Next-Generation Strategies for Crop Improvement

Yogesh Ahlawat • Darshna Chaudhary
Pawan K. Jaiwal
Editors

Next-Generation Strategies for Crop Improvement

Springer

Editors
Yogesh Ahlawat
University Center for Research
and Development, Chandigarh University
Mohali, Punjab, India

Darshna Chaudhary
Centre for Biotechnology
Maharshi Dayanand University
Rohtak, Haryana, India

Pawan K. Jaiwal
Centre for Biotechnology
Maharshi Dayanand University
Rohtak, Haryana, India

ISBN 978-981-95-0308-7 ISBN 978-981-95-0309-4 (eBook)
https://doi.org/10.1007/978-981-95-0309-4

This Springer imprint is published by the registered company Springer Nature Singapore Pte Ltd.
The registered company address is: 152 Beach Road, #21-01/04 Gateway East, Singapore 189721, Singapore

Preface

The rapid advancements in agricultural sciences, coupled with the increasing demand for food security, have emphasized the need for innovative and sustainable crop improvement strategies. As the world grapples with challenges such as climate change, diminishing arable land, and a growing global population, the urgency to develop next-generation crop breeding and genetic enhancement techniques has never been more pressing.

Crop improvement is undergoing a paradigm shift, driven by state-of-the-art technologies like genome editing, molecular markers, systems biology, and bioinformatics. These breakthroughs provide unparalleled insights into plant biology, enabling the creation of crops that are more resilient, higher-yielding, and nutritionally enriched. These next-generation breeding techniques promise to revolutionize crop production, offering solutions to meet the nutritional needs of future generations while preserving the environment.

This book, *Next-Generation Strategies for Crop Improvement*, aims to encapsulate these transformative strategies, offering a thorough understanding of the various tools and approaches shaping the future of agriculture. The chapters cover a broad range of topics, from CRISPR and other genome-editing technologies to the integration of omics-based techniques for trait discovery and improvement. The book also explores precision agriculture, synthetic biology, and plant-microbe interactions in enhancing crop resilience to environmental stresses.

Each chapter provides an in-depth overview of the current state of the art, highlighting both challenges and opportunities. The expertise of leading scientists in plant breeding, genetics, and agricultural biotechnology contributes to advancing scientific knowledge while inspiring future research that promotes sustainable agricultural practices. This book aims to be a valuable resource for researchers, practitioners, and policymakers working toward a food-secure and sustainable future.

Mohali, India
Rohtak, India

Yogesh Ahlawat
Darshna Chaudhary
Pawan K. Jaiwal

Acknowledgments

We would like to express our sincere gratitude to all those who contributed to the creation and success of this book, *Next-Generation Strategies for Crop Improvement*. This project would not have been possible without the support, dedication, and expertise of our renowned authors and reviewers.

First and foremost, we are deeply grateful to the distinguished contributors who have shared their knowledge and research in the chapters of this book. Their cutting-edge work, insights, and commitment to advancing the field of crop improvement have been the backbone of this publication. Each chapter reflects the authors' expertise and passion for their respective areas, and we extend our heartfelt thanks to them for their collaboration.

A special acknowledgment goes to the editorial and production teams at Springer for their tireless efforts in bringing this book to fruition. Their professionalism, attention to detail, and efficient handling of the manuscript ensured a smooth and timely publication process.

Finally, we would like to extend our deepest appreciation to our families for their understanding, patience, and encouragement during the long hours spent on this work. Their unwavering support has been a constant source of strength.

This book is the product of collective effort and we are proud to share it with the global community of researchers, practitioners, and policymakers dedicated to advancing sustainable crop improvement strategies. We hope that it serves as both a comprehensive resource and an inspiration for future innovations in crop science.

Contents

Editors and Contributors

About the Editors

Yogesh Ahlawat earned his PhD from Michigan Technological University, USA, focusing on enhancing biofuel production by manipulating plant cell wall biosynthesis. He later completed postdoctoral research at the University of Florida, specializing in extending postharvest shelf life in horticultural crops. His core research involves modifying lignin biosynthesis in *Populus trichocarpa* to improve biofuel yields, using cutting-edge tools like tissue-specific promoters, gene overexpression/silencing, and chemical pretreatments to overcome biomass recalcitrance. He has developed hundreds of transgenic poplar lines targeting key genes and transcription factors related to bioenergy. His work also spans the use of multi-omics strategies to prolong the postharvest life of fruits and vegetables. With over 13 years of experience in functional genomics and crop trait enhancement, he has made notable contributions to improving quantitative and qualitative traits across various crops. He has published extensively, holds patents, and has received prestigious honors including the Doctoral Finishing Fellowship, Young Scientist Award, Young Biotechnologist Award, and Best Graduate Teaching Award. He serves on editorial boards of reputed journals and has received multiple national and international travel grants for his research excellence. His work bridges bioenergy innovation and sustainable agriculture through advanced molecular approaches.

Darshna Chaudhary is an Associate Professor at the Centre for Biotechnology, Maharshi Dayanand University, Rohtak (India). She received postgraduate and doctorate in Biotechnology from M. D. University, Rohtak (India). Her key areas of interest include edible vaccine development in plants, oil quality improvement in *Brassica juncea* using CRISPR-Cas9, RNAi-based pest management, and genetic transformation of *Vigna* species for crop improvement. She has guided over 70 MSc dissertation students and 8 PhD students. She has published more than 50 research papers in nationally and internationally reputed peer-reviewed journals. She has been awarded seven national research projects funded by agencies such as University Grant Commission (UGC), Department of Biotechnology (DBT), Govt. of India, Department of Science and Technology-Science and Engineering Research Board (DST-SERB), and Haryana State Council for Science, Innovation and

Technology (HSCSIT), Govt. of Haryana, and out of which, she has completed five successfully. She has edited a book published by Springer Nature. She is a life member of several scientific societies. She has been awarded DST Young Scientist and university incentives for research achievements. In addition to her research and teaching roles, she actively contributes to academic administration, curriculum development, and institutional biosafety at M. D. University.

Pawan K. Jaiwal is a Professor Emeritus at the Centre for Biotechnology, M. D. University, Rohtak (India). He has worked as the Dean of the Faculty of Life Sciences and the Director of the Centre for Biotechnology, M. D. University, Rohtak. He graduated in Botany (Honors) at the University of Delhi, Delhi; and completed post-graduation and doctorate in Botany at the Kurukshetra University, Kurukshetra (India). He has been meritorious throughout his studies. He has served at the M. D. University in various positions as an Assistant Professor (1986–1998), Associate Professor (1998–2006), Professor (July 2006–December 2017), Director, and Dean, and he has participated in multiple committees and statutory bodies of the University from time to time as chairman or member. He has nearly 35 years of postgraduate teaching experience in molecular biology, genetic engineering, and plant biotechnology. He has contributed 16 books. He has also published 28 book chapters and 112 research papers in nationally and internationally reputed and referred research journals. He is a member of several academic bodies and is on the editorial board of national and international journals. He has guided 27 PhD students. He has completed 12 major research projects funded by various national and international agencies, such as University Grant Commission (UGC), Council of Scientific & Industrial Research (CSIR), Department of Biotechnology (DBT), Govt. of India, Department of Science and Technology (DST), Haryana State Council for Science, Innovation and Technology (HCST), Govt. of Haryana, and Japan Society for the Promotion of Science (JSPS), Japan. He has been awarded the DST Young Scientist, Indian National Science Academy (INSA) visiting Fellow, DBT overseas Associate, Prof H.S. Srivastava Gold medal, and Basic Scientific Research (BSR) Faculty Fellow. He has worked and trained at the Institute of Plant Sciences, ETH, Zurich (Switzerland); the National Research Centre on Plant Biotechnology, New Delhi; and the University of Ghent, Belgium. He visited National Institute of Agrobiological Sciences (NIAS), Tsukuba, Japan; the Chinese Academy of Agriculture Sciences, Beijing, China; and the University of Massachusetts, Amherst, USA, for collaborative research. He has significantly contributed to the genetic improvement of legumes, oil crops, and cereals.

Contributors

Ankita Ahlawat Centre for Biotechnology, Maharshi Dayanand University, Rohtak, Haryana, India

Yogesh Ahlawat University Centre for Research and Development, Chandigarh University, Mohali, Punjab, India
Allied Health Sciences, Datta Meghe Institute of Higher Education and Research, Wardha, Maharashtra, India

Mohd Anas Department of Botany, Aligarh Muslim University, Aligarh, India

Sowmiya Arumugam Tamil Nadu Agricultural University, Coimbatore, Tamil Nadu, India

Ashish H. Bankar Department of Agricultural Botany, SVKM's College of Agriculture, Tardi, Shirpur, Dhule, Maharashtra (Affiliated to MPKV, Rahuri), India

P. K. Bhati Borlaug Institute for South Asia (BISA) India, DPS Marg, NASC Complex, New Delhi, India

Neha Boora West Virginia State University, Institute, WV, USA

Walunjkar Babasaheb Changdeo Shri Vaishnav Institute of Agriculture, SVVV (Shri Vaishnav Vidyapeeth Vishwavidyalaya), Indore, India

Darshna Chaudhary Centre for Biotechnology, Maharshi Dayanand University, Rohtak, Haryana, India

Vineeta Dixit Department of Botany, Nilamber-Pitamber University, Medininagar, Palamu, Jharkhand, India

J. Essemine Bioinformatics Section, Department of Informatics, University of Quebec at Montreal, Pavillon President-Kennedy, Montreal, QC, Canada

Jaya Gangwar Department of Life Sciences, Christ University, Bangalore, Karnataka, India

Mutian Gao Key Laboratory of the Ministry of Education for Crop Physiology and Molecular Biology, College of Agronomy, Hunan Agricultural University, Changsha, China

Mokhtar Guerfel Applied College, Shaqra University, Afif, Saudi Arabia

Narayan Pandit Gurav Department of Agricultural Botany, SVKMs College of Agriculture, Tardi, Shirpur, Dhule, Maharashtra (Affiliated to MPKV, Rahuri), India

Sarahani Harun Centre for Bioinformatics Research, Institute of Systems Biology (INBIOSIS), Universiti Kebangsaan Malaysia, Bangi, Selangor, Malaysia

Minza Hasan Department of Botany, Aligarh Muslim University, Aligarh, India

Cheng Huang Key Laboratory of the Ministry of Education for Crop Physiology and Molecular Biology, College of Agronomy, Hunan Agricultural University, Changsha, China
Yuelushan Laboratory, Changsha, China

Pawan K. Jaiwal Centre for Biotechnology, Maharshi Dayanand University, Rohtak, Haryana, India

Ranjana Jaiwal Department of Zoology, Maharshi Dayanand University, Rohtak, Haryana, India

K. P. Jaseem Department of Botany, Aligarh Muslim University, Aligarh, India

Kadanthottu Sebastian Joseph Department of Life Sciences, Christ University, Bangalore, Karnataka, India

Samwel Muiruri Kariuki International Institute of Tropical Agriculture, Nairobi, Kenya

Sharma Shanti Kumar Krishi Anusandhan Bhavan, ICAR, New Delhi, India

Manish Kumar Vishwakarma Borlaug Institute for South Asia (BISA) India, DPS Marg, NASC Complex, New Delhi, India

Ankit Kumari Centre for Biotechnology, Maharshi Dayanand University, Rohtak, Haryana, India

Shilpi Kundu Department of Agricultural Extension & Information System, Faculty of Agriculture, Sher-e-Bangla Agricultural University, Dhaka, Bangladesh

Anurag Malik Division of Research and Innovation, Uttaranchal University, Dehradun, Uttarakhand, India

Suganthy Mariappan Tamil Nadu Agricultural University, Coimbatore, Tamil Nadu, India

Shikha Nara Centre for Biotechnology, Maharshi Dayanand University, Rohtak, Haryana, India

Anu Naruka Shri Vaishnav Institute of Agriculture, SVVV (Shri Vaishnav Vidyapeeth Vishwavidyalaya), Indore, India

Ravisankar Natesan Indian Institute of Farming System Research, Modipuram, Meerut, Uttar Pradesh, India

Duncan Njora Waweru International Institute of Tropical Agriculture, Nairobi, Kenya

P. M. S. Noor Mohamed Ibrahim Department of Botany, Aligarh Muslim University, Aligarh, India

Janaki Ponnusamy Tamil Nadu Agricultural University, Coimbatore, Tamil Nadu, India

Himani Punia Saraswati Group of Colleges, SGC Group, Gharuan, Mohali, Punjab, India

Natalia Repkina Institute of Biology of Karelian Research Centre, Russian Academy of Sciences, Petrozavodsk, Russian Federation Pushkinskaya

Suman Roy ICAR-National Bureau of Plant Genetic Resources, New Delhi, India

Pratik Satya ICAR-Central Research Institute for Jute and Allied Fibres, Kolkata, India

Laxmi Sharma ICAR-National Bureau of Plant Genetic Resources, New Delhi, India

Varun Shende Shri Vaishnav Institute of Agriculture, SVVV (Shri Vaishnav Vidyapeeth Vishwavidyalaya), Indore, India

Ekta Shukla Department of Botany, Faculty of Science and Technology, Sunbeam College for Women, Varanasi, India

Shruti Sinha ICAR-National Bureau of Plant Genetic Resources, New Delhi, India

Svetla Sofkova-Bobcheva School of Agriculture and Environment, Massey University, Palmerston North, New Zealand

Jaindra Nath Tripathi International Institute of Tropical Agriculture, Nairobi, Kenya

Leena Tripathi International Institute of Tropical Agriculture, Nairobi, Kenya

Priti Upadhyay Division of Vegetable Sciences, Indian Agricultural Research Institute, New Delhi, India

Manish K. Vishwakarma Borlaug Institute for South Asia (BISA)-India, DPS Marg, NASC Complex, New Delhi, India

Nashra Waqar Department of Biotechnology, Faculty of Science, Era University, Lucknow, India

Ying Xu Key Laboratory of the Ministry of Education for Crop Physiology and Molecular Biology, College of Agronomy, Hunan Agricultural University, Changsha, China
Yuelushan Laboratory, Changsha, China

Sheel Yadav ICAR-National Bureau of Plant Genetic Resources, New Delhi, India

Rabiatul-Adawiah Zainal-Abidin Biotechnology and Nanotechnology Research Centre, Malaysian Agricultural Research and Development Institute (MARDI), Serdang, Malaysia

A Historical Perspective of Crop Improvement

1

Vineeta Dixit, Priti Upadhyay, Ekta Shukla, and Manish K. Vishwakarma

Abstract

Since ancient times, agriculture has been the foundation of human civilization, guiding humans from hunter-gatherer tribes to settled agricultural communities. Early efforts concentrated on domesticating wild plants like wheat, barley, rice, maize, etc., to maintain production. Crop rotation, irrigation systems, and polyculture were employed. Advances in crop rotation techniques, ploughing technology, and crop exchange during the Columbian Exchange in the Middle Ages and Renaissance improved food security and diversified diets. The Green Revolution of the twentieth century was a revolutionary period that greatly enhanced the world's food output by creating high-yielding crop varieties, applying synthetic agrochemicals, irrigation infrastructure, and mechanization. These developments also highlighted the necessity of sustainable farming methods to lessen environmental problems. Crop rotation, agroforestry, and cover crops are sustainable farming methods prioritizing biodiversity and environmental health. Modern crop modification techniques have transformed agriculture in the last few decades, including mutation breeding. Crop resilience, yield, and nutritional value have increased due to precise genetic alterations made possible by

V. Dixit
Department of Botany, Nilamber-Pitamber University, Medininagar, Palamu, Jharkhand, India

P. Upadhyay
Division of Vegetable Sciences, Indian Agricultural Research Institute, New Delhi, India

E. Shukla
Department of Botany, Faculty of Science and Technology, Sunbeam College for Women, Varanasi, India

M. K. Vishwakarma (✉)
Borlaug Institute for South Asia (BISA)-India, DPS Marg, NASC Complex, New Delhi, India
e-mail: M.Vishwakarma@cgiar.org

Y. Ahlawat et al. (eds.), *Next-Generation Strategies for Crop Improvement*,
https://doi.org/10.1007/978-981-95-0309-4_1

marker-assisted selection, genomic selection, tissue culture, and Clustered Regularly Interspaced Short Palindromic Repeats (CRISPR)/Cas gene editing. The intervention of Global Positioning System (GPS), remote sensing, artificial intelligence, machine learning, and data analytics enables precision agriculture which guarantees the focused, effective usage of resources.

Keywords

Agriculture · Crop · Green revolution · Yield · Mutation

1.1 Introduction

Agriculture has been the foundation of human civilization for thousands of years. The introduction of crop cultivation signalled a dramatic change from nomadic hunter-gatherer communities to permanent agricultural civilizations (Bhargava and Srivastava 2019). Over the past century, the world has witnessed remarkable advancements in crop productivity, largely driven by a suite of technological innovations, including the development of higher-yielding crop varieties, improved farm management practices, the introduction of synthetic agrochemicals, and the widespread adoption of agricultural mechanization. This 'Green Revolution' has been credited with reducing famine and malnutrition globally (Pingali 2012). This decade has witnessed various techniques which have immense potential to revolutionize agriculture (Boomer 2023). Advanced molecular approaches like marker-assisted selection, genomic selection, and gene stacking have contributed to the development of crops with improved tolerance to biotic and abiotic stresses (Khan et al. 2024). Techniques such as CRISPR/Cas genome editing have also enabled precise genetic modifications, enhancing crop resilience, yield, and nutritional value (Feng 2024). This chapter delves into the historical perspectives of crop improvement, tracing the journey from ancient practices to modern advancements, which charts the development of crop improvement from conventional techniques to state-of-the-art technologies and highlights the inventiveness of human endeavours in agriculture. These developments are essential to tackling global food security issues and promoting agriculture's sustainability in the era of artificial intelligence technology.

1.2 Crop Domestication and Early Improvement Efforts

1.2.1 Early Beginnings of Crop Cultivation and Ancient Agriculture Practices

The advent of agriculture is one of the most critical events in human history; it signifies substantial transition of human civilization from the stage of nomadic hunter-gatherer societies to sedentary agricultural civilizations. This monumental

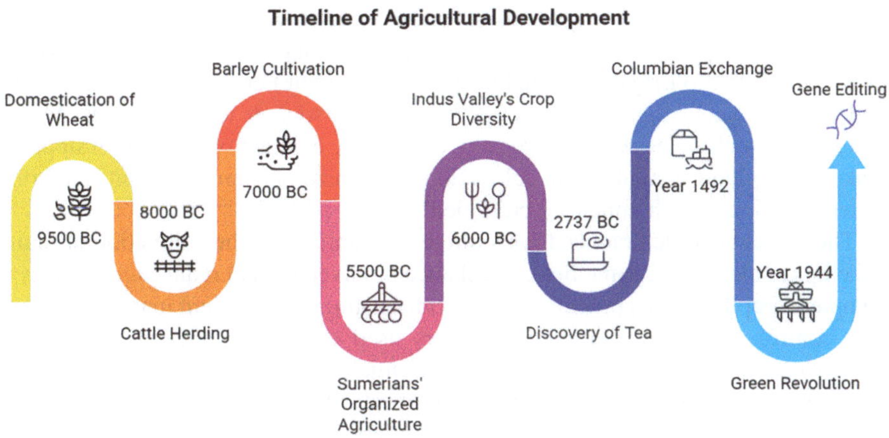

Fig. 1.1 A descriptive illustration of milestones achieved in agriculture evolution

shift is the Neolithic Revolution occurring around 10,000–12,000 BCE) (Fig. 1.1). Early humans started with the domestication of wild plants in this period, which marked a foundational step toward modern agriculture (Larsen 1981; Haas 2018). Domestication of plants required screening wild plant communities and selecting a specific one for a particular character, such as seed size, good yield, and desirable traits. In addition, selecting a specific part of the plant has been done precisely to meet various human needs. All these were taken over many generations in gradual artificial selection to develop the first domesticated crops. Among the earliest domesticated plants were wheat and barley in the Fertile Crescent, a region now often referred to as the cradle of civilization; it comprises parts of today's Iraq, Syria, Lebanon, Israel, and Jordan. The fertile soil conditions and the favourable climate helped contribute to supporting early agricultural activities in this region (Haas 2018). At the same time, other parts of the world were witnessing similar agrarian advancements. Rice cultivation began in the Yangtze River basin in East Asia, where early farmers selected rice varieties that were more productive and adaptable to different environments. Rice became a staple food within a short period, supporting the growth of complex societies in China and Southeast Asia (Fuller 2011). Maize was domesticated in the Balsas River Valley of southwestern Mexico around 9000–8600 years ago (Piperno et al. 2009). It spread to Central and South America 4000–3000 years ago (Iriarte 2007), reached North America around 2000 years ago (Fritz 2007), and has since spread globally over the past 500 years. Indigenous people then cultivated maize from a wild grass called teosinte by selecting plants with more and larger kernels.

These early agricultural practices were much more than planting crops; intricate knowledge about the local environment, patterns of the seasons, and behaviour of different plant species involved. Crop rotation was another practice of ancient farmers in maintaining soil fertility, in which different types of crops are grown in the same field across different seasons to help prevent soil depletion and reduce the risk of pests and diseases. Ingenious irrigation systems were developed to ensure a reliable water supply to crops. The Egyptians, for instance, used the annual flooding of the Nile River to irrigate their fields, while the ancient Persians developed qanats-underground canals that brought water from distant sources to arid regions. Mountain regions such as the Andes were cultivated by the creation of terraces on slopes to make arable land available for cultivation, minimize soil erosion, and maximize water retention, which enhanced crop yields.

The ancient people practiced polyculture (Slotten 2020), the cultivation of multiple crops at once in the same field, to enhance biodiversity and reduce the chances of complete crop loss. This practice also provides natural soil nutrient management and pest control. The most common fertilizers used during the ancient period were animal manure and compost, providing essential nutrients to the soil for healthy plant growth. Alongside cultivation, the domestication of cattle, sheep, and goats allowed for a regular supply of manure for fertilization purposes and labour for tilling the fields.

The development of primary agricultural tools, including ploughs, sickles, and hoes, transformed agriculture into a more efficient means of farming. Most of these were made from wood, stone, and later, metal. Ancient societies were highly religious and culturally inclined towards agriculture. Many cultures had rituals and ceremonies to ensure a good harvest and to honour deities associated with fertility and agriculture. Ancient agricultural practices were a testament to human ingenuity and adaptability. They laid the ground for modern farming techniques to grow civilizations and develop more complex societies. Understanding these practices holds valuable insights into agriculture's evolutionary history and the relationship between humans and the crops that sustain them.

1.2.2 Medieval and Renaissance Innovations

Agricultural developments, especially during the Middle Ages and the Renaissance, greatly influenced farming practices in Europe. Their crop cultivation improved remarkably. The most outstanding innovation developed during the Middle Ages was the three-field system of crop rotation. The three-field system divided the arable into three parts; one had a winter crop, and the other had a spring crop. The remaining was kept fallow. Rotating crops in this manner helped maintain soil fertility and reduce the risk of soil exhaustion, allowing for more efficient use of land and increased agricultural output (Simon-Miquel and Plaza-Bonilla 2021).

Through other cultures' interactions in the process of trade and conquest, the Middle Ages witnessed several new crops introduced into Europe. For instance, durum wheat, which hails from the Middle East, was introduced and proved very important in Mediterranean areas (Woźniak et al. 2021). Rye, oats, and barley also assumed new importance as staple crops throughout Europe (Leff et al. 2004). The heavy Plow, which had been developed at this time with the ability to cut further into heavy and dense Northern European soils, was in use during this period. This allowed for deeper cutting through the soil; it is made of metal and allows for better digging than other wooden or even lighter versions of the earlier Plows. This improved the cultivation of previously uncultivable lands, thus expanding agricultural production.

The Renaissance period was characterized by renewed interest in science and observation. Scholars and agriculturists began to systematically look at plants and their growth, which led to improved knowledge of crop physiology and soil science. During that time, agricultural manuals became available, compiling knowledge and best practices for crop improvement (Stone 2005). The Age of Exploration allowed for the transfer of crops between the Old and New Worlds, which is known as the Columbian Exchange (Maurya et al. 2010). European explorers brought back crops such as maize, potatoes, tomatoes, and beans from the Americas. These crops were gradually integrated into European agriculture, diversifying diets and improving food security (Hobhouse 1999).

During the Renaissance, more deliberate and systematic approaches to selective breeding emerged. Farmers and gardeners experimented with cross-breeding different plant varieties to enhance desirable traits like yield, disease resistance, and adaptability to local climates (Lantinga and Rabbinge 1996). These practices paved the way for the scientific plant breeding techniques that would emerge later. Both practical innovations and an increasing spirit of scientific inquiry characterized the Middle Ages and the Renaissance, with all these elements propelling further advancements in crop improvement. Crop rotation systems, ploughing technologies, new crops, and a systematic approach to plant growth contributed towards increased agricultural productivity and ensured food security. All of these were historical advancements that would shape the agricultural revolutions that eventually led to the modern ways of crop cultivation today (Pingali 2010).

1.2.3 The Green Revolution

A series of innovations and improvements in agriculture between the 1940s and the late twentieth century marked a transformative shift in global agriculture, increasing food production with the view of combating hunger, which is the concept of the green revolution (Pingali 2012). It is characterized by the development and diffusion of high-yielding crop varieties with chemical fertilizers and pesticides, advancements in irrigation infrastructure, and the mechanization of farming. This is regarded as a landmark period in agricultural history. High levels of food production increase and food security improvement for millions worldwide followed it. While the Green Revolution came with its challenges, it provided a basis for modern agriculture and continues to impact work toward sustainable and resilient food systems (Pingali 2012).

One of the most remarkable outcomes of the Green Revolution was the creation of high-yielding varieties (HYVs) of staple crops like wheat, rice, and maize. These varieties were developed to be much more productive compared to conventional varieties. For instance, the dwarf wheat variety Norman Borlaug, known by many as the 'Father of the Green Revolution,' has developed smaller stalks (semi-dwarf) and greater resistance to diseases and pests in wheat varieties. This was able to carry the increased weight of the grain due to its productivity. Borlaug's work earned him the Nobel Peace Prize in 1970 for his contributions to global food production (Phillips 2013).

High-yielding rice varieties were also developed to transform rice cultivation in Asia. Among these varieties was IR8, also referred to as 'Miracle Rice'. It was bred at the International Rice Research Institute in the Philippines, which gave it a relatively short growth period so that farmers could harvest several crops within a year. The crop was also resistant to pests and diseases, contributing to its adoption and subsequent successes in raising rice yields (Peng et al. 2010; John and Babu 2021).

The use of chemical fertilizers and pesticides was the major input that contributed to the Green Revolution. These inputs improved the fertility of the soil as well as the protection of crops from pests and diseases, thus leading to higher agricultural productivity (Chakwanda et al. 2024). For instance, the application of nitrogen-based fertilizers boosted wheat and rice's growth to achieve their full yield potential. Improvements in irrigation infrastructure were also crucial to the Green Revolution. The construction of dams, canals, and other irrigation systems ensured a consistent water supply for crops, even in arid and semi-arid regions. This increased irrigated farming led to more consistent yields and higher crop yield. For example, the building of the Indira Gandhi Canal in India turned large parts of the Thar Desert into fertile agricultural lands, allowing crops such as wheat and mustard to be grown on them (Roy et al. 2015).

Another crucial component of the Green Revolution was mechanization in agriculture (Rosset and Altieri 1997). With the advent of tractors, harvesters, and irrigation pumps, farmers increased their farming efficiency. The machinery made the work much less laborious for them, enabling more lands to be farmed on larger scales. More significantly, machinery made planting and harvesting possible within timely seasons, thereby raising crop yields (Chakwanda et al. 2024). The Green Revolution of the mid-twentieth century marked a turning point in agricultural techniques across the globe, particularly in countries like India. Visionaries like Dr M.S. Swaminathan led this movement, developing revolutionary technology to increase production and reduce poverty in rural communities (Ullah et al. 2024). Over time, the Green Revolution's legacy has grown, with modern innovations and technologies further altering farming operations.

A more significant part of the developing world mainly Asia and Latin America was touched by The Green Revolution. India turned itself into a food-efficient state that transformed into one of the leading agricultural producers in its turn, and Mexico experienced that kind of change as its high-yielding varieties of maize started to fuel up maize production, aiding that country's population growth and feeding its growing people (Evenson and Gollin 2002).

Despite its many successes, the Green Revolution also brought out the need for sustainable agricultural practices to address environmental challenges. The overuse of chemical inputs, soil degradation, and water resource depletion are issues that continue to be addressed through the development of sustainable farming practices and technologies (Chakwanda et al. 2024).

1.3 Modern Crop Improvement Techniques

Modern crop improvement uses scientific research and other technological advancement to increase yields, enhance the nutritional quality of crops, and increase resilience to environmental stresses, diseases, and pests (Khan et al. 2024). A few of the techniques are discussed below:

1.3.1 Mutation Breeding and Crop Improvement

Mutation breeding, also known as variation breeding, was initiated in the early twentieth century and is a method of developing new crop varieties by causing genetic changes. Mutations can be induced with physical agents like radiation or chemical mutagens. The idea is to cause beneficial variations that can be selected for higher yields, disease resistance, and environmental adaptability. One of the landmark events was in 1927 when Hermann J. Muller showed that X-rays could cause mutation in the fruit fly, and later, this method was used with plants. This discovery helped in using radiation to cause mutations in crops, hence leading to the development of new and improved varieties.

1.3.1.1 Mutation Breeding and Its Procedure

1. Induction of mutations: Physical or chemical agents induce mutations in the seeds or plant tissues. Physical mutagens include X-rays, gamma rays, and fast neutrons, while chemical mutagens include substances like ethyl methane sulfonate (EMS) and sodium azide (Fig. 1.2).

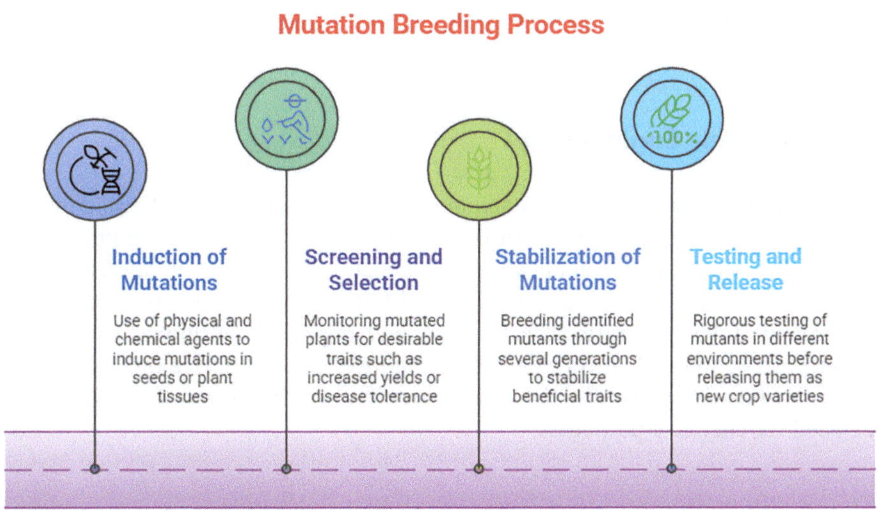

Fig. 1.2 Mutation breeding pipeline

2. Screening and selection: Mutated plants are grown and closely monitored for any changes that display desirable traits. Desirable traits can include, but not be limited to, increased yields, more excellent nutritional value, or tolerance to diseases and insects.
3. Stabilization of mutations: Identified mutants with desirable traits are bred/selected through several generations for stabilizing the beneficial characteristics passed onto subsequent generations.
4. Testing and release: The selected mutants undergo rigorous testing in different environments to ensure their performance and stability. Successful mutants are then released as new crop varieties for commercial cultivation.

Mutation breeding has contributed significantly to crop improvement by creating varieties with enhanced traits. All the information about the number of leading countries in the release of mutant variety, the number of leading crop that has more mutant variety so far, etc., is available on the website of Food and Agriculture Organization (FAO)/International Atomic Energy Agency (IAEA) at the link given here, https://nucleus.iaea.org/sites/mvd/SitePages/Home.aspx.

1.3.2 Marker-Assisted Selection (MAS)

Marker-assisted selection is a technique in plant breeding that uses molecular markers to select desirable plants with concerned traits. This method allows the easy and accurate selection of plants with the desired traits, which, in turn, accelerates the conventional breeding process. Using MAS, the breeder can identify or select plants with certain specific characteristics, such as disease resistance, improved yield, or drought tolerance, without waiting for the plants to mature. For instance, MAS has been applied to breed wheat varieties resistant to rust diseases. The detailed study has been found in another chapter of this book.

1.3.3 Genetic Engineering

Genetic engineering is the direct manipulation of an organism's DNA using biotechnology. This technique allows scientists to introduce or enhance new traits in crops. Among the most known applications of genetic engineering in agriculture are GMOs. Genetically Modified Organism (GMO) may possess characteristics such as herbicide tolerance, resistance to pests, and improved nutritional content. For instance, Bt cotton and Bt corn have been engineered to produce a protein from the bacterium *Bacillus thuringiensis*, which is toxic to specific pests, thereby reducing the need for chemical pesticides.

1.3.4 Genomic Selection

Genomic selection is an advanced breeding technique. This method uses genome-wide markers to predict the performance of plants, which the genetic makeup would otherwise achieve. Genomic selection helps more efficiently select the best candidates for breeding programmes. It increases the accuracy and speed of breeding programmes and could help develop superior crop varieties. For example, genomic selection has been used for the improvement of dairy cattle, and this is similar to what has been happening with crops like maize and wheat.

1.3.5 Tissue Culture and Micropropagation

Plant tissue culture is an assemblage of techniques that describe growing or maintaining plant cells, tissues, or organs under strictly hygienic laboratory conditions using a nutrient culture medium. It takes advantage of totipotency, where each plant cell carries the potential to regenerate back into a whole plant. Micropropagation is one of the specific applications of plant tissue culture, involving the rapid multiplication of plant material to produce many progeny plants, often called clones, from a single explant. Plant tissue culture as a concept began early in the twentieth century with the great pioneering work of Gottlieb Haberlandt, the man proudly titled the 'father of plant tissue culture'. Haberlandt laid the foundation with his experiments way back in 1902. Still, important developments did not occur until the 1950s and 1960s; by this time, the applications were finally put into practical use with plant tissue culture and micropropagation techniques. These are strong tools for rapid, efficient plant propagation, genetic modification, disease elimination, and conservation of plant genetic resources (Khatik 2024). Despite their challenges, they are still crucial in modern agriculture, horticulture, and plant research to achieve sustainable, innovative global food security and biodiversity conservation solutions.

1.3.5.1 Techniques and Processes

Explants and sterilization: The tissue culture process begins with the selection of explants, which are small pieces of plant tissue taken from various parts of a plant, such as leaves, stems, or roots. These explants are then sterilized to eliminate microorganisms that could contaminate the culture medium.

Culture media: The sterilized explants are placed on a culture medium providing essential nutrients, vitamins, and hormones for growth and development. Murashige and Skoog medium is one of plant tissue culture's most commonly used culture medium.

Callus formation: The explants often form an unorganized mass of cells known as a callus, which can be induced to develop into roots and shoots under specific conditions. Growth regulators like auxins and cytokinins play a crucial role in this process.

Organogenesis and somatic embryogenesis: Organogenesis refers to the formation of organs such as roots and shoots from the callus. Somatic embryogenesis is the process of somatic cells (nonreproductive cells) developing into embryos, which can then grow into complete plants.

1.3.5.2 Micropropagation Stages

Stage I—initiation: Establishment of aseptic cultures from the explants

Stage II—multiplication: Rapid multiplication of shoots or embryos through subculturing

Stage III—rooting: Induction of root formation in the regenerated shoots

Stage IV—acclimatization: Gradual adaptation of tissue-cultured plants to external conditions, followed by transfer to soil

1.3.5.3 Applications

Clonal propagation: Micropropagation is widely used for the clonal propagation of plants, ensuring the production of genetically uniform and disease-free plants. This technique is especially valuable for propagating ornamental plants, fruit trees, and commercially important crops like bananas, orchids, and potatoes.

Genetic engineering: Plant tissue culture serves as a fundamental tool in genetic engineering, allowing for the introduction of new genes into plants through techniques such as Agrobacterium-mediated transformation and biolistics (gene gun).

Germplasm conservation: Tissue culture techniques are employed for the conservation of germplasm, enabling the long-term storage of plant genetic resources. Cryopreservation is a method used to store plant tissues at ultralow temperatures, preserving their viability for future use.

Disease elimination: Tissue culture can eliminate viruses and other pathogens from infected plant material, producing healthy, disease-free plants. Techniques like meristem culture are particularly effective in achieving this goal.

Conventional breeding for a specific trait has existed since plant species were domesticated. Therefore, crop improvement techniques are not new (Georges and Ray 2017). Molecular developments led to the random integration of transgenes, or gene with a defined function, often achieved using *Agrobacterium*-mediated transformation (Que et al. 2010). Both genome engineering and recombinant technologies have made it simpler to generate precise and focused genomic alterations for crop improvement, despite their many uses in both plant and animal systems (Schiml et al. 2014; Kamthan et al. 2016; Ryu et al. 2018). The public's perception, laws, and intricacy surrounding genetically modified animals have led to the increased application and rapid commercialization of GE technology in plant systems.

1.3.6 Precision Agriculture

Precision agriculture, also called precision farming or site-specific crop management (SSCM), is a high-tech approach in agricultural practice that uses technology to monitor, measure, and react to changes in crop conditions and field environments. It aims to achieve efficiency in the use of inputs, increase crop production, and reduce environmental impact through precision and control in agricultural operations (European Commission 2017). Precision agriculture is a paradigm shift in modern farming. It uses technology to make agriculture more efficient, productive, and sustainable. With Global Positioning System (GPS), remote sensing, Geographic Information Systems (GIS), variable rate technology (VRT), and advanced data analytics, farmers can optimize their practices and produce more with fewer resources. As technology continues to evolve, precision agriculture will play a crucial role in shaping the future of global agriculture (Maurya et al. 2024).

1.3.6.1 Major Technologies in Precision Agriculture

Global positioning system (GPS): GPS technology gives farmers precise location data, allowing them to map their fields and monitor crop conditions with great precision. GPS-guided tractors and machinery ensure the precise planting, fertilizing, and harvesting of fields with less overlap and gaps in field operations.

Remote sensing: Satellite imagery, drones, and aerial photography are remote sensing technologies that provide valuable data on soil properties, crop health, and field variability. These technologies enable farmers to identify stress, disease, or nutrient deficiency and apply interventions accordingly.

Geographic information systems (GIS): GIS technology integrates and analyses spatial data to create detailed maps of fields, showing the variations in soil, crop health, and environmental conditions. GIS tools help farmers in planning and decision-making and enable them to implement site-specific management practices.

Variable rate technology (VRT): VRT enables the effective application of inputs, such as fertilizers, pesticides, and seeds, at all points in a field based on specific needs. This technology aims to ensure that resources are used efficiently, waste has been minimized, and cropping has been optimized.

Soil and crop sensors: Soil sensors measure soil moisture, temperature, pH, and nutrient levels in real time, providing critical information for irrigation and fertilization management. Crop sensors monitor the health of plants, growth rates, and chlorophyll levels, allowing inputs to be applied precisely according to plant needs.

Automated machinery and robotics: Tractors, harvesters, and drones perform high accuracy and consistency in field operations. Robotics are used for planting, weeding, and crop monitoring, which reduces labour costs and increases efficiency.

Unmanned agriculture vehicle (UAV)-based decision-making: Temperature, soil composition, and weather are important external factors that affect agricultural output. Farmers may adapt to certain environmental peculiarities and make well-informed decisions with the help of agricultural drones (Vishwakarma et al. 2023). UAVs collect the data, and using the information they gather, they may manage crop health, carry out accurate crop scouting, apply focused crop treatments, control irrigation, and carry out in-depth field soil analysis and crop loss assessments. Drone surveys, therefore, improve crop yields while streamlining operations and cutting down on farming-related expenses and time.

Data analytics, machine learning and AI: Advanced data analytics and machine learning algorithms analyse large datasets from multiple sources, identify patterns, and predict crop performance.

1.3.7 Sustainable Agriculture: Omics and Epigenomics

Omics approaches encompassing various fields such as genomics, proteomics, metabolomics, metagenomics, phenomics, transcriptomics along with epigenomics have played a pivotal role in deciphering the complex interplay of genetic, molecular, and environmental factors on crop plants. These insights have been effectively utilized to influence and optimize crop traits. Over the past few decades, plant-omics and system biology have been the primary stream of plant biology research.

Genomics: It focuses on complete or partial set of genome to identify various genes controlling higher yield, stress resistance, and/or improved nutritional qualities in plants. Gene *DREB1A*, identified with drought resistance, was introduced into rice through genetic engineering ensuring in the development of rice varieties that thrive in water-scarce environments, ensuring stable yields despite climate challenges.

Proteomics: Systemic identification and quantification of various underlying mechanisms of protein synthesis and overall proteome is key to understand functional proteins that control growth and development in plants. Development of highly nutritive 'quality protein maize' was achieved through proteomic studies enhancing the content of essential amino acids such as lysine. Alterations in wheat proteomic profile in response to abiotic stresses have been successfully achieved (Setsuko et al. 2014).

Metabolomics: It analyses metabolite development and thus is utilized in enhancing crop flavour, nutrient content, and resistance to environmental challenges. Metabolomics helped researchers modify the major pathways responsible for flavour and sweetness in tomatoes.

Metagenomics: Metagenomics focuses on analysing collective genomes of microorganisms directly from environmental samples. It helps in studying microbial communities contributing in nutrient cycling, degradation of pollutants and thereby soil fertility. Metagenomic analysis of rhizospheric microbiomes helped in identifying nitrogen-fixing bacteria, uncovered beneficial microbes (like mycorrhizal fungi or endophytes) that promote plant growth or protect against pathogens.

Phenomics: This branch of study integrates data on physical and physiological characteristics of plants to establish coordination between genotype and phenotype to assist as moderator in precision breeding programmes. Phenomics has been crucial in breeding stress-tolerant barley. Advanced imaging systems have been utilized to generate high-resolution phenotypic data, assisting workers in connecting genetic and environmental interactions to identify the best-performing varieties.

Transcriptomics: This focuses on study of total RNA or mRNA transcripts in plants. This assists in deciphering the expression of genes under specific conditions, such as pathogen attack or heat stress. Transcriptomic studies have identified genes involved in disease resistance in wheat. Researchers have successfully bred wheat varieties resistant to rust diseases by targeting these specific gene expressions, ensuring a significant reduction in crop losses.

Epigenomics: Epigenomics focuses on heritable changes in gene activity that are not caused by alterations in the DNA sequence but rather by modifications, such as DNA methylation and histone modification. In Arabidopsis (a model plant), researchers observed that plants exposed to drought retained 'stress memory' through epigenetic changes, allowing them to better survive subsequent drought events. This principle is now being studied in crops like maize and wheat for real-world applications.

1.4 Conclusion

The history of crop improvement is a testament to human ingenuity and adaptability. From the early days of seed selection to the cutting-edge genetic technologies of today, each era has contributed to the advancement of agriculture. Understanding these historical perspectives provides a solid foundation for appreciating the complexities and challenges of modern crop improvement efforts. Modern crop improvement techniques transform agriculture by leveraging cutting-edge technologies and scientific advancements. Modern crop improvement also emphasizes sustainable farming practices that promote environmental health and biodiversity. Cover cropping, crop rotation, and agroforestry help maintain soil fertility, reduce erosion, and enhance pest and disease management. Genetic engineering, CRISPR-Cas9, marker-assisted selection, genomic selection, tissue culture, precision agriculture, and sustainable farming practices all contribute to the development of more productive, resilient, and nutritious crops. These innovations promise to address global food security challenges and ensure a sustainable and prosperous future for agriculture.

References

Bhargava A, Srivastava S (2019) Human civilization and agriculture. In: Participatory plant breeding: concept and applications. Springer, Singapore, pp 1–27. https://doi.org/10.1007/978-981-13-7119-6_1

Boomer J (2023) Advances in crop improvement: a comprehensive review. Int Res J Agric Sci Soil Sci 1(1):1–3. https://www.interesjournals.org/agricultural-science-soil-science.html

Chakwanda MA, Chenge PT, Baloyi N, Tamburayi GL, Muchangana M, Jarbah P, Rera A, Sahoo JP (2024) Green revolution: the catalyst for agricultural transformation. Vigyan Varta 5(4): 294–302. https://doi.org/10.13140/RG.2.2.16649.45920

European Commission (2017) Precision farming: sowing the seeds of a new agricultural revolution. The community research and development information service (CORDIS). Publications Office of the European Union, Luxembourg. https://cordis.europa.eu/article/id/401847-precision-farming-sowing-the-seeds-of-a-new-agricultural-revolution

Evenson R, Gollin D (2002) The green revolution: an end of century perspective (Research Memorandum No. 17). Williams College. Retrieved from https://core.ac.uk/download/pdf/6315369.pdf

Feng CL (2024) Recent advances in CRISPR-based genome editing for enhancing staple crops. Front Plant Sci 1(1):1–21. https://doi.org/10.3389/fpls.2024.1478398

Fritz GJ (2007) Keepers of Louisiana's levees: early mound builders and forest managers. In: Denham TP, Iriarte J, Vrydaghs L (eds) Rethinking agriculture: archaeological and ethnoarchaeological perspectives. Left Coast Press, Walnut Creek, pp 189–209

Fuller DQ (2011) Pathways to Asian civilizations: tracing the origins and spread of rice and rice cultures. Rice 4(2):78–92. https://doi.org/10.1007/s12284-011-9078-7

Georges F, Ray H (2017) Genome editing of crops: a renewed opportunity for food security. GM Crops & Food 8(1):1–12. https://doi.org/10.1080/21645698.2016.1270489

Haas MS (2018) Domestication and crop evolution of wheat and barley: genes, genomics, and future directions. J Integr Plant Biol 60(3):197–377. https://doi.org/10.1111/jipb.12737

Hobhouse H (1999) Seeds of change: six plants that transformed mankind. Shoemaker and Hoard, New York, pp 1–381. Macmillan Publishing Ltd

Iriarte J (2007) New perspectives on plant domestication and the development of agriculture in the New World. In: Denham TP, Iriarte J, Vrydaghs L (eds) Rethinking agriculture: archaeological and ethnoarchaeological perspectives. Left Coast Press, Walnut Creek, pp 167–188

John DA, Babu GR (2021) Lessons from the aftermaths of green revolution on food system and health. Front Sustain Food Syst 5:644559. https://doi.org/10.3389/fsufs.2021.644559

Kamthan A, Chaudhuri A, Kamthan M, Datta A (2016) Genetically modified (GM) crops: milestones and new advances in crop improvement. Theor Appl Genet 129(9):1639–1655. https://doi.org/10.1007/s00122-016-2747-6

Khan AA, Iqbal B, Jalal A et al (2024) Advanced molecular approaches for improving crop yield and quality: a review. J Plant Growth Regul 43(8):2091–2103. https://doi.org/10.1007/s00344-024-11253-7

Khatik N (2024) Plant tissue culture and its applications in horticultural crop improvement. In: Biotechnological advances in horticultural crops. CRC Press, pp 1–22. https://doi.org/10.1201/9781003343707

Lantinga E, Rabbinge R (1996) The renaissance of mixed farming systems: a way towards sustainable agriculture. In: Jarvis SC, Pain BF (eds) Gaseous nitrogen emissions from grasslands. CAB International, Wallingford, pp 408–410

Larsen CS (1981) Skeletal and dental adaptations to the shift to agriculture on the Georgia coast. Curr Anthropol 22(4):422–423. https://www.jstor.org/stable/2742237

Leff B, Ramankutty N, Foley JA (2004) Geographic distribution of major crops across the world. Glob Biogeochem Cycles 18(1):GB1009. https://doi.org/10.1029/2003GB002108

Manish K. Vishwakarma, K. Thiyagarajan and P. K. Bhati (2023) Application of Unmanned Aerial Vehicles/Drones in Precision Agriculture. GRAIN ISSUE - 06 VOL – 02

Maurya DK, Maurya SK, Kumar M, Chaubey C, Gupta D, Patel KK, Mehta AK, Nunn N, Qian N (2010) The Columbian exchange: a history of disease, food, and ideas. J Econ Perspect 24(2): 163–188. https://doi.org/10.1257/jep.24.2.163

Peng S, Huang J, Cassman K, Laza MR, Visperas R, Khush G (2010) The importance of maintenance breeding: a case study of the first miracle rice variety—IR8. Field Crop Res 119(3):342–347. https://doi.org/10.1016/j.fcr.2010.08.003

Phillips RL (2013) Norman Ernest Borlaug: 25 March 1914–12 September 2009. Biogr Mem Fellows R Soc 59:59–72. https://doi.org/10.1098/rsbm.2013.0012

Pingali P (2010) Agriculture renaissance: making "agriculture for development" work in the 21st century. Handb Agric Econ 4:3867–3894. https://doi.org/10.1016/S1574-0072(09)04065-7

Pingali PL (2012) Green revolution: impacts, limits, and the path ahead. Proc Natl Acad Sci 109(31):12302–12308. https://doi.org/10.1073/pnas.0912953109
Piperno DR, Ranere AJ, Holst I, Iriarte J, Dickau R (2009) Starch grain and phytolith evidence for early ninth millennium B.P. maize from the Central Balsas River Valley, Mexico. Proc Natl Acad Sci 106(13):5019–5024. https://doi.org/10.1073/pnas.0812525106
Que G, Chilton MD, De Fontes CM, He C, Nuccio M, Zhu T et al (2010) Trait stacking in transgenic crops: challenges and opportunities. GM Crops 1(4):220–229. https://doi.org/10.4161/gmcr.1.4.13439
Rosset P, Altieri MA (1997) Agroecology versus input substitution: a fundamental contradiction of sustainable agriculture. Soc Nat Resour 10(3):283–295. https://doi.org/10.1080/08941929709381027
Roy PS, Roy A, Joshi PK, Kale MP, Srivastava VK, Srivastava SK, Dwevidi RS, Joshi C, Behera MD, Meiyappan P, Sharma Y, Jain AK, Singh JS, Palchowdhuri Y, Ramachandran RM, Pinjarla B, Chakravarthi V, Babu N, Gowsalya MS, Kushwaha D (2015) Development of decadal (1985–1995–2005) land use and land cover database for India. Remote Sens 7(3):2401–2430. https://doi.org/10.3390/rs70302401
Ryu J, Prather RS, Lee K (2018) Use of gene-editing technology to introduce targeted modifications in pigs. J Anim Sci Biotechnol 9(1):5–10. https://doi.org/10.1186/s40104-017-0228-7
Schiml S, Fauser F, Puchta H (2014) CRISPR/Cas system can be used as nuclease for in planta gene targeting and as paired nickases for directed mutagenesis in Arabidopsis resulting in heritable progeny. Plant J 80(6):1139–1150. https://doi.org/10.1111/tpj.12704
Setsuko K, Kamal AH, M., & Zahed, H. (2014) Wheat proteomics: proteome modulation and abiotic stress acclimation. Front Plant Sci 5:Article 684. https://doi.org/10.3389/fpls.2014.00684
Simon-Miquel G, Plaza-Bonilla D (2021) Crop rotation. In: Recarbonizing global soils: a technical manual of recommended management practices, Food and Agriculture Organization of the United Nations (FAO) and Intergovernmental Technical Panel on Soils (ITPS), vol 3. FAO, Rome, Italy, pp 26–37. https://www.fao.org/documents/en/
Slotten V (2020) Landscape management and polyculture in the ancient gardens and fields at Joya de Cerén, El Salvador. J Anthropol Archaeol 59:Article 101203. https://doi.org/10.1016/j.jaa.2020.101203
Stone D (2005) The use of agricultural techniques in medieval England. In: Decision-making in medieval agriculture. Oxford University Press, pp 191–207. https://doi.org/10.1093/acprof:oso/9780199247769.003.0008
Ullah Z et al (2024) The green revolution: promoting environmental stewardship and plant growth. In: Fahad S, Saud S, Nawaz T, Gu L, Ahmad M, Zhou R (eds) Environment, climate, plant, and vegetation growth. Springer, Cham, pp 425–469. https://doi.org/10.1007/978-3-031-69417-2_15
Woźniak A, Nowak A, Gawęda D (2021) The effect of the three-field crop rotation system and cereal monoculture on grain yield and quality and the economic efficiency of durum wheat production. Pol J Environ Stud 30(6):5297–5305. https://doi.org/10.15244/pjoes/134640

Advancements in Genomics: Applications in Crop Improvement, Genetic Mapping, and Germplasm Characterization

2

Laxmi Sharma, Suman Roy, Shruti Sinha, Sheel Yadav, and Pratik Satya

Abstract

Genomics, a baby born about 4.5 min ago in the 24-h timeline of crop improvement, is a collection of tools for inspection and dissection of genome structure and function. The technology boom in DNA sequencing has provided plant breeders with vast data on the structure, function and variation of genomic content in breeding populations, which are efficiently utilized in many major crop species for high-resolution genetic map construction, alignment of genetic and physical map, genomic variant identification, association analysis, mutant population generation and selection of superior genotypes using genome information. This chapter provides an overview of major areas of genomics and current applications of genomics tools in germplasm characterization, genetic mapping and quantitative trait loci (QTL) identification, mutant population development, genome-wide association analysis and genomic selection. We carefully avoided genome editing and transgenic technologies as these are separate technological branches that use gene sequence information as input but use an engineering approach to alter genetic makeup. Finally, current and possible future applications of genomics tools in evaluating crop wild relatives are discussed.

Keywords

Genomics · QTL · GBS · GWAS · SNP · Genomic selection · Crop wild relatives

L. Sharma · S. Roy · S. Sinha · S. Yadav
ICAR-National Bureau of Plant Genetic Resources, New Delhi, India

P. Satya (✉)
ICAR-Central Research Institute for Jute and Allied Fibres, Kolkata, India

Y. Ahlawat et al. (eds.), *Next-Generation Strategies for Crop Improvement*,
https://doi.org/10.1007/978-981-95-0309-4_2

2.1 Introduction

Crop improvement through the timeline of a civilization spanning approximately 12,000 years has witnessed the evolution, selection and domestication of about 7000 plants for serving food, fibre and energy to humanity. Throughout this period, new crop genotypes have been developed by domestication and long-term selection by the farmers. The era of artificial hybridization started with the discovery of the two sex forms in plants from the works of Rudolf Jakob Camerarius in 1694 and the establishment of a classification of plant kingdom based on floral structure by Carolus Linnaeus in 1735. Artificial hybridization was first extensively examined by Joseph Gottlieb Koelreuter during 1761–1766 in 54 species, indicating that hybridization success depends on the resemblance of the parents, and the hybrids are mostly intermediate in performance. Early selection principles for crop improvement were laid down by John Le Couteur in 1836 and Louis de Vilmorin in 1856. However, the scientific basis of plant breeding was only established in the early 1900s, with the rediscovery of Gregor Johanne Mendel's laws of heredity. The development of understanding of the science of crop improvement shows a high association with the yield performance of crop plants. For example, before 1900, the productivity of wheat around the world was around 1 tonne/hectare for about 1000 years, while in the past 100 years, the productivity increased to 3.6 tonne/hectare (Ritchie et al. 2022). Though crop productivity has increased by 300%–400% in the major food crops during the past 100 years, the world population has escalated over 500%, from 1.6 billion in 1900 to 8.16 billion in 2024. We must produce 50%–70% more food by 2050 to sustain the population. The upcoming constraints are manifold, including limited area, degraded soil, scarcer water, deteriorating climate, rising temperature and hazards of intensive agriculture. Hence, crop improvement needs to harness the benefits of modern scientific tools to enhance the yield potential of the major crops.

Genomics, the structural and functional genome analysis, can augment crop breeding programmes in multiple ways (Fig. 2.1). Genomics can help in the identification of molecular markers to construction of genetic and physical maps, identification of desired genes for obtaining desired change in the crop, understanding the biological functional processes that contribute to higher productivity, screening of genotypes at genomic and functional levels, creation of new variability and select superior genotypes for developing ideotypes for specific climates. For the past 20 years, genomics has contributed significantly to the primary crop improvement programmes worldwide, delving deeper into the genomic variation in the gene pool and helping breeders improve precision during selection. Genomics has revolutionized plant sciences by enabling comprehensive analysis of plant genomes, leading to significant advancements in understanding genetic diversity, gene function, and evolutionary biology. This field has facilitated the identification of genes responsible for important agronomic traits, such as disease resistance, drought tolerance, and yield improvement, with advances in sequencing technologies and molecular biology (Kumar et al. 2024). From the commencement of Sanger sequencing to the advent of next-generation sequencing (NGS) platforms, each technological

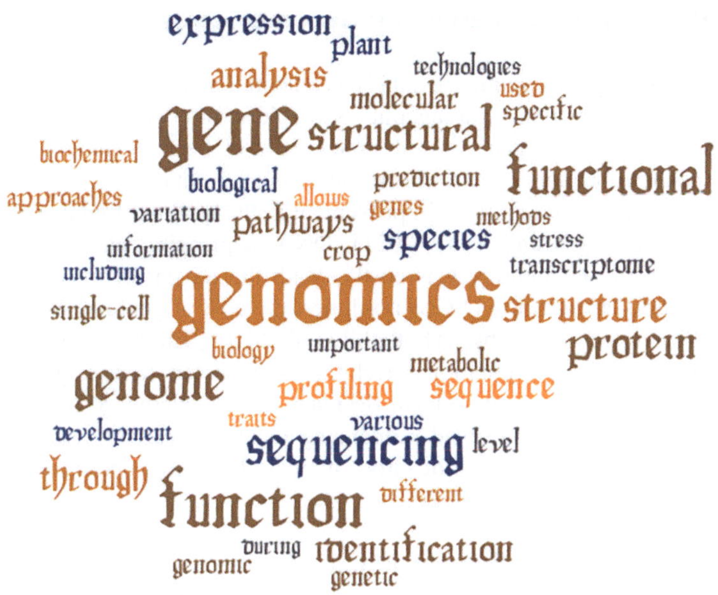

Fig. 2.1 A word cloud map of genomics

leap has increased the speed, accuracy, and affordability. In plant genomics, NGS has been instrumental in genome sequencing, gene expression profiling, the identification of genetic variants and structural changes, and the markers responsible for them (Everett et al. 2011). NGS technologies, such as Illumina and Ion Torrent, have enabled high-throughput sequencing of large genomes, uncovering genetic variations and facilitating the discovery of molecular markers.

2.2 Genomics: Classification

Genome refers to the total content of DNA or genetic material of the nucleus or the organelle. The term 'Genomics' was coined by Thomas Roderick in 1986 while deciding the name of an upcoming journal. The journal *Genomics* appeared first in September 1987, and the term became popular instantly, followed by the birth of a series of 'omics' terminologies like proteomics, transcriptomics, metabolomics, epigenomics, and metagenomics. The scope of genomics has been expanded from the initial structural and functional study of the genome to much broader applications, including evolutionary studies (phylogenomics), population analysis (population genomics), gene-environment interaction, genome regulations at multiple levels, medical studies and drug discovery (medical genomics, personal genomics, and pharmacogenomics), system biology study (system genomics), and analysis of model organisms.

2.2.1 Prerequisite of Genomics: Genome Sequencing

The generation of quality sequence data is an essential prerequisite for the success of genomics. The 'garbagein—garbageout' principle is also appropriate for genome sequence information; hence, several strategies have been developed to generate accurate genome sequences and check the sequence data quality. Over the past 20 years, there has been a notable improvement in the quality of genome sequences and assemblies (Soltis and Soltis 2021). DNA sequencing based on short-read Sanger sequencing technology was costly. With the advent of massively parallel 'next-generation' sequencing (NGS) machines, the cost of genome sequencing has dramatically dropped over time (Thottathil et al. 2016). The developments in NGS technology and new computational techniques have led to a large increase in sequencing capacity and a significant decrease in sequencing costs over the past years. Developments in long-read sequencing technologies (for example, PacBio and Oxford Nanopore) significantly increased contig N50 length from 124.6 to 4033.4 Kb (Marks et al. 2021).

2.2.1.1 Whole Genome Sequencing

Because of the large percentage of repeated sequences, widespread polyploidy, and heterozygosity, plant genomes are incredibly complicated and difficult to assemble (Jackson et al. 2011; Chen et al. 2019). They contain a variety of transposons, which are repeating sequences known to have a role in the wide variations in size between the collinear genome segments of diverse plant species. These factors may account for at least 50% of the variation in total genome size between species. The nuclear genome sizes of flowering plants, or angiosperms, vary by more than 1000 times; the smallest is *Genlisea tuberosa* (61 Mb), and the largest is *Paris japonica* (150 Gb) (Paajanen et al. 2019). As high as 70%–80% of flowering plants are the result of at least one polyploidization event (Barnes 2002); many economically significant plant species, including corn, wheat, potatoes, and oats, are either ancient or recent polyploids; wheat (*Triticum aestivum*) has three distinct homologous genomes. Consequently, most sequenced genomes have $< 0.8\%$ heterozygosity and $< 60\%$ repetitive sequences (Kong et al. 2023). Despite these obstacles, the genomes of over 798 land plant species have been sequenced (Marks et al. 2021); however, this number only represents a small portion (~0.16%) of the approximately 350,000 extant land plants, so more genomes must be sequenced to fill this gap. The highest numbers of genome assemblies are available for orders that are composed of agriculturally important species like Brassicales, Cucurbitales, Fagales, Malvales, Rosales, and Solanales (Fig. 2.2a).

2.2.1.2 Whole Genome Resequencing (WGRS)

With the availability of reference genome assemblies for 54 different crop species, genome resequencing initiatives have been undertaken in many of these species (Song et al. 2023) (Fig. 2.2b). Whole-genome resequencing recognizes the genomic variation in a population using a reference genome (Martin et al. 2018). It involves sequencing at a higher depth. A total of 54,413 accessions belonging to 187 plant

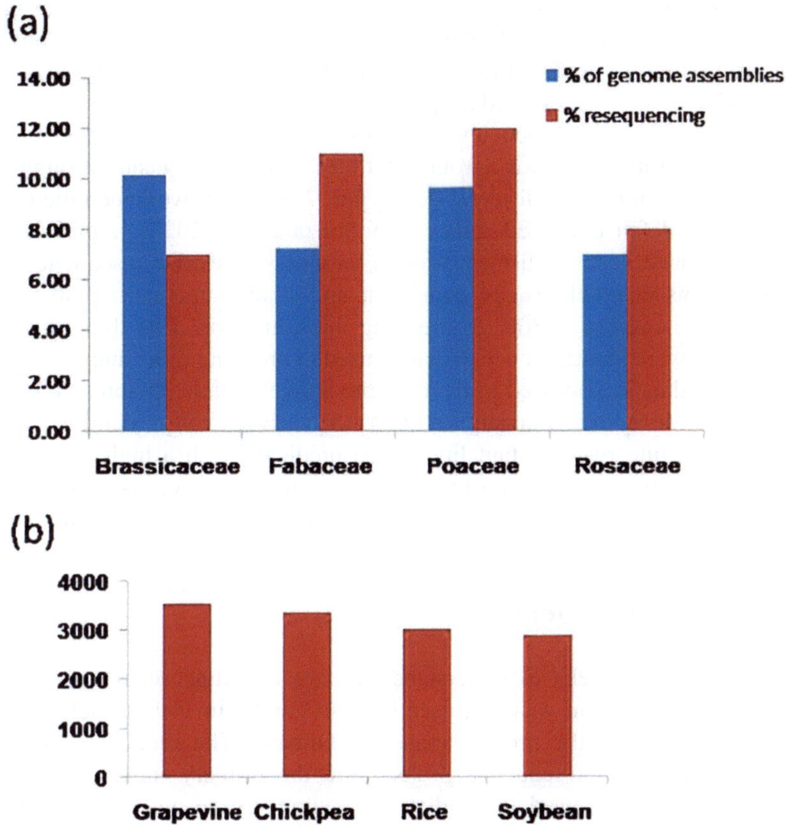

Fig. 2.2 (**a**) Approximate share (%) of genomes sequenced for the species belonging to the four major taxonomic orders. (**b**) The resequencing efforts undertaken in different crops with the number of accessions resequenced shown on the y-axis. (Data taken from Marks et al. 2021; Song et al. 2023)

species have been resequenced. These have provided novel insights into crop domestication, genome and phenotype associations, and candidate genes associated with various traits, which can be used for future research for crop improvement. The most noteworthy among these species with the number of accessions resequenced are given in Fig. 2.1b. Structural variants (SVs) in the form of single-nucleotide polymorphism (SNPs), InDels, etc., have been mined using resequencing approaches. This helped to study the genes under domestication sweep in vegetable soybean (Liu et al. 2020), tomato (Causse et al. 2013), eggplant (Gramazio et al. 2019), *Brassica napus* (Lu et al. 2019). A total of 588 *B. napus* accessions were resequenced, and the possible origins of the A and C subgenomes were deciphered (Lu et al. 2019). The narrow genetic diversity in the cultivated eggplant accessions (*Solanum melongena*) compared to the wild accessions (*S. incanum*) was identified through the resequencing of eight cultivated and wild accessions of eggplant

(Gramazio et al. 2019). This analysis led to the identification of introgressed genomic regions from the wild relatives into the cultivated accessions, providing crucial information to unravel the evolutionary history of *Solanum* species. Similar studies in diverse species reported a significant loss in gene content during evolution (Bayer et al. 2022). Resequencing has also allowed the identification of genes involved in controlling various agronomic traits. In common bean, through whole genome resequencing of 40 different genotypes, 7481 SNPs were identified in 85% of the disease-defence related genes (De Souza et al. 2023). Applications of resequencing have been confirmed through various studies. By resequencing over 3000 accessions of chickpea, 24 consistent and stable superior haplotypes were identified to be missing in 80% of breeding lines, and based on this information, 56 donor lines were identified which can be used in breeding (Varshney et al. 2021). Resequencing data is also used to develop predictive models. In spinach, based on the genome-wide association studies (GWAS)-derived SNPs identified through resequencing, white rust-resistant lines were predicted with a high correlation of 75% (Shi et al. 2022). Hence, resequencing can provide novel insights and robust predictionsfor designing breeding systems to achieve higher genetic gains.

2.2.2 Structural Genomics

Structural genomics seeks to understand the structure-function relationship of a DNA sequence. The term was first used by Eric Lander in 1996 to define the task of identification of all the protein shapes in humans. The sequence information generated from genome sequence projects are the primary inputs for structural genomics, which were used to predict the coding region and protein sequence followed by prediction of protein folding and three-dimensional structure of the protein. The development of robust computational biology programmes is an essential, integral part of structural genomics. The results of structural genomics studies during the past decades have culminated in elucidating protein structural and functional roles. The resemblances of the predicted structure, especially the folding pattern, with its homologues, if available, further allow the prediction of its molecular function. The molecular function, in turn, allows the prediction of its cellular function through its interaction with other genes and proteins using different genomic and proteomic approaches. The final validation is done through its biochemical assay after the prediction of the protein structure and its molecular and cellular function through different bioinformatics and modelling approaches. Moreover, the presence of ligands in the structure often provides additional clues that make the prediction of the biochemical function easier. Thus, structural genomics is a better approach for predicting the genome structure to function than comparing sequences. Protein Data Bank (PDB), the largest protein structure repository, has grown at an astonishing rate from a mere 13 entries in 1976 to 22,969 entries in 2024. These structures, determined by X-ray diffraction, nuclear magnetic resonance (NMR) spectroscopy, electron microscopy, or multiple methods serve as the basic repository for structural genomics analysis. However, the scope of structural genomics is not limited to experimental validation of protein structure, and 'omics'-based analysis of protein structures is also being gradually included in the domain of structural genomics.

2.2.3 Functional Genomics

Functional genomics clearly defines all the aspects of gene function at the transcriptome, proteome or metabolome level derived from its DNA structure. The first definition of functional genomics was proposed by Hieter and Boguski (1997) to denote the ideas and approaches for understanding the biological mechanism in the context of whole genome structure. The starting point of functional genomics projects may be the collection of mutant libraries or collections, either already available or generated through insertional mutagenesis and knockdown of gene function, as well as reading frames of sequenced genomes whose function has not been assigned. Transposable elements, T-DNA insertions, RNA interference-mediated knockdown or mutagenesis through random nucleotide insertions and substitutions are extensively used to generate mutant libraries. Insertion of these mutated sequences into expression systems through transgenesis is one of the popular approaches for the determination of the function of a gene. The multi-dimensional dynamic interaction and association of genes, transcription factors, proteins and metabolites during growth and development, as well as on exposure to environmental perturbations, are modulated by those regulatory elements coded in the genetic sequence. Functional genomics offers an incredible potential to generate precise metabolic models that can accurately predict gene functions and their interaction with other components in the biological system. Functional genomics integrates multi-omics information, including 'epigenomics', 'epi transcriptomics', 'proteomics' and 'metabolomics'. Integration of all these large-scale datasets is computationally challenging. Hence, robust algorithms are required to extract meaningful information from functional genomics data. New machine learning methods and artificial intelligence, including deep learning methods, are currently being developed to increase the efficiency of functional genomics (Caudai et al. 2021).

2.2.4 Transcriptomics

The emergence of NGS and advanced in silico analytical tools has established global transcriptome analysis as a fundamental method for elucidating the connections between genotype and phenotype. RNA sequencing represents a groundbreaking technique that enables extensive transcriptome profiling, thereby facilitating the identification of genes and alleles linked to various gene regulatory pathways, including those related to stress tolerance and sensitivity (Tyagi et al. 2022). The core principle of transcriptome analysis is founded on the notion that the depth of sequence coverage correlates directly with the expression level of the respective gene. Consequently, by mapping and quantifying the sequenced reads against a specific transcript, researchers can assess gene expression, transforming raw sequence data into biologically relevant insights. This advancement has rendered previous methodologies, such as microarrays and serial analysis of gene expression (SAGE), obsolete in gene expression analysis. Two new innovative technologies, namely single-cell RNA sequencing (scRNA-seq) technology and spatial transcriptome (ST) technology, have furthered the scope of transcriptomics in crop improvement (Wang et al. 2024). The distinguishing feature of these techniques over previously used bulk RNA sequencing lies in their capacity to detect RNA molecules within individual cells at a high resolution and across a genomic scale. The capabilities of single-cell transcriptomics, especially single-cell RNA sequencing, in elucidating the cell-specific responses to pathogen invasion have been eloquently articulated by Cole et al. in 2021. The authors illustrate how single-cell RNA sequencing facilitates the identification of unique transcriptomic responses at the cellular level during pathogen interactions. It is understood that only specific localized plant cells engage directly with most pathogens, and disease progression is regulated spatially and temporally. Employing single-cell RNA sequencing can significantly improve the signal-to-noise ratio, enhancing the detection of specific gene expression alterations against the backdrop of general gene expression fluctuations, thereby offering comprehensive insights into the dynamics of disease progression. ST utilizes combining advancements in situ RNA capture and ex situ sequencing methods and has the potential to address the limitations associated with earlier approaches, including low throughput and inefficient capture rates. These innovations have significantly enhanced sequencing resolution and capacity (Hu et al. 2024a, b).

2.2.5 Metabolomics

Metabolomics allows the profiling of a whole set of metabolites in the plant species, and thus, its integration with transcriptomics has been elemental in highlighting several important biochemical pathways like phenylpropanoid pathway, saponin pathway, cell wall biosynthesis, photosynthesis, and many other biosynthetic pathways. Moreover, metabolic profiling in horticultural crops at different growth stages, accessions, or mutants has elucidated various fruit ripening pathways or development pathways. Apart from these small molecules, lipid profiling has also assisted in bringing out the mechanisms of biotic and abiotic stress tolerance as plant lipid composition changes in response to stress. Thus, as an important part of functional genomics, metabolic profiling adds information to explain the gene function at the biochemical or phenotype level. Further, the association of metabolome with the structural genomics or the DNA sequence in diverse accessions through phenotyping helps in genome-wide association studies (GWAS). Previously, quantitative trait loci (QTL) mapping was such an approach to target a genomic region and the candidate gene with the assistance of phenotyping. QTL approach for allelic regions has been used in tomato and *Arabidopsis* by profiling metabolites (Barrero-Sicilia et al. 2017; Alseekh et al. 2015). GWAS has been popular for locating the candidate gene in various crops for traits of interest (discussed later). Recently, metabolome, centric GWAS (mGWAS) has emerged as a potential tool in functional genomics to associate the metabolite, both primary and secondary, with the predicted candidate gene related to various biosynthetic pathways, secondary metabolite pathways, amino acid biosynthesis pathways and many more. On the other hand, computational biology, with its powerful technologies, has increased the efficiency of genes and their function prediction for the next generation of functional genomics. Genome-scale metabolic models are now possible through deep learning-assisted genomics, and these theoretical models accurately define the metabolic profile and network of biological systems.

2.2.6 Pangenomics

A species' whole gene pool is known as the pangenome, which is made up of the 'dispensable' genome and the 'core' genome. A'core genome' comprises the minimum required DNA sequences for survival and carrying essential functions and is shared by all species members (Tahir et al. 2024). The best part of the pangenome is that it records the allelic variation in the species. Obviously, one reference genome assembly is inadequate to represent the gene pool of a species (Hu et al. 2025). Pangenomics has allowed the identification of structural variations (SVs) like presence-absence variation (PAV), copy number variation (CNV), and chromosomal rearrangements, which largely are responsible for gene content variation within a species. These SVs are implicated in governing various traits like the determination of environmental adaptation of plants to abiotic and biotic stress, flowering time domestication traits, etc. (Nitcher et al. 2013; Würschum et al. 2015; Della Coletta et al. 2021). The noteworthy and major crop species with pangenomes available are depicted in Fig. 2.3. The super-pangenome, which is a pangenome created out of the pangenomes of several species for a specific genus (Khan et al. 2020), is a new idea that allows investigating genome properties at a higher biological order.

The availability of pangenomes has bolstered the association studies to estimate marker-trait associations. This was because pangenomes could overcome the reference bias generated by using a single reference genome while mapping the sequence reads to the reference genome (Gage et al. 2019). This was validated in *Brassica napus*, where GWAS conducted based on eight whole-genome assemblies, led to the identification of PAVs associated with traits like silique length, seed weight, and flowering time, which were not identified typically by SNP-GWAS (Liu et al. 2020). Another example where the higher efficacy of pangenome over single reference genome-based mapping and deriving marker-trait association comes from wheat. The gene *YR63* of wheat that provides broad-spectrum resistance against multiple races of stripe rust that could not be captured in the first two reference genomes was correctly mapped in the pangenome, allowing its successful cloning (Mackenzie et al. 2023). The utility of pangenomes in developing effective genomic prediction models has also been demonstrated. It was observed that the predictive ability for determining plant height under low nitrogen conditions significantly increased by adding a few hundred CNVs to the existing ~20 k SNPs in maize (Lyra et al. 2019).

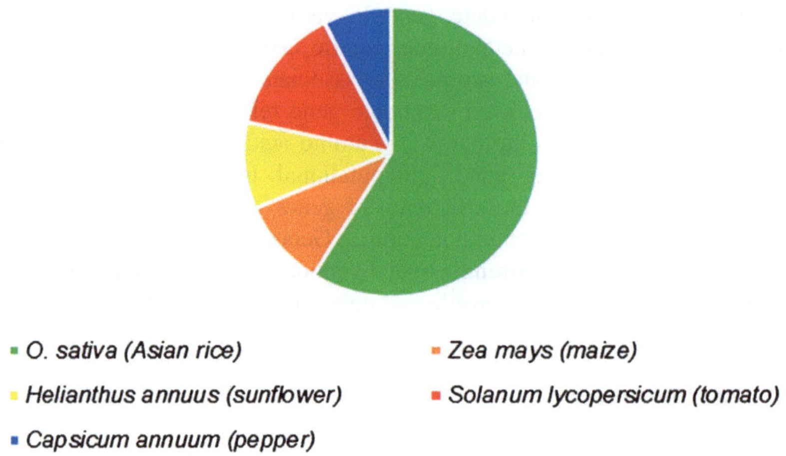

Fig. 2.3 Five major crop species with available pangenomes. (Data from Della Coletta et al. 2021)

2.3 Applications of Genomics Technologies in Crop Improvement

Crop improvement refers to the selection of superior genotypes from a population having variation, either generated naturally through mutation or open pollination or created by genetic recombination through artificial hybridization. Genomics can augment crop improvement programmes in multiple phases, from germplasm characterization, identification of molecular markers, integration of genetic and physical maps, fine mapping and gene identification associated with traits/QTLs, and selection of superior genotypes.

2.3.1 Germplasm Evauation and Genetic Diversity Analysis

Plant germplasms or genetic resources are the fundamental inputs for genetic improvement. The characterization and utilization of these resources are thus vital for crop breeding against the target of interest. These represent the total genetic diversity present in nature that has economic value in the present or for the future and include genetic stocks, landraces, weedy relatives, farmers' varieties, obsolete and modern varieties, breeding material and potential domesticates. Thus, these genetic resources are assigned a specific role as the donor of the desired trait for utilization in crop improvement programmes to achieve food security and sustainability. The use of molecular or genomic tools is gaining importance to characterize the germplasm owing to the advantage of generation of additional information like evolutionary origin, generation of inter-accessional and geographical relationship, identification of genetic diversity and duplicates, conservation of germplasm based on genetic diversity, designing of core collections and gene discovery (Jia et al. 2017). Over the years, SNPs have been a choice to study the genetic diversity in large accessions or populations. It provides an additional advantage of association of linkage disequilibrium that directly provides the association of genotype with the phenotype (Kilian and Graner 2012). Progresses in NGS have provided the opportunity to resequence thousands of individuals within a species to harness information on several candidate genes and the genomic regions, which will further derive the genetic and functional diversity of the germplasm pool. With the arrival of pangenomics, almost complete species diversity representation provides a convenient and robust platform for QTL mapping and gene identification. Pangenome clearly has prospect to investigate the evolutionary history, origin, domestication, and gene flow through the genetic variations observed between the cultivated and wild species (Qiao et al. 2021), functional gene mapping through QTLs and GWAS, mechanisms underlying the genome evolution within a species mainly the origin of structural variations like transposons, polyploidy, outcrossings, etc., and its impact on chromatin organization and expression patterns (Panchy et al. 2016; Dunning et al. 2019; Zhang et al. 2019) and opportunities for molecular breeding for crop improvement (Li et al. 2022).

2.3.2 Development of Molecular Markers

Molecular markers are DNA sequences identified by hybridization of small probes or primer sequences to the genomic DNA, generating differential hybridization or amplification patterns, which can be used to distinguish two genotypes. They have wide applications in plant breeding, from genetic diversity assessment, identification of heterotic parental combinations, genetic map construction QTL mapping, and marker-assisted selection. Some of these markers are species independent, where the primer or probe sequences are designed from random sequences or prominent genomic features common in eukaryotic genomes, such as repetitive elements, transcription start sites, highly conserved gene sequences or transposable elements. While these markers are excellent choices for diversity analysis, the marker loci on the linkage map may not correspond well with their physical proximity to the genome. A vast majority of DNA markers are species-specific, where the probe or primer sequences are designed from the genomic information of a species. When genome or transcriptome sequences are available for two or more genotypes, the sequences can be directly compared to identify allelic variants at a single nucleotide position. Whole genome sequences and transcriptome sequences allow the discovery of these species-specific markers, such as simple sequence repeat (SSR), diversity array technology (DArT) and single nucleotide polymorphism (SNP) that are abundant in the genome. The linkage maps constructed by these markers have genotype-specific allelic variants ordered according to their relative positions. Therefore, these markers are excellent choices for aligning with the genomic contigs for physical maps. Besides these small nucleotide differences, large-scale changes to the genome, including insertions, deletions, duplications, and inversions, also exist in the population, collectively known as structural variants (SVs) that significantly impact gene function and phenotypic characteristics. They can now be easily identified due to advances in bioinformatics analysis from NGS data and comparative genomic studies. Duplications or deletions of DNA segments that alter the number of copies of a specific gene or area are known as copy number variations or CNVs. Small-scale changes in the DNA sequence are known as indels (insertions and deletions), which are also responsible for variations and polymorphism in different individuals, such as in rice (Huynh et al. 2021). The integration of NGS in plant research has accelerated the discovery of marker polymorphism at a higher pace and precision (Imelfort et al. 2009).

2.3.2.1 Mining of Simple Sequence Repeat (SSR) from Genome/Transcriptome Sequence

Simple sequence repeats (SSRs) are abundant, polymorphic tandem repeats of one to six nucleotide sequences randomly spread in various species' nuclear, chloroplastic or mitochondrial genomes (Amiteye 2021). Previously, SSRs were identified by constructing genomic or expressed sequence tags (EST) libraries through cloning and Sanger sequencing. With the advances in NGS, SSR discovery has become an integral part of genome and transcriptome sequencing studies. Using the basic features, the SSR sequences can be extracted from the sequence database, followed by the design of primers flanking the SSR sequences. Finally, the markers are validated in more populations to determine the extent of SSR polymorphism. The proficiency of polymorphic SSR identification from sequence data depends on the sequence quality and the extent of variation in the genotypes sequenced. Mining of SSRs from 18 sorghum genomes revealed the presence of 39,120 to 64,667 SSRs in different genomes (An et al. 2023), indicating that the success of SSR identification depends extensively on sequence quality. Several SSR databases have been developed from genomics studies, such as TriticeaeSSRdb for wheat and its relatives, LegumeSSRdb for legumes, Genomic SSR Millets Database, JuteMarkerdb for jute, etc. In addition, crop genome databases like RiceGDB and MaizeGDB also include information on SSR markers. SSR markers can also be developed from genes (genic SSRs), intronic regions, regulatory regions and other genomic landmarks. For example, SSRs have been mined from microRNA coding regions in wheat that are linked to heat tolerance (Sihag et al. 2021). In jute, genes associated with the phenylpropanoid pathway and fibre development were mined to identify genic SSRs with high genetic diversity (Satya et al. 2017). It has been observed in maize that the diversity of SSRs was much higher in the genic and intronic regions compared to other genomic regions, indicating the evolution of SSRs may not be neutral (Zhao et al. 2023).

2.3.2.2 Identification of SNPs

SNPs are biallelic variants found in the genome with the highest frequency. Like SSRs, SNPs are principally identified from genomic or transcriptomic sequences by mapping genomic or transcriptomic sequences of a genotype on a reference genome by a method called variant calling. The calling may be performed without a reference genome using RNA-seq, genotyping by sequencing (GBS), or reduced-representation-based sequences like RAD-seq and targeted amplification. Openly available RNA-seq data are a good starting point for SNA identification that has been extensively utilized in SNP identification of major crops. Although RNA-seq is a potent method for discovering SNPs, there are ongoing challenges, especially in the areas of read length selection, the choice of assemblers, and the tools used for SNP calling, all of which have a considerable impact on the accuracy of detection (Costa et al. 2010). GBS offers more comprehensive genome coverage for SNP identification over RNA-seq. Millions of SNPs have been identified in major crops (Table 2.1) using data generated from various sources and sequencing strategies. These sequences have been used to construct genetic maps and identify QTLs/genes. SNP can also be identified using microarray-based genotyping, and the two technologies can be combined for optimization. The more recently developed and advanced DNA-based marker systems based on SNP genotyping include Kompetitive Allele-Specific PCR SNP genotyping system (KASP) developed by KBiosciences, Restriction site-associated DNA (RAD), Illumina Goldengate and multiplex shotgun genotyping (MSG) (Guo et al. 2014; Deschamps et al. 2010). Further, advanced DNA-based molecular marker techniques utilized in genome analysis of crops lacking reference genome sequence include double digest RAD (ddRAD) and genotyping by sequencing (GBS) (Heinrich et al. 2020).

2.3.3 Genetic and Physical Map Construction

The primary objective of genetic mapping is to identify markers located near the genetic determinants or genes responsible for complex quantitative traits (Abdurakhmonov and Abdukarimov 2008). There are primarily two methods for genetic mapping; the first method relies on recombination in breeding populations generated from controlled crosses to develop a linkage map. The second method depends on the historical association of loci in diverse genotypes that are related ancestrally and is known as linkage disequilibrium (LD) mapping or 'association mapping' (Xu et al. 2017). Classical linkage mapping is the most robust approach to developing a genetic map. The construction of a high-resolution linkage map relies on several factors, including the nature of the mapping population, marker polymorphism, and the breeding behaviour of the crop. The main limitation of mapping by utilizing a biparental population is that single recombination in the female F_1 plant generates the population, which may not be sufficient for high-resolution mapping of QTL. To overcome this limitation, in many crops, multi-parent mapping populations (nested association mapping (NAM) and multi-parent advanced generation intercross populations (MAGIC)) have been developed (Scott et al. 2020; Satya et al. 2023). On the other hand, association mapping provides for the optimal use of ex-situ conserved natural genetic diversity found within global crop germplasm resources. Linkage disequilibrium (LD) denotes a historically diminished level of recombination among specific alleles at various loci that govern particular genetic variations within a population (Goldstein 2001). SNP markers are available as chips

Table 2.1 Examples of molecular marker development using genomics in crops

Sl No.	Crop	Sequencing platform	Number of markers	DOI
1	Chickpea	Illumina HiSeq 2000	4.97 million SNPs	10.1038/s41588-019-0401-3
2	Cotton	Roche 454	151,712 SNPs	10.1038/s41588-019-0401-3
3	Cotton	lluminaHiSeq 2000	40,503 SNPs	10.1534/g3.114.012542
4	Faba bean	Illumina HiSeq X Ten	1,579,411 SNPs	10.3389/fpls.2021.618133
5	Finger millet	Roche 454 and Illumina	10,327 SSRs and 23,285 SNPs	10.1371/journal.pone.0159437
6	Finger millet	Illumina NextSeq	17,245 SNPs	10.1186/s12870-018-1316-3
7	Jute	Illumina HiSeq X	3269,808 SNPs and 1179,601 InDels.	10.1016/j.indcrop.2023.117247
8	Jute	Illumina HiSeq™ 2000	1115 SNPs	10.1007/s00438-018-1526-2
9	Lentil	Illumina NovaSeq 6000	734,716 InDels	10.1371/journal.pone.0302870
10	Maize	Roche 454 Sequencing	4543 SNPs	10.1111/j.1365-313X.2007.03193.x
11	Pea	Illumina HiSeq	419,024 SNPs	10.1111/j.1365-313X.2007.03193.x
12	Pigeonpea	Illumina Hiseq X-ten platform	1,419,904 SNPs; 292,347 InDels; 92,903 SVs	10.1016/j.ijbiomac.2023.128559
13	Pigeonpea	PacBio Sequel II and BGI Tech MGISEQ-2000	3.97 million SNPs and 0.97 million InDels	10.1093/hr/uhae201
14	Potato	Illumina HiSeq 2500	204,397 SNPs and 6217 indels	10.1007/s00122-024-04651-8
15	Potato	Illumina	19,560,602 SNPs	10.1038/s41598-024-55669-3
16	Rice	Amplicon sequencing, Sanger sequencing	54,465 SNPs	10.1007/s00122-011-1633-5
17	Rice	lluminaHiSeq 2000	2,040,30 SNPs and 184,111 InDels	10.14719/pst.2021.8.3.1047
18	Sugarcane	Illumina HiSeq 2000	76,251 SNPs, 5642 InDels, 6380 PAVs, and 826 CNVs	10.1186/s12864-017-3980-3
19	Sugarcane	Illumina Hiseq2500	6,064,835 SNPs	10.1111/tpj.16861
20	Tobacco	Illumina Novaseq6000	580,942 SNPs	10.3389/fpls.2021.618133

(continued)

Table 2.1 (continued)

Sl No.	Crop	Sequencing platform	Number of markers	DOI
21	Tobacco	Illumina Novaseq6000	17.9 million SNPs	10.1007/s11756-024-01853-9
22	Tomato	5500xl SOLiD sequencer	1,473,798 SNPs	10.1093/dnares/dst033
23	Wheat	Illumina HiSeq2000	3.3 million SNPs	10.1371/journal.pone.0186329

and arrays, which have enabled high-throughput genotyping in many crops. In barley, a 50 K single nucleotide polymorphism (SNP) genotyping array was used to map and precisely identify alleles controlling plant height and flowering time (Cosenza et al. 2024). It has been estimated that prior to NGS, QTL resolutions were low (10–30 cM) that has increased substantially (0.5–10 cM) due to NGS (Jaganathan et al. 2020). The various techniques followed for fine mapping, which now constitute a vast majority of mapping studies, include QTL-seq, BSR-seq (Bulked segregant-RNA Seq), WGRS (whole genome resequencing), RAD-seq (restriction site-associated DNA sequencing), GBS (genotyping by sequencing), etc. Because near-finished, reference-based genomes are available for many cereal crops like rice, wheat, maize, sorghum, barley, and pearl millet, most of the above-cited techniques have been used in these crops and, therefore, monumental progress in the area of trait mapping has taken place in these crops. An open-access high-resolution platform was developed in rice containing a wide collection of diverse germplasm, high-density SNP marker data and highly sophisticated bioinformatics tools to facilitate genome-wide association mapping (McCouch et al. 2016). In wheat, high-density linkage mapping utilizing markers derived from NGS has facilitated the precise mapping of significant traits, including resistance to fungi (Cockram et al. 2015), resistance to powdery mildew (Liang et al. 2016), characteristics of awn shape and length (Yoshioka et al. 2017), traits related to flag leaves (Hussain et al. 2017), and resistance to stripe rust (Ma et al. 2019). Similar success stories are available in many other species. Thus, it can be concluded that the projects that previously required a significant duration of 10–15 years for fine mapping, cloning quantitative trait loci (QTL), and identifying a candidate gene can now, with NGS and bioinformatic analysis tools, be accomplished in a timeframe of 10–20 months.

2.3.4 Genotyping by Sequencing (GBS)

Genotyping by sequencing (GBS) is perhaps the most applied genomic technology in crop improvement. It employs a reduced-representation sequencing technique based on restriction enzyme digestion where a barcode system was developed to allow the discovery of genome-wide SNPs with a lower error rate and cost (Elshire et al. 2011). The restriction enzyme Ape KI has been frequently used in GBS, although later, multiple enzymes were more effective. AGBS protocol employing two enzymes (PstI/MspI) was developed in wheat and barley for complexity reduction. It was more effective for uniform library preparation than the ApeKI-based GBS (Poland and Rife 2012). To further make GBS an economically viable tool for genomic selection (GS), a new approach called 3D-GBS was formulated using three restriction enzymes (PstI/NsiI/MspI) (de Ronne et al. 2024). The protocol reduced genome complexity to fourfold, with a simultaneous reduction of ~ 92% cost for GS and 86% for QTL mapping. The high-throughput SNPs detected by GBS are widely used for genetic diversity analysis, genome-wide association studies (GWAS), QTL mapping, and genomic prediction (GP) in many plant species (Kim et al. 2016; He et al. 2014; Scheben et al. 2017) (Table 2.2). The advantage of GBS lies in the fact that it does not require any a priori sequence information of the desired species, as it simultaneously conducts SNP discovery and genotyping (Kim et al. 2016).

Table 2.2 Application of GBS in crop improvement

Crop	Variants/SNPs	Application	DOI
Rice	30,984 SNPs	QTLs for leaf width and aluminium tolerance identified	10.1007/s00122-013-2166-x
	8145 SNPs	Four QTLs for number of tillers	10.1534/g3.116.038190
	10,836 SNPs	114 QTLs associated with cold tolerance under Brassinosteroid response	10.3389/fpls.2022.938339
Maize	1.4 million SNPs and 200,000 indels	Divergent haplotypes	10.1126/science.1177837
	681,257 SNP markers	SNPs associated with kernel colour, sweetness, and flowering time were identified	10.1186/gb-2013-14-6-r55
	One million SNPs	Marker generation for maize breeding	10.1038/s41598-020-73,321-8
Wheat	28,644 SNPs	Three rust-resistant genes and 15 QTLs were validated	10.1186/s12864-015-1424-5
	16,506 SNPs	Genetic diversity in Iranian wheat	10.3389/fpls.2017.01293
Soybean	205,614 SNPs	Genetic diversity and evolutionary relationship	10.1038/ng.715
	47,000 SNPs	Marker-trait association for five different traits	10.1111/pbi.12249
Yellow mustard	828 SNPs	Genetic diversity in germplasm	10.1007/s10722-013-0058-1
Brassica	8092 SNPs	Seven QTLs for disease resistance	10.1007/s00122-023-04483-y
Potato	30,363 SNP	189 QTLs	10.1007/s00122-024-04651-8
Chickpea	3237 SNPs	QTLs for drought tolerance traits identified.	10.1038/s41598-023-44990-y
Pea	95,740 SNPs	Genomic selection models for grain yield under severe terminal drought	10.3835/plantgenome2016.07.0072

2.3.5 Genome-Wide Association Studies (GWAS)

The primary goal of genomics is to understand the genotypic basis of the phenotypic variation. Various molecular tools have been used to generate genetic maps and identify the genomic regions or candidate genes responsible for the trait of interest. Continuous development in sequencing technologies and efforts from the scientific community to dissect complex traits using large natural populations have made genome-wide association studies (GWAS) a tool of choice. GWAS have emerged as an increasingly potential and powerful genomic tool to study the association of genetic variants with the trait of interest using high throughput genotyping and high throughput phenotyping technologies (Dhondt et al. 2013; Ellegren 2014; Gill et al. 2022). The existing variants in the natural population or cultivars are used as samples associating to the phenotype with the genetic variant. Genetic variants associated with desirable traits for selective breeding (Albert et al. 2016; Yano et al. 2016), variants associated with various biotic stresses (Todesco et al. 2010; Demirjian et al. 2023), abiotic stresses (Josephs et al. 2017; Rees et al. 2020) and resolving biological pathways (Atwell et al. 2010) have been identified using GWAS studies. The basis of GWAS underlies employing the SNPs, identified through genotyping, to mark a genomic region carrying the gene of interest, followed by the association of the marker with the trait and the identification of the candidate gene responsible for the trait expression. The association between the genetic variants and the phenotype of the diverse natural population is based on linkage disequilibrium (LD). A lower LD is expected on a diverse panel, which will further generate a higher recombination frequency between the genetic loci and the trait, allowing for a finer map with higher polymorphism. GWAS is conducted on the set up of experimental design, capturing the variation, genotyping technologies, association tests, and statistical data analysis through replication and meta-analysis (Bush and Moore 2012). The GWAS analysis protocol comprises a population selection, sample collection, genotyping and data processing (quality control), association studies and data interpretation.

2.3.5.1 Selection of Population and Sample Selection

A sizable population is required as a fundamental and essential input for conducting GWAS to identify the genetic variant associated with the trait of interest. However, population diversity is more important than population size. Genetic studies must determine the location and density of commonly occurring SNPs in particular genomic regions, and the correlation among the common genetic variants must be drawn to avoid redundancy. Moreover, the experimental design set-up is based on the trait and is set on *case* versus *control* or quantitative design. The quantitative design is set on populations where trait measurement is required. At the same time, the case versus control is carried out on qualitative traits, where the data is recorded via scoring. For instance, in *case* versus *control*, if an individual plant has the presence of the trait, it is the control. In contrast, it will be the case when the individual/accession does not express the trait or has a deviation of expression from the *control*. The quantitative design is set up when a gradient of variation with a wide range is observed on a specific trait. Thus, the design set-up and sample selection depend on the population size, a trait of interest, attributes of the population (geographical area, morphological character, etc.), availability of pre-existing data and the ease of data collection.

2.3.5.2 Genotyping Technologies

Genotyping for GWAS relies on the SNP identification technologies, as discussed earlier. Whole genome sequencing, genotype by sequencing, RNA-seq or RAD-seq technologies are extensively used for GWAS.

2.3.5.3 Quality Control and Genotype Imputation

Before proceeding to the association studies, the quality control of the generated sequence is essential to remove the false signals in the genotype-phenotype test. Monomorphic variants and genotyping errors are eliminated by using filters. Quality control mainly depends on the population structure, evolutionary relationship, and duplication (Pavan et al. 2020). After the false signals are filtered, the information of the missing genotype is derived against the reference genome through genotype imputations. It is an indirect method of assigning SNP to the genotype that has not been sequenced. It is done to improve the efficiency of meta-analysis by enhancing the power and association of fine mapping.

2.3.5.4 Phenotyping

The phenotypic data for the trait(s) of interest are to be recorded accurately to reflect the actual variation among the population. Many GWAS studies use high-throughput phenomics platforms for precision phenotyping.

2.3.5.5 Data Analysis

On the availability of the phenotypic and genotypic data, the association between them is drawn through linear or logistic regression models, depending on the design of the experiment. The spurious association is one of the major problems associated with the genetic variant with the phenotype. A mixed linear model (MLM) or multi-trait GWAS is used in advanced GWAS modes. MLM GWAS accounts for two levels of genetic relatedness [population structure (Q) and kinship (K)] as well. It also accounts for the LD of the population, which reduces the chances of spurious association for further fine mapping to identify the gene(s) for the trait of interest (Fig. 2.4).

2.3.5.6 Applications of GWAS

GWAS targets to understand the genetic mechanism underlying the traits of interest for crop improvement under variable environmental conditions. GWAS has been used to study the genotype to phenotype in several crops like rice, wheat, maize, barley, soybean, cotton, tomato, cucumber, sesame, peanut, peach, melon, tea, and lettuce (Alseekh et al. 2021). Several studies have highlighted GWAS mediated yield attributes in several crops. For example, flowering time, panicle architecture and grain attributes have been key target traits of GWAS in rice for the identification of genes for crop improvement in several studies. A study used 49 panicle

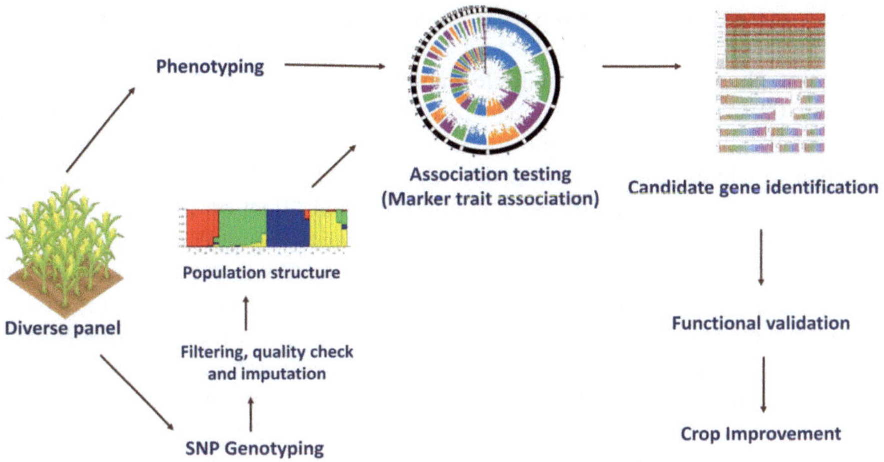

Fig. 2.4 Stepwise schematic representation of genome-wide association studies

phenotypic traits in a diverse panel of 246 tropical rice accessions and identified 10 robust genomic locations that revealed a relationship between panicle size and yield (Crowell et al. 2016). Similarly, in 809 soybean accessions, 84 agronomic traits were assessed to underline the multifariousness of yield traits (Fang et al. 2017). Moreover, analysing the traits with metabolic profile also provides better information on the genetic basis underlying the regulation of metabolome with sequential studies like screening of mutants, identification of gene family and comparison of sister families to the genetic map to dissect the metabolic pathways underlying the trait of interest. Using 368 maize inbred lines with 83% phenotypic variation for oil content, GWAS identified 74 loci associated with kernel oil content and fatty acid composition. Out of these, 24 loci were conserved for enzymes of oil biosynthesis (Li et al. 2019). Similarly, tomato fruit quality has been evaluated to characterize the metabolic profile and fruit taste panel (Tieman et al. 2017). Combining metabolic and transcriptome profiling traits preferred by domestication has been addressed (Zhu et al. 2018). Similar studies have also been carried out to approach the genetic mechanisms for biotic and abiotic stresses. Applying single- and multiple-trait GWAS, the leaf resistance to stripe and stem rust has been investigated in 2300 wheat accessions (Joukhadar et al. 2020). Haplotype analysis followed by GWAS identified SEMIDWARF1 as a potential gene for imparting flood tolerance in rice. Several examples of biotic and abiotic stress are highlighted in the literature. GWAS has been successful in identifying several loci for traits of interest. Using a simple and robust statistical model, the classical or standard GWAS method relates the genotype to phenotype. Since the association in standard GWAS is made between a single SNP and a single phenotype, the chance of spurious association increases (Clauw et al. 2024). As a result, promising loci or valuable genetic associations may be missed. The association may also be challenging to validate. Consequently, standard GWAS models may not represent well complex natural variation. The primary reasons for the spurious association include the existence of a complicated pattern of correlation among the trait as well as the trait and the environment, imprecise phenotyping, and multiple haplotypes (Clauw et al. 2024). Thus, the standard GWAS using a simple model restricts the clear understanding of the genetic basis of the trait. New multiple model GWAS have been developed along the timeline to address the problem (Tibbs Cortes et al. 2021) to reveal accurate and efficient associations. To address the environmental effect on the trait, the multi-locus GWAS is used to assess the pleiotropic relationship directly. The multi-trait GWAS thus addresses the genetic, environmental and G x E effect easily by linking the multivariate regression with the population structure control through multi-trait mixed model (MTMM) GWAS (Korte et al. 2012). The G x E effect using MTMM has been extended to many crops, such as *Arabidopsis* and tomato under drought (El-Soda et al. 2015; Albert et al. 2016) and flowering time of Arabidopsis under high temperature (Sasaki et al. 2015). The multi-locus models are efficient and often result in significant marker-trait associations and candidate gene identification.

2.3.6 Genomics in Mutational Analysis: TILLING and Its Variants

Identification of functional mutants and their underlying causes has long been an interest to plant breeders. There are two general approaches for gene function identification: forward or reverse genetic. In a forward genetics approach, researchers identify phenotypic traits of interest, such as disease resistance, drought tolerance, or high yield, and then work to determine the underlying genetic basis. This is traditionally carried out by strategies that involve mutagenesis and transgenic approaches for genetic mapping and quantitative trait locus (QTL) analysis. Another approach, reverse genetics, starts with a gene of unknown phenotype and searches for its functionality by introducing mutations or altering its expression to observe phenotypic changes. Techniques such as RNA interference (RNAi), CRISPR-Cas systems, and Targeting Induced Local Lesions in Genomes (TILLING) are common reverse genetics and a precise tool for gene-function studies and trait improvement. With developments in molecular biology techniques, the identification of changes in nucleotide sequences from plants was realized. At this time, the classical mutagenesis approach was combined with molecular biology technique to screen a population for desired mutational changes in the nucleotide sequences (McCallum et al. 2000). The TILLING process involves three major steps- mutant population creation, targeted screening of these mutants and analysis of the phenotype of those mutants. TILLING started with the generation of an ethylmethanne sulphonate (EMS)-induced mutant population. TILLING was first used for the development of a mutant population using EMS in Arabidopsis thaliana and the identification of mutants by detecting changes in heteroduplex analysis using denaturing high-performance liquid chromatography (DHPLC) (McCallum et al. 2000). Classically generated mutant populations of crops like maize and rice were screened through a high-throughput method for genome-wide identification of variations point mutations or small insertions and deletions in targeted regions of DNA (Till et al. 2004; Till et al. 2007).

TILLING is potentially helpful to identify functional gene-gene variants in any crop capable of antagonization regardless of its mating/pollinating system, ploidy level, or genome size (Chen et al. 2014). It can detect allelic series of mutations causing changes in nucleotide sequence such as silent, missense or nonsense that generates sub-lethal and sub-sterile mutants (Greene et al. 2003). It is a non-transgenic approach and thus is exempted from GM regulations capable of detecting even single nucleotide polymorphism (SNP) in plants (Colbert et al. 2001). These SNPs are the causatives for mutations in the gene or its regulatory regions, leading to changes in the gene function. The identification of historic SNPs that have been inherited stably in the population for a very long and lead to phenotypic plasticity of the species is equally essential for breeders and evolutionary biologists. Such changes exist in the naturally occurring alleles of a gene in a diverse population of the species. Thus, EcoTILLING, a variant of TILLING, was developed by Comai et al. in 2004 to examine natural genetic variation in populations. Since then, it has been successfully utilized in plants to discover SNPs, including rare ones. Therefore, the variation detected by Eco-TILLING is not limited to functional analysis but expands to study phylogenetic diversity and allele mining in crops. Thus, TILLING has applications in rare mutation detection and trait improvement while generating a genomic resource for the developed mutant population as a seed bank or its nucleic acid sequence as an essential resource for the research community, also known as the TILLING Platform. TILLING mutant population resources are available for several crops like maize, wheat, rice, soybean, tomato, tobacco, pea, barley, and peanut (Chen et al. 2014). The seeds or DNA of 3000–5000 M2 population and up to 10,000 M2 in some crops are archived in the repository and are called TILLING Platforms. Multiple crop-specific TILLING platforms have been developed CoTIP for cotton (Aslam et al. 2016), BRACHYTIL for *Brachypodium distachyon* (Dalmais et al. 2013), HorTILLUS in Barley (Szurman-Zubrzycka et al. 2018), Micro-Tom TILLING platform (Okabe et al. 2013), its database TOMATOMA (Shikata and Ezura 2016) and Red Setter TILLING platform for Tomato along with a LycoTILL phenotype database (Minoia et al. 2010), Piel de Sapo TILLING platform for *Cucumis melo* L. non-climacteric *cantalupensis* group of the subspecies *melo* (González et al. 2011).

Multiple variants of TILLING, such as high-resolution melting (HRM), assist TILLING, double-stranded EcoTILLING (DEcoTILLING), Self-EcoTILLING, Individualized TILLING (tilling), Deletion-TILLING (De-TILLING), PolyTILLING, and VeggieTILLING; In silico TILLING platforms have been reviewed by Singh et al. (2024). The recent one is TILLING by Sequencing (TbyS), developed by Tsai et al. 2011 who used NGS coupled with multidimensional pooling to identify rare alleles in populations using rice (*Oryza sativa*) and wheat (*Triticum durum* and *Triticum aestivum*). TbyS is a highly efficient, robust and cost-effective approach that is being increasingly utilized to discover rare mutations in crops like polyploid soybean where a detected mutation in stearoyl-acyl carrier protein desaturase gene caused sub-functionalization of homeologous genes for specific roles in unsaturated fatty acid biosynthesis (Lakhssassi et al. 2020), 16 mutations in 21 flowering-related genes in sunflower (Fanelli et al. 2021), in stress resistance gene phospholipase D (PLD) and lipoxygenase (LOX) gene for their role in drought tolerance in peanut (Guo et al. 2015). EcoTILLING of the sucrose synthase genes (Sus) gene in 277 accessions of cotton identified 24 allelic variations for fibre yield and quality (Zeng et al. 2016). EcoTILLING in the Starch synthase gene in wheat identified haplotypes for yield attributing factor thousand grain weight (Irshad et al. 2019). The TbyS approach has led to the identification of causative SNPs for traits like phytic acid content in rice (Kim and Tai 2014), crop yield by regulating kernel number in wheat (Bommert et al. 2013), abiotic stress tolerance in barley (Xia et al. 2013) and many more. ECoTILLING of diverse rice landraces and varieties led to the identification of markers for drought tolerance. The 95 rice varieties were scanned with EcoTILLING, and the polymorphic loci were then used to study the association of the phenotype and drought tolerance. The study led to the identification of variation in promoter sequences of five transcription factor families, namely Alfin-like, AP2/EREBP, GRF, Tify, and CCAAT-HAP3, correlated with the drought tolerance of rice genotypes (Yu et al. 2012). Raja et al. (2017) used the ETbyS approach (EcoTILLING by sequencing) of six candidate starch synthase family genes with a potential role in regulating glycaemic levels in rice. Using 512 *indica* rice germplasm, they uncovered variants with deletions in starch synthase homologs.

2.3.7 Genomics for Increasing Selection Efficiency (Genomic Selection)

While GWAS identifies SNPs associated with traits, genomic selection (GS) or genomic prediction (GP), a variation of marker-assisted selection targets to utilize genome-wide marker information to derive genotype-phenotype relationship. GS can quickly improve traits with low heritability and high trait complexity and significantly reduce breeding costs (Crossa et al. 2017). The genetic gain depends on the ratio of the additive genetic variance to the square root of phenotypic variance at a given selection intensity. Since additive genetic variance is the variance of breeding value (BV), attempts have been made to increase true breeding value (BV) by better estimating the genetic value of the progeny, followed by increasing the accuracy of the estimated breeding value (EBV). The framework of genomic selection in plants was developed by Meuwissen et al. (2001) in a simulated population for estimating the relationship between true breeding values and estimated breeding values (EBV). The target population, also termed as breeding population (BP) or test population (TST), is genotyped using the same set of SNPs. The TP is phenotyped and genotyped to derive the EBV, while the BP need not be phenotyped. The allelic association of marker loci with the phenotypes in the TP is used to predict the phenotypic value of individuals having similar allelic associations in the BP. This is particularly useful in early generation evaluation, such as in the F_2 or F_3 population, where estimation of additive genetic variance using phenotypic observation is less accurate.

Several statistical methods have been developed to increase the accuracy of estimated EBV, including the best linear unbiased prediction (BLUP) and genomic-BLUP (G-BLUP) (Crossa et al. 2017). Best linear unbiased prediction (BLUP) is a technique for estimating the genetic value of a trait. Initially developed by Henderson in 1950 (Henderson 1975), the method has been extensively used in animal breeding to identify the best progenies. While testing varieties over locations, the genotypic value can be considered a fixed effect and estimated by BLUE. A BLUP estimate can be obtained when genotypes are regarded as random, such as progenies in a mass/recurrent selection programme. It is possible to obtain a BLUP estimate for components of genetic variances using an appropriate plant breeding design. In structural populations, such as pedigree-based selection, BLUP values can be estimated from pedigree information, which is measured by the coefficient of kinship (K). In cross-pollinated crops, BLUP can be used to estimate hybrid performance, general and specific combining ability. GBLUP/RR-BLUP is used to estimate the association between marker data and phenotype data in TP and to select superior individuals based on genomic estimated breeding value (GEBV) in the BP. SNP polymorphism can be integrated with BLUP to predict EBV more accurately. The derived breeding value obtained from GBLUP is the genomic estimated breeding value (GEBV). Ridge-regression BLUP (RR-BLUP), random regression BLUP or SNP BLUP is a method of determination of GEBV that employs regression technique. For traits with very low heritability and a significant influence on the environment, optimal statistical model selection is an essential prerequisite. Machine learning (ML) tools have introduced more flexibility in genomic selection model development without strict assumptions. A plethora of ML tools like partial least square regression, BayesA, BayesB, BayesC, BayesU, BayesHP, BayesHE, least absolute shrinkage and selection operator (LASSO), adaptive LASSO, Bayesian LASSO, reproducing kernel Hilbert space (RKHS) model, gradient boosting machine, random forest, support vector machine, convolutional neural network, etc., have been used in genomic selection (Alemu et al. 2024).

GS is currently extensively used to predict complex traits for its advantage over phenotypic selection (PS). Apart from selection, GS can be very helpful in the estimation of heterosis. GS-based heterosis estimates are available for a large number of crops for yield and associated traits, biotic and abiotic stress tolerance and nutrient use efficiency (Lv et al. 2020; Saxena et al. 2021). Das et al. (2020) observed that GS was advantageous over PS under drought conditions, but under waterlogging conditions, PS was at par or advantageous over GS. Accuracy of GP is associated with heritability, marker density and population size (Liu et al., 2018). Recently, crop modelling data have been integrated to improve the accuracy of GS under diverse environmental conditions (Zhen et al. 2025). Another novel metabolomics-based GS approach, metabolic marker-assisted genomic prediction (MM_GP), has resulted in a 5% improvement in predictive ability in maize hybrids (Xu et al. 2024).

2.4 Genomics of Crop Wild Relatives (CWRs): Future Applications

Genomics is being used as an integral part of crop genetics and breeding, focussing on the cultivated species. The crop wild relatives (CWRs) or landraces carry valuable traits favoured by natural selection, particularly under extreme weather conditions and poor nutrient availability. However, they have lost these important traits during domestication at the cost of some desirable traits, and thus, the genetic diversity also has been narrowed down. Moreover, some wild relatives were also treated as weeds and therefore rejected. Thus, accessing the genetic diversity of the crop's wild relatives can diversify the existing breeders' sources. CWRs have an additional source of genetic diversity, which, when used in crop breeding programmes, can increase yield by up to 20% (Brozynska et al. 2016). Abiotic stresses may incur a global yield loss of up to 60% in crop plants (Vogel et al. 2019). The potential for climate change resilience for wild relatives of *Lens* sp. has been reviewed by Rajpal et al. (2023). Similarly, the potential has been explored in the wild relatives of several globally important crops like rice, soybean, barley, *Brassica,* etc. (Pourkheirandish et al. 2020). Advances in sequence technologies expanding to pangenomes comprising sequences of wild relatives are ideal for comparative genomics that will land in the identification of key genes and novel traits, as discussed in previous sections. De novo domestication for the domestication of a completely new crop using wild species, engineering polyploidy for genome duplication or bridging the genomes of two related species and harnessing the plant-microbe interaction for choosing suitable crop to suitable geographic location are some other areas where the crop wild relatives can be used to impart abiotic stress tolerance (Pourkheirandish et al. 2020). While germplasm characterization and GWAS panels often include some CWRs, their application in genetic mapping by GBS or in genomic selection is increasing rapidly. Large numbers of QTLs for different traits have been located using the populations derived from interspecific crosses in major and minor crop species (Table 2.3). These QTLs will be valuable in breeding programmes for marker-assisted QTL introgression. In addition, GWAS panels developed from interspecific hybridization have helped to detect QTL-SNP linkages and structural variants (SV).

2.5 Conclusion

In a 24-h timescale of crop improvement, genomics is a baby born only 4.56 min ago. Despite this short span, genomic tools have generated huge amounts of information and products for helping the genetic improvement of cultivated crops. The current focus of major genomics programmes is on maize, wheat, rice mustard, canola, cotton and soybean, where quality reference genome sequences are available. Many genetic resources have been developed, including mapping populations and mutant lines. One should not ignore the basic conventional breeding approaches for broadening the genetic base of crops. Nevertheless, draft genome sequences of

Table 2.3 Use of CWRs in genomic analysis

Interspecific cross/ genotypes	Genomic study design	Findings	DOI
Triticum aestivm × *Leymusmollis 4Ns (4D) alien disomic substitution line*	Transcriptome sequencing	16 markers transferred resistance to stripe rust and Fusarium head blight	10.3389/fpls.2022.1012939
Oryza sativa × *O. glaberrima*	QTL mapping using SSR	27 QTLs for nine characters	10.1007/s00122-004-1668-y
O. sativa × *O. rufipogon*	QTL mapping using SSR and KASP	Six QTLs for mineral nutrition	10.3390/agronomy13010076
O. sativa × *O. rufipogon*	Using Cornell 6 K SNP array	Diversity and genetic analysis of Oryza species	10.1186/s12284-017-0181-2
Solanum lycopersicum × *Solanum pennellii*	SNP genotyping of backcross inbred lines	Yield-associated introgressions and QTLs	10.1073/pnas.2205787119
Lycopersicon esculentum × *Lycopersicon pennellii*	QTL mapping using RFLP	74 QTLs	10.1093/genetics/134.2.585
Capsicum chinense × *Capsicum annuum*	QTL mapping using specific length amplified fragment sequencing (SLAF-seq)	6 QTLs for the flowering time and flower number per node	10.1038/s41598-018-38370-0
Capsicum annuum × *wild Capsicum frutescens*	GBS	Genetic map with 3708 SNPs	10.1080/13102818.2020.1787863
E. guineensis × *E. oleifera*	Genetic mapping using SNP and SSR	1814 SNPs and 149 SSRs, QTLs for multiple characters	10.1186/s12870-020-02563-5
Sorghum bicolor × *Sorghum propinquum*	QTL mapping using SNP	19 QTLs	10.1016/j.stress.2021.100024
Gossypium tomentosum × *Gossypium hirsutum*	QTL mapping using SSR	11 QTLs for salt tolerance	10.1007/s10681-016-1674-6
E. guineensis (dura × *pisifera)*	GBS with SNP	Linkage map with 1085 SNPs, 3 QTLs	10.1016/j.ygeno.2015.02.002
Populus deltoides × *Populus simonii*	RAD sequencing	Linkage map with 2097 SNPs, 39 QTLs	10.1093/jhered/esaa039
B. napus × *B. oleracea*	Genetic mapping using SNP	Nine QTLs	10.1002/csc2.20824
Castanea mollissima × *C. Henryi*	GBS with SNP	Linkage map with 4578 SNPs, 71 QTLs	10.3390/f14081684

(continued)

Table 2.3 (continued)

Interspecific cross/ genotypes	Genomic study design	Findings	DOI
Bi-parental F_1 families with at least six Vitis species in their ancestry	GBS with SNP	Maternal genetic maps consisting of 1077 and 1641 SNPs, two QTLs	10.1007/s11032-016-0586-4
Asparagus maritimus × *Asparagus acutifolius*	GWAS of 378 genotypes	First example of GWAS and GP in asparagus	10.3389/fpls.2023.1187663
Elaeis oleifera × *Elaeis guineensis*	GBS and GWAS with SNP	Twelve SNPs for tissue growth, cellular trafficking, and physiological processes	10.1186/s12870-019-2153-8
C. echinospermum × *C. reticulatum*	GWAS with diversity arrays technology (DArT) markers	11 QTLs	10.1002/tpg2.20408
Gosypium spp.	Pangenome analysis	182,593 non-redundant SVs	10.1016/j.molp.2023.02.004

most of the crops are published now, which need to be converted into quality reference genome assembly for improving the accuracy of the application of genomics in other crops. Future generation sequencing (FGS) technologies would enable faster and more productive integration of genomics in crop improvement. Cutting-edge informatics tools like machine learning and artificial intelligence are increasingly integrated into various facets of genomic technologies to improve decision-making. However, in the biological world of complex interactions, the environment is a key determinant of genotypic expression; hence, precise phenotyping facilities are required to be developed for the evaluation of promising breeding populations on a large scale for QTL detection, gene identification, and predictive model development for genomic selection.

References

Abdurakhmonov IY, Abdukarimov A (2008) Application of association mapping to understanding the genetic diversity of plant germplasm resources. Int J Plant Genom 1:574927

Albert E, Segura V, Gricourt J, Bonnefoi J, Derivot L, Causse M (2016) Association mapping reveals the genetic architecture of tomato response to water deficit: focus on major fruit quality traits. J Exp Bot 67:6413–6430

Alemu A, Åstrand J, Montesinos-López OA, Isidro Y, Sánchez J, Fernández-Gónzalez J, Tadesse W, Vetukuri RR, Carlsson AS, Ceplitis A, Crossa J, Ortiz R, Chawade A (2024) Genomic selection in plant breeding: key factors shaping two decades of progress. Mol Plant 17(4):552–578

Alseekh S, Tohge T, Wendenberg R, Scossa F, Omranian N, Li J, Kleessen S, Giavalisco P, Pleban T, Mueller-Roeber B, Zamir D (2015) Identification and mode of inheritance of quantitative trait loci for secondary metabolite abundance in tomato. Plant Cell 27(3):485–512

Alseekh S, Kostova D, Bulut M, Fernie AR (2021) Genome-wide association studies: assessing trait characteristics in model and crop plants. Cell Mol Life Sci 78:5743–5754

Amiteye S (2021) Basic concepts and methodologies of DNA marker systems in plant molecular breeding. Heliyon 7(10):e08093

An Y, Xia X, Zheng H, Yu S, Jing T, Zhang F (2023) Multi-genome comprehensive identification of SSR/SV and development of molecular markers database to serve *Sorghum bicolor* (L.) breeding. BMC Genom Data 24(1):62

Aslam U, Cheema HM, Ahmad S, Khan IA, Malik W, Khan AA (2016) COTIP: cotton TILLING platform, a resource for plant improvement and reverse genetic studies. Front Plant Sci 7:1863

Atwell S, Huang YS, Vilhjálmsson BJ, Willems G, Horton M, Li Y, Meng D, Platt A, Tarone AM, Hu TT, Jiang R (2010) Genome-wide association study of 107 phenotypes in *Arabidopsis thaliana* inbred lines. Nature 465(7298):627–631

Barnes S (2002) Comparing Arabidopsis to other flowering plants. Curr Opin Plant Biol 5: 128–134

Barrero-Sicilia C, Silvestre S, Haslam RP, Michaelson LV (2017) Lipid remodelling: unravelling the response to cold stress in Arabidopsis and its extremophile relative *Eutremasalsugineum.* Plant Sci 263:194–200

Bayer PE, Fraser MW, Martin BC, Petereit J, Severn-Ellis AA, Sinclair EA, Batley J, Kendrick GA, Edwards D (2022, 2022) Not all pathways are the same–unique adaptations to submerged environments emerge from comparative seagrass genomics. bioRxiv 24:–11

Bommert P, Nagasawa NS, Jackson D (2013) Quantitative variation in maize kernel row number is controlled by the FASCIATED EAR2 locus. Nat Genet 45(3):334–337

Brozynska M, Furtado A, Henry RJ (2016) Genomics of crop wild relatives: expanding the gene pool for crop improvement. Plant Biotechnol J 14(4):1070–1085

Bush WS, Moore JH (2012) Chapter 11: genome-wide association studies. PLoS Comput Biol 8(12):e1002822

Caudai C, Galizia A, Geraci F, Le Pera L, Morea V, Salerno E, Via A, Colombo T (2021) AI applications in functional genomics. Comput Struct Biotechnol J 19:5762–5790

Causse M, Desplat N, Pascual L, Le Paslier MC, Sauvage C, Bauchet G, Bérard A, Bounon R, Tchoumakov M, Brunel D, Bouchet JP (2013) Whole genome resequencing in tomato reveals variation associated with introgression and breeding events. BMC Genom 14:1–4

Chen K, Shan Q, Gao C (2014) An efficient TALEN mutagenesis system in rice. Methods 69(1): 2–8

Chen F, Song Y, Li X, Chen J, Mo L, Zhang X, Lin Z, Zhang L (2019) Genome sequences of horticultural plants: past, present, and future. Hortic Res 1:6

Clauw P, Ellis TJ, Liu HJ, Sasaki E (2024) Beyond the standard GWAS – a guide for plant biologists. Plant Cell Physiol:pcae079

Cockram J, Scuderi A, Barber T, Furuki E, Gardner KA, Gosman N, Kowalczyk R, Phan HP, Rose GA, Tan KC, Oliver RP (2015) Fine-mapping the wheat Snn1 locus conferring sensitivity to the *Parastagonosporanodorum* necrotrophic effector SnTox1 using an eight founder multiparent advanced generation inter-cross population. G3: genes, genomes. Genetics 5(11):2257–2266

Colbert T, Till BJ, Tompa R, Reynolds S, Steine MN, Yeung AT, McCallum CM, Comai L, Henikoff S (2001) High-throughput screening for induced point mutations. Plant Physiol 126: 480–484

Cole B, Bergmann D, Blaby-Haas CE, Blaby IK, Bouchard KE, Brady SM, Ciobanu D, Coleman-Derr D, Leiboff S, Mortimer JC, Nobori T (2021) Plant single-cell solutions for energy and the environment. Commun Biol 4(1):962

Comai L, Young K, Till BJ, Reynolds SH, Greene EA, Codomo CA, Enns LC, Johnson JE, Burtner C, Odden AR, Henikoff S (2004) Efficient discovery of DNA polymorphisms in natural populations by Ecotilling. Plant J 37(5):778–786

Cosenza F, Shrestha A, Van Inghelandt D, Casale FA, Wu PY, Weisweiler M, Li J, Wespel F, Stich B (2024) Genetic mapping reveals new loci and alleles for flowering time and plant height using the double round-robin population of barley. J Exp Bot 75(8):2385–2402

Costa V, Angelini C, De Feis I, Ciccodicola A (2010) Uncovering the complexity of transcriptomes with RNA-Seq. Biomed Res Int 1:853916

Crossa J, Pérez-Rodríguez P, Cuevas J, Montesinos-López O, Jarquín D, de Los CG, Burgueño J, González-Camacho JM, Pérez-Elizalde S, Beyene Y, Dreisigacker S, Singh R, Zhang X, Gowda M, Roorkiwal M, Rutkoski J, Varshney RK (2017) Genomic selection in plant breeding: methods, models, and perspectives. Trends Plant Sci 22(11):961–975

Crowell S, Korniliev P, Falcao A, Ismail A, Gregorio G, Mezey J, McCouch S (2016) Genome-wide association and high-resolution phenotyping link *Oryza sativa* panicle traits to numerous trait-specific QTL clusters. Nat Commun 7(1):10527

Dalmais M, Antelme S, Ho-Yue-Kuang S, Wang Y, Darracq O, d'Yvoire MB, Cezard L, Légée F, Blondet E, Oria N, Troadec C (2013) A TILLING platform for functional genomics in Brachypodium distachyon. PLoS One 8(6):e65503

Das RR, Vinayan MT, Patel MB, Phagna RK, Singh SB, Shahi JP, Sarma A, Barua NS, Babu R, Seetharam K, Burgueño JA, Zaidi PH (2020) Genetic gains with rapid-cycle genomic selection for combined drought and waterlogging tolerance in tropical maize (*Zea mays* L.). Plant. Genome 13(3):e20035

de Ronne M, Abed A, Légaré G, Laroche J, Boucher St-Amour VT, Fortier É, Beattie A, Badea A, Khanal R, O'Donoughue L, Rajcan I (2024) Integrating targeted genetic markers to genotyping-by-sequencing for an ultimate genotyping tool. Theor Appl Genet 137(10):247

De Souza IP, De Azevedo BR, Coelho AS, De Souza TL, Valdisser PA, Gomes-Messias LM, Funicheli BO, Brondani C, Vianello RP (2023) Whole-genome resequencing of common bean elite breeding lines. Sci Rep 13(1):12721

Della Coletta R, Qiu Y, Ou S, Hufford MB, Hirsch CN (2021) How the pan-genome is changing crop genomics and improvement. Genome Biol 22:1–9

Demirjian C, Vailleau F, Berthomé R, Roux F (2023) Genome-wide association studies in plant pathosystems: success or failure? Trends Plant Sci 28:471–485

Deschamps S, la Rota M, Ratashak JP, Biddle P, Thureen D, Farmer A, Luck S, Beatty M, Nagasawa N, Michael L, Llaca V (2010) Rapid genome-wide single nucleotide polymorphism discovery in soybean and rice via deep resequencing of reduced representation libraries with the Illumina genome analyzer. Plant Genome 3(1):53–68

Dhondt S, Wuyts N, Inzé D (2013) Cell to whole-plant phenotyping: the best is yet to come. Trends Plant Sci 18:428–439

Dunning LT, Olofsson JK, Parisod C, Choudhury RR, Moreno-Villena JJ, Yang Y, Dionora J, Quick WP, Park M, Bennetzen JL, Besnard G (2019) Lateral transfers of large DNA fragments spread functional genes among grasses. Proc Nat Acad Sci 116(10):4416–4425

Ellegren H (2014) Genome sequencing and population genomics in non-model organisms. Trend Ecol Evol 29:51–63

Elshire RJ, Glaubitz JC, Sun Q, Poland JA, Kawamoto K, Buckler ES, Mitchell SE (2011) A robust, simple genotyping-by-sequencing (GBS) approach for high diversity species. PLoS One 6(5): 19379

El-Soda M, Kruijer W, Malosetti M, Koornneef M, Aarts MG (2015) Quantitative trait loci and candidate genes underlying genotype by environment interaction in the response of *Arabidopsis thaliana* to drought. Plant Cell Environ 38(3):585–599

Everett MV, Grau ED, Seeb JE (2011) Short reads and nonmodel species: exploring the complexities of next-generation sequence assembly and SNP discovery in the absence of a reference genome. Mol Ecol Resour 11:93–108

Fanelli V, Ngo KJ, Thompson VL, Silva BR, Tsai H, Sabetta W, Montemurro C, Comai L, Harmer SL (2021) A tilling by sequencing approach to identify induced mutations in sunflower genes. Sci Rep 11(1):9885

Fang C, Ma Y, Wu S, Liu Z, Wang Z, Yang R, Hu G, Zhou Z, Yu H, Zhang M, Pan Y (2017) Genome-wide association studies dissect the genetic networks underlying agronomical traits in soybean. Genome Biol 18:1–4

Gage JL, Vaillancourt B, Hamilton JP, Manrique-Carpintero NC, Gustafson TJ, Barry K, Lipzen A, Tracy WF, Mikel MA, Kaeppler SM, Buell CR (2019) Multiple maize reference genomes impact the identification of variants by genome-wide association study in a diverse inbred panel. Plant Genome 12(2):180069

Gill HS, Halder J, Zhang J, Rana A, Klcinjan J, Amand PS, Bernardo A, Bai G, Sehgal SK (2022) Whole-genome analysis of hard winter wheat germplasm identifies genomic regions associated with spike and kernel traits. Theor Appl Genet 135(9):2953–2967

Goldstein DB (2001) Islands of linkage disequilibrium. Nat Genet 29(2):109–111

González M, Xu M, Esteras C, Roig C, Monforte AJ, Troadec C, Pujol M, Nuez F, Bendahmane A, Garcia-Mas J, Picó B (2011) Towards a TILLING platform for functional genomics in Piel de Sapo melons. BMC Res Notes 4:1–1

Gramazio P, Yan H, Hasing T, Vilanova S, Prohens J, Bombarely A (2019) Whole-genome resequencing of seven eggplant (*Solanum melongena*) and one wild relative (*S. incanum*) accessions provides new insights and breeding tools for eggplant enhancement. Front Plant Sci 10:1220

Greene EA, Codomo CA, Taylor NE, Henikoff JG, Till BJ, Reynolds SH, Enns LC, Burtner C, Johnson JE, Odden AR, Comai L (2003) Spectrum of chemically induced mutations from a large-scale reverse-genetic screen in Arabidopsis. Genetics 164(2): 731–740

Guo G, Dondup D, Zhang L, Hu S, Yuan X, Zhang J (2014) Identification of SNPs in barley (*Hordeum vulgare* L.) by deep sequencing of six reduced representation libraries. Crop J 2(6): 419–425

Guo Y, Abernathy B, Zeng Y, Ozias-Akins P (2015) TILLING by sequencing to identify induced mutations in stress resistance genes of peanut (*Arachis hypogaea*). BMC Genom 16:1–3

He J, Zhao X, Laroche A, Lu ZX, Liu H, Li Z (2014) Genotyping-by-sequencing (GBS), an ultimate marker-assisted selection (MAS) tool to accelerate plant breeding. Front Plant Sci 5: 484

Heinrich F, Wutke M, Das PP, Kamp M, Gültas M, Link W, Schmitt AO (2020) Identification of regulatory SNPs associated with vicine and convicine content of *Vicia faba* based on genotyping by sequencing data using deep learning. Genes 11(6):614

Henderson CR (1975) Best linear unbiased estimation and prediction under a selection model. Biometrics 31(2):423–447

Hieter P, Boguski M (1997) Functional genomics: it's all how you read it. Science 278(5338): 601–602

Hu H, Wang J, Nie S, Zhao J, Batley J, Edwards D (2024a) Plant pangenomics, current practice and future direction. Agric Commun 30:100039

Hu Z, Zhu Z, Cai L, Zhan Y, Yan X, Chen J, Sun B, Du S, Jiang S, Wang H, Zhang Y (2024b) STEP: deciphering spatial atlas at single-cell level with whole-transcriptome coverage. bioRxiv 2024:2024-11

Hu G, Wang Z, Tian Z, Wang K, Ji G, Wang X, Zhang X, Yang Z, Liu X, Niu R, Zhu D (2025) A telomere-totelomere genome assembly of cotton provides insights into centromere evolution and short-season adaptation. Nat Genet 17:1–3

Hussain W, Baenziger PS, Belamkar V, Guttieri MJ, Venegas JP, Easterly A, Sallam A, Poland J (2017) Genotyping-by-sequencing derived high-density linkage map and its application to QTL mapping of flag leaf traits in bread wheat. Sci Rep 7(1):16394

Huynh K, Van Quoc G, Thanh TNC, Loc HN, Thanh VC (2021) Whole-genome sequencing of three local rice varieties (*Oryza sativa L.*) in Vietnam. Plant Sci Today 8(3):437–444

Imelfort M, Duran C, Batley J, Edwards D (2009) Discovering genetic polymorphisms in next-generation sequencing data. Plant Biotechnol J 7(4):312–317

Irshad A, Guo H, Zhang S, Gu J, Zhao L, Xie Y, Xiong H, Zhao S, Ding Y, Ma Y, Liu L (2019) EcoTILLING reveals natural allelic variations in starch synthesis key gene TaSSIV and its haplotypes associated with higher thousand-grain weight. Genes 10(4):307

Jackson SA, Iwata A, Lee SH, Schmutz J, Shoemaker R (2011) Sequencing crop genomes: approaches and applications. New Phytol 191(4):915–925

Jaganathan D, Bohra A, Thudi M, Varshney RK (2020) Fine mapping and gene cloning in the post-NGS era: advances and prospects. Theor Appl Genet 133(5):1791–1810

Jia J, Li H, Zhang X, Li Z, Qiu L (2017) Genomics-based plant germplasm research (GPGR). Crop J 5(2):166–174

Josephs EB, Stinchcombe JR, Wright SI (2017) What can genome-wide association studies tell us about the evolutionary forces maintaining genetic variation for quantitative traits? New Phytol 214(1):21–33

Joukhadar R, Hollaway G, Shi F, Kant S, Forrest K, Wong D, Petkowski J, Pasam R, Tibbits J, Bariana H, Bansal U (2020) Genome-wide association reveals a complex architecture for rust resistance in 2300 worldwide bread wheat accessions screened under various Australian conditions. Theor Appl Genet 133:2695–2712

Khan AW, Garg V, Roorkiwal M, Golicz AA, Edwards D, Varshney RK (2020) Super-pangenome by integrating the wild side of a species for accelerated crop improvement. Trends Plant Sci 25(2):148–158

Kilian B, Graner A (2012) NGS technologies for analyzing germplasm diversity in genebanks. Brief Funct Genom 11(1):38–50

Kim SI, Tai TH (2014) Identification of novel rice low phytic acid mutations via TILLING by sequencing. Mol Breed 34:1717–1729

Kim C, Guo H, Kong W, Chandnani R, Shuang LS, Paterson AH (2016) Application of genotyping by sequencing technology to a variety of crop breeding programs. Plant Sci 242:14–22

Kong W, Wang Y, Zhang S, Yu J, Zhang X (2023) Recent advances in assembly of complex plant genomes. Genom Proteom Bioinform 21(3):427–439

Korte A, Vilhjálmsson BJ, Segura V, Platt A, Long Q, Nordborg M (2012) A mixed-model approach for genome-wide association studies of correlated traits in structured populations. Nat Genet 44:1066–1071

Kumar R, Das SP, Choudhury BU, Kumar A, Prakash NR, Verma R, Chakraborti M, Devi AG, Bhattacharjee B, Das R, Das B (2024) Advances in genomic tools for plant breeding: harnessing DNA molecular markers, genomic selection, and genome editing. Biol Res 57(1):80

Lakhssassi N, Zhou Z, Liu S, Piya S, Cullen MA, El Baze A, Knizia D, Patil GB, Badad O, Embaby MG, Meksem J (2020) Soybean TILLING-by-Sequencing+ reveals the role of novel GmSACPD members in unsaturated fatty acid biosynthesis while maintaining healthy nodules. J Exp Bot 71(22):6969–6987

Li H, Thrash A, Tang JD, He L, Yan J, Warburton ML (2019) Leveraging GWAS data to identify metabolic pathways and networks involved in maize lipid biosynthesis. Plant J 98(5):853–863

Li W, Liu J, Zhang H, Liu Z, Wang Y, Xing L, He Q, Du H (2022) Plant pan-genomics: recent advances, new challenges, and roads ahead. J Genet Genomics 49(9):833–846

Liang J, Fu B, Tang W, Khan NU, Li N, Ma Z (2016) Fine mapping of two wheat powdery mildew resistance genes located at the Pm1 cluster. Plant Genome 9(2):2015–2009

Liu X, Wang H, Wang H, Guo Z, Xu X, Liu J, Wang S, Li WX, Zou C, Prasanna BM, Olsen MS (2018) Factors affecting genomic selection revealed by empirical evidence in maize. Crop J 6: 341–352

Liu Y, Du H, Li P, Shen Y, Peng H, Liu S, Zhou GA, Zhang H, Liu Z, Shi M, Huang X (2020) Pan-genome of wild and cultivated soybeans. Cell 182(1):162–176

Lu K, Wei L, Li X, Wang Y, Wu J, Liu M, Zhang C, Chen Z, Xiao Z, Jian H, Cheng F (2019) Whole-genome resequencing reveals Brassica napus origin and genetic loci involved in its improvement. Nat Commun 10(1):1154

Lv Q, Li W, Sun Z, Ouyang N, Jing X, He Q, Wu J, Zheng J, Zheng J, Tang S, Zhu R, Tian Y, Duan M, Tan Y, Yu D, Sheng X, Sun X, Jia G, Gao H, Zeng Q, Li Y, Tang L, Xu Q, Zhao B,

Huang Z, Lu H, Li N, Zhao J, Zhu L, Li D, Yuan L, Yuan D (2020) Resequencing of 1,143 indica rice accessions reveals important genetic variations and different heterosis patterns. Nat Commun 11(1):4778

Lyra DH, Galli G, Alves FC, Granato ÍS, Vidotti MS, Bandeira e Sousa M, Morosini JS, Crossa J, Fritsche-Neto R (2019) Modeling copy number variation in the genomic prediction of maize hybrids. Theor Appl Genet 132:273–288

Ma J, Qin N, Cai B, Chen G, Ding P, Zhang H, Yang C, Huang L, Mu Y, Tang H, Liu Y (2019) Identification and validation of a novel major QTL for all-stage stripe rust resistance on 1BL in the winter wheat line 20828. Theor Appl Genet 132:1363–1373

Mackenzie A, Norman M, Gessese M, Chen C, Sørensen C, Hovmøller M, Ma L, Forrest K, Hickey L, Bariana H, Bansal U (2023) Wheat stripe rust resistance locus YR63 is a hot spot for evolution of defence genes–a pangenome discovery. BMC Plant Biol 23(1):590

Marks RA, Hotaling S, Frandsen PB, Van Buren R (2021) Representation and participation across 20 years of plant genome sequencing. Nat Plant 7(12):1571–1578

Martin J, Schackwitz W, Lipzen A (2018) Genomic sequence variation analysis by resequencing. Fung Genom Method Prot 229–239

McCallum CM, Comai L, Greene EA, Henikoff S (2000) Targeting induced local lesions IN genomes (TILLING) for plant functional genomics. Plant Physiol 123(2):439–442

McCouch SR, Wright MH, Tung CW, Maron LG, McNally KL, Fitzgerald M, Singh N, DeClerck G, Agosto-Perez F, Korniliev P, Greenberg AJ (2016) Open access resources for genome-wide association mapping in rice. Nat Commun 7(1):10532

Meuwissen TH, Hayes BJ, Goddard ME (2001) Prediction of total genetic value using genome-wide dense marker maps. Genetics 157(4):1819–1829

Minoia S, Petrozza A, D'Onofrio O, Piron F, Mosca G, Sozio G, Cellini F, Bendahmane A, Carriero F (2010) A new mutant genetic resource for tomato crop improvement by TILLING technology. BMC Res Notes 3:1–8

Nitcher R, Distelfeld A, Ta C, Yan L, Dubcovsky J (2013) Increased copy number at the HvFT1 locus is associated with accelerated flowering time in barley. Mol Gen Genom 288:261–275

Okabe Y, Ariizumi T, Ezura H (2013) Updating the micro-tom TILLING platform. Breed Sci 63(1): 42–48

Paajanen P, Kettleborough G, López-Girona E, Giolai M, Heavens D, Baker D, Lister A, Cugliandolo F, Wilde G, Hein I, Macaulay I (2019) A critical comparison of technologies for a plant genome sequencing project. Gigascience 8(3):giy163

Panchy N, Lehti-Shiu M, Shiu SH (2016) Evolution of gene duplication in plants. Plant Physiol 171:2294–2316

Pavan S, Delvento C, Ricciardi L, Lotti C, Ciani E, D'Agostino N (2020) Recommendations for choosing the genotyping method and best practices for quality control in crop genome-wide association studies. Front Genet 11:447

Poland JA, Rife TW (2012) Genotyping-by-sequencing for plant breeding and genetics. Plant Genom 5(3)

Pourkheirandish M, Golicz AA, Bhalla PL, Singh MB (2020) Global role of crop genomics in the face of climate change. Front Plant Sci 11:922

Qiao Q, Edger PP, Xue L, Qiong L, Lu J, Zhang Y, Cao Q, Yocca AE, Platts AE, Knapp SJ (2021) Evolutionary history and pan-genome dynamics of strawberry (*Fragaria* spp.). Proc Natl Acad Sci USA 118:e2105431118

Raja RB, Agasimani S, Jaiswal S, Thiruvengadam V, Sabariappan R, Chibbar RN, Ram SG (2017) EcoTILLING by sequencing reveals polymorphisms in genes encoding starch synthases that are associated with low glycemic response in rice. BMC Plant Biol 17:1–3

Rajpal VR, Singh A, Kathpalia R, Thakur RK, Khan MK, Pandey A, Hamurcu M, Raina SN (2023) The prospects of gene introgression from crop wild relatives into cultivated lentil for climate change mitigation. Front Plant Sci 14:1127239

Rees JS, Castellano S, Andrés AM (2020) The genomics of human local adaptation. Trend Genet 36:415–428

Ritchie H, Rosado P, Roser M (2022) Crop yields. Data adapted from the food and agriculture Organization of the United Nations. Broadberry et al., Brassley, Bayliss-Smith &Wanmali. Retrieved from https://ourworldindata.org/grapher/wheat-yields. Retrieved on 12 Feb 2025

Sasaki E, Zhang P, Atwell S, Meng D, Nordborg M (2015) 'Miss-ing' G x E variation controls flowering time in *Arabidopsis thaliana*. PLoS Genet 11:e1005597

Satya P, Chakraborty A, Jana S, Majumdar S, Karan M, Sarkar D, Datta S, Mitra J, Kar CS, Karmakar PG, Singh NK (2017) Identification of genic SSRs in jute (*Corchorus capsularis*, Malvaceae) and development of markers for phenylpropanoid biosynthesis genes and regulatory genes. Plant Breed 136(5):784–797

Satya P, Das D, Mandal NA, Kundu A, Sarkar D, Kar CS, Mitra J, Kar G, Singh NK (2023) Development and characterization of a multiparent advanced generation inter-cross (MAGIC) population of jute (*Corchorus olitorius*). Agric Res 12(3):266–276

Saxena RK, Jiang Y, Khan AW, Zhao Y, Kumar Singh V, Bohra A, Sonappa M, Rathore A, Kumar CVS, Saxena K, Reif J, Varshney RK (2021) Characterization of heterosis and genomic prediction-based establishment of heterotic patterns for developing better hybrids in pigeonpea. Plant Genom 14(3):e20125

Scheben A, Batley J, Edwards D (2017) Genotyping-by-sequencing approaches to characterize crop genomes: choosing the right tool for the right application. Plant Biotechnol J 15(2):149–161

Scott MF, Ladejobi O, Amer S, Bentley AR, Biernaskie J, Boden SA, Clark M, Dell'Acqua M, Dixon LE, Filippi CV, Fradgley N (2020) Multi-parent populations in crops: a toolbox integrating genomics and genetic mapping with breeding. Heredity 125(6):396–416

Shi A, Bhattarai G, Xiong H, Avila CA, Feng C, Liu B, Joshi V, Stein L, Mou B, du Toit LJ, Correll JC (2022) Genome-wide association study and genomic prediction of white rust resistance in USDA GRIN spinach germplasm. Hortic Res 9:uhac069

Shikata M, Ezura H (2016) Micro-tom tomato as an alternative plant model system: mutant collection and efficient transformation. Plant Signal Trans Method Prot 2016:47–55

Sihag P, Sagwal V, Kumar A, Balyan P, Mir RR, Dhankher OP, Kumar U (2021) Discovery of miRNAs and development of heat-responsive miRNA-SSR markers for characterization of wheat germplasm for terminal heat tolerance breeding. Front Genet 12: 699420

Singh D, Chaudhary P, Taunk J, Singh CK, Chinnusamy V, Sevanthi AM, Singh VJ, Pal M (2024) Targeting induced local lesions in genomes (TILLING): advances and opportunities for fast-tracking crop breeding. Crit Rev Biotechnol 44(5):817–836

Soltis PS, Soltis DE (2021) Plant genomes: markers of evolutionary history and drivers of evolutionary change. Plant People Planet 3(1):74–82

Song B, Ning W, Wei D, Jiang M, Zhu K, Wang X, Edwards D, Odeny DA, Cheng S (2023) Plant genome resequencing and population genomics: current status and future prospects. Mol Plant. incomplete

Szurman-Zubrzycka ME, Zbieszczyk J, Marzec M, Jelonek J, Chmielewska B, Kurowska MM, Krok M, Daszkowska-Golec A, Guzy-Wrobelska J, Gruszka D, Gajecka M (2018) Hor TILLUS—a rich and renewable source of induced mutations for forward/reverse genetics and pre-breeding programs in barley (*Hordeum vulgare* L.). Front Plant Sci 9:216

Tahir M, Ren Y, Wu B, Li M, Refaiy M, Cao M, Kong D, Pang X (2024) Parental reconstruction from a half-sib population of stoneless Jujube *Ziziphus jujuba* Mill. Based on individual specific SNP markers using multiplex PCR. Plants 13(22):3163

Thottathil GP, Jayasekaran K, Othman AS (2016) Sequencing crop genomes: a gateway to improve tropical agriculture. Trop Life Sc Res 27(1):93

Tibbs Cortes L, Zhang Z, Yu J (2021) Status and prospects of genome-wide association studies in plants. Plant Genom 14(1):e20077

Tieman D, Zhu G, Resende MF Jr, Lin T, Nguyen C, Bies D, Rambla JL, Beltran KS, Taylor M, Zhang B, Ikeda H (2017) A chemical genetic roadmap to improved tomato flavor. Science 355(6323):391–394

Till BJ, Reynolds SH, Weil C, Springer N, Burtner C, Young K, Bowers E, Codomo CA, Enns LC, Odden AR, Greene EA (2004) Discovery of induced point mutations in maize genes by TILLING. BMC Plant Biol 4:1–8

Till BJ, Cooper J, Tai TH, Colowit P, Greene EA, Henikoff S, Comai L (2007) Discovery of chemically induced mutations in rice by TILLING. BMC Plant Biol 7:1–2

Todesco M, Balasubramanian S, Hu TT, Traw MB, Horton M, Epple P, Kuhns C, Sureshkumar S, Schwartz C, Lanz C, Laitinen RA (2010) Natural allelic variation underlying a major fitness trade-off in Arabidopsis thaliana. Nature 465(7298):632–636

Tsai H, Howell T, Nitcher R, Missirian V, Watson B, Ngo KJ, Lieberman M, Fass J, Uauy C, Tran RK, Khan AA (2011) Discovery of rare mutations in populations: TILLING by sequencing. Plant Physiol 156(3):1257–1268

Tyagi P, Singh D, Mathur S, Singh A, Ranjan R (2022) Upcoming progress of transcriptomics studies on plants: an overview. Front Plant Sci 13:1030890

Varshney RK, Roorkiwal M, Sun S, Bajaj P, Chitikineni A, Thudi M, Singh NP, Du X, Upadhyaya HD, Khan AW, Wang Y (2021) A chickpea genetic variation map based on the sequencing of 3366 genomes. Nature 599(7886):622–627

Vogel E, Donat MG, Alexander LV, Meinshausen M, Ray DK, Karoly D, Meinshausen N, Frieler K (2019) The effects of climate extremes on global agricultural yields. Environ Res Lett 14(5): 054010

Wang Y, Li Z, Lu J (2024) Single-cell RNA sequencing reveals the epithelial cell, fibroblast, and key gene alterations in chronic rhinosinusitis with nasal polyps. Sci Rep 14(1):2270

Würschum T, Boeven PH, Langer SM, Longin CFH, Leiser WL (2015) Multiply to conquer: copy number variations at Ppd-B1 and Vrn-A1 facilitate global adaptation in wheat. BMC Genet 16: 1–8

Xia Y, Li R, Ning Z, Bai G, Siddique KH, Yan G, Baum M, Varshney RK, Guo P (2013) Single nucleotide polymorphisms in HSP17. 8 and their association with agronomic traits in barley. PLoS One 8(2):e56816

Xu Y, Li P, Yang Z, Xu C (2017) Genetic mapping of quantitative trait loci in crops. Crop J 5: 175–184

Xu Y, Yang W, Qiu J, Zhou K, Yu G, Zhang Y, Wang X, Jiao Y, Wang X, Hu S, Zhang X, Li P, Lu Y, Chen R, Tao T, Yang Z, Xu Y, Xu C (2024) Metabolic marker-assisted genomic prediction improves hybrid breeding. Plant Commun 29:101199

Yano K, Yamamoto E, Aya K, Takeuchi H, Lo PC, Hu L, Yamasaki M, Yoshida S, Kitano H, Hirano K, Matsuoka M (2016) Genome-wide association study using whole-genome sequencing rapidly identifies new genes influencing agronomic traits in rice. Nat Genet 48(8):927–934

Yoshioka M, Iehisa JC, Ohno R, Kimura T, Enoki H, Nishimura S, Nasuda S, Takumi S (2017) Three dominant awnless genes in common wheat: fine mapping, interaction and contribution to diversity in awn shape and length. PloS One 12(4):0176148

Yu S, Liao F, Wang F, Wen W, Li J, Mei H (2012) Luo L (2012) identification of rice transcription factors associated with drought tolerance using the ecotilling method. PLoS One 7(2):e30765

Zeng YD, Sun JL, Bu SH, Deng KS, Tao T, Zhang YM, Zhang TZ, Du XM, Zhou BL (2016) EcoTILLING revealed SNPs in GhSus genes that are associated with fiber-and seed-related traits in upland cotton. Sci Rep 6(1):29250

Zhang L, Ren Y, Yang T, Li G, Chen J, Gschwend AR, Yu Y, Hou G, Zi J, Zhou R, Wen B (2019) Rapid evolution of protein diversity by de novo origination in *Oryza*. Nat Ecol Evol 3(4): 679–690

Zhao M, Shu G, Hu Y, Cao G, Wang Y (2023) Pattern and variation in simple sequence repeat (SSR) at different genomic regions and its implications to maize evolution and breeding. BMC Genomics 24(1):136

Zhen X, Luo J, Xiao Y, Yan J, Cordoba BC, Batchelor WD (2025) Integrating genomics with crop modelling to predict maize yield and component traits: towards the next generation of crop models. Eur J Agron 162:127391

Zhu G, Wang S, Huang Z, Zhang S, Liao Q, Zhang C, Lin T, Qin M, Peng M, Yang C, Cao X (2018) Rewiring of the fruit metabolome in tomato breeding. Cell 172(1):249–261

Genome Editing in Banana: Tools, Techniques, and Challenges

3

Jaindra Nath Tripathi, Samwel Muiruri Kariuki, Duncan Njora Waweru, and Leena Tripathi

Abstract

Bananas (*Musa* spp.) are one of the most important staple crops and fruit sources, providing livelihoods for millions of farmers in tropical and subtropical regions worldwide. It is cultivated in over 145 countries and small island nations, crucial to food security. However, banana production faces significant challenges due to various biotic and abiotic factors. The bananas' narrow genetic diversity and long generation cycle make conventional breeding methods difficult and time-consuming. Genome editing offers a powerful approach to developing high-yield, disease-resistant banana varieties. Several genome editing technologies enable precise and targeted modifications within the plant genome. Among these, Clustered Regularly Interspaced Short Palindromic Repeats (CRISPR)-based tools stand out for their significant advantages, though they also present regulatory and technical challenges. In this chapter, we describe the importance of banana cultivation, examine the benefits and limitations of various CRISPR tools, explore advancements in banana genome editing, and cover the technical and regulatory aspects of implementing genome editing for banana improvement.

Keywords

Banana · CRISPR/Cas9 · Genome editing · Base editing · Prime editing

J. N. Tripathi (✉) · S. M. Kariuki · D. Njora Waweru · L. Tripathi (✉)
International Institute of Tropical Agriculture, Nairobi, Kenya
e-mail: j.tripathi@cgiar.org; l.tripathi@cgiar.org

Y. Ahlawat et al. (eds.), *Next-Generation Strategies for Crop Improvement*,
https://doi.org/10.1007/978-981-95-0309-4_3

3.1 Introduction to Banana

3.1.1 Overview of Banana Cultivation

Bananas are one of the world's most important staple crops and fruits, playing a critical role in food security, nutrition, and livelihoods, especially in tropical and subtropical regions globally. They serve as a food source for over 500 million people worldwide, particularly in Asia, Africa, Latin America, and the Caribbean, and are cultivated in more than 145 countries and island nations (FAOSTAT 2023). Bananas are a significant energy source due to their enrichment with carbohydrates, minerals, and vitamins, including potassium, vitamin B6, vitamin C, and dietary fibers. They are vital in combating malnutrition and micronutrient deficiencies in developing regions. Both ripe and unripe bananas are consumed in various modes, from fresh fruit to processed products like flour, chips, and fermented beverages. Bananas are the most exported fruit worldwide, with an annual trade value exceeding 15 billion USD (FAOSTAT 2023). Major exporting countries include those in Latin America and the Caribbean, such as Ecuador, Colombia, and the Philippines. The cultivation of bananas supports millions of smallholder farmers and plantation workers, providing vital income and employment opportunities for smallholder farmers. In many regions, bananas are not just a fruit but a staple food. Cooking bananas, such as plantains, are the cornerstone of diets in Africa and Latin America. Bananas are used in various traditional diets, rituals, and cuisines. Bananas are grown in multiple agroecological zones and require relatively low inputs, and they can be harvested year-round, offering a continuous food supply. However, banana cultivation faces significant challenges from pests and diseases, such as Fusarium wilt (also known as Panama disease) and bacterial wilt, which threaten global production (Tripathi et al. 2024). Bananas are a lifeline for millions of producers worldwide. Their role in combating hunger, malnutrition, and poverty highlights their importance as a staple crop.

In Hinduism, bananas are considered sacred, and the banana plant symbolizes prosperity, and its leaves are used in several rituals. Banana green and new leaves are used as plates during traditional feasts and ceremonies, symbolizing purity in South India and Southeast Asia. In the Philippines, bananas are the fruit of love. In Caribbean cultures, bananas are celebrated in traditional dances and culinary practices, reflecting their historical and cultural ties to plantation life. Banana production is hindered by abiotic and biotic constraints that need to be tackled to increase production several fold and fulfill sustainable output due to increasing demand globally. Significant biotic production constraints include bacterial, fungal, and viral diseases, including pests like nematodes and weevils. Banana production can be improved by applying sustainable practices and innovative technologies, such as genetic transformation, genome editing, and disease management. It is crucial to ensure their ongoing contribution to global food security.

3.1.2 Banana Genetic Complexity

Bananas (*Musa* ssp.) are large monocarpic herbaceous perennial plants that originated in Southeast Asia and were the first crops deemed to be domesticated by human dwellers about 7000 years ago in Southeast Asian countries (D'hont et al. 2012). The genus *Musa* belongs to the monocotyledon Musaceae family. All the edible bananas belong to the Eumusa section and are divided into dessert, cooking, and brewing bananas based on their application. Furthermore, bananas of this section are distinguished based on their genetic background as *Musa acuminata* (A-genome, $2n = 2x = 22$; n represents the haploid chromosome number), *Musa balbisiana* (B-genome, $2n = 2x = 22$), *Musa schizocarpa* (S-genome, $2n = 2x = 22$), and *Australimusa* species (T-genome, $2n = 2x = 20$)2. Most edible cultivated bananas originated from intraspecific or interspecific hybridization between wild diploid *M. acuminata* (A-genome) and *M. balbisiana* (B-genome) species. Additionally, the hybridization of A- and B-genomes has resulted in various polyploidy of cultivated edible bananas, including diploid (AA, BB, and AB), triploid (AAA, AAB, and ABB), and tetraploid (AAAB, AABB, and ABBB) varieties (Simmonds and Shepherd 1955). The triploid genotypes constitute the predominant cultivated varieties that are grown globally. The banana genome is bigger (~430–525 Mb) with significant repetitive DNA sequences (transposable elements and retrotransposons), making genome assembly and annotation challenging. Recent genome sequencing projects, such as the reference genome for *M. acuminata* (AA), have provided insights into genome structure and function. D'Hont et al. described the draft sequence of the 523-megabase genome of an *M. acuminata* AA genotype, providing a breakthrough for genome editing of bananas. Similarly, Wang et al. (2019) presented a high-quality draft genome assembly of *M. balbisiana* with 430 Mb (87%) collected into 11 chromosomes.

3.1.3 Challenges Posed by Polyploidy and Complex Genetics

Most bananas are clonally propagated through suckers derived from mother plants, leading to minimal genetic variation within cultivated varieties. Due to the clonal nature of propagation, bananas are particularly susceptible to pests and diseases attributed to their narrow genetic diversity (Heslop-Harrison and Schwarzacher 2007). Genetic diversity in banana is highly challenging but offers potential for innovative strategies like genome editing to meet food and nutrition security challenges and improve sustainable agriculture.

3.2 Fundamentals of Genome Editing and Its Tools

The innovation of genome editing tools was driven by the limitations associated with transgenesis, which involves the transfer of genetic information across organisms. The earliest forms of genome editing involved the use of radiation- and chemical-induced mutagenesis in the mid-twentieth century (Shelake et al. 2019). After the initial application in *Drosophila* (Muller 1927), X-rays were successfully employed in cereal crops (barley and maize) to induce mutations (Stadler 1928), thereby laying the foundation for induced mutagenesis in plants. The technique was later applied in tobacco, leading to the release of the first mutant cultivar, Chlorina, in the late 1930s (Tollenaar 1938). In the 1940s, Auerbach and Robson (1944) documented the role of chemicals in inducing mutations, leading to the development of various chemical mutagens, including alkylating agents and azides. Over the years, radiation-induced and chemical-induced mutagenesis techniques have been widely applied to investigate gene functions, develop new agronomic traits, and explore DNA repair processes (Shu et al. 2012). However, although these techniques generated random genetic alterations, they lacked specificity, were not eco-friendly, and were hazardous to humans (Shelake et al. 2019). Significant progress was achieved after the introduction of recombinant DNA (rDNA) technology in the 1970s, which marked the first step toward precise genetic manipulation (Jackson et al. 1972). The technique pioneered through the work on bacterial plasmids and viral genomes enabled researchers to insert or delete DNA sequences at targeted loci (Cohen et al. 1973). A pivotal development in genome editing was the discovery of meganucleases, which induce double-strand breaks (DSBs) that are subsequently repaired by the organism's endogenous repair mechanism through non-homologous end joining (NHEJ) or homology-directed repair (HDR). The meganucleases were first identified in yeast in 1985 and are characterized by a recognition sequence usually 20–30 bp. The initial studies induced DSBs in the mouse genome using meganuclease I-SCeI (Choulika et al. 1995; Cohen-Tannoudji et al. 1998). However, their cutting frequency was low and had a fixed recognition sequence, which limited their use. The discovery of zinc finger nucleases (ZFN) significantly enhanced the ease of inducing DSBs, resulting in their widespread application in biotechnology (Bibikova et al. 2003; Urnov et al. 2005). Zinc fingers, small structural protein domains that bind to DNA, RNA, or other proteins, were initially discovered in 1985, but their application in genome editing was only feasible after researchers fused them with a DNA cleavage domain from Fok1 restriction enzyme (Kim et al. 1996), thereby generating an easily programmable nuclease. ZFNs allowed for site-specific genetic modifications but were challenging to design and costly to produce. In the early 2000s, transcription activator-like effector nucleases (TALENs) were developed, offering a more flexible and modular approach to targeted genome editing through customizable DNA recognition domains (Christian et al. 2010; Miller et al. 2011). A breakthrough occurred in D'hont et al. 2012 with the adaptation of the clustered regularly interspaced short palindromic repeats (CRISPR)-Cas9 system, derived from a bacterial adaptive immune mechanism (Fig. 3.1d) (Jinek et al. 2012). CRISPR-Cas9 revolutionized genome editing due to its simplicity and target genome modifications. This technology is a breakthrough in modern molecular biology, facilitating advances in diverse fields, including biotechnology, functional genomics, and therapeutic development (Molla et al. 2020; Songstad et al. 2017; Wang and Doudna 2023).

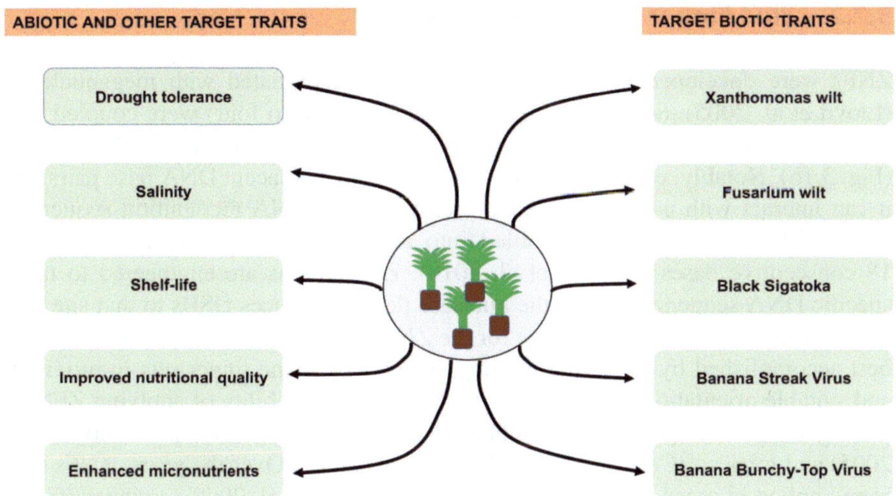

Fig. 3.1 The major genome editing tools. (**a**) Meganucleases, (**b**) Zinc finger nuclease (**c**) TALENs, (**d**) CRISPR/Cas9, (**e**) Base editing using cytidine deaminase, (**f**) Base editing using adenine deaminase, and (**g**) Prime editing

3.2.1 Meganucleases

Meganucleases, also referred to as engineered endonucleases, are highly specific endonucleases that recognize a specific sequence, typically 20–30 bp, known as the homing site (Fig. 3.1a). Although there are over a hundred meganucleases, I-SCeI and I-CreI, characterized by the LAGLIDADG motif, are the commonly utilized nucleases (Aglawe et al. 2018). I-SCeI, isolated from yeast, was the first nuclease-based genome editing technology developed. It was initially applied in mammalian cells to induce double-stranded breaks (DSB) that were repaired through a cellular mechanism called homology-directed repair (HDR) (Choulika et al. 1995; Cohen-Tannoudji et al. 1998). In plants, the efficiency of I-SCeI was first tested in the model plant *Arabidopsis thaliana*, where a double-strand break was induced when transgenic plants expressing the I-SCeI nuclease were crossed with plants with the I-SCeI recognition site (Orel et al. 2003). Next, I-SCeI was utilized for targeted mutagenesis in the progeny of maize (*Zea mays*) transgenic plants (Yang et al. 2009). Moreover, I-CreI meganuclease was successfully applied to create targeted mutations at the liguleless1 (*LG1*) gene in maize (Gao et al. 2010). However, despite the greater specificity exhibited by meganucleases, their application in genome editing is hampered by single homing sites in most organisms and the overlap between DNA binding domains and cleavage sites, thereby complicating the design of the DNA binding domain.

3.2.2 Zinc Finger Nucleases (ZFNs)

ZNFs were developed to counter the challenges associated with meganucleases (Lloyd et al. 2005). Several zinc finger proteins (three to four) were coupled with the cleavage domain of the Fok1 restriction enzyme to create a ZFN DNA domain (Fig. 3.1b). Notably, each zinc finger (ZF) binds three adjacent DNA base pairs, and it can interact with a fourth base. To create specific DNA recognition sequences, several ZF modules were assembled into arrays capable of targeting 9, 12, or 18 consecutive bases (Aglawe et al. 2018). ZF domains are engineered to target specific DNA sequences, while the nuclease domain induces DSBs to that site. The cleavage domain needs to dimerize for the ZNFs to function as a nuclease, which is best accomplished by arranging binding sites for two zinc finger sets in proximity and suitable orientations (Bibikova et al. 2003). The feasibility of applying ZFNs in plant genome editing was first explored in model plants, Arabidopsis (Lloyd et al. 2005) and tobacco (Cai et al. 2009; Wright et al. 2005). Over the years, ZNFs have been applied in other plants, including maize (Shukla et al. 2009), soybean (Curtin et al. 2011), and rice (Jung et al. 2018). However, their application is hampered by several limitations, such as limited availability of target sites, difficulties in designing efficient zinc finger arrays, and low targeting precision, leading to persistent off-target effects (Maeder et al. 2008).

3.2.3 Transcription Activator-Like Effector Endonucleases (TALENs)

TALENs, a family of type III effector proteins secreted by bacteria in the *Xanthomonas* genus, are virulence factors that bind to effector-specific DNA sequences in host plants, altering gene expression to favor infection and disease development (Fig. 3.1c) (Boch et al. 2009; Bogdanove et al. 2010). The interaction between TALEs and target DNA is facilitated by a central region consisting of up to 30 tandem repeats of a sequence motif comprising 33 to 35 amino acids. While most of the amino acid sequence remains constant, two adjacent amino acids, referred to as the repeat variable di residue (RVD), differ among the repeats. These RVDs determine the specificity of DNA recognition, with each RVD corresponding to a specific DNA base pair. The predictable DNA-binding specificity of both native and engineered TALEs opens numerous possibilities for their use in precise genome editing. The DNA recognition domain can guide an attached nuclease to a designated DNA sequence. To this effect, Christian et al. (2010) proposed replacing ZF arrays with the DNA recognition domain of TALEs to develop TALENs (TALEs + Fok1 nuclease) capable of recognizing and cleaving specific DNA targets. These groundbreaking experiments utilized two well-characterized TALENs: AvrBs3 from the pepper pathogen *Xanthomonas campestris* pv. *vesicatoria* and PthXo1 from the rice pathogen *Xanthomonas oryzae* pv. *oryzae*. In plants, TALENs were first applied in the model plants Arabidopsis (Cermak et al. 2011; Christian et al. 2013) and tobacco (Mahfouz et al. 2011), followed by rice (Li et al. 2012; Zhang et al. 2013), barley (Wendt et al. 2013), maize (Liang et al. 2014), and soybean (Haun et al.

2014), among others. The ease of assembly, access to robust resources, cross-species adaptability, and high success rates collectively made TALEN technology a preferred tool for genome editing applications.

3.2.4 CRISPR/Cas9

The GE tools described above rely on the ability of protein molecules with DNA-binding motifs to recognize specific DNA sequences. However, designing DNA-binding motifs to match target sequences requires extensive expertise in protein biochemistry and engineering (Fig. 3.1d) (Aglawe et al. 2018). The best alternative would be having a protein with programmable DNA-binding capabilities, like TALE and zinc finger proteins, but without the need for the labor-intensive process of creating a new protein domain. Ideally, such a protein would naturally function as a nuclease, eliminating the need for extensive engineering. Unknowingly, a system with these characteristics had already been discovered in the late 1980s in *Escherichia coli* (Ishino et al. 1987), now popularly known as the CRISPR/Cas system, a form of acquired immunity in prokaryotes (Sapranauskas et al. 2011). CRISPRs (clustered regularly interspaced short palindromic repeats) are commonly linked to Cas (CRISPR-associated) genes, which encode proteins that function as natural endonucleases. The system utilizes small RNA sequences called CRISPR RNA (crRNA) to guide the Cas protein to its target (Jansen et al. 2002). Over the years, researchers have identified eleven such systems categorized into three groups. The type II CRISPR/Cas system has been particularly adapted for genome editing due to the presence of the protospacer adjacent motif (PAM, NGG, where N can be any nucleotide) sequence and a second RNA molecule, known as trans-activating CRISPR RNA (tracrRNA) (Tröder and Zevnik 2022). The tracrRNA works with CRISPR RNA (crRNA) to facilitate crRNA maturation and recruit the Cas9 nuclease to the target DNA (Brouns et al. 2008). To enhance efficiency, the natural three-component system is streamlined by combining crRNA and tracrRNA into a single synthetic chimeric "guide" RNA (sgRNA or gRNA) (Jinek et al. 2012). Using an RNA molecule to guide the nuclease to a specific DNA target is a significant advantage of the CRISPR/Cas system over other SDNs largely because RNA is more cost-effective and easier to synthesize than the protein domains used in ZFN and TALEN methods. This makes the CRISPR/Cas system simpler, more accessible, and highly efficient (Nasrallah et al. 2022; Zaidi et al. 2020). Moreover, the system allows for the simultaneous introduction of mutations in multiple genes using multiple gRNAs. Given these advantages, the CRISPR/Cas system has been widely applied in plants to introduce crucial agricultural traits. The system's applicability in plants was first tested using model plants: Arabidopsis (Li et al. 2013) and tobacco (Nekrasov et al. 2013). Over the years, CRISPR/Cas technology has been utilized in numerous plants, including rice, maize, wheat, banana, tomato, potato, soybean, and cotton (Chen et al. 2019). However, despite the huge success attributed to CRISPR/Cas9 technology in agriculture, the frequent editing in off-target regions is a significant challenge associated with the technology. Researchers have explored various

Cas nucleases in a bid to mitigate this challenge. For instance, CRISPR-Cas12a (previously known as Cpf1), a type V-A class 2 CRISPR-Cas system, was the first Cas12 nuclease employed for editing plant genomes (Anzalone et al. 2020). Unlike Cas9, Cas12a recognizes T-rich (TTTN) PAM sequences and functions with only crRNA (lacking tracrRNA) (Zetsche et al. 2015). It generates a staggered double-stranded DNA break, producing a 4- or 5-nt 5′ overhang. Moreover, Cas12a exhibits ribonuclease activity, enabling it to process crRNA into its mature form. Although Cas9 and Cas12a exhibit distinct characteristics, both have been shown to mediate plant genome editing with high specificity (Tang et al. 2018). Overall, both nucleases have proven to be highly efficient and precise sequence-specific nucleases (SSNs) in plant systems (Wang et al. 2022b Tang et al. 2017). Several other Cas12 variants, including Cas12b, Cas12e, Cas12g, and Cas12i, all possess the ability to cleave double-stranded DNA (Qi et al. 2020). Cas12 and Cas13 proteins are part of the CRISPR system, used for precise nucleic acid detection in diagnostic applications, leveraging their ability to cleave target RNA or DNA sequences with rapid and accurate detection of viral diseases (Hillary and Ceasar 2024). This technique is called "collateral cleavage," where they disrupt nearby non-target nucleic acids upon binding to a target sequence. This quality makes these proteins useful for developing sensitive diagnostic assays like Specific High-sensitivity Enzymatic Reporter Unlocking (SHERLOCK) for detecting various diseases, including viral infections (Hillary and Ceasar 2024). The Cas12 protein can be used as an effector RNA-guided DNA endonuclease that becomes an alternative to the Cas9 protein for genome editing in human cells (Zetsche et al. 2015). Similarly, CRISPR-Cas13 can target RNA knocking down with high specificity to study RNA in mammals and plants (Abudayyeh et al. 2017). Another class of Cas proteins applied in genome editing is Cas13, a multifunctional enzyme containing two higher eukaryotes and prokaryotes nucleotide-binding (HEPN) domains that confer RNase activity (Kavuri et al. 2022). When combined with crRNA, Cas13 forms an RNA-guided complex capable of recognizing and cleaving single-stranded RNA (ssRNA) targets. Given that the Cas13 protein specifically targets ssRNA, the CRISPR/Cas13 system shows great potential for engineering plant immunity against a wide array of RNA viruses, which comprise most plant viruses (Kavuri et al. 2022). There are several Cas13 proteins (Cas13a, Cas13b, Cas13c, Cas13d, and Cas13X), with all exhibiting two distinct RNase activities: the processing of pre-crRNA into mature functional crRNA and the degradation of target RNA mediated by their HEPN domains (higher eukaryotes and prokaryotes nucleotide-binding domains) (Li et al. 2024; Perez-Sanjose et al. 2022). Recently, researchers have developed a novel high-fidelity RNA-guided endonucleases, called Cas-CLOVER (Madison et al. 2022). The system comprises a catalytically inactive SpCas9 (dCas9) fused to the nuclease domain of a *Clostridium* Clo051 type IIs restriction endonuclease. Its functionality relies on the dimerization of the Clo051 nuclease domain, triggered by RNA-guided recognition of two neighboring 20-nt target sequences separated by a spacer region (11–31 nt). The Cas-CLOVER technique enables precise DNA cleavage with minimal risk of off-target effects, producing larger, sticky-ended overhangs than the smaller blunt-ended deletions generated by

CRISPR/Cas9 (Madison et al. 2022; Tripathi et al. 2023). CRISPR interference (CRISPRi) is divided into two groups, especially CRISPR activation and repression, based on the overexpression or repression of endogenous genes in the plant genome. CRISPR activation (CRISPRa) is a gene regulation tool that increases the overexpression of specific endogenous genes without double-stranded breaks in genomic sequences. It uses dead or inactivated Cas9 protein and acts through a transcription system with minimal cell manipulation (Heidersbach et al. 2023). CRISPR/dCas9 has also been used to visualize genome and nuclear systems, leading to a novel concept named 4D nucleosome (Dekker et al. 2017).

Base editing is a precise, targeted genome editing technique that allows a single nucleotide's direct, irreversible replacement into another without causing double-strand breaks (DSBs) or requiring donor DNA templates. It is a powerful tool for making targeted single-nucleotide changes in the genome, and it is helpful for therapeutic applications and agriculture research. Base editing is a genome-editing strategy that employs specialized "base editors" to precisely convert one nucleotide into another at a specific target site. Base editors enable precise C-to-T or A-to-G substitutions in a target genome sequence without generating DSBs or triggering repair pathways like NHEJ or HDR (Komor et al. 2016). A base editor combines catalytically inactive or dead CRISPR/Cas9 variants (such as dCas9) or Cas9 nickase and a cytosine or adenosine deaminase domain capable of converting one nucleotide into another (Fig. 3.1e, f) (Qi et al. 2020; Mishra et al. 2020). Editing occurs when the sgRNA guides the nCas9-deaminase fusion protein to the target site, initiating deamination and resulting in a base mismatch that is then corrected by the cell's DNA repair mechanism (Porto et al. 2020; Rees and Liu 2018). By enabling single-base changes, base editing can introduce elite trait variations in crop plants, significantly enhancing crop improvement efforts to solve food insecurity (Runo et al. 2024; Tripathi et al. 2022). Currently, base editors are classified into three main groups: cytosine base editors (CBEs), adenine base editors (ABEs), and RNA base editors (adenosine deaminases that act on RNA, or ADARs) (Bharat et al. 2019). CBEs convert C-G base pairs to T-A pairs by initially deaminating cytosine to uracil, creating a U-G mismatch. U is treated as T during repair or replication, resulting in a T-A pair. Unlike CBEs, ABEs facilitate the conversion of A-T base pairs to G-C in genomic DNA. RNA base editors facilitate the conversion of adenine (A) to inosine (I) in RNA (Molla and Yang 2019).

A significant hurdle for BEs is that they are limited to inducing base transitions and cannot facilitate base transversions. Researchers have sought to develop a versatile genome engineering approach capable of introducing all types of genetic modifications to address this technological limitation. A significant breakthrough was achieved when Anzalone et al. (2020) coupled a Cas9 nickase with a reverse transcriptase domain (nCas9-RT), thereby creating a form of "search-and-replace" genome editing tool termed prime editor (PE) (Yang et al. 2019).

Prime editing is a more advanced genome editing technology that allows precise and targeted modifications to DNA, including base substitutions, insertions, and deletions, without requiring double-strand breaks (DSBs) or significant donor DNA templates (Fig. 3.1g). It expands the scope of genome editing beyond the capabilities of traditional CRISPR/Cas9 and base editing. Prime editing has the potential to perform all 12 types of base-to-base conversions, as well as desired insertions or deletions of the targeted genome, without inducing double-strand breaks and requiring a donor DNA template (Tingting et al. 2023; Vu et al. 2024). The nCas9-RT is fused to a prime-editing guide RNA (pegRNA), which consists of two components: a guide sequence that directs the nCas9-RT to a specific genomic target and a 3′-extended template for reverse transcription. Upon binding, the nCas9-RT induces a single-stranded DNA nick, which serves as a primer for the reverse transcription of the template sequence (Anzalone et al. 2020). Vu et al. (2024) successfully optimized PE in tomato and Arabidopsis with high efficiency and specificity. Engineered plant prime editor (ePPE) with two modifications synergistically enhanced the efficiency of base substitutions, deletions, and insertions at various endogenous sites (Zong et al. 2022). A significant bottleneck in prime editing is low efficiency, and several researchers are trying to resolve this issue by using an engineered Prime editor. Li et al. (2022) worked on rice to increase the efficiency several folds.

3.3 Tools and Techniques of Genome Editing in Banana

Bananas are one of the most important food crops globally and are a significant income source for millions of resource-poor farmers in Africa. Predominantly cultivated by smallholder farmers, the crop is mainly used for domestic consumption and local or regional markets, with less than 15% reaching international markets (Pillay and Tripathi 2007). The average annual yield for bananas is 12 tons/ha [1], significantly lower than their potential yield of 70 tons/ha/year. This yield gap is primarily due to the crops' susceptibility to various diseases (bacterial, fungal, and viral), particularly when multiple diseases coexist in the same region (Pillay and Tripathi 2007). Fortunately, genome editing offers a viable tool for developing banana varieties with durable disease resistance through targeted gene knockouts and knock-ins. The CRISPR/Cas9 system was first successfully implemented in bananas through targeted editing of the *phytoene desaturase* (*PDS*) gene (Kaur et al. 2018; Naim et al. 2018; Ntui et al. 2020).

3.3.1 CRISPR-Cas9 Genome Editing in Banana

Banana improvement using conventional methods is hampered by long breeding cycles, poor seed setting, and sterility, especially in farmer-preferred varieties. The CRISPR-Cas9 genome editing, a precise tool for establishing mutations, could be used to develop new traits within existing and difficult-to-breed farmer-preferred varieties. Significant milestones have been achieved in the genome editing of bananas, starting with protocol establishment by targeted knockout of the Phytoene desaturase (*PDS*) gene and developing disease-resistant bananas (Table 3.1)—the *PDS* gene is preferred in establishing plant genome editing protocol due to its

Table 3.1 Advances of genome editing in banana for crop improvement

S. No	Trait	Editing system	Delivery method	Explant	Target	References
1.	*Stress resistance*	CRISPR-Cas9-RNP	*PEG method*	Protoplasts	*STP13 gene*	Leh et al. (2024)
2.	*Albino phenotype*	Cas-Clover	*Agrobacterium tumefaciens*	Embryogenic cells	*MusaPDS*	Tripathi et al. (2023)
3.	*Xanthomonas* wilt resistance	CRISPR-Cas9	*Agrobacterium tumefaciens*	Embryogenic cells	*MusaENOD*3 genes	Ntui et al. (2024)
4.	*Xanthomonas* wilt resistance	CRISPR-Cas9	*Agrobacterium tumefaciens*	Embryogenic cells	*MusaDMR6* gene	Tripathi et al. (2021)
5.	*Albino phenotype*	CRISPR-Cas9	*Agrobacterium tumefaciens*	Embryogenic cells	*MusaPDS*	Ntui et al. (2020)
6.	*Delayed riping and plant height*	CRISPR-Cas9	*Agrobacterium tumefaciens*	Embryogenic cells	*MaACO1* gene	Hu et al. (2023)
7.	Increase β-carotene	CRISPR-Cas9	*Agrobacterium tumefaciens*	Embryogenic cells	*Musa lycopene epsilon-cyclase gene*	Kaur et al. (2020)
8.	Shorter height	CRISPR-Cas9	*Agrobacterium tumefaciens*	Embryogenic cells	*Gibberellin 20ox2* (*MaGA20ox2*) gene	Shao et al. (2020)
9.	Banana streak virus resistance	CRISPR-Cas9	*Agrobacterium tumefaciens*	Embryogenic cells	*Endogenous banana streak virus in the B genome of banana*	Tripathi et al. (2019a)
10.	*Albino phenotype*	CRISPR-Cas9	*Agrobacterium tumefaciens*	Embryogenic cells	*MusaPDS*	Kaur et al. (2018)
11.	*Albino phenotype*	CRISPR-Cas9	*Agrobacterium tumefaciens*	Embryogenic cells	*MusaPDS*	Naim et al. (2018)

visually identifiable albino phenotype upon successful knockout. Several studies on bananas targeting different cultivars have been accomplished so far. Kaur et al. 2018 reported efficient editing of the *PDS* gene in the cultivar Rasthali using the CRISPR/Cas9 system, where a single guide RNA (gRNA) was used to target conserved regions. Another study by Mascarenhas et al. (2024) developed targeted knockout of the *PDS* gene in the variety Prata-Anã. We have also successfully edited the *PDS* gene in a dessert banana 'SukaliNdiizi' and Plantain 'GonjaManjaya' using the CRISPR/Cas9 system (Ntui et al. 2020). A recent Hu et al. 2023 investigated *PDS* gene knocking out establishing a transgene-free genome editing system.

Besides the protocol established by *PDS* knockout, CRISPR/Cas9 editing also targets genes associated with specific traits, including enhancing biotic and abiotic stress tolerance. Studies by Tripathi et al. (2019b) and (2021), for example, have targeted two known disease susceptibility genes Mildew Locus O (*MLO*) and downy mildew resistance 6 (*DMR6*), respectively, which have conferred resistance to Banana Xanthomonas Wilt (BXW) caused by *Xanthomonas campestris* pv. *musacearum* (Xcm). Previously, our laboratory used CRISPR/Cas9 to knock out an endogenous Banana Streak Virus (BSV) that is naturally integrated into the B-genome and prevents germplasm movement (Tripathi et al., 2019a). A recent study showed that editing the early nodulin-like 3 (*ENOD3*) gene conferred resistance against BXW in GonjaManjaya (Ntui et al. 2024). Moreover, a study by Kaur et al. (2020) targeted editing of the lycopene epsilon-cyclase (*LCYε*) gene for enhanced carotene levels in the Cavendish banana variety.

3.3.2 Cas-Clover Genome Editing in Banana

The Cas-CLOVER system is an innovative high-fidelity genome-editing technology that employs two gRNAs, unlike CRISPR/Cas9, which relies on a single gRNA. The use of two gRNAs significantly enhances editing precision. The dual gRNA design also ensures that the Cas-CLOVER system only induces double-stranded breaks at the intended target sites, minimizing the risk of off-target effects (Alvarez et al. 2023). One of the key components of the Cas-CLOVER system is the use of a catalytically inactive Cas protein, often referred to as "dead Cas" or dCas, fused to the Clo051 endonuclease domain. This fusion protein requires the dimerization of the Clo051 domains, which is facilitated by binding the two guide RNAs to their respective target sequences (Madison et al. 2022). This dimerization is essential for the nuclease activity, ensuring that DNA cleavage only occurs when both guide RNAs are correctly positioned. This mechanism provides an additional layer of specificity, making Cas-CLOVER a highly reliable tool for precise genome editing. The usability of the Cas-CLOVER system has already been successfully tested in banana genome editing, where the *PDS* gene was knocked out, resulting in albino plants (Tripathi et al. 2023).

3.3.3 Methods of Delivery of Genome Editing Components in Plant Cells

Delivering genome editing components into plant cells is essential for achieving precise genetic modifications. One of the traditional methods is *Agrobacterium*-mediated transformation, which uses the natural ability of *Agrobacterium tumefaciens* to transfer DNA into plant genomes. This method is particularly effective and allows for stable integration of editing tools, although it is limited by plant species specificity and tissue culture requirements (Doyle-Prestwich et al. 2023). Most studies have relied on using *Agrobacterium* to deliver genome-editing components to bananas. The significant challenge for using *Agrobacterium* in the delivery of genome editing components in bananas is its nature of propagation being asexual and, therefore, no means of segregating out the transgene. Thus, methods that can result in transgene-free edited plants could benefit vegetatively propagated crops like bananas.

Biolistic delivery, also known as particle bombardment, is an alternative technique for editing components, whether DNA or protein. This method involves using high-velocity microprojectiles, typically gold or tungsten particles, coated with plasmid DNA or ribonucleoprotein (RNP) complexes. These particles are physically propelled into plant cells, which penetrate the cell walls and membranes to deliver the genetic material directly into the nucleus. This technique is particularly advantageous for bananas, which are often challenging to transform due to their polyploidy and sterility. Biolistic delivery has been successfully used to deliver plasmid for genetic engineering in bananas as an alternative to *Agrobacterium*-mediated transformation, which can be less efficient and more time-consuming (Dong et al. 2020). Therefore, this method could successfully deliver genome editing components in the future, given that no successful studies have been reported.

Another alternative method of delivering genome editing components into plant cells is stripping off the cell wall to form protoplasts, which can then be electroporated or porous by chemicals. The use of protoplasts typically involves isolating protoplasts from banana tissues, such as leaves or immature male flower buds, using enzymatic digestion. Once isolated, these protoplasts can be transformed using polyethylene glycol (PEG)-mediated transfection or electroporation to deliver CRISPR/Cas9 plasmids or preassembled ribonucleoprotein (RNP) complexes directly into the cells. This approach allows efficient and targeted gene editing without integrating foreign DNA, making it a promising tool for developing improved banana varieties (Wu et al. 2020). One of the key benefits of using protoplast genome editing in bananas is the ability to achieve transgene-free modifications. By delivering preassembled CRISPR/Cas9 RNPs directly into the protoplasts, researchers can induce targeted mutations without integrating foreign DNA into the plant genome. This reduces the regulatory hurdles associated with genetically modified organisms (GMOs) and increases the acceptance of edited plants. Studies have demonstrated that protoplast genome editing can effectively target genes such as the *phytoene desaturase* (*PDS*) gene, resulting in observable phenotypes like albinism, which are markers for successful gene editing. The continued refinement of protoplast transformation techniques holds great potential for advancing banana breeding programs and developing varieties with enhanced traits such as disease resistance, improved yield, and better adaptability to environmental stresses (Leh et al. 2024).

Though not yet used in bananas, transient delivery methods like virus-mediated delivery could be exploited in future genome editing. These methods exploit plant viruses as vectors to carry genome editing components. Positive-strand RNA viruses like Potato Virus X (PVX) and engineered Tobacco Rattle Virus (TRV) have already demonstrated systemic delivery of compact Cas proteins or guide RNAs across plant tissues. These methods bypass tissue culture, enabling efficient mutagenesis and transgene-free genome editing in crops (Mengstie 2022).

Each delivery method has strengths and limitations, and the choice depends on factors such as plant species, tissue type, and the desired outcome (stable integration or transient expression). Continued advancements in delivery technologies, including integrating multiple approaches, will enhance the efficiency and accessibility of genome editing in diverse plant species.

3.3.4 Regeneration of Edited Plants

Banana tissue culture and plant regeneration of edited cells present significant challenges and promising opportunities for banana biotechnology. One of the primary challenges is the inherent difficulty in generating banana plants from tissue culture embryogenesis. Bananas are known for their recalcitrance to tissue culture, specifically somatic embryogenesis. This is particularly true for certain banana varieties, which are highly valued for their commercial traits but are less amenable to tissue culture techniques. Inducing somatic embryogenesis, crucial for regenerating whole plants from edited cells, can be highly genotype-dependent and requires optimization for each specific variety (Tripathi et al. 2015).

The polyploid nature of most cultivated banana varieties complicates genetic transformation and regeneration because it involves multiple sets of chromosomes, making it more difficult to achieve uniform gene editing across all genome copies. This, coupled with the fact that most cultivated bananas are vegetatively propagated and there is no way of segregating out the transgene, makes editing more complex. This complexity can lead to chimeric plants, where only some cells are edited, resulting in inconsistent phenotypes. Additionally, the sterility of many banana cultivars poses a challenge to traditional breeding methods, making tissue culture and genetic transformation essential for introducing new traits (Wang et al. 2022a; b). However, despite these challenges, there are significant opportunities in banana tissue culture and plant regeneration. Advances in somatic embryogenesis and the development of embryogenic cell suspension (ECS) cultures have improved the efficiency of banana transformation. ECS cultures provide a reliable source of totipotent cells that can be genetically transformed and regenerated into whole plants. This technique has been successfully introduced to introduce disease-resistance genes into a banana, offering a potential solution to significant threats such as Fusarium wilt, Xanthomonas wilt, and banana bunchy top virus (Tripathi et al. 2020). Using CRISPR/Cas9 and other genome-editing technologies in conjunction with tissue culture techniques offers exciting opportunities for banana improvement. These technologies allow for precise modifications to the banana genome, enabling the development of varieties with enhanced traits such as improved yield, disease resistance, and stress tolerance. The ability to create transgene-free edited plants through biolistic delivery or protoplast transformation further enhances the potential for regulatory approval and public acceptance of genetically edited bananas. Continued research and optimization of tissue culture protocols, along with the integration of advanced genome-editing tools, hold the potential to revolutionize banana breeding and cultivation.

3.4 Applications of Genome Editing in Banana

Banana diseases pose a significant threat to production and consequently make the development of disease-resistant banana varieties one of the primary applications of the genome. Bananas are susceptible to devastating diseases like Fusarium wilt (Panama disease) and banana Xanthomonas wilt (BXW). Using CRISPR/Cas9 technology, researchers have successfully edited genes associated with BXW disease susceptibility, thereby enhancing the plant's resistance. For instance, targeting susceptibility genes has conferred resistance to BXW, a primary bacterial disease affecting banana production in Africa (Tripathi et al. 2020). The gene editing approaches can be considered for various traits like biotic, abiotic constraints, nutritional enhancement with high-yield varieties (Fig. 3.2).

Another important application is improving bananas' nutritional content. Genome editing has been used to enhance the levels of essential nutrients such as provitamin A (β-carotene) and iron. By editing specific genes involved in the biosynthesis pathways of these nutrients, researchers have developed banana varieties with increased nutritional value. This is particularly important in regions where bananas are a staple food and can help address micronutrient deficiencies (Bansal et al. 2022).

The potential of genome editing in banana improvement is enormous, with possible application in improving agronomic traits. Traits such as drought tolerance, early flowering, and enhanced yield could be targeted through precise genetic modifications. For example, editing genes related to the plant's response to water stress can result in banana varieties that are more resilient to drought conditions. Furthermore, genome editing could also be used to develop bananas with reduced post-harvest losses. Such innovations could reduce food waste and extend banana shelf life, making bananas more appealing to consumers and reducing farmers' economic losses. Gene editing has also been applied to develop bananas with enhanced plant architecture, where the gibberellin 20ox2 (MaGA20ox2) gene was edited, resulting in bananas' shorter height (Shao et al. 2020).

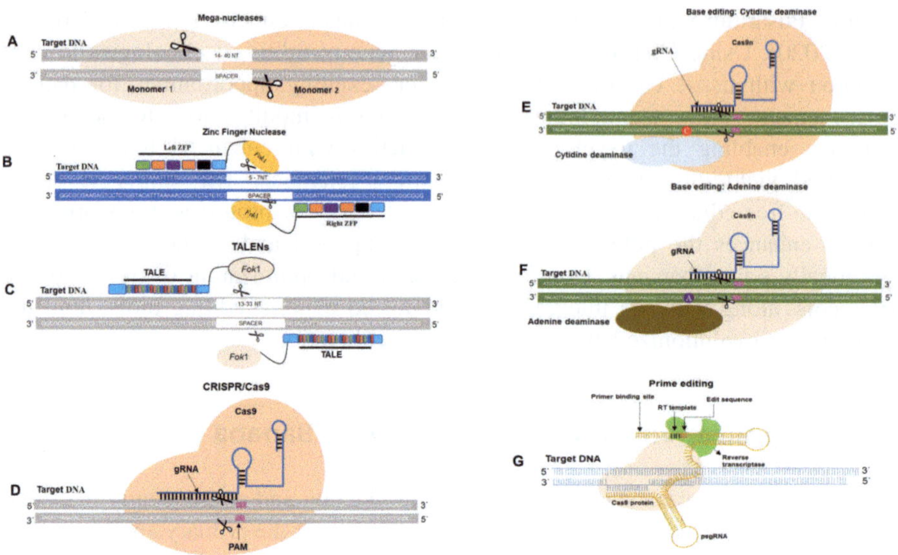

Fig. 3.2 A schematic presentation of various traits that can be targeted for genome editing in bananas

3.4.1 Genome Editing of Bananas for Disease Resistance

Of the significant banana diseases, Xanthomonas wilt has had the most genome editing research. Based on a comparative transcriptome study that we did (Tripathi et al., 2019c), a diverse number of genes were observed to be downregulated in the resistant progenitor species *Musa balbisiana* but not in the susceptible variety, Pisang Awak. Using CRISPR/Cas9 as the significant approach, we have targeted to knock out a vast number of susceptibility genes, including *DMR6* (Tripathi et al. 2021) and early nodulin-like 3 (*ENOD3*) (Ntui et al. 2024).

BSV is a significant constraint to banana production, causing severe yield losses and limiting the international movement of banana germplasm. Traditional breeding methods have been largely ineffective in managing BSV due to the complex nature of the virus and the polyploidy of banana genomes. However, recent advancements in CRISPR/Cas9-based genome editing have provided a powerful tool to develop BSV-resistant banana varieties. Researchers have successfully used CRISPR/Cas9 to target and inactivate endogenous BSV (eBSV) sequences integrated into the banana genome, thereby preventing virus activation and reducing disease incidence.

3.4.1.1 Resistance to Fusarium Wilt

Fusarium wilt is another primary disease of bananas, which can cause up to 100% losses. Fusarium wilt, also known as Panama disease, is a devastating disease that affects banana plants. It is caused by the soil-borne fungus *Fusarium oxysporum* f. sp. *cubense* (Foc). This pathogen enters the banana plant through the roots and spreads through the vascular system, blocking the flow of water and nutrients. This disease is complicated by the presence of multiple races, including the devastating tropical race-4 (TR4). Genome editing could allow the precise modification of genes to enhance the banana's immune response to Foc. Functional genomic studies have revealed insights into genes governing immune signaling and plant-pathogen interactions, aiding the identification of key targets for editing. For example, enhancing the expression of pathogenesis-related (PR) genes and manipulating hormonal pathways involved in plant defense can bolster resistance hurdles (Wang et al., 2022c). Targeting susceptibility genes for knockout could also be used as an alternative strategy. Despite these advancements, integrating genome-edited bananas into agricultural systems faces practical and regulatory challenges. Field trials are necessary to validate the efficacy of edited cultivars against TR4 under diverse environmental conditions, but the absence of TR4 in most countries limits extensive research. Furthermore, regulatory frameworks governing genetically modified organisms (GMOs) vary worldwide, influencing the deployment of genome-edited bananas in areas with TR4 already reported. Research into transgene-free editing methods, such as transient delivery of CRISPR components, is gaining momentum to mitigate public concerns and regulations.

3.4.1.2 Resistance to Black Sigatoka

Black Sigatoka, caused by *Pseudocercospora fijiensis*, is another devastating fungal disease of bananas that can lead to significant yield losses or increased dependency on fungicides. The pathogen's high evolutionary potential and ability to develop resistance to fungicides make it a formidable challenge for sustainable agriculture. Genome editing offers a promising avenue to enhance resistance by targeting genes involved in the plant-pathogen interaction. For example, host-induced gene silencing (HIGS) can suppress fungal pathogenicity genes, while CRISPR/Cas9-mediated editing can modify susceptibility genes in bananas, conferring durable resistance. Identifying resistant genes in wild banana species has further facilitated genome editing applications. Studies have revealed that specific diploid wild bananas exhibit natural resistance to Black Sigatoka, providing a genetic reservoir for resistance traits. Genome-wide analyses of these accessions have highlighted several candidate genes involved in defense pathways, including those regulating jasmonic acid and ethylene signaling, which could be targeted for editing (Nascimento et al. 2023). Genome editing has the potential to combat Black Sigatoka in bananas. Researchers can develop resistant cultivars, reduce fungicide dependency, and ensure sustainable banana production by leveraging advancements in genetic tools and understanding plant-pathogen interactions.

3.4.1.3 Resistance to Banana Xanthomonas Wilt

Banana Xanthomonas Wilt (BXW), caused by *Xanthomonas campestris* pv. *musacearum* (Xcm), is the leading threat to banana production in East Africa, causing up to 100% yield losses. This disease affects all banana varieties except the progenitor diploid species *M. balbisiana*, which shows natural resistance. At the International Institute of Tropical Agriculture (IITA), extensive efforts have been made to establish bananas resistant to BXW. Multiple susceptibility genes have been targeted for knockout to confer resistance in various bananas and plantains (Tripathi et al. 2021). The S-genes targeted at IITA have previously been observed to be differentially expressed in susceptible variety and naturally resistant species (Shao et al. 2020). Based on the knowledge from previous studies within IITA and from literature that the target genes were indeed S-genes, the genes were targeted for knockout using CRISPR/Cas9.

Some major S-genes already targeted at IITA include *MusaDMR6* (Tripathi et al. 2021). The *DMR6* gene negatively regulates plant disease resistance by modulating the salicylic acid (SA) pathway and suppressing immune responses. Knocking out *MusaDMR6* through multiplexed gRNAs in the cultivar 'SukaliNdiizi' resulted in enhanced resistance to BXW without adverse effects on the growth and morphology of the events.

CRISPR/Cas has also been used to knock out the *MusaENOD*3 gene in the BXW-susceptible plantain 'GonjaManjaya', resulting in significant resistance to Xcm (Ntui et al. 2024). Nodulin-like genes were initially identified in legumes for their role in symbiotic nitrogen fixation but have been found to play additional roles in plant defense in different plant species. These genes transport nutrients, solutes, amino acids, and hormones, which are crucial for plant development and stress responses. In non-leguminous plants such as Arabidopsis and rice, nodulin-like proteins have been shown to participate in defense mechanisms against pathogens (Denancé et al. 2014). By targeting specific genes associated with susceptibility and immune response, researchers can develop resistant cultivars that reduce dependency on chemical controls and enhance productivity. Continued collaboration among researchers, policymakers, and stakeholders is essential to translate these scientific advancements into practical agricultural solutions. Immune pathways, including pattern-triggered immunity (PTI) and effector-triggered immunity (ETI), are critical for plant defense against pathogens. By editing the susceptibility gene promoter, researchers can enhance the detection and response mechanisms of these pathways, thereby improving disease resistance. Recent advances include the modification of receptor-like kinases (RLKs) and nucleotide-binding leucine-rich repeat (NLR) proteins to strengthen immune responses against broad-spectrum pathogens (Choudhary et al. 2024).

3.5 Key Challenges in Genome Editing of Bananas

3.5.1 Polyploidy and Genetic Diversity

Banana is a polyploid crop, which presents several challenges for genome editing, mainly when target genes exist in multiple copies (Tripathi et al. 2020). For a desired outcome, all gene copies must be edited simultaneously (Schaart et al. 2021). If some copies remain functional, they may compensate for the mutation, reducing the phenotypic impact. CRISPR/Cas9 genome editing tools can sometimes recognize and cut similar sequences elsewhere in the genome, leading to unintended mutations.

Additionally, different alleles may be edited at varying efficiencies, resulting in mosaicism within the same plant. Partial editing during regeneration can lead to genetic variability among regenerated plants, making obtaining stable, uniform lines difficult.

Furthermore, tissue culture-based regeneration often introduces additional genetic changes, further complicating trait evaluation. Achieving transgene-free genome editing in polyploids requires transient expression or RNP delivery, which is particularly complex in bananas. Screening for desired mutations without selectable markers is also challenging, especially in large genomes. Polyploids often exhibit epigenetic modifications that influence gene expression, making it difficult to predict the functional outcome of edits.

Additionally, redundancy in regulatory elements may buffer the effects of gene edits, requiring multiple modifications to achieve a significant impact. In vegetatively propagated crops like bananas, polyploidy can lead to unpredictable segregation of edited alleles (Hu et al. 2023), further complicating breeding programs. Regulatory frameworks for genome-edited crops vary across countries, and polyploidy adds another layer of complexity in distinguishing edited from non-edited plants.

3.5.2 Transformation Efficiency

Embryogenic cells are preferred explants for plant transformation and genome editing in the case of banana. These cells are developed from multiple buds or immature male flowers embedded within the inner part of the inflorescence (Tripathi et al. 2015). However, developing embryogenic cells is labor-intensive and time-consuming, often taking up to a year. A major limiting factor in this process is its cultivar specificity, as some banana cultivars are highly recalcitrant to producing friable calli.

3.5.3 Off-Target Effects

Off-target effects in plant genome editing using CRISPR technology refer to unintended mutations or indels occurring at sequences other than the specific target site of the CRISPR-Cas9 system. These unintended cuts can lead to nonspecific genetic alterations, potentially impacting the desired trait modifications. Off-target effects are a significant regulatory concern due to their potential adverse consequences on plant genomes. However, several advanced tools are now available to design highly specific gRNAs, minimizing or eliminating off-target effects (Guo et al. 2023). Additionally, the availability of Cas9 variants with improved specificity further enhances the precision of genome editing.

3.5.4 Ethical and Regulatory Considerations

Ethical concerns regarding the genetic modification (GM) of staple crops comprise social, environmental, and moral considerations. The main problem is the safety of GM foods for both human and animal health, as they may pose toxicity, allergenicity, or potential long-term health implications (Qaim 2010). However, it is worth noting that GM foods undergo rigorous testing before reaching consumers. Nevertheless, concerns remain regarding the adequacy and reliability of current food safety assessment protocols. In addition to health-related issues, environmental concerns include the potential for gene flow from GM crops to conventional or wild relatives, the loss of natural biodiversity, and unintended effects on non-target beneficial insects and soil microorganisms. Economic and social concerns associated with GM crops are more challenging to regulate. Major seed companies may dominate the food system, potentially limiting the choices of resource-poor farmers, who may be compelled to purchase expensive seeds. In addition, misinformation about GM foods, often propagated by opponents, contributes to public skepticism. There is an urgent need for science-based communication to address misconceptions, provide fact-based justifications, and draw upon past experiences. Over the past two decades, several GM foods have been introduced to the market, yet no scientifically validated evidence has demonstrated adverse health effects. Furthermore, ethical considerations have regulatory implications, particularly for the biosafety authorities, to ensure that small-scale farmers can express their preferences regarding traits selected by plant breeders and to establish mechanisms for cultivating GM crops based on risk-based evaluations (Weale 2010).

3.5.5 Regulatory Challenges in Different Countries for Genome-Edited Crops

Most concerns surrounding gene-edited products are like those associated with GM crops, particularly in areas such as human health, environmental impact, off-target mutations, and intellectual property rights. Modern breeding tools like gene editing are more precise and targeted than genetically modified techniques, but off-target effects can still occur, especially in complex genomes. Advances in gene-editing technology and bioinformatics tools are continually improving the accuracy of edits, minimizing the frequency of off-target mutations (Young et al. 2019). Gene editing tools are frequently presented as a naturally occurring mutation resulting from traditional breeding if there is no foreign gene integration into the plant genome, which could help gain public acceptance. Regulatory frameworks vary across countries, with some distinguishing gene-edited crops from GMOs based on the presence or absence of foreign DNA in plants, the plant genome of mutated events. Regulatory authorities distinguish three types of genome editing based on site-directed nucleases (SDNs). SDN-1 introduces small changes at the target site. SDN-2 employs template-guided repair via homologous recombination to teach a specific DNA sequence replacement in the genome. Moreover, SDN-3 replaces more considerable genetic material, like SDN-2 (Friedrichs et al. 2019). Most biosafety regulators across countries agree that crops edited using SDN-1 and SDN-2, which do not contain exogenous DNA, are considered non-GM crops. The USA, Canada, and four South American countries have implemented regulations that treat genome-edited crops (SDN-1, -2) as equivalent to conventionally bred crops (Schmidt et al. 2020).

In contrast, countries such as China, Ethiopia, Ghana, India, Kenya, Nigeria, and Malawi have biosafety guidelines that determine regulatory decisions on a case-by-case basis, depending on the presence or absence of foreign DNA in the edited crops (Buchholzer and Frommer 2023). Transparency regarding gene-edited products is essential for ensuring social acceptance among researchers, biosafety regulators, and farmers to fully realize the benefits of this revolutionary technology (Gordon et al. 2021). All stakeholders must be involved in the process to maximize the potential of genome editing for global food security.

3.6 Future Directions and Prospects

Epigenome editing is a genetic engineering technology that modifies the epigenome at specifically targeted locations using epigenome effectors comprising zinc finger nucleases, transcription activator-like (TAL) effectors, CRISPR/Cas9 complexes, and chromatin markers. This approach does not induce double-strand breaks in plant DNA, minimizing damage and relying on repair pathways (McCutcheon et al. 2024). Modifying the epigenome can lead to gene overexpression, facilitating the development of disease-resistant and high-yielding crop varieties.

3.6.1 Nanoparticles for Delivery

Nanotechnology presents a promising method for efficiently and precisely delivering CRISPR/Cas proteins to target cells. This approach is particularly appealing for gene editing applications that require temporary delivery of CRISPR components without integrating foreign DNA into the host genome (Li et al. 2021). Various types of nanoparticles have been designed to deliver CRISPR/Cas proteins and guide RNA (gRNA) into cells. Among these, lipid nanoparticles (LNPs) are commonly used due to their biocompatibility and capacity to encapsulate Cas proteins and gRNA (Wei et al. 2020). Other polymers, such as polyethyleneimine (PEI), poly (lactic-co-glycolic acid) (PLGA), and dendrimers, can also encapsulate CRISPR/Cas components. These nanoparticles facilitate controlled release and protect the cargo from degradation. Gold nanoparticles can be conjugated with CRISPR/Cas proteins and gRNA using chemical linkers.

Additionally, their surfaces can be functionalized to target specific cells and enable controlled release. Magnetic particles allow guided delivery through external magnetic fields, making them particularly useful for localized delivery in vivo. Silica nanoparticles are chemically stable and can be modified to carry CRISPR components efficiently. DNA-based nanocarriers can deliver CRISPR/Cas proteins and gRNA with high precision due to their programmability. Nanoparticles enable the delivery of CRISPR/Cas proteins as ribonucleoprotein (RNP) complexes, which reduces the risk of off-target effects compared to plasmid DNA delivery. They also protect CRISPR components from enzymatic degradation in the extracellular environment. These nanoparticles can achieve cell or tissue-specific targeting by employing surface modifications, thus minimizing off-target effects. Moreover, nanoparticle systems are relatively easy to scale for clinical applications in human or animal models, avoiding the risks associated with viral vectors, such as insertional mutagenesis and immunogenicity. However, achieving high delivery efficiency to specific cells like embryogenic cells or meristematic tissues remains challenging, and maintaining the stability of CRISPR/Cas proteins and gRNA within nanoparticles is critical. While nanoparticles are generally biocompatible, some formulations may trigger immune responses. Manufacturing nanoparticle formulations at a large scale poses technical challenges. Applying nanotechnology in gene editing may enhance plant traits, such as drought tolerance and disease resistance.

3.7 Conclusion

Banana breeding programs solely depend on access to diploid wild-type parent populations with sufficient genomic variability, which can contribute significantly to the development of bananas with enhanced yield and nutritional content. This variability can be induced through chemical mutagens or physical irradiation. However, not all types of genetic variations provide the expected results, such as higher yield, improved nutritional value, disease resistance, and climate resilience. As a result, breeding efforts can be time-consuming and labor-intensive, requiring significant effort to incorporate desirable traits into elite banana varieties. In contrast, genome editing offers the ability to produce precisely targeted modifications. The genome editing of bananas has become a fully established technology facilitated by the availability of whole genome sequences and robust protocols for generating complete plants through embryogenic cells. This technology plays a crucial role in developing elite varieties of bananas by applying genome editing tools, especially CRISPR/Cas9. Several new tools, such as Cas-Clover and PiggyBac, are currently being researched to develop DNA-free disease-resistant banana plants. Genome editing technology is advancing rapidly, potentially creating high-yielding crops with multiple beneficial traits through multiplexing. Since bananas are clonally propagated crops, transgenes do not segregate as they do in seed-propagated crops, making removing transgenes through conventional breeding challenging and time-consuming. Developing foreign gene-free plants remains a significant bottleneck, but researchers are working toward solving this challenge. To facilitate the adoption of genome-edited bananas, science-based regulations are required to ensure safe and responsible deployment. Public perception must be carefully managed to support large-scale adoption, with ongoing communication between scientists, media, biosafety regulators, and farmers.

Overall, banana genome editing will play a pivotal role in crop improvement by enabling the development of high-yielding crops. This will contribute significantly to global food security and sustainable agriculture, especially in banana-producing regions.

References

Abudayyeh OO, Gootenberg JS, Essletzbichler P (2017) RNA targeting with CRISPR–Cas13. Nature 550(7675):280–284. https://doi.org/10.1038/nature24049

Aglawe SB, Barbadikar KM, Mangrauthia SK et al (2018) New breeding technique "genome editing" for crop improvement: applications, potentials and challenges. 3 Biotech, 8(8):1–20. https://doi.org/10.1007/s13205-018-1355-3

Alvarez O, Homa L, Kearney S et al (2023) Cas-Clover editing efficiency and off-target activity in human hepatocytes at the KLKB1 locus. Blood 142:1753. https://doi.org/10.1182/blood-2023-

Anzalone AV, Koblan LW, Liu DR (2020) Genome editing with CRISPR–Cas nucleases, base editors, transposases and prime editors. Nat Biotechnol 38(7):824–844. https://doi.org/10.1038/s41587-020-0561-9

Auerbach C, Robson JM (1944) Production of mutations by allyl isothiocyanate. Nature 154(3898): 81–81. https://doi.org/10.1038/154081a0

Bansal KC, Molla KA, Chinnusamy V (2022) Genome editing: a boon for plant biologists, breeders and farmers. Curr Sci 123(1):5

Bharat SS, Li S, Li J et al (2019) Science direct base editing in plants: current status and challenges. The Crop J 8(3):384–395. https://doi.org/10.1016/j.cj.2019.10.002

Bibikova M, Beumer K, Trautman JK et al (2003) Enhancing gene targeting with designed zinc finger nucleases. Science 300(5620):764. https://doi.org/10.1126/science.1079512

Boch J, Scholze H, Schornack S et al (2009) Breaking the code of DNA binding specificity of TAL-type III effectors. Science 326(5959):1509–1512. https://doi.org/10.1126/science.1178811

Bogdanove AJ, Schornack S, Lahaye T (2010) TAL effectors: finding plant genes for disease and defense. Curr Opin Plant Biol 13(4):394–401. https://doi.org/10.1016/j.pbi.2010.04.010

Brouns SJJ, Davies K, Harbor CS et al (2008) Small CRISPR RNAs guide antiviral defense in prokaryotes. Science 321(15):960–964. https://doi.org/10.1126/science.1159689

Buchholzer M, Frommer WB (2023) An increasing number of countries regulate genome editing in crops. New Phytol 237:12–15. https://doi.org/10.1111/nph.18333

Cai CQ, Doyon Y, Ainley WM et al (2009) Targeted transgene integration in plant cells using designed zinc finger nucleases. Plant Mol Biol 69(6):699–709. https://doi.org/10.1007/s11103-008-9449-7

Cermak T, Doyle EL, Christian M et al (2011) Efficient design and assembly of custom TALEN and other TAL effector-based constructs for DNA targeting. Nucleic Acids Res 39(12):1–11. https://doi.org/10.1093/nar/gkr218

Chen K, Wang Y, Zhang R et al (2019) CRISPR/Cas genome editing and precision plant breeding in agriculture. Ann Rev 70:667–697. https://doi.org/10.1146/annurev-arplant-050718-100049

Choudhary S, Prasanna NL, Ashwini S et al (2024) A review on molecular mechanism of plant immunity against fungal pathogens. J Adv Biol Biotechnol 27(10):71–81. https://doi.org/10.9734/jabb/2024/v27i101431

Choulika A, Perrin A, Dujon B et al (1995) Induction of homologous recombination in mammalian chromosomes by using the I- Sce I system of *Saccharomyces cerevisiae*. Mol Cell Biol 15(4): 1968–1973. https://doi.org/10.1128/mcb.15.4.1968

Christian M, Cermak T, Doyle EL et al (2010) Targeting DNA double-strand breaks with TAL effector nucleases. Genetics 186(2):756–761. https://doi.org/10.1534/genetics.110.120717

Christian M, Qi Y, Zhang Y, Voytas DF (2013) Targeted mutagenesis of *Arabidopsis thaliana* using engineered TAL effector nucleases. G3: genes, genomes. Genetics 3(9):1697–1705. https://doi.org/10.1534/g3.113.007104

Cohen SN, Chang ACY, Boyer HW et al (1973) Construction of biologically functional bacterial plasmids in vitro. Proc Natl Acad Sci USA 70(11):3240–3244. https://doi.org/10.1073/pnas.70.11.3240

Cohen-Tannoudji M, Robine S, Choulika A et al (1998) I- Sce I-induced gene replacement at a natural locus in embryonic stem cells. Mol Cell Biol 18(3):1444–1448. https://doi.org/10.1128/mcb.18.3.1444

Curtin SJ, Zhang F, Sander JD et al (2011) Targeted mutagenesis of duplicated genes in soybean with zinc-finger nucleases. Plant Physiol 156(2):466–473. https://doi.org/10.1104/pp.111.172981

D'hont A, Denoeud F, Aury JM et al (2012) The banana (Musa acuminata) genome and the evolution of monocotyledonous plants. Nature 488(7410):213–217. https://doi.org/10.1038/nature11241

Dekker J, Belmont AS, Guttman M et al (2017) The 4D nucleome project. Nature 549(7671): 219–226. https://doi.org/10.1038/nature23884

Denancé N, Szurek B, Noël LD (2014) Emerging functions of nodulin-like proteins in non-nodulating plant species. Plant Cell Physiol 55(3):469–474. https://doi.org/10.1093/pcp/pct198

Dong T, Bi FC, Huang YH et al (2020) Highly efficient biolistic transformation of embryogenic cell suspensions of banana via a liquid medium selection system. Hortic Sci 55(5):703–708. https://doi.org/10.21273/HORTSCI14884-20

Doyle-Prestwich B, Cardi T, Bakhsh A et al (2023) Novel delivery methods for CRISPR-based plant genome editing. In: Doyle Prestwich B, Cardi T, Bakhsh A, Nicolia A, Bhati KK (eds) A roadmap for plant genome editing. Springer. https://doi.org/10.1007/978-3-031-46150-7_3

FAOSTAT (2023) FAO statistical database. http://www.fao.org. Accessed 18 Dec 2024

Friedrichs S, Takasu Y, Kearns P et al (2019) An overview of regulatory approaches to genome editing in agriculture. Biotechnol Res Innov 3(2):208–220. https://doi.org/10.1016/j.biori.2019.07.001

Gao H, Smith J, Yang M et al (2010) Heritable targeted mutagenesis in maize using a designed endonuclease. Plant J 61(1):176–187. https://doi.org/10.1111/j.1365-313X.2009.04041.x

Gordon DR, Jaffe G, Doane M et al (2021) Responsible governance of gene editing in agriculture and the environment. Nat Biotechnol 39:1055–1057. https://doi.org/10.1038/s41587-021-01023-1

Guo C, Ma X, Gao F et al (2023) Off-target effects in CRISPR/Cas9 gene editing. Front Bioeng Biotech 9(11):1143157. https://doi.org/10.3389/fbioe.2023.1143157

Haun W, Coffman A, Clasen BM et al (2014) Improved soybean oil quality by targeted mutagenesis of the fatty acid desaturase 2 gene family. Plant Biotechnol J 12(7):934–940. https://doi.org/10.1111/pbi.12201

Heidersbach AJ, Dorighi KM, Gomez JA et al (2023) A versatile, high-efficiency platform for CRISPR-based gene activation. Nat Commun 14(1):902. https://doi.org/10.1038/s41467-023-36452-w

Heslop-Harrison JS, Schwarzacher T (2007) Domestication, genomics and the future for banana. Ann Bot 100(5):1073–1084. https://doi.org/10.1093/aob/mcm191

Hillary VE, Ceasar SA (2024) CRISPR/Cas system-mediated base editing in crops: recent developments and future prospects. Plant Cell Rep 43(11):1–17. https://doi.org/10.1007/s00299-024-03346-0

Hu C, Liu F, Sheng O et al (2023) Efficient and transgene-free genome editing in banana using a REG-2 promoter-driven gene-deletion system. Mol Hortic 3(1):16. https://doi.org/10.1186/s43897-023-00065-0

Ishino Y, Shinagawa H, Makino K et al (1987) Nucleotide sequence of the iap gene, responsible for alkaline phosphatase isoenzyme conversion in Escherichia coli, and identification of the gene product. J Bacteriol 169(12):5429–5433. https://doi.org/10.1128/jb.169.12.5429-5433.1987

Jackson DA, Symons RH, Berg P (1972) Biochemical method for inserting new genetic information into DNA of Simian Virus 40: circular SV40 DNA molecules containing lambda phage genes and the galactose operon of *Escherichia coli*. Proc Natl Acad Sci USA 69(10):2904–2909. https://doi.org/10.1073/pnas.69.10.2904

Jansen R, Van Embden JDA, Gaastra W et al (2002) Identification of genes that are associated with DNA repeats in prokaryotes. Mol Microbiol 43(6):1565–1575. https://doi.org/10.1046/j.1365-2958.2002.02839.x

Jinek M, Chylinski K, Fonfara I et al (2012) A programmable dual-RNA-guided DNA endonuclease in adaptive bacterial immunity. Science 337(6096):816–821. https://doi.org/10.1126/science.1225829

Jung YJ, Nogoy FM, Lee SK et al (2018) Application of ZFN for site directed mutagenesis of rice SSIVagene. Biotechnol Bioprocess Eng 23(1):108–115. https://doi.org/10.1007/s12257-017-0420-9

Kaur N, Alok A, Shivani K et al (2018) CRISPR/Cas9-mediated efficient editing in phytoene desaturase (PDS) demonstrates precise manipulation in banana cv. Rasthali genome. Funct Integr Genomics 18(1):89–99. https://doi.org/10.1007/s10142-017-0577-5

Kaur N, Alok A, Kumar P et al (2020) CRISPR/Cas9 directed editing of lycopene epsilon-cyclase modulates metabolic flux for β-carotene biosynthesis in banana fruit. Metab Eng 59:76–86. https://doi.org/10.1016/j.ymben.2020.01.008

Kavuri NR, Ramasamy M, Qi Y et al (2022) Applications of CRISPR/Cas13-based RNA editing in plants. Cell 11(17):1–11. https://doi.org/10.3390/cells11172665

Kim YG, Cha J, Chandrasegaran S (1996) Hybrid restriction enzymes: zinc finger fusions to Fok I cleavage domain. Proc Natl Acad Sci USA 93(3):1156–1160. https://doi.org/10.1073/pnas.93.3.1156

Komor AC, Kim YB, Packer MS et al (2016) Programmable editing of a target base in genomic DNA without double-stranded DNA cleavage. Nature 533(7603):420–424. https://doi.org/10.1038/nature17946

Leh LS, Mazumdar P, Tan BC et al (2024) Establishment of a PEG-assisted protoplast transfection system in Musa acuminata cv. Berangan (AAA) using a CRISPR/Cas9 ribonucleoprotein complex. Biol Bull 50:S298–S309. https://doi.org/10.1134/S1062359023600010

Li T, Liu B, Spalding MH et al (2012) High-efficiency TALEN-based gene editing produces disease-resistant rice. Nat Biotechnol 30(5):390–392. https://doi.org/10.1038/nbt.2199

Li JF, Norville JE, Aach J et al (2013) Multiplex and homologous recombination–mediated genome editing in Arabidopsis and Nicotiana benthamiana using guide RNA and Cas9. Nat Biotechnol 31(8):688–691. https://doi.org/10.1038/nbt.2654

Li D, Kan O, Xiao X et al (2021) Nanoparticle delivery of CRISPR/Cas9 for genome editing. Front Genet 12:1–10. https://doi.org/10.3389/fgene.2021.673286

Li J, Chen L, Liang J et al (2022) Development of a highly efficient prime editor 2 system in plants. Genome Biol 23:161. https://doi.org/10.1186/s13059-022-02730-x

Li Z, Cheng X, Wang S et al (2024) Intrinsic targeting of host RNA by Cas13 constrains its utility. Nat Biomed Eng 8(2):177–192. https://doi.org/10.1038/s41551-023-01109-y

Liang Z, Zhang K, Chen K et al (2014) Targeted mutagenesis in Zea mays using TALENs and the CRISPR/Cas system. J Genet Genomics 41(2):63–68. https://doi.org/10.1016/j.jgg.2013.12.001

Lloyd A, Plaisier CL, Carroll D et al (2005) Targeted mutagenesis using zinc-finger nucleases in Arabidopsis. Proc Natl Acad Sci 102(6):2232–2237. https://doi.org/10.1073/pnas.0409339102

Madison BB, Patil D, Richter M et al (2022) Cas-CLOVER is a novel high-fidelity nuclease for safe and robust generation of TSCM-enriched allogeneic CAR-T cells. Mol Ther Nuc Acids 29(9):979–995. https://doi.org/10.1016/j.omtn.2022.06.003

Maeder ML, Thibodeau-Beganny S, Osiak A et al (2008) Rapid "Open-Source" engineering of customized Zinc-Finger Nucleases for highly efficient gene modification. Mol Cell 31(2): 294–301. https://doi.org/10.1016/j.molcel.2008.06.016

Mahfouz MM, Li L, Shamimuzzaman M et al (2011) De novo-engineered transcription activator-like effector (TALE) hybrid nuclease with novel DNA binding specificity creates double-strand breaks. Proc Nalt Acad Sci USA 108(6):2623–2628. https://doi.org/10.1073/pnas.1019533108

Mascarenhas MS, Nascimento FDS, Schittino LMP et al (2024) Construction and validation of CRISPR/Cas vectors for editing the PDS gene in Banana (*Musa* spp.). Curr Issues Mol Biol 46(12):14422–14437. https://doi.org/10.3390/cimb46120865

McCutcheon SR, Rohm D, Iglesias N et al (2024) Epigenome editing technologies for discovery and medicine. Nat Biotechnol 42:1199–1217. https://doi.org/10.1038/s41587-024-02320-1

Mengstie MA (2022) Viral vectors for the in vivo delivery of CRISPR components: advances and challenges. Front Bioeng Biotech 10:895713. https://doi.org/10.3389/fbioe.2022.895713

Miller JC, Tan S, Qiao G et al (2011) A TALE nuclease architecture for efficient genome editing. Nat Biotechnol 29(2):143–150. https://doi.org/10.1038/nbt.1755

Mishra R, Joshi RK, Zhao K (2020) Base editing in crops: current advances, limitations and future implications. Plant Biotechnol J 18:20–31. https://doi.org/10.1111/pbi.13225

Molla KA, Yang Y (2019) CRISPR/Cas-Mediated Base editing: technical considerations and practical applications. Trends Biotechnol 37(10):1121–1142. https://doi.org/10.1016/j.tibtech.2019.03.008

Molla KA, Karmakar S, Islam MT (2020) Wide horizons of CRISPR-Cas-derived technologies for basic biology, agriculture, and medicine. In: Islam MT, Bhowmik PK, Molla KA (eds) CRISPR-Cas methods. Springer Protocols Handbooks. Humana, New York. https://doi.org/10.1007/978-1-0716-0616-2_1

Muller HJ (1927) Artificial transmutation of the gene. Science 66(1699):84–87. https://doi.org/10.1038/285058b0

Naim F, Dugdale B, Kleidon J et al (2018) Gene editing the phytoene desaturase alleles of Cavendish banana using CRISPR/Cas9. Transgenic Res 27(5):451–460. https://doi.org/10.1007/s11248-018-0083-0

Nascimento FDS, Rocha ADJ, Soares JMDS et al (2023) Gene editing for plant resistance to abiotic factors: a systematic review. Planta 12(2):305. https://doi.org/10.3390/plants12020305

Nasrallah A, Sulpice E, Kobaisi F et al (2022) CRISPR-Cas9 technology for the creation of biological avatars capable of modeling and treating pathologies: from discovery to the latest improvements. Cell 11(22):3615. https://doi.org/10.3390/cells11223615

Nekrasov V, Staskawicz B, Weigel D et al (2013) Targeted mutagenesis in the model plant Nicotiana benthamiana using Cas9 RNA-guided endonuclease. Nat Biotechnol 31(8): 691–693. https://doi.org/10.1038/nbt.2655

Ntui VO, Tripathi JN, Tripathi L (2020) Robust CRISPR/Cas9 mediated genome editing tool for banana and plantain (Musa spp.). Cur Plant Biol 21(10):100128. https://doi.org/10.1016/j.cpb.2019.100128

Ntui VO, Tripathi JN, Shah T et al (2024) Targeted knockout of early nodulin-like 3 (MusaENODL3) gene in banana reveals its function in resistance to Xanthomonas wilt disease. Plant Biotechnol J 22(5):1101–1112. https://doi.org/10.1111/pbi.14248

Orel N, Kyryk A, Puchta H (2003) Different pathways of homologous recombination are used for the repair of double-strand breaks within tandemly arranged sequences in the plant genome. Plant J 35(5):604–612. https://doi.org/10.1046/j.1365-313X.2003.01832.x

Perez-Sanjose D, de la Fuente MA, Pérez JS et al (2022) Crispr/casrx proof-of-concept for rna degradation: a future tool against rna viruses? Pharmaceuticals 15(1):1–11. https://doi.org/10.3390/ph15010032

Pillay M, Tripathi L (2007) Banana. In: Fruits and nuts. Springer, Berlin/Heidelberg, pp 281–301

Porto EM, Komor AC, Slaymaker IM et al (2020) Base editing: advances and therapeutic opportunities. Nat Rev Drug Discov 19:839–859. https://doi.org/10.1038/s41573-020-0084-6

Qaim M (2010) Benefits of genetically modified crops for the poor: household income, nutrition, and health. Nat Biotechnol 27(5):552–557. https://doi.org/10.1016/j.nbt.2010.07.009

Qi Y, Gürel AF, Zhang Y (2020) CRISPR-Cas nucleases and base editors for plant genome editing. aBiotech 1(1):74–87. https://doi.org/10.1007/s42994-019-00010-0

Rees HA, Liu DR (2018) Base editing: precision chemistry on the genome and transcriptome of living cells. Nat Rev Genet 19(12):770–788. https://doi.org/10.1038/s41576-018-0059-1

Runo S, Karembu M, Nan'gayo F et al (2024) Africanizing genome editing for food sustainability. Glob Food Sec 42(6): 100785. https://doi.org/10.1016/j.gfs.2024.100785

Sapranauskas R, Gasiunas G, Fremaux C et al (2011) The Streptococcus thermophilus CRISPR/Cas system provides immunity in Escherichia coli. Nucl Acids Res 39(21):9275–9282. https://doi.org/10.1093/nar/gkr606

Schaart JG, van de Wiel CCM, Smulders MJM (2021) Genome editing of polyploid crops: prospects, achievements and bottlenecks. Transgenic Res 30(4):337–351. https://doi.org/10.1007/s11248-021-00251-0

Schmidt SM, Belisle M, Frommer WB (2020) The evolving landscape around genome editing in agriculture: many countries have exempted or moveto exempt forms of genome editing from GMO regulation of crop plants. EMBO Rep 21(6):e50680. https://doi.org/10.15252/embr.202050680

Shao X, Wu S, Dou T et al (2020) Using CRISPR/Cas9 genome editing system to create MaGA20ox2 gene-modified semi-dwarf banana. Plant Biotechnol J 18:17–19. https://doi.org/10.1111/pbi.13216

Shelake RM, Pramanik D, Kim JY (2019) Evolution of plant mutagenesis tools: a shifting paradigm from random to targeted genome editing. Plant Biotechnol Rep 13(5):423–445. https://doi.org/10.1007/s11816-019-00562-z

Shu QY, Forster BP, Nakagawa H (eds) (2012) Plant mutation breeding and biotechnology. CabinetMaker. https://doi.org/10.1079/9781780640853.0000

Shukla VK, Doyon Y, Miller JC et al (2009) Precise genome modification in the crop species Zea mays using zinc-finger nucleases. Nature 459(7245):437–441. https://doi.org/10.1038/nature07992

Simmonds NW, Shepherd K (1955) The taxonomy and origins of the cultivated bananas. Bot J Linn Soc 55:302–312. https://doi.org/10.1111/j.1095-8339.1955.tb00015.x

Songstad DD, Petolino JF, Voytas DF et al (2017) Genome editing of plants. Crit Rev Plant Sci 36(1):1–23. https://doi.org/10.1080/07352689.2017.1281663

Stadler LJ (1928) Genetic effects of X-rays in maize. Proc Natl Acad Sci 14(1):69–75. https://doi.org/10.1073/pnas.14.1.69

Tang X, Lowder LG, Zhang T et al (2017) A CRISPR-Cpf1 system for efficient genome editing and transcriptional repression in plants. Nat Plant 3(2):1–5. https://doi.org/10.1038/nplants.2017.18

Tang X, Liu G, Zhou J et al (2018) A large-scale whole-genome sequencing analysis reveals highly specific genome editing by both Cas9 and Cpf1 (Cas12a) nucleases in rice. Genome Biol 19(1):1–13. https://doi.org/10.1186/s13059-018-1458-5

Tingting L, Jinpeng Z, Xi Y et al (2023) Development and application of prime editing in plants. Rice Sci 30(6):509–522. https://doi.org/10.1016/j.rsci.2023.07.005

Tollenaar D (1938) Investigations on mutation in tobacco II. Some artificially induced chromosome mutants. Genetics 20:285–294

Tripathi JN, Oduor RO, Tripathi L et al (2015) A high-throughput regeneration and transformation platform for production of genetically modified Banana. Front. Plant Sci 6:1025. https://doi.org/10.3389/fpls.2015.01025

Tripathi JN, Ntui VO, Ron M et al (2019a) CRISPR/Cas9 editing of endogenous banana streak virus in the B genome of Musa spp. overcomes a major challenge in banana breeding. Commun Biol 2(1):46. https://doi.org/10.1038/s42003-019-0288-7

Tripathi L, Ntui VO, Tripathi JN. Application of CRISPR/Cas-based gene-editing for developing better banana. Front Bioeng Biotechnol. 2024 Aug 16;12:1395772. doi: 10.3389/fbioe.2024.1395772.

Tripathi L, Tripathi JN, Shah T et al (2019b) Molecular basis of disease resistance in banana progenitor *Musa balbisiana* against *Xanthomonas campestris* pv. *musacearum.* Sci rep 9(1): 7007. https://doi.org/10.1038/s41598-019-43421-1

Tripathi L, Ntui VO, Tripathi JN (2020) CRISPR/Cas9 based genome editing of banana for disease resistance. Curr Opin Plant Biol 56:118–126. https://doi.org/10.1016/j.pbi.2020.05.003

Tripathi JN, Ntui VO, Shah T et al (2021) CRISPR/Cas9-mediated editing of DMR6 orthologue in banana (Musa spp.) confers enhanced resistance to bacterial disease. Plant Biotechnol J 19(7): 1291–1293. https://doi.org/10.1111/pbi.13614

Tripathi L, Dhugga KS, Ntui VO et al (2022) Genome editing for sustainable agriculture in Africa. Front Genome Ed 4(5):1–20. https://doi.org/10.3389/fgeed.2022.876697

Tripathi L, Ntui VO, Tripathi JN et al (2023) A new and novel high-fidelity genome editing tool for banana using Cas-CLOVER. Plant Biotechnol J 21(9):1731–1733. https://doi.org/10.1111/pbi.14100

Tripathi JN, Ntui VO, Tripathi L (2024) Precision genetics tools for genetic improvement of banana. Plant Genome 27:e20416. https://doi.org/10.1002/tpg2.20416

Tröder SE, Zevnik B (2022) History of genome editing: from meganucleases to CRISPR. Lab An 56(1):60–68. https://doi.org/10.1177/0023677221994613

Urnov FD, Miller JC, Lee YL et al (2005) Highly efficient endogenous human gene correction using designed zinc-finger nucleases. Nature 435(7042):646–651. https://doi.org/10.1038/nature03556

Vu TV, Nguyen NT, Kim J et al (2024) Optimized dicot prime editing enables heritable desired edits in tomato and Arabidopsis. Nat Plant 10:1502–1513. https://doi.org/10.1038/s41477-024-01786-w

Wang JY, Doudna JA (2023) CRISPR technology: a decade of genome editing is only the beginning. Science 379(6629):eadd8643. https://doi.org/10.1126/science.add8643

Wang Z, Miao H, Liu J et al (2019) *Musa balbisiana* genome reveals subgenome evolution and functional divergence. Nat Plants 5:810–821. https://doi.org/10.1038/s41477-019-0452-6

Wang J, Gan S, Zheng Y et al (2022a) Banana somatic embryogenesis and biotechnological application. Trop Plant 1(1):1–13. https://doi.org/10.48130/TP-2022-0012

Wang Y, Zafar N, Ali Q et al (2022b) CRISPR/Cas genome editing technologies for plant improvement against biotic and abiotic stresses: advances, limitations, and future perspectives. Cell 11(23):3928. https://doi.org/10.3390/cells11233928

Wang Y, Zhang X, Wang T et al (2022c) The small, secreted protein FoSsp1 elicits plant defenses and negatively regulates pathogenesis in Fusarium oxysporum f. sp. cubense (Foc4). Front Plant Sci 13:873451. https://doi.org/10.3389/fpls.2022.873451

Weale A (2010) Ethical arguments relevant to the use of GM crops. Nat Biotechnot 27(5):582–587. https://doi.org/10.1016/j.nbt.2010.08.013

Wei T, Cheng Q, Min YL et al (2020) Systemic nanoparticle delivery of CRISPR-Cas9 ribonucleoproteins for effective tissue-specific genome editing. Nat Commun 11:3232. https://doi.org/10.1038/s41467-020-17029-3

Wendt T, Holm PB, Starker CG et al (2013) TAL effector nucleases induce mutations at a pre-selected location in the genome of primary barley transformants. Plant Mol Biol 83(3): 279–285. https://doi.org/10.1007/s11103-013-0078-4

Wright DA, Townsend JA, Winfrey RJ et al (2005) High-frequency homologous recombination in plants mediated by zinc-finger nucleases. Plant J 44(4):693–705. https://doi.org/10.1111/j.1365-313X.2005.02551.x

Wu S, Zhu H, Liu J et al (2020) Establishment of a PEG-mediated protoplast transformation system based on DNA and CRISPR/Cas9 ribonucleoprotein complexes for banana. BMC Plant Biol 20: 1–10. https://doi.org/10.1186/s12870-020-02609-8

Yang M, Djukanovic V, Stagg J et al (2009) Targeted mutagenesis in the progeny of maize transgenic plants. Plant Mol Biol 70(6):669–679. https://doi.org/10.1007/s11103-009-9499-5

Yang L, Yang B, Chen J (2019) Previews one prime for all editing. Cell 179(7):1448–1450. https://doi.org/10.1016/j.cell.2019.11.030

Young J, Zastrow-Hayes G, Deschamps S et al (2019) CRISPR-Cas9 editing in maize: systematic evaluation of off-target activity and its relevance in crop improvement. Sci Rep 9:6729. https://doi.org/10.1038/s41598-019-43141-6

Zaidi SSA, Mahas A, Vanderschuren H et al (2020) Engineering crops of the future: CRISPR approaches to develop climate-resilient and disease-resistant plants. Genome Biol 21(1):1–19. https://doi.org/10.1186/s13059-020-02204-y

Zetsche B, Gootenberg JS, Abudayyeh OO et al (2015) Cpf1 is a single RNA-guided endonuclease of a class 2 CRISPR-Cas system. Cell 163(3):759–771. https://doi.org/10.1016/j.cell.2015.09.038

Zhang Y, Shan Q, Wang Y et al (2013) Rapid and efficient gene modification in rice and brachypodium using TALENs. Mol Plant 6(4):1365–1368. https://doi.org/10.1093/mp/sss162

Zong Y, Liu Y, Xue C et al (2022) An engineered prime editor with enhanced editing efficiency in plants. Nat Biotechnol 40:1394–1402. https://doi.org/10.1038/s41587-022-01254-w

Breeding for Biotic and Abiotic Stress Tolerance: Engineering Resilient Crop Varieties, Enhancing Plant Defense Mechanisms

4

Ashish H. Bankar, Anu Naruka, Varun Shende, Walunjkar Babasaheb Changdeo, Anurag Malik, Himani Punia, and Natalia Repkina

Abstract

Agriculture crop yield and productivity have been affected by both human activity and the changing worldwide environment. Frequent ecological stressors affect plants, which depresses and limits the yield of agricultural crops. Advanced methods are desired to generate robust crop varieties that can assurance sustainable agricultural production and food security in bright of the rising complications obtainable by biotic and abiotic stressors. This chapter covered the processes underpinning plant defense against biotic and abiotic stressors. This chapter also emphasizes how vital it is to syndicate conventional understanding with state-of-the-art approaches in order to produce thorough breeding plants. By taking into interpretation the social, economic, and environmental sides of agriculture, these all-encompassing tactics make sure that the formation of crops that can withstand stress is consistent with the thoughts of sustainable growth. To

A. H. Bankar
Department of Agricultural Botany, SVKM's College of Agriculture, Tardi, Shirpur, Dhule, Maharashtra (Affiliated to MPKV, Rahuri), India

A. Naruka · V. Shende · W. B. Changdeo
Shri Vaishnav Institute of Agriculture, SVVV (Shri Vaishnav Vidyapeeth Vishwavidyalaya), Indore, India

A. Malik (✉)
Division of Research and Innovation, Uttaranchal University, Dehradun, Uttarakhand, India
e-mail: anuragmalikseed@hau.ac.in

H. Punia
Saraswati Group of Colleges, SGC Group, Gharuan, Mohali, Punjab, India
e-mail: puniahimani@hau.ac.in

N. Repkina
Institute of Biology of Karelian Research Centre, Russian Academy of Sciences, Petrozavodsk, Russian Federation Pushkinskaya

Y. Ahlawat et al. (eds.), *Next-Generation Strategies for Crop Improvement*,
https://doi.org/10.1007/978-981-95-0309-4_4

produce robust crop kinds that can embellishment in challenging environments, it is crucial to breed for tolerance to both biotic and abiotic stress. To generate resilient crop types that can flourish in cruel environments, breeding for biotic and abiotic stress tolerance is important.

Keywords

Biotic stress · Abiotic stress · Breeding strategies · Marker-assisted selection · Insect-pest resistance · CRISPR/Cas9 · Biotechnological interventions

4.1 Introduction

In order to preserve food safety, ecological constancy, and agricultural production, plants must be resistant to stress. Various biotic and abiotic stresses, including pathogens, pests, drought, salt, and severe temperatures, are experienced by plants (Umar et al. 2021). These stresses have a harmful effect on the manufacture, progress, and development of plants. These stressors have a crucial economic impact. Substantial yield losses can result from biotic stressors (Rahman et al. 2021). For example, the rice blast disease, which is carried on by the fungus *Magnaporthe oryzae*, has the potential to wipe out sufficient rice respectively year to feed more than 60 million individuals (Wilson and Talbot 2009). Abiotic stressors are also unsafe for the economy. Conferring to the National Drought Mitigation Center (2020), the United States experiences up to $6 billion in economic dropping due to drought alone, which affects around 55 million people yearly (Seleiman et al. 2021). These strains place a heavy monetary burden on farmers and the agricultural industry, which makes the creation of resilient crop types and applied management practices essential (Malhi et al. 2021).

Environment trouble and biodiversity damage can result from biotic and abiotic ecological stressors. When chemical pesticides are used extremely to counter biotic pressures, the atmosphere may become contaminated, non-target plants may be harmed, and resistant pest strains may emerge (Gill et al. 2014). Abiotic stressors can limit biodiversity and harm natural ecosystems by reducing arable land, deteriorating soil quality, and rushing desertification. The distribution and cruelty of both biotic and abiotic stress factors are being altered by rising temperatures and shifting precipitation patterns brought on by climate change, which is also rising these pressures (Nawaz et al. 2023). Improving plant tolerance to stress is vital for reducing the undesirable impacts of climate change, as these stressors are prophesied to become more recurrent and intense (Rivero et al. 2022).

Strengthening agricultural sustainability, output, and resilience in contradiction of the mounting complications posed by climate change and environmental degradation needs breeding plants for stress tolerance (Acosta-Motos et al. 2024). Crop varieties that are resistant to biotic and abiotic stressors have been established via the use of both conventional breeding methods and cutting-edge biotechnological techniques together with genetic engineering and marker-assisted selection (MAS)

(Parmar et al. 2017). Securing food source is largely reliant on these breeding activities, especially in zones that are vulnerable to harsh environmental conditions. The formation of crops resistant to stress has an important financial impact (Çakmakçı et al. 2023). In water-scarce areas, drought-tolerant maize varieties have verified yield profits of up to 25% above conventional varieties, offering farmers monetary relief and attractive food security (Cairns et al. 2012). In a similar vein, breeding for disease and insect resistance reduces environmental pollution and production costs by dipping the need for chemical pesticides. Breeding for stress resistance has profits for society (Swarnapriya and Swarnapriya 2021). By strengthening crop resilience, smallholder farmers in emerging countries who are often the most susceptible to environmental stresses can increase yields and earnings. This plays a part to poverty reduction and rural development. Resilient crops also pay to global food security by safeguarding a steady production in the face of shifting weather patterns and rising populations (Kapari et al. 2023).

4.2 Biotic Stress Resistance

In agriculture, plant resistance to biotic stress, with the aim of strengthening plant defenses in contradiction of a variety of live harm to crop yield and strength. A wide range of living things, including fungus, viruses, bacteria, nematodes, and insects, can induce biotic stress to plants (Fahad et al. 2017). Dropping crop production is the outcome of these biotic stress factors encouraging various illnesses, infections, and damage to crop plants. However, via research approaches, numerous ways have been developed to get the better of biotic stresses. By researching the genetic mechanism of the agents manufacturing these stressors, biotic challenges in plants can be alleviated (Rollins et al. 2013). Although plant-parasitic nematodes can take nourishment on any part of the plant and swallow the insides of plant cells, their primary reaction is on the root system and soil-borne illnesses. Many pathogens examples are mentioned in Table 4.1. They cause stunting and wilting, which are indications of inadequate nutrition. Although viruses seldom slay their hosts, they can cause deformities, stunting, and chlorosis in many parts of the plant, in addition to confined harm (Savchenko et al. 2014). When pests eat or place their eggs on plants, it troubles the plants. Insects that pierce plants by means of their stylets can transmit viruses to plants. Fungal parasites occur in two varieties: necrotrophs, which produce toxins that kill their host cells, and biotrophs, which do not damage the living cells of their host. When attached with bacteria, they can contaminate different sections of the plant and cause vascular wilts, leaf spots, and cankers, among other signs (Schlenker and Roberts 2009).

4.2.1 Mechanisms of Biotic Stress Resistance in Plants

A crowd of biotic stress conditions can affect plants. When stress is sensed by the plant, a series of molecular and cellular progressions is initiated (Lamers et al. 2020). This sets off a multifaceted web of molecular, physiological, and biochemical

Table 4.1 A comprehensive analysis of diseases and effects of biotic stress on diverse plant species

Types of biotic stress	Pathogens	Diseases triggered	Affected plants	References
Viruses	Barley yellow dwarf virus (BYDV)	Yellow dwarf disease	Barley, wheat, oats, rye	D'Arcy and Domier (2000)
	Tomato spotted wilt virus (TSWV)	Wilt, stunted growth, necrotic spots	Tomatoes, peppers, peanuts	Kikkert et al. (1999)
	Banana streak virus (BSV)	Streak disease	Bananas	PDF (n.d.)
	Lettuce mosaic virus (LMV)	Mosaic disease	Lettuce	German-Retana et al. (2008)
Bacteria	*Ralstonia solanacearum*	Bacterial wilt	Potatoes, tomatoes, eggplants, bananas	Hayward (1991)
	Agrobacterium tumefaciens	Crown gall disease	Grapes, stone fruits, roses	Escobar and Dandekar (2003)
	Xylella fastidiosa	Pierce's disease, leaf scorch	Grapes, olives, almond trees	Hopkins and Purcell (2007)
	Erwinia amylovora	Fire blight	Apples, pears, quince	Zhao et al. (2019)
Fungi	*Monilinia fructicola*	Brown rot	Stone fruits (peaches, plums, cherries)	Byrde and Willetts (1977)
	Venturia inaequalis	Apple scab	Apples, crabapples	Singh (2019)
	Gaeumannomyces graminis	Take-all disease	Wheat, barley, rye	Palma-Guerrero et al. (2021)
	Thielaviopsis basicola	Black root rot	Cotton, tobacco, legumes	Bucki et al. (2020)
Nematodes	*Pratylenchus penetrans*	Root lesion nematode infestation	Potatoes, carrots, strawberries	Wilson et al. (2021)
	Bursaphelenchus xylophilus	Pine wilt disease	Pine trees	Donald et al. (2003)
	Rotylenchulus reniformis	Reniform nematode infestation	Cotton, soybeans, pineapple	Palomares-Rius et al. (2021)
	Ditylenchus dipsaci	Stem and bulb nematode infestation	Onions, garlic, ornamental plants	French et al. (2017)

(continued)

Table 4.1 (continued)

Types of biotic stress	Pathogens	Diseases triggered	Affected plants	References
Insect pests	Vine weevil (*Otiorhynchus sulcatus*)	Root and foliage damage	Strawberries, ornamentals, vines	Fezza et al. (2022)
	Glassy-winged sharpshooter (*Homalodisca vitripennis*)	Pierce's disease vector	Grapes, citrus, almonds	Redak et al. (2004)
	Coffee borer beetle (*Hypothenemus hampei*)	Coffee berry borer	Coffee	Johnson et al. (2020)
	Mango mealybug (*Rastrococcus invadens*)	Fruit and leaf damage	Mango, citrus, guava	Williams and Denno (2014)
Weeds	*Striga* spp.	Parasitic plant, nutrient extraction	Maize, sorghum, millet	Parker (2012)
	Orobanche spp.	Parasitic plant, nutrient extraction	Sunflowers, tomatoes, tobacco, legumes	Nosratti et al. (2020)
	Parthenium hysterophorus	Competitive weed, allelopathic effects	Various crops, especially in disturbed soils	Adkins and Shabbir (2014)
	Cuscuta spp.	Parasitic plant, nutrient extraction	Alfalfa, clover, cranberries, ornamentals	Kim and Westwood (2015)

mechanisms mentioned in Fig. 4.1 (Saijo et al. 2020). In order to sense biotic invasion and avoid it from disturbing growth, production, and survival, plants have changed sophisticated sensing systems (Lamers et al. 2020). As a result, in order to defend themselves from a extensive range of pests and diseases, such as viruses, nematodes, bacteria, fungus, and herbivorous insects, plants have progressed an abundance of defense mechanisms (Saijo et al. 2020).

Appropriate to region off infections and insects, plants have a passive first line of defense that involves physical barriers like trichomes, wax, and cuticles. Furthermore, plants have the skill to produce chemical defenses in contradiction of diseases. Besides, plants use two layers of pathogen recognition to recruit defense against biotic invaders (Saijo et al. 2020). Pattern recognition receptors (PRRs), which identify pathogen-associated molecular patterns (PAMPs), are part of the initial level of pathogen identification. We state to this category of plant immunity as PAMP-triggered immunity (PTI). Plant resistance (R) proteins, which distinguish certain pathogen receptors (Avr proteins), are intricate in the second stage of pathogen identification (Malik et al. 2020). Maximum R genes encode proteins

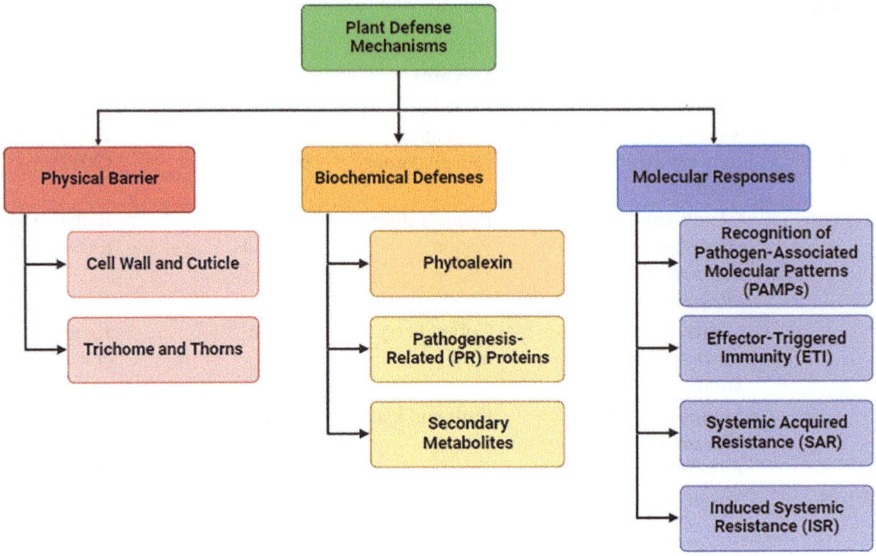

Fig. 4.1 The way plants overcome biotic stress

with a convinced domain as well as a conserved nucleotide-binding site (NBS). Leucine-rich repeat (LRR) is the second important domain (Santamaria et al. 2013). NB-LRR receptors recognize pathogen effectors either directly (physical association) or indirectly (association of an attachment protein) (Dodds and Rathjen 2010). Systemic acquired resistance (SAR) is the term used to define the superior level of defense that a plant's response to a hostile pathogen can occasionally elicit through the R gene. Whole-plant systemic resistance to an extensive range of diseases is produced via SAR. Through intraplane message, a resident encounter in SAR stimulates resistance to the extra plant organs (Fu and Dong 2013). Both kinds of plant immune responses often cause the similar response, although ETI is thought to be further severe in response to pathogen contagion (Tao et al. 2003). So, in order to respond to the detrimental effect on their survival, plants often treasure a balance between their response and biotic stress. A significant deal of information has been exposed about the molecular processes behind plant defensive responses (Wang et al. 2019). However, it is still uncertain how and why many signaling pathways come composed to produce biotic stress responses (Iqbal et al. 2021).

4.2.2 Breeding Strategies for Biotic Stress Resistance

Toward discourse the serious risks that illnesses and pests bring to global agriculture, breeding practices for biotic stress tolerance are vital. Pathogens may cause important agricultural losses, which can consequence in food shortages and unbalanced economies (Savary et al. 2019). In addition to existence highly expensive, traditional chemical controller methods like pesticides and herbicides also enough for health and environmental hazards. By 2050, it was prophesied that around 9 billion people on the planet, rising at a pace of 0.75% each year. By the end of 2050, an excess of 1 billion tons of grains must be manufactured in order to feed this increasing human population alone (Food and Agriculture Organization of the United Nations [Internet] n.d.). On the assumption that a practical and long-lasting response, breeding resistant crop varieties can support in reducing these challenges. This approach defends biodiversity and encourages environmental sustainability by dipping the necessity for chemical controls. It is extensively acknowledged that in order to attain these objectives, novel, combined dealings that decrease net time and cost per unit produced and speed up the breeding cycle must be working in tandem with traditional breeding programs (Collard and Mackill 2008; Platten et al. 2019).

4.2.2.1 Conventional Breeding Approaches

Emerging healthy crops and livestock that can bear pests, infections, and pathogens need the use of conventional breeding approaches for biotic stress tolerance. In order to select and spread individuals with chosen resistance qualities, these methods take benefit of the intrinsic genetic heterogeneity within species, assuring agricultural productivity and sustainability (Ahmar et al. 2020). The first step in the process is to identify and select parent plants that have a high level of acceptance to particular biotic stressors. Phenotypic monitoring and, genetic markers related with resistance qualities help as the basis for this variety. The next phase is to crossbreed in order to assort these resistance qualities from several parents. For plants, this means controlled cross-pollination (Stetter et al. 2016). The advantage of conventional plant breeding is in growing the pool of genetic resources accessible for improving crops by familiarizing required features. However, certain plants run the hazard of losing their genetic variety and becoming more vulnerable to environmental stressors (Basey et al. 2015). Hence, existing farming practices are inadequate to address the difficulties with global food security. It has been usually detecting that combining numerous phenotypic traits within a sole plant variation can positively boost harvest; nevertheless, new breeding actions will enable speedier manufacture of hereditarily altered crops at a lesser price (Ahmar et al. 2020).

4.2.2.2 Marker-Assisted Selection (MAS)

Modern age of molecular breeding initiates with the possible use of marker-assisted selection (MAS), which is defined as the alteration of genomic areas implicated in the desirable characteristic of interest using DNA markers (Gupta et al. 2010). Because the characteristic of interest is related with a molecular marker that enhancements the targeted trait's selection effectiveness, the MAS has an advantage over graphic phenotypic selection (Jiang and Jiang 2013). In disparity to conventional breeding practices, which are contingent on phenotypic selection and can be laborious and imprecise, genetic markers associated with certain genes of interest are used by MAS to identify plants that have those genes.

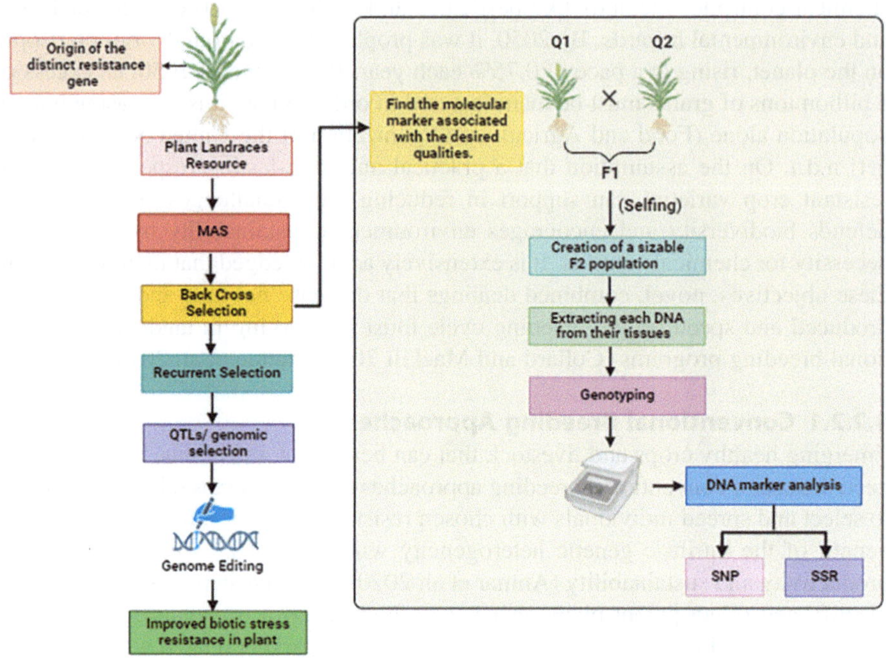

Fig. 4.2 Marker-assisted selection

This saves period, money, and effort by permitting breeders to select for characteristics at the seedling stage (Kumawat et al. 2020). Discovery molecular markers related to the desired trait, thorough the breeding population for these markers, and selecting individuals that possess the beneficial alleles are the main phases in the MAS procedure. These markers may be DNA sequences that are nearly associated with the genes leading to the chosen features, such as simple sequence repeats (SSRs) or single-nucleotide polymorphisms (SNPs) (Fig. 4.2) (Wani et al. 2022; Kamaluddin et al. 2022). The application of MAS enables the exact and effective selection of complex characteristics, including those that are interesting to evaluate phenotypically, such as drought tolerance and disease resistance (Collard and Mackill 2008).

4.2.2.3 Genetic Engineering and Transgenic Approaches

Powerful methods for producing biotic stress tolerance in plants are provided by genetic engineering and transgenic techniques. In contrast to conventional breeding, which be contingent on genetic variety that occurs naturally, these methods directly modify a plant's DNA to introduce novel features (Parmar et al. 2017). One process of genetic engineering is the straight insertion of particular genes into the genome of the envisioned plant part, which confers resistance to illnesses, pests, or infections (Health NRC (US) C on I and AUE of GEF on H 2004). Many methods, including clustered regularly interspaced short palindromic repeats (CRISPR)/Cas9 gene editing, gene gun technology, and *Agrobacterium*-mediated transformation, can be used to achieve this. Crops like *Btcorn* and *Bt cotton*, which communicate a gene from the *Bacillus thuringiensis* bacterium that produces an insecticidal protein and effectively guard against several insect pests, have been positively made by transgenic techniques (Abbas 2018).

Insect-Pest Resistance

Various crop plants currently are mainly weak in insect-pest resistance. Chemical control approaches are not globally sustainable and are established to be dangerous for clients. In the least, hereditary advancement that may lessen the expense of applying chemicals to regulate pests would be very beneficial from the cultivator's point of view (Parmar et al. 2017). The *Bt (Cry)* gene, which was gained from the soil bacterium *Bacillus thuringiensis*, has been shown to be successful in regulating a variety of *lepidopteran* insects in a range of crops. In 1987, the *Bt* gene was used to statement tomato insect resistance for the first time (Transgenic crops 2006) (Table 4.2).

Table 4.2 Transgenic Crops with Bt Genes for Insect Pest Resistance

Transgenic plant name	Gene name	Target plant	Insect pest controlled	References
Bt corn	cry1Ab	Maize	European corn borer (*Ostrinia nubilalis*)	James (2000)
Bt cotton	cry2Ab	Cotton	Bollworm (*Helicoverpa zea*)	Naranjo (2011)
Bt potato	cry3A	Potato	Colorado potato beetle (*Leptinotarsa decemlineata*)	Mi et al. (2015)
Bt soybean	cry1Ac	Soybean	Soybean looper (*Pseudoplusia includens*)	Sugimoto et al. (2012)
Bt eggplant	cry1Fa2	Eggplant	Fruit and shoot borer (*Leucinodes orbonalis*)	Hautea et al. (2016)
Bt rice	cry1Ab	Rice	Yellow stem borer (*Scirpophaga ceratellus*)	Ye et al. (2001)
Bt tomato	cry1Ac	Tomato	Tomato fruit borer (*Helicoverpa armigera*)	Bakhsh et al. (2018)
Bt brinjal	cry1Ac	Brinjal (eggplant)	Fruit and shoot borer (*Leucinodes orbonalis*)	The Development and Regulation of Bt Brinjal in India (2024)
Bt mustard	cry1Ab	Mustard	Diamondback moth (*Plutella xylostella*)	Sudha et al. (1999)
Bt alfalfa	cry1C	Alfalfa	Alfalfa weevil (*Hypera postica*)	Austin et al. (1995)

Disease Resistance and Virus Resistance

Decreasing tolerance or presenting resistance to various illnesses in plants is one of the main areas of genetic transformation. Plant genetic engineering for disease resistance has grown in popularity and value in terms of both cost and effectiveness (Buziashvili and Yemets 2023). Several genes, such as chitinase, glucanase, osmotin, and defensin, have been inserted into various horticultural crops worldwide to confer resistance against bacterial and fungal diseases. Using genes related to systemic acquired resistance (SAR) is one of the most significant genetic engineering methods for disease resistance. Long-lasting SAR is regularly linked to the accumulation of salicylic acid (SA) both locally and systemically (Parmar et al. 2017). It also induces the countenance of several genes, including those linked to pathogenesis (PR). The coat protein-mediated technique of virus engineering has been used in fruit plants to generate resistance in opposed a range of viral infections. Numerous tropical countries cultivate papaya, but the pathogen Papaya Ring Spot Virus (PRSV) is intimidating the plant and suggestively dipping efficiency. The virus's coat protein gene has been introduced into papaya by biotechnological means, consulting PRSV resistance. Hawaii, USA has been producing genetically modified papayas since 1998, and these plants have been verified to have strong resistance to PRSV (Azad et al. 2014) (Table 4.3).

4.2.2.4 CRISPR/Cas9 and Editing Genomes

The most exact method of editing a cell's genome in any plant is through the use of CRISPR (clustered regularly interspaced short palindromic repeats) and the protein that they are connected with (Cas-9) (Ibrahim et al. 2019). This method is applied in numerous fields. The two important parts of the CRISPR/Cas-9 system are director RNA (gRNA) and CRISPR-associated (Cas-9) proteins. There are three phases in the CRISPR/Cas-9 genome editing mechanism: recognition, cleavage, and repair. The genome-editing technique CRISPR/Cas-9 has many uses in biotechnology, agriculture, and medicine, among other fields (Mengstie and Wondimu 2021). Plant breeding has been severely transformed by CRISPR/Cas9 and genome editing methods, which deliver before unheard-of precision and competence in the growth of plant resistant to biotic stress. Instead of these technologies, it is feasible to surely adjust genes linked to disease and pest resistance at definite loci within the plant genome (Erdoğan et al. 2023). The ability of CRISPR/Cas9 to remove defenseless genes or recover resistance genes deprived of introducing foreign DNA is a big advantage, and this size can be a crucial factor for gaining regulatory clearance and public acceptability. For example, OsSWEET14 gene alterations have been created in rice using CRISPR/Cas9, providing resistance to *Xanthomonas oryzae pv. oryzae*-caused bacterial blight (Zeng et al. 2020) (Table 4.4).

Table 4.3 Advanced plant protection strategies and yield stability through resistance to specific viruses and diseases

Transgenic plant name	Gene name	Source organism	Target plant	Disease resistance	References
PRSV-resistant papaya	PRSV CP (coat protein)	Papaya ringspot virus (PRSV)	Papaya	Papaya ringspot virus (PRSV)	Azad et al. (2014)
Late blight-resistant potato	RB (Rpi-blb1)	*Solanum bulbocastanum*	Potato	Late blight (*Phytophthora infestans*)	Song et al. (2003)
Fungal-resistant wheat	Chitinase	*Trichoderma harzianum*	Wheat	Fusarium head blight (*Fusarium graminearum*)	Gil-Humanes et al. (2011)
Virus-resistant squash	ZYMV CP (coat protein)	Zucchini yellow mosaic virus (ZYMV)	Squash	Zucchini yellow mosaic virus (ZYMV)	Tricoli et al. (1995)
Bacterial blight-resistant rice	Xa21	*Oryza longistaminata*	Rice	Bacterial blight (*Xanthomonas oryzae pv. oryzae*)	Song et al. (1995)
Powdery mildew-resistant grapevine	PGIP (polygalacturonase-inhibiting protein)	*Phaseolus vulgaris*	Grapevine	Powdery mildew (*Erysiphe necator*)	Agüero et al. (2005)
Nematode-resistant tomato	Mi-1	*Solanum peruvianum*	Tomato	Root-knot nematode (*Meloidogyne* spp.)	Milligan et al. (1998)
Virus-resistant plum	PPV CP (coat protein)	Plum pox virus (PPV)	Plum	Plum pox virus (PPV)	Ilardi and Tavazza (2015)
PRSV-resistant papaya	PRSV CP (coat protein)	Papaya ringspot virus (PRSV)	Papaya	Papaya ringspot virus (PRSV)	Gonsalves (1998)
TMV-resistant tobacco	TMV CP (coat protein)	Tobacco mosaic virus (TMV)	Tobacco	Tobacco mosaic virus (TMV)	Beachy (1999)
PVY-resistant potato	PVY CP (coat protein)	Potato virus Y (PVY)	Potato	Potato virus Y (PVY)	Lawson et al. (1990)
CMV-resistant tomato	CMV CP (coat protein)	Cucumber mosaic virus (CMV)	Tomato	Cucumber mosaic virus (CMV)	Mochizuki and Ohki (2012)
TSWV-resistant peanut	N gene	Tomato spotted wilt virus (TSWV)	Peanut	Tomato spotted wilt virus (TSWV)	MacKenzie and Ellis (1992)
WSMV-resistant wheat	WSMV CP (coat protein)	Wheat streak mosaic virus (WSMV)	Wheat	Wheat streak mosaic virus (WSMV)	Borisjuk et al. (2019)

Table 4.4 Examples of Genetically Modified Crops with Target Genes, and Agronomic Benefits

Plant	Use	Gene target	Benefits	References
Tomato	Improved drought resistance	SlMAPK3	Improved yield and development under water-limited conditions	Huang et al. (2023a)
	Powdery mildew resistance	SlMlo1	Reduced need for fungicides	Jiang et al. (2016)
Rice	Increased disease resistance	OsSWEET14	Reduced crop losses due to bacterial blight	Yang et al. (2022)
	Improved blast resistance	OsERF922	Higher yield stability	Wang et al. (2016)
Wheat	Gluten reduction	TaGW2	Development of gluten-free wheat varieties	Zhou et al. (2023)
	Powdery mildew resistance	TaMLO	Reduced need for fungicides	Wang et al. (2014)
Potato	Enhanced resistance to late blight	StNRL1	Decreased dependency on chemical fungicides	Nourozi et al. (2023)
	Altered starch composition	StGBSS	Improved industrial processing qualities	Mali et al. (2023)
Soybean	Increased oil content	GmFAD2-1A	Improved nutritional value and economic value	Silva et al. (2022)
	Improved nitrogen fixation	GmPAP1	Enhanced growth and yield	Sun et al. (2024)
Maize	Increased yield and biomass	ZmNAC111	Higher productivity and efficiency in biofuel production	Gesteiro et al. (2023)
	Improved phosphorus utilization	ZmIPK	Enhanced growth in low-phosphorus soils	Sparvoli and Cominelli (2015)
Banana	Resistance to Panama disease	RGA2	Preservation of banana crops globally	Dale et al. (2017)
Cucumber	Parthenocarpy (seedless fruit production)	CpCAL112	Higher quality and consumer preference	Martínez et al. (2014)
Grapevine	Powdery mildew resistance	VvMLO7	Reduced need for fungicides and improved grape quality	Fedorina et al. (2022)
Barley	Enhanced nitrogen use efficiency	HvPSTOL1	Improved sustainability and reduced need for fertilizers	Papathanasiou et al. (2024)
	Powdery mildew resistance	HvMLO	Reduced need for fungicides	Acevedo-Garcia et al. (2017)

(continued)

Table 4.4 (continued)

Plant	Use	Gene target	Benefits	References
Apple	Fire blight resistance	MdDIPM1	Reduced losses from bacterial infection	Pompili et al. (2020)
Cassava	Enhanced starch quality	MeSBEII	Improved industrial and food processing qualities	Muiruri et al. (2023)
Cotton	Increased fiber quality	GhMYB25	Better textile quality and higher economic returns	Walford et al. (2011)

4.3 Abiotic Stress Resistance

Abiotic stress resistance describes a plant's strength to bear severe environmental factors such salt, drought, high temperatures, and heavy metal pollution. Worldwide food safety and agricultural production are importantly having an influence by these stressors. Improving crop resistance to abiotic stress is becoming a serious issue of research as the difficulties posed by climate alteration develop. One major environmental factor restrictive plant manufacture is water scarcity. Due to the seriousness of both stress severity and period, losses from agricultural yield triggered by drought are expected to exceed losses from all other causes (Gull et al. 2019). The quantity of rainfall and its spreading, evaporative requirements, and soil moisture-storing capability all affect how severe the drought is (Hazel 1995). Another abiotic stressor that depresses agricultural yield efficiency is cold stress, which alters crop quality and post-harvest life. It has been revealed that freezing remarkably inhibits the growth of many agricultural plant species (Jiang et al. 2011). Due to their sessile nature, plants have established special modifications to deal with temperature variations in their environment (Jones 2021). Frosty temperatures are faced by plants and these situations may be immensely detrimental to them as stressors. Plants that are visible to heat stress exhibit decreased manufacture, photosynthetic efficiency, and seed germination rate. Rising industrial development, a better dependence on chemical fertilizers, irrigation of sewage effluent, and extra factors have played a part in toxic metals to agricultural soils, which has an adverse impact on the soil-plant environment structure (Nievola et al. 2017). Eventually, these metals bioaccumulate and increasingly make their way into plants through the food chain, the air, and water (Rauwane and Ntushelo 2020) (Table 4.5).

4.3.1 Mechanisms of Abiotic Stress Resistance

Plants hold a various series of morphological, physiological, biochemical, and molecular mechanisms recognized as abiotic stress resistance. These mechanisms combine efforts to help plants endure and adjust to severe environmental conditions, counting but not limited to drought, salinity, high temperatures, and heavy metal

Table 4.5 Examples of abiotic stress—resistant crops with key adaptive traits

Type of abiotic stress	Examples of resistant plants	Key characteristics	References
Drought resistance	Sorghum	Deep root systems, condensed leaf area, osmotic adjustment	Blum (2011)
	Pearl millet	Efficient water use, early maturation	Srivastava et al. (2022)
	Cowpea	Stomatal closure, production of osmo-protectants	Hall et al. (2003)
Salinity resistance	Barley	Ion compartmentalization, salt exclusion	Munns and Tester (2008)
	Sugar beet	Production of compatible solutes, efficient Na + transport	Hasegawa et al. (2000)
	Tomato	Salt-tolerant rootstocks, antioxidant production	Bonarota et al. (2022)
Heat resistance	Wheat	Heat shock proteins, efficient photosynthesis	Wahid et al. (2007)
	Rice	Altered membrane fluidity, antioxidant enzymes	Zhao et al. (2016)
	Maize	Variations in leaf location, heat-avoidance mechanisms	Cairns et al. (2012)
Cold resistance	Winter rye	Gathering of cryoprotectants, membrane stabilization	Rihan et al. (2017)
	Canola	Antifreeze proteins, cold acclimation	Thakur et al. (2010)
	Potato	Altered lipid composition, enhanced antioxidant activity	Nasir and Toth (2022)
Heavy metal resistance	*Brassica juncea* (Indian mustard)	Hyperaccumulation, chelation of heavy metals	Rascio and Navari-Izzo (2011)
	Sunflower	Phytoremediation capabilities, metal-binding proteins	Pilon-Smits (2005)
	Poplar	Improved detoxification pathways, metal sequestration	Tuskan et al. (2006)
Flooding resistance	Rice	Formation of aerenchyma, anaerobic respiration	Xu et al. (2006)
	Mangroves	Pneumatophores (breathing roots), salt excretion	Parida and Jha (2010)
	Sago palm	Tolerance to waterlogged conditions, oxidative stress management	Ehara et al. (2018)

pollution. In morphology, plants frequently grow bigger, deeper root systems to advance water intake during droughts. They can also produce aerenchyma tissues in their roots to assist in conveyance of oxygen in soggy soils (Blum 2011). Depletions in the zone of the leaf nearby for transpiration, such as enlarged leaf width and reduced leaf range, assist in limiting water damage (Srivastava et al. 2022). One of the core physiological procedures is the regulator of stomatal aperture. The minute

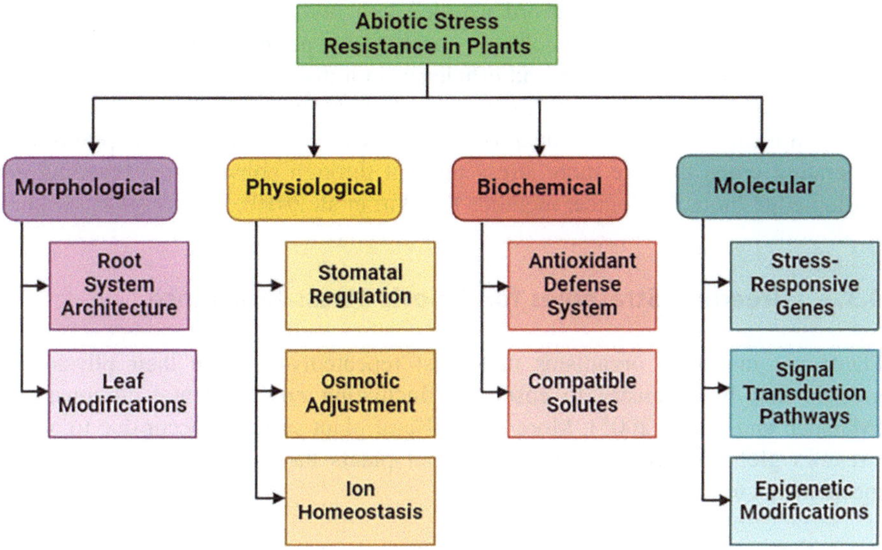

Fig. 4.3 Mechanisms of abiotic stress resistance in plants

holes named stomata on the external of leaves regulate both water loss and gas exchange. Plants often shield their stomata to diminish transpiration while below drought stress, but doing so also decreases CO_2 uptake, which has an influence on photosynthesis (Hall et al. 2003). Osmotic adjustment, in which plants stock osmo-protectants such proline, glycine betaine, and carbohydrates, is alternative vital physiological response (Fig. 4.3).

Plants make use of their antioxidant protection mechanisms biochemically to reduce the destructive effects of reactive oxygen species (ROS) generated under stressful circumstances. Superoxide dismutase (SOD), catalase (CAT), and peroxidases (POD) are significant antioxidant enzymes that hunt reactive oxygen species (ROS) and protection cellular components from oxidative damage (Wahid et al. 2007). Furthermore, under stressful situations, the manufacture of appropriate solutes counting glycine betaine, proline, and trehalose benefits stabilizes proteins and cellular membranes, maintaining their structure and function (Bonarota et al. 2022).

Plants practice a multiple of molecular methods to regulate a gene expression and activate stress-responsive pathways. Important transcription factors, like heat shock factors (HSF) and dehydration-responsive element-binding proteins (DREB), are upregulated in retaliation to stress, which initiate the transcription of genes related with stress tolerance (Huang et al. 2023b). Signal transduction paths that include molecules like calcium ions (Ca^{2+}), abscisic acid (ABA), and nitric oxide (NO) are crucial for recognizing stress signals and transmitting them to the applicable cellular machinery, leading to adaptive responses (Cairns et al. 2012). Lastly, epigenetic modifications, such as alterations in DNA methylation, histone modifications, and the action of small RNAs, afford a quick and reversible mechanism for stress tolerance. All collected, these cohesive mechanisms permit plants to develop more resilient and assurance growth and efficiency in a diversity of strict environmental circumstances. Investigate in this range is still enduring and is serving to elucidate the complications of these adaptive responses, which will be supportive in generating crops that are more stress tolerant through conventional breeding and cutting-edge biotechnological methods (Thakur et al. 2010).

4.3.2 Breeding Strategies for Abiotic Stress Resistance

Plants are immobile organisms that must repeatedly regulate their physiology, growth, and biochemistry to ensemble fluctuating environmental circumstances (Mahajan and Tuteja 2005). More than 50% of plant sufferers occur due to abiotic stressors global (Alcázar et al. 2006), and plants have a diversity of defensive mechanisms to endure these burdens.

4.3.2.1 Conventional Breeding Techniques

These approaches select and crossbreed plants that have necessary traits like heat, salinity, or drought tolerance. By gathering beneficial alleles through hybridization and recurrent selection, breeders can recover crop diversities' stress resistance over generations (Blum 2011). Although this approach is productive, it takes an extended period to find the best execution genotypes.

4.3.2.2 MAS

Genetic markers linked to stress tolerance characteristics, genomics-assisted breeding and molecular breeding have altered conventional breeding. By permitting the assortment of plants bearing certain genetic markers connected to necessary qualities deprived of coming for the whole phenotypic expression, MAS accelerates the breeding process (Collard and Mackill 2008). By choosing plants with better stress tolerance using whole-genome estimate models, genomic selection (GS) advances breeding precision and productivity (Ahmar et al. 2020).

4.3.2.3 Biotechnological Interventions

Alteration genes connected to stress tolerance and biotechnological interventions use genome editing and genetic engineering techniques. Through genetic engineering, genes from other plant that offer resistance to abiotic stressors can be comprised. For occurrence, frequent crops have importantly amplified their resistance to salt and drought when genes encoding DREB, transcription factors were introduced (Agarwal et al. 2006). Accurate genome editing is made conceivable by CRISPR-Cas9 technology, which permits targeted modifications such as the elimination of stress-susceptible genes or the supplementation of beneficial alleles to raise stress resilience (Zhang et al. 2018a).

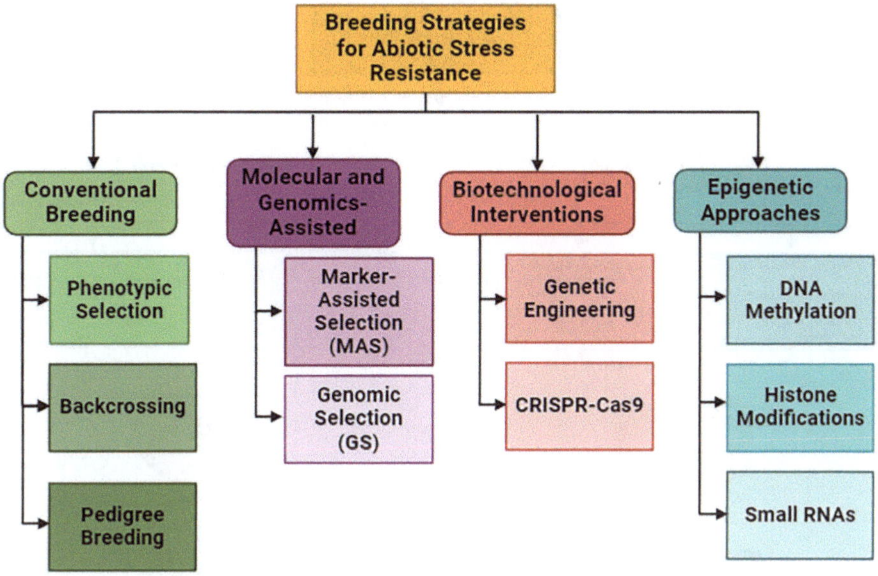

Fig. 4.4 The different breeding strategies for abiotic stress resistance

4.3.2.4 Epigenetic Approaches

The importance of epigenetic approaches is on inherited alterations to gene expression that do not comprise variations to the DNA sequence. Through procedures like DNA methylation, histone modification, and the action of short RNAs, these changes which can be carried on by environmental factors might recover stress tolerance. Epigenetic alterations can enhance genetic and molecular breeding efforts by contribution a rapid and rescindable way for plant to respond to stress (Zhang et al. 2018b) (Fig. 4.4).

4.4 Integrating Biotic and Abiotic Stress Resistance

4.4.1 Synergistic Effects of Combined Stress Resistance

Most of the plants are subjected to several stressors at once, such as salinity and pathogen invasions or drought and insect invasions. If the plant's reactions are not appropriately combined, these immediate burdens may degenerate the injury. However, plants have established multifaceted defense systems to grip these mixed loads, leading to synergistic effects in which a plant's reaction to one stress can strengthen its resistance to another (Atkinson and Urwin 2012). Studies have verified that plants have established multifaceted systems to syndicate responses to these coexisting stresses, ensuing in synergistic effects where resistance to one stress can be

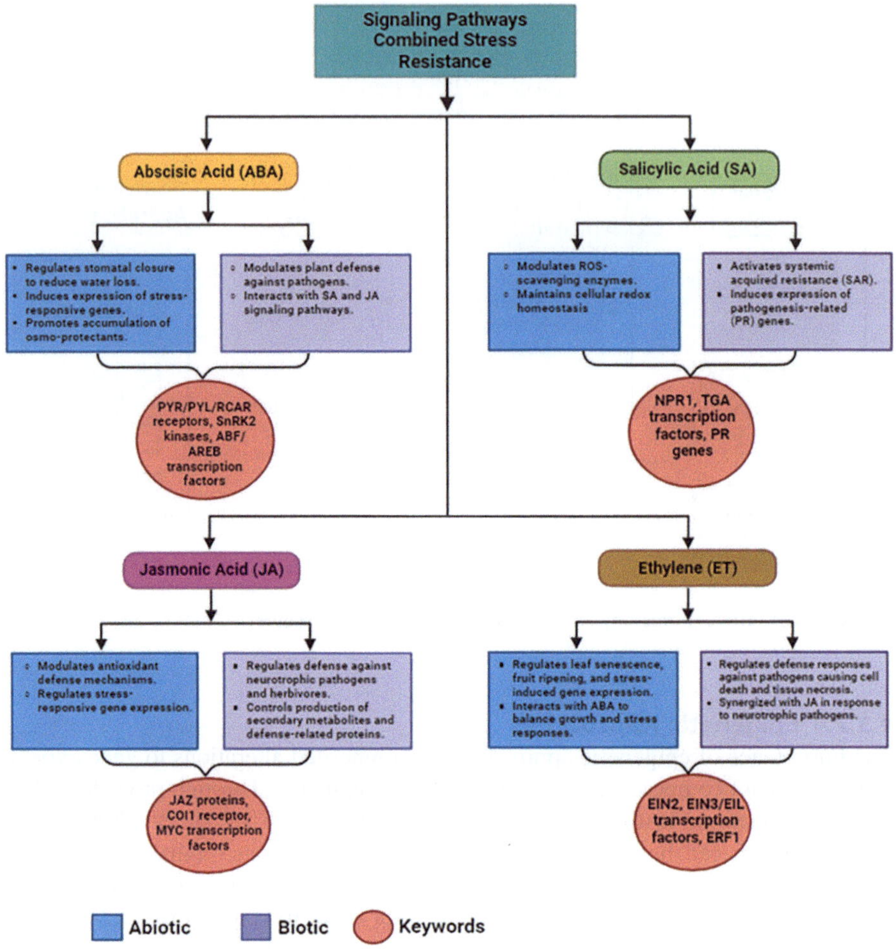

Fig. 4.5 Biotic and abiotic stress signaling pathways+

improved by resistance to another. A major element of this synergy is the shared signaling pathways that arbitrate responses to abiotic and biotic stresses; stress hormones like ethylene (ET), salicylic acid (SA), ABA, and jasmonic acid (JA) are significant players in these pathways illustrated in Fig. 4.5 (Fujita et al. 2006).

Agriculture will be greatly impacted by understanding and using the synergistic benefits of mutual biotic and abiotic stress resistance. In the face of climate alteration and other environmental problems, consistent harvests may be guaranteed by generating crop types that are resistant to various pressures. Using significant regulatory genes and common signaling pathways, integrated breeding techniques provide a viable option to raise agricultural productivity and resilience.

4.4.2 Integrated Pest and Stress Management Strategies (IPSM)

Integrate pest control and abiotic stress management techniques to provide inclusive resolutions that recover crop production and resilience. Sustainable agriculture needs this integrated strategy, particularly in light of increasing pest pressure and climate alteration. In order to handle biotic and abiotic pressures parallel, IPSM techniques make use of resistant crop types, cultural practices, biological control agents, and minimum chemical inputs.

4.4.2.1 Resistance Varieties

In order to sustain agricultural productivity and food security in the face of biotic (drought, salt, and temperature extremes) and abiotic (diseases and pests) encounters, resistant crop types are essential. To generate plants that can resist a variety of hindrances, these cultivars combine traditional breeding methods, MAS, and cutting-edge genetic engineering. Enhancing crop resilience has been made possible by the application of traditional breeding techniques including hybridization and phenotypic selection. For example, via selective breeding, drought-resistant maize cultivars have been generated, improving yield stability in dry locations (Maazou et al. 2016).

Drought-tolerant maize cultivars were created by the Water Efficient Maize for Africa (WEMA) initiative by merging genetic engineering and traditional breeding. In Africa during droughts, these cultivars have significantly enlarged yields (Sheoran 2022).

Traditional Indian rice cultivars have been shown to convey the Sub1 gene, which imparts submersion resistance. High-yielding rice cultivars were bred with the gene to produce harvests resistant to extended floods, a typical occurrence in several regions of Asia (Xu et al. 2006).

In order to treat vitamin A insufficiency, this genetically modified rice type is supplemented with beta-carotene, or pro-vitamin A. It is advantageous for crop resilience and nutrition since it also possesses features for stress tolerance and insect resistance (Paine et al. 2005).

4.4.2.2 Cultural Practices

Agronomic techniques including conservation digging, crop rotation, and intercropping support control pests and advance soil health, which raises plant resistance to abiotic pressures. By upsetting the life cycles of pests, these approaches lessen their numbers and improve soil moisture retention and nutrient availability, which helps plants adapt to nutritional stress and drought (Sharma et al. 2021).

4.4.2.3 Biological Control

A crucial element of IPSM is the use of microbiological agents, parasitoids, and natural predators to manage pest populations. In addition to decreasing insect populations, biological control agents cause plants to develop systemic resistance, which improves the plants' tolerance to abiotic stressors. As to Harman et al. (2004), *Trichoderma* spp. are helpful fungi that inhibit soil-borne diseases and enhance plant development under stressful environments.

4.4.2.4 Minimal Chemical Inputs

In order to lessen the influence on the environment and sluggish the appearance of pest resistance, IPSM prioritizes the use of biopesticides and the targeted application of synthetic pesticides when chemical interventions are compulsory. In order to accomplish sustainable pest and stress management, integrated techniques ensure that chemical treatments are applied as a last option and in concert with other management strategies (Peshin 2014).

4.5 Challenges and Future Prospects

4.5.1 Limitations of Current Breeding Techniques

Certain breeding and hybridization are two examples of traditional breeding approaches that have made a substantial contribution to agricultural output. Unfortunately, these techniques take a long time to produce desirable features, sometimes taking many generations. Moreover, they can accidentally introduce unwanted features and are constrained by the genetic variety found within a species. Furthermore, several significant characteristics, including yield and drought resistance, are complexly polygenic, which presents further difficulties for conventional breeding techniques.

4.5.2 Emerging Technologies and Innovations

Plant breeding is being transformed by recent advances in genetics and biotechnology. With the use of methods like CRISPR-Cas9, one may precisely edit the genome and familiarize or modify particular features with beforehand unheard-of precision. Furthermore, breeding operations are becoming more efficient because to genomic selection, which makes use of genetic information to forecast the performance of breeding prospects. Additional developments include synthetic biology, which has the ability to create totally new plant kinds with specific features, and high-throughput phenotyping, which uses cutting-edge imaging and sensor technology to quickly analyze plant properties.

4.5.3 Regulatory and Policy Aspects

Worldwide varying governmental and regulatory outlines impact the adoption of novel breeding methods. Genome-edited crops are subject to strict regulatory oversight in many areas, much like genetically modified organisms (GMOs). The rate of innovation and breeders' access to new technology may be impacted by these rules. Harmonized and scientifically grounded regulatory strategies are required to support the ethical application of modern breeding technologies while maintaining public safety and confidence.

4.5.4 Future Paths for Research

In order to get above the existing constraints, future plant breeding research is probably going to concentrate on merging numerous modern technologies. For example, combining gene editing with genomic selection might hasten the creation of crops that are more resilient to climate change.

4.6 Conclusion

It is crucial to breed for biotic and abiotic stress tolerance in order to protect global food production, improve plant defense systems, and maintain agricultural sustainability. Researchers may accurately target and tweak certain genes to generate crops that are more resistant to diseases, pests, drought, salt, and other environmental challenges by combining genome editing with modern breeding techniques like CRISPR/Cas9. This leads to more environmentally friendly agriculture practices by increasing crop resilience and yields while lowering the demand for chemical inputs. To fully use the promise of new technology and get over the constraints of existing breeding techniques, more research and development in this field is necessary. The capability of the globe to meet the growing food demands of an expanding population and adapt to the challenges posed by climate change will be greatly impacted by the adoption of these state-of-the-art breeding procedures. By encouraging collaboration between researchers, decision-makers, and farmers, we can ensure the development of resilient crop varieties that sustain and enhance agricultural productivity worldwide.

Author Contributions Conceptualization, methodology, investigation, data curation, writing, original draft preparation, A.H.B., A.N., V.S., W.B.C., and A.M.; study design, supervision, validation, H.P, N.R and A.M; literature search, data collection, writing review and editing, H.P., A.M., A.H.B. All authors have read and agreed to the published version of the manuscript.

Conflicts of Interest The authors declare no conflict of interest.

Funding This research received no external funding.

References

Abbas MST (2018) Genetically engineered (Modified) crops (bacillus thuringiensis crops) and the world controversy on their safety. Egypt J Biol Pest Control [Internet] 28(1):1–12. Available from: https://ejbpc.springeropen.com/articles/10.1186/s41938-018-0051-2 [cited 2024 Jul 27]

Acevedo-Garcia J, Spencer D, Thieron H, Reinstädler A, Hammond-Kosack K, Phillips AL et al (2017) mlo-based powdery mildew resistance in hexaploid bread wheat generated by a non-transgenic TILLING approach. Plant Biotechnol J [Internet] 15(3):367. Available from: /pmc/articles/PMC5316926/ [cited 2024 Jul 30]

Acosta-Motos JR, Franco-Navarro JD, Gómez-Bellot MJ, Álvarez S (2024) Editorial: crop resistance mechanisms to alleviate climate change-related stress. Front Plant Sci [Internet] 15: 1368573. Available from: http://www.fao.org/3/a-i6030e.pdf [cited 2024 Jul 25]

Adkins S, Shabbir A (2014) Biology, ecology and management of the invasive parthenium weed (Parthenium hysterophorus L.). Pest Manag Sci [Internet] 70(7):1023–1029. Available from: https://pubmed.ncbi.nlm.nih.gov/24430973/ [cited 2024 Jul 29]

Agarwal PK, Agarwal P, Reddy MK, Sopory SK (2006) Role of DREB transcription factors in abiotic and biotic stress tolerance in plants. Plant Cell Rep [Internet] 25(12):1263–1274. Available from: https://pubmed.ncbi.nlm.nih.gov/16858552/ [cited 2024 Jul 31]

Agüero CB, Uratsu SL, Greve C, Powell ALT, Labavitch JM, Meredith CP et al (2005) Evaluation of tolerance to Pierce's disease and Botrytis in transgenic plants of Vitis vinifera L. expressing the pear PGIP gene. Mol Plant Pathol [Internet] 6(1):43–51. Available from: https://pubmed.ncbi.nlm.nih.gov/20565637/ [cited 2024 Jul 29]

Ahmar S, Gill RA, Jung KH, Faheem A, Qasim MU, Mubeen M et al (2020) Conventional and molecular techniques from simple breeding to speed breeding in crop plants: recent advances and future outlook. Int J Mol Sci [Internet] 21(7):2590. Available from: /pmc/articles/PMC7177917/ [cited 2024 Jul 27]

Alcázar R, Marco F, Cuevas JC, Patron M, Ferrando A, Carrasco P et al (2006) Involvement of polyamines in plant response to abiotic stress. Biotechnol Lett [Internet] 28(23):1867–1876. Available from: https://pubmed.ncbi.nlm.nih.gov/17028780/ [cited 2024 Jul 31]

Atkinson NJ, Urwin PE (2012) The interaction of plant biotic and abiotic stresses: from genes to the field. J Exp Bot [Internet] 63(10):3523–3543. Available from: https://doi.org/10.1093/jxb/ers100 [cited 2024 Jul 31]

Austin S, Bingham ET, Mathews DE, Shahan MN, Will J, Burgess RR (1995) Production and field performance of transgenic alfalfa (Medicago sativa L.) expressing alpha-amylase and manganese-dependent lignin peroxidase. Euphytica Int J Plant Breed [Internet] 85(1): 381–393. Available from: https://www.researchgate.net/publication/226858148_Production_and_field_performance_of_transgenic_alfalfa_Medicago_sativa_L_expressing_alpha-amylase_and_manganese-dependent_lignin_peroxidase_Euphytica [cited 2024 Jul 29]

Azad MAK, Amin L, Sidik NM (2014) Gene technology for Papaya Ringspot Virus disease management. Sci World J [Internet] 2014. Available from: /pmc/articles/PMC3976845/ [cited 2024 Jul 29]

Bakhsh A, Dınç T, Hussain T, Demırel U, Aasim M, Çalişkan ME (2018) Development of transgenic tobacco lines with pyramided insect resistant genes. Turkish J Biol [Internet] 42(2):174. Available from: /pmc/articles/PMC6353294/ [cited 2024 Jul 29]

Basey AC, Fant JB, Kramer AT (2015) Producing native plant materials for restoration: 10 rules to collect and maintain genetic diversity. Nativ Plants J 16(1):37–53

Beachy RN (1999) Coat-protein-mediated resistance to tobacco mosaic virus: discovery mechanisms and exploitation. Philos Trans R Soc Lond B Biol Sci [Internet] 354(1383): 659–664. Available from: https://pubmed.ncbi.nlm.nih.gov/10212946/ [cited 2024 Jul 29]

Blum A (2011) Plant breeding for water-limited environments. Springer, New York, pp 1–255

Bonarota MS, Kosma DK, Barrios-Masias FH (2022) Salt tolerance mechanisms in the Lycopersicon clade and their trade-offs. AoB Plants [Internet] 14(1) Available from: /pmc/articles/PMC8782609/ [cited 2024 Jul 31]

Borisjuk N, Kishchenko O, Eliby S, Schramm C, Anderson P, Jatayev S et al (2019) Genetic modification for wheat improvement: from transgenesis to genome editing. Biomed Res Int [Internet] 2019. Available from: /pmc/articles/PMC6431451/ [cited 2024 Jul 29]

Bucki P, Qing X, Castillo P, Gamliel A, Dobrinin S, Alon T et al (2020) The genus pratylenchus (Nematoda: Pratylenchidae) in Israel: from taxonomy to control practices. Plants [Internet] 9(11):1475. Available from: https://www.mdpi.com/2223-7747/9/11/1475/htm [cited 2024 Jul 29]

Buziashvili A, Yemets A (2023) Lactoferrin and its role in biotechnological strategies for plant defense against pathogens. Transgenic Res [Internet] 32(1–2):1. Available from: /pmc/articles/PMC9761627/ [cited 2024 Jul 29]

Byrde RJW, Willetts HJ (1977) The brown rot fungi of fruit: their biology and control [Internet]. Pergamon Press, 171 p. Available from: http://www.sciencedirect.com:5070/book/9780080197401/the-brown-rot-fungi-of-fruit [cited 2024 Jul 29]

Cairns JE, Sonder K, Zaidi PH, Verhulst N, Mahuku G, Babu R et al (2012) Maize production in a changing climate: impacts, adaptation, and mitigation strategies. Adv Agron 114:1–58

Çakmakçı R, Salık MA, Çakmakçı S (2023) Assessment and principles of environmentally sustainable food and agriculture systems. Agric [Internet] 13(5):1073. Available from: https://www.mdpi.com/2077-0472/13/5/1073/htm [cited 2024 Jul 25]

Collard BCY, Mackill DJ (2008) Marker-assisted selection: an approach for precision plant breeding in the twenty-first century. Philos Trans R Soc Lond B Biol Sci [Internet] 363(1491):557–572. Available from: https://pubmed.ncbi.nlm.nih.gov/17715053/ [cited 2024 Jul 26]

D'Arcy CJ, Domier LL (2000) Barley yellow dwarf. Plant Heal Instr [Internet]. Available from: https://www.apsnet.org/edcenter/disandpath/viral/pdlessons/Pages/BarleyYelDwarf.aspx [cited 2024 Jul 29]

Dale J, James A, Paul JY, Khanna H, Smith M, Peraza-Echeverria S et al (2017) Transgenic Cavendish bananas with resistance to Fusarium wilt tropical race 4. Nat Commun [Internet] 8(1):1496. Available from: /pmc/articles/PMC5684404/ [cited 2024 Jul 30]

Dodds PN, Rathjen JP (2010) Plant immunity: towards an integrated view of plant–pathogen interactions. Nat Rev Genet [Internet] 11(8):539–548. Available from: https://www.nature.com/articles/nrg2812 [cited 2024 Jul 26]

Donald PA, Stamps WT, Linit MJ (2003) Pine wilt. Plant Heal Instr [Internet]. Available from: https://www.apsnet.org/edcenter/disandpath/nematode/pdlessons/Pages/PineWilt.aspx [cited 2024 Jul 29]

Ehara H, Toyoda Y, Johnson DV (2018) Sago palm: multiple contributions to food security and sustainable livelihoods. Sago Palm Mult Contrib Food Secur Sustain Livelihood 15:1–330

Erdoğan İ, Cevher-Keskin B, Bilir Ö, Hong Y, Tör M (2023) Recent developments in CRISPR/Cas9 genome-editing technology related to plant disease resistance and abiotic stress tolerance. Biology (Basel) [Internet] 12(7):1037. Available from: /pmc/articles/PMC10376527/ [cited 2024 Jul 30]

Escobar MA, Dandekar AM (2003) Agrobacterium tumefaciens as an agent of disease. Trends Plant Sci 8(8):380–386

Fahad S, Bajwa AA, Nazir U, Anjum SA, Farooq A, Zohaib A et al (2017) Crop production under drought and heat stress: plant responses and management options. Front Plant Sci [Internet] 8: 265598. Available from: www.frontiersin.org [cited 2024 Jul 25]

Fedorina J, Tikhonova N, Ukhatova Y, Ivanov R, Khlestkina E (2022) Grapevine gene systems for resistance to gray mold botrytis cinerea and powdery Mildew Erysiphe necator. Agron [Internet] 12(2):499. Available from: https://www.mdpi.com/2073-4395/12/2/499/htm [cited 2024 Jul 30]

Fezza E, Roberts JM, Bruce TJA, Walsh LE, Gaffney MT, Pope TW (2022) Optimising vine weevil, Otiorhynchus sulcatus F. (Coleoptera: Curculionidae), monitoring tool design. Insects [Internet] 13(1):80. Available from: https://www.mdpi.com/2075-4450/13/1/80/htm [cited 2024 Jul 29]

(n.d.) World agriculture: towards 2030/2050 – the 2012 revision. Global perspectives studies. Food and Agriculture Organization of the United Nations [Internet]. Available from: https://www.fao.org/global-perspectives-studies/resources/detail/en/c/411108/ [cited 2024 Jul 26]

French JM, Beacham J, Garcia A, Goldberg NP, Thomas SH, Hanson SF (2017) First report of stem and bulb nematode ditylenchus dipsaci on Garlic in New Mexico. Plant Heal Prog [Internet] 18(2):91–92. Available from: https://apsjournals.apsnet.org/doi/10.1094/PHP-12-16-0069-BR [cited 2024 Jul 29]

Fu ZQ, Dong X (2013) Systemic acquired resistance: turning local infection into global defense. Annu Rev Plant Biol [Internet] 64:839–863. Available from: https://pubmed.ncbi.nlm.nih.gov/23373699/ [cited 2024 Jul 26]

Fujita M, Fujita Y, Noutoshi Y, Takahashi F, Narusaka Y, Yamaguchi-Shinozaki K et al (2006) Crosstalk between abiotic and biotic stress responses: a current view from the points of convergence in the stress signaling networks. Curr Opin Plant Biol 9(4):436–442

German-Retana S, Walter J, Le Gall O (2008) Lettuce mosaic virus: from pathogen diversity to host interactors. Mol Plant Pathol [Internet] 9(2):127–136. Available from: https://onlinelibrary.wiley.com/doi/full/10.1111/j.1364-3703.2007.00451.x [cited 2024 Jul 29]

Gesteiro N, Butrón A, Santiago R, Gomez LD, López-Malvar A, Álvarez-Iglesias L et al (2023) Breeding dual-purpose maize: grain production and biofuel conversion of the Stover. Agronomy [Internet] 13(5):1352. Available from: https://www.mdpi.com/2073-4395/13/5/1352/htm [cited 2024 Jul 30]

Gil-Humanes J, Ozuna CV, Marín S, León E, Barro F, Pistón F et al (2011) Genetic transformation of wheat: advances in the transformation method and applications for obtaining lines with improved bread-making quality and low toxicity in relation to celiac disease. Genet Transform [Internet]. Available from: https://www.intechopen.com/chapters/18817 [cited 2024 Jul 29]

Gill HK, Garg H, Gill HK, Garg H (2014) Pesticides: environmental impacts and management strategies. Pestic Tox Asp [Internet]. Available from: https://www.intechopen.com/chapters/46083 [cited 2024 Jul 25]

Gonsalves D (1998) Control of papaya ringspot virus in papaya: a case study. Annu Rev Phytopathol [Internet] 36:415–437. Available from: https://pubmed.ncbi.nlm.nih.gov/15012507/ [cited 2024 Jul 29]

Gull A, Lone AA, Wani NUI, Gull A, Lone AA, Wani NUI (2019) Biotic and abiotic stresses in plants. Abiotic Biot Stress Plant [Internet]. Available from: https://www.intechopen.com/chapters/66714 [cited 2024 Jul 31]

Gupta PK, Langridge P, Mir RR (2010) Marker-assisted wheat breeding: present status and future possibilities. Mol Breed 26(2):145–161

Hall AE, Cisse N, Thiaw S, Elawad HOA, Ehlers JD, Ismail AM et al (2003) Development of cowpea cultivars and germplasm by the Bean/Cowpea CRSP. Field Crop Res 82(2–3):103–134

Harman GE, Howell CR, Viterbo A, Chet I, Lorito M (2004) Trichoderma species – opportunistic, avirulent plant symbionts. Nat Rev Microbiol [Internet] 2(1):43–56. Available from: https://pubmed.ncbi.nlm.nih.gov/15035008/

Hasegawa PM, Bressan RA, Zhu JK, Bohnert HJ (2000) Plant cellular and molecular responses to high salinity. Annu Rev Plant Physiol Plant Mol Biol [Internet] 51:463–499. Available from: https://pubmed.ncbi.nlm.nih.gov/15012199/ [cited 2024 Jul 31]

Hautea DM, Taylo LD, Masanga APL, Sison MLJ, Narciso JO, Quilloy RB et al (2016) Field performance of Bt eggplants (Solanum melongena L.) in The Philippines: Cry1Ac expression and control of the eggplant fruit and shoot borer (Leucinodes orbonalis Guenée). PLoS One [Internet] 11(6) Available from: /pmc/articles/PMC4913932/ [cited 2024 Jul 29]

Hayward AC (1991) Biology and epidemiology of bacterial wilt caused by pseudomonas solanacearum. Annu Rev Phytopathol [Internet] 29:65–87. Available from: https://pubmed.ncbi.nlm.nih.gov/18479193/ [cited 2024 Jul 29]

Hazel JR (1995) Thermal adaptation in biological membranes: is homeoviscous adaptation the explanation? Annu Rev Physiol [Internet] 57:19–42. Available from: https://pubmed.ncbi.nlm.nih.gov/7778864/ [cited 2024 Jul 31]

Health NRC (US) C on I and AUE of GEF on H (2004) Methods and mechanisms for genetic manipulation of plants, animals, and microorganisms. Available from: https://www.ncbi.nlm.nih.gov/books/NBK215771/ [cited 2024 Jul 27]

Hopkins DL, Purcell AH (2007) Xylella fastidiosa: cause of Pierce's disease of grapevine and other emergent diseases. [Internet] 86(10):1056–1066. Available from: https://apsjournals.apsnet.org/doi/10.1094/PDIS.2002.86.10.1056 [cited 2024 Jul 29]

Huang X, Wei JM, Feng WZ, Luo Q, Tan GF, Li YZ (2023a) Interaction between SlMAPK3 and SlASR4 regulates drought resistance in tomato (Solanum lycopersicum L.). Mol Breed [Internet] 43(10). Available from: https://www.researchgate.net/publication/374417851_Interaction_between_SlMAPK3_and_SlASR4_regulates_drought_resistance_in_tomato_Solanum_lycopersicum_L [cited 2024 Jul 30]

Huang D, Zhang Z, Fan Y, Tang S, Zhuang J, Zhu Y (2023b) Detection of QTL for high-temperature tolerance in rice using a high-density bin map. Agronomy [Internet] 13(6):1582. Available from: https://www.mdpi.com/2073-4395/13/6/1582/htm [cited 2024 Jul 31]

Ibrahim AU, Özsöz M, Saeed Z, Tirah G, Gideon O (2019) Genome engineering using the CRISPR Cas9 system. J Biomed Pharm Sci [Internet] 2(2):1–7. Available from: https://www.hilarispublisher.com/abstract/genome-engineering-using-the-crispr-cas9-system-44436.html [cited 2024 Jul 29]

Ilardi V, Tavazza M (2015) Biotechnological strategies and tools for Plum pox virus resistance: trans-, intra-, cis-genesis, and beyond. Front Plant Sci [Internet] 6(June) Available from: /pmc/articles/PMC4458569/ [cited 2024 Jul 29]

Iqbal Z, Iqbal MS, Hashem A, Abd_Allah EF, Ansari MI (2021) Plant defense responses to biotic stress and its interplay with fluctuating dark/light conditions. Front Plant Sci [Internet] 12. Available from: /pmc/articles/PMC7982811/ [cited 2024 Jul 26]

James C (2000) Transgenic crops worldwide: current situation and future outlook. Agric Biotechnol Dev Ctries [Internet]. 11–23. Available from: https://link.springer.com/chapter/10.1007/978-1-4757-3178-1_2 [cited 2024 Jul 29]

Jiang G-L, Jiang G-L (2013) Molecular markers and marker-assisted breeding in plants. Plant Breed from Lab Field [Internet]. Available from: https://www.intechopen.com/chapters/40178 [cited 2024 Jul 26]

Jiang W, Jin YM, Lee J, Lee KI, Piao R, Han L et al (2011) Quantitative trait loci for cold tolerance of rice recombinant inbred lines in low temperature environments. Mol Cells [Internet] 32(6): 579. Available from: /pmc/articles/PMC3887680/ [cited 2024 Jul 31]

Jiang P, Chen Y, Wilde HD (2016) Reduction of MLO1 expression in petunia increases resistance to powdery mildew. Sci Hortic (Amsterdam) [Internet] 201:225–229. Available from: https://www.researchgate.net/publication/294421845_Reduction_of_MLO1_expression_in_petunia_increases_resistance_to_powdery_mildew [cited 2024 Jul 30]

Johnson MA, Ruiz-Diaz CP, Manoukis NC, Rodrigues JCV (2020) Coffee berry borer (Hypothenemus hampei), a global pest of coffee: perspectives from historical and recent invasions, and future priorities. Insects [Internet] 11(12):882. Available from: https://www.mdpi.com/2075-4450/11/12/882/htm [cited 2024 Jul 29]

Jones RAC (2021) Global plant virus disease pandemics and epidemics. Plants [Internet] 10(2):233. Available from: https://www.mdpi.com/2223-7747/10/2/233/htm [cited 2024 Jul 31]

Kamaluddin SP, Sharma V, Kumar H, Kumar M, Negi HS et al (2022) Marker-assisted selection for value addition in crop plants. Technol Plant Biotechnol Breed F Crop:23–39

Kapari M, Hlophe-Ginindza S, Nhamo L, Mpandeli S (2023) Contribution of smallholder farmers to food security and opportunities for resilient farming systems. Front Sustain Food Syst 20: 1149854

Kikkert M, Van LJ, Storms M, Bodegom P, Kormelink R, Goldbach R (1999) Tomato spotted wilt virus particle morphogenesis in plant cells. J Virol [Internet] 73(3):2288. Available from: /pmc/articles/PMC104474/ [cited 2024 Jul 29]

Kim G, Westwood JH (2015) Macromolecule exchange in Cuscuta–host plant interactions. Curr Opin Plant Biol 26:20–25

Kumawat G, Kumawat CK, Chandra K, Pandey S, Chand S, Mishra UN et al (2020) Insights into marker assisted selection and its applications in plant breeding. Plant Breed Curr Futur Views [Internet]. Available from: https://www.intechopen.com/chapters/74259 [cited 2024 Jul 26]

Lamers J, Der Meer T, Van TC (2020) How plants sense and respond to stressful environments. Plant Physiol [Internet] 182(4):1624–1635. Available from: https://pubmed.ncbi.nlm.nih.gov/32132112/ [cited 2024 Jul 26]

Lawson C, Kaniewski W, Haley L, Rozman R, Newell C, Sanders P et al (1990) Engineering resistance to mixed virus infection in a commercial potato cultivar: resistance to potato virus X and potato virus Y in transgenic Russet Burbank. Biotechnology (NY) [Internet] 8(2):127–34. Available from: https://pubmed.ncbi.nlm.nih.gov/1366358/ [cited 2024 Jul 29]

Maazou A-RS, Tu J, Qiu J, Liu Z, Maazou A-RS, Tu J et al (2016) Breeding for drought tolerance in maize (Zea mays L.). Am J Plant Sci [Internet] 7(14):1858–1870. Available from: http://www.scirp.org/journal/PaperInformation.aspx?PaperID=70987 [cited 2024 Aug 1]

MacKenzie DJ, Ellis PJ (1992) Resistance to tomato spotted wilt virus infection in transgenic tobacco expressing the viral nucleocapsid gene. Mol Plant Microbe Interact [Internet] 5(1): 34–40. Available from: https://pubmed.ncbi.nlm.nih.gov/1600236/ [cited 2024 Jul 29]

Mahajan S, Tuteja N (2005) Cold, salinity and drought stresses: an overview. Arch Biochem Biophys [Internet] 444(2):139–158. Available from: https://pubmed.ncbi.nlm.nih.gov/16309626/ [cited 2024 Jul 31]

Malhi GS, Kaur M, Kaushik P (2021) Impact of climate change on agriculture and its mitigation strategies: a review. Sustainability [Internet] 13(3):1318. Available from: https://www.mdpi.com/2071-1050/13/3/1318/htm [cited 2024 Feb 9]

Mali S, Dutta M, Zinta G (2023) Genome editing advancements in potato (Solanum tuberosum L.): operational challenges and solutions. J Plant Biochem Biotechnol [Internet] 32(4):730–742. Available from: https://www.researchgate.net/publication/366372446_Genome_editing_advancements_in_potato_Solanum_tuberosum_L_operational_challenges_and_solutions [cited 2024 Jul 30]

Malik NAA, Kumar IS, Nadarajah K (2020) Elicitor and receptor molecules: orchestrators of plant defense and immunity. Int J Mol Sci [Internet] 21(3). Available from: /pmc/articles/PMC7037962/ [cited 2024 Jul 26]

Martínez C, Manzano S, Megías Z, Garrido D, Picó B, Jamilena M (2014) Sources of parthenocarpy for Zucchini breeding: relationship with ethylene production and sensitivity. Euphytica [Internet] 200(3):349–362. Available from: https://link.springer.com/article/10.1007/s10681-014-1155-8 [cited 2024 Jul 30]

Mengstie MA, Wondimu BZ (2021) Mechanism and applications of CRISPR/Cas-9-mediated genome editing. Biol Int 15:353. Available from: /pmc/articles/PMC8388126/ [cited 2024 Jul 29]

Mi X, Ji X, Yang J, Liang L, Si H, Wu J et al (2015) Transgenic potato plants expressing cry3A gene confer resistance to Colorado potato beetle. C R Biol [Internet] 338(7):443–550. Available from: https://pubmed.ncbi.nlm.nih.gov/26025753/ [cited 2024 Jul 29]

Milligan SB, Bodeau J, Yaghoobi J, Kaloshian I, Zabel P, Williamson VM (1998) The root knot nematode resistance gene Mi from tomato is a member of the leucine zipper, nucleotide binding, leucine-rich repeat family of plant genes. Plant Cell [Internet] 10(8):1307–1319. Available from: https://pubmed.ncbi.nlm.nih.gov/9707531/ [cited 2024 Jul 29]

Mochizuki T, Ohki ST (2012) Cucumber mosaic virus: viral genes as virulence determinants. Mol Plant Pathol [Internet] 13(3):217. Available from: /pmc/articles/PMC6638793/ [cited 2024 Jul 29]

Muiruri KS, Fathima AA, Muiruri KS, Fathima AA (2023) Advances in cassava trait improvement and processing technologies for food and feed. Cassava Recent Updat Food Feed Ind [Internet]. Available from: https://www.intechopen.com/chapters/86172 [cited 2024 Jul 30]

Munns R, Tester M (2008) Mechanisms of salinity tolerance. Annu Rev Plant Biol [Internet] 59: 651–681. Available from: https://pubmed.ncbi.nlm.nih.gov/18444910/ [cited 2024 Jul 31]

Naranjo SE (2011) Impacts of Bt transgenic cotton on integrated pest management. J Agric Food Chem [Internet] 59(11):5842–5851. Available from: https://pubmed.ncbi.nlm.nih.gov/20942488/ [cited 2024 Jul 29]

Nasir MW, Toth Z (2022) Effect of drought stress on potato production: a review. Agron [Internet] 12(3):635. Available from: https://www.mdpi.com/2073-4395/12/3/635/htm [cited 2024 Jul 31]

Nawaz M, Sun J, Shabbir S, Khattak WA, Ren G, Nie X et al (2023) A review of plants strategies to resist biotic and abiotic environmental stressors. Sci Total Environ 900:165832

Nievola CC, Carvalho CP, Carvalho V, Rodrigues E (2017) Rapid responses of plants to temperature changes. Temp Multidiscip Biomed J [Internet] 4(4):371. Available from: /pmc/articles/PMC5800372/ [cited 2024 Jul 31]

Nosratti I, Mobli A, Mohammadi G, Yousefi AR, Sabeti P, Chauhan BS (2020) The problem of Orobanche spp. and Phelipanche spp. and their management in Iran. Weed Sci [Internet] 68(6): 555–564. Available from: https://www.cambridge.org/core/journals/weed-science/article/abs/problem-of-orobanche-spp-and-phelipanche-spp-and-their-management-in-iran/6D9E16EF0A7881D2FA0FE8E3B93EDB2D [cited 2024 Jul 29]

Nourozi M, Nazarain-Firouzabadi F, Ismaili A, Ahmadvand R, Poormazaheri H (2023). CRISPR/Cas StNRL1 gene knockout increases resistance to late blight and susceptibility to early blight in potato. Front Plant Sci [Internet] 14:1278127. Available from: www.endmemo.com/bio/gc [cited 2024 Jul 30]

Paine JA, Shipton CA, Chaggar S, Howells RM, Kennedy MJ, Vernon G et al (2005) Improving the nutritional value of Golden Rice through increased pro-vitamin A content. Nat Biotechnol [Internet] 23(4):482–487. Available from: https://pubmed.ncbi.nlm.nih.gov/15793573/ [cited 2024 Aug 1]

Palma-Guerrero J, Chancellor T, Spong J, Canning G, Hammond J, McMillan VE et al (2021) Take-all disease: new insights into an important wheat root pathogen. Trends Plant Sci [Internet] 26(8):836–848. Available from: http://www.cell.com/article/S1360138521000388/fulltext [cited 2024 Jul 29]

Palomares-Rius JE, Clavero-Camacho I, Archidona-Yuste A, Cantalapiedra-Navarrete C, León-Ropero G, Miyara SB et al (2021) Global distribution of the reniform nematode genus rotylenchulus with the synonymy of rotylenchulus macrosoma with rotylenchulus borealis. Plants [Internet] 10(1):7. Available from: https://www.mdpi.com/2223- 7747/10/1/7/htm [cited 2024 Jul 29]

Papathanasiou F, Ninou E, Olšovská K, Sytar O, Kováčik P (2024) Optimizing nitrogen application for enhanced barley resilience: a comprehensive study on drought stress and nitrogen supply for sustainable agriculture. Sustain [Internet] 16(5):2016. Available from: https://www.mdpi.com/2071-1050/16/5/2016/htm [cited 2024 Jul 30]

Parida AK, Jha B (2010) Salt tolerance mechanisms in mangroves: a review trees. Struct Funct [Internet] 24(2):199–217. Available from: https://link.springer.com/article/10.1007/s00468-010-0417-x [cited 2024 Jul 31]

Parker C (2012) Parasitic weeds: a world challenge. Weed Sci [Internet] 60(2):269–276. Available from: https://www.cambridge.org/core/journals/weed-science/article/abs/parasitic-weeds-a-world-challenge/5C43CF33C514FC159DBC214B2554CAC3 [cited 2024 Jul 29];

Parmar N, Singh KH, Sharma D, Singh L, Kumar P, Nanjundan J et al (2017) Genetic engineering strategies for biotic and abiotic stress tolerance and quality enhancement in horticultural crops: a comprehensive review. 3 Biotech [Internet] 7(4) Available from: /pmc/articles/PMC5507805/ [cited 2024 Jul 25]

(PDF) (n.d.) Banana streak virus: a highly diverse plant pararetrovirus [Internet]. Available from: https://www.researchgate.net/publication/43480621_Banana_streak_virus_A_highly_diverse_plant_pararetrovirus [cited 2024 Jul 29]

Peshin R (2014) Integrated pest management: pesticide problems. Integr Pest Manag Pestic Probl 3: 1–474

Pilon-Smits E (2005) Phytoremediation. Annu Rev Plant Biol [Internet] 56:15–39. Available from: https://pubmed.ncbi.nlm.nih.gov/15862088/ [cited 2024 Jul 31]

Platten JD, Cobb JN, Zantua RE (2019) Criteria for evaluating molecular markers: comprehensive quality metrics to improve marker-assisted selection. PLoS One [Internet] 14(1) Available from: /pmc/articles/PMC6333336/ [cited 2024 Jul 26]

Pompili V, Dalla Costa L, Piazza S, Pindo M, Malnoy M (2020) Reduced fire blight susceptibility in apple cultivars using a high-efficiency CRISPR/Cas9-FLP/FRT-based gene editing system. Plant Biotechnol J [Internet] 18(3):845. Available from: /pmc/articles/PMC7004915/ [cited 2024 Jul 30]

Rahman NSNA, Hamid NWA, Nadarajah K (2021) Effects of abiotic Stress on soil microbiome. Int J Mol Sci [Internet] 22(16):22. Available from: /pmc/articles/PMC8396473/ [cited 2024 Jul 25]

Rascio N, Navari-Izzo F (2011) Heavy metal hyperaccumulating plants: how and why do they do it? And what makes them so interesting? Plant Sci 180(2):169–181

Rauwane M, Ntushelo K (2020) Understanding biotic stress and hormone signalling in cassava (Manihot esculenta): potential for using hyphenated analytical techniques. Appl Sci [Internet] 10(22):8152. Available from: https://www.mdpi.com/2076-3417/10/22/8152/htm 18 [cited 2024 Jul 31]

Redak RA, Purcell AH, Lopes JRS, Blua MJ, Mizell RF, Andersen PC (2004) The biology of xylem fluid-feeding insect vectors of Xylella fastidiosa and their relation to disease epidemiology. Annu Rev Entomol [Internet] 49:243–270. Available from: https://pubmed.ncbi.nlm.nih.gov/14651464/ [cited 2024 Jul 29]

Rihan HZ, Al-Issawi M, Fuller MP. Advances in physiological and molecular aspects of plant cold tolerance. J Plant Interact [Internet] 2017 ;12(1):143–157. Available from: https://www.tandfonline.com/doi/abs/10.1080/17429145.2017.1308568 [cited 2024 Jul 31]

Rivero RM, Mittler R, Blumwald E, Zandalinas SI (2022) Developing climate-resilient crops: improving plant tolerance to stress combination. Plant J [Internet] 109(2):373–389. Available from: https://pubmed.ncbi.nlm.nih.gov/34482588/ [cited 2024 Jul 25]

Rollins JA, Habte E, Templer SE, Colby T, Schmidt J, Von Korff M (2013) Leaf proteome alterations in the context of physiological and morphological responses to drought and heat stress in barley (Hordeum vulgare L.). J Exp Bot [Internet] 64(11):3201–3212. Available from: https://pubmed.ncbi.nlm.nih.gov/23918963/ [cited 2024 Jul 25]

Saijo Y, Loo EP, iian. (2020) Plant immunity in signal integration between biotic and abiotic stress responses. New Phytol [Internet] 225(1):87–104. Available from: https://pubmed.ncbi.nlm.nih.gov/31209880/ [cited 2024 Jul 26]

Santamaria ME, Martínez M, Cambra I, Grbic V, Diaz I (2013) Understanding plant defence responses against herbivore attacks: an essential first step towards the development of sustainable resistance against pests. Transgenic Res [Internet] 22(4):697–708. Available from: https://pubmed.ncbi.nlm.nih.gov/23793555/ [cited 2024 Jul 26]

Savary S, Willocquet L, Pethybridge SJ, Esker P, McRoberts N, Nelson A (2019) The global burden of pathogens and pests on major food crops. Nat Ecol Evol [Internet] 3(3):430–439. Available from: https://pubmed.ncbi.nlm.nih.gov/30718852/ [cited 2024 Jul 26]

Savchenko TV, Zastrijnaja OM, Klimov VV (2014) Oxylipins and plant abiotic stress resistance. Biochemistry (Mosc) [Internet] 79(4):362–375. Available from: https://pubmed.ncbi.nlm.nih.gov/24910209/ [cited 2024 Jul 25]

Schlenker W, Roberts MJ (2009) Nonlinear temperature effects indicate severe damages to U.S. crop yields under climate change. Proc Natl Acad Sci U S A [Internet] 106(37): 15594–15598. Available from: https://pubmed.ncbi.nlm.nih.gov/19717432/ [cited 2024 Jul 25]

Seleiman MF, Al-Suhaibani N, Ali N, Akmal M, Alotaibi M, Refay Y et al (2021) Drought stress impacts on plants and different approaches to alleviate its adverse effects. Plants [Internet] 10(2):1–25. Available from: /pmc/articles/PMC7911879/ [cited 2024 Jul 25]

Sharma G, Shrestha S, Kunwar S, Tseng TM (2021) Crop diversification for improved weed management: a review. Agric [Internet] 11(5):461. Available from: https://www.mdpi.com/2077-0472/11/5/461/htm [cited 2024 Aug 1]

Sheoran S (2022) Recent advances for drought stress tolerance in maize (Zea mays L.): present status and future prospects. Front. Plant Sci 13:872566

Silva LCC, Mayrink DB, Bueno RD, Piovesan ND, Ribeiro C, Dal-Bianco M (2022) Reference genes and expression analysis of seed desaturases genes in soybean mutant accessions. Biochem Genet [Internet] 60(3):937–952. Available from: https://pubmed.ncbi.nlm.nih.gov/34554351/ [cited 2024 Jul 30]

Singh KP (2019) Aerobiology, epidemiology and management strategies in apple scab: science and its applications. Indian Phytopathol [Internet] 72(3):381–408. Available from: https://link.springer.com/article/10.1007/s42360-019-00162-5 [cited 2024 Jul 29]

Song WY, Wang GL, Chen LL, Kim HS, Pi LY, Holsten T et al (1995) A receptor kinase-like protein encoded by the rice disease resistance gene, Xa21. Science [Internet] 270(5243):1804. Available from: https://pubmed.ncbi.nlm.nih.gov/8525370/ [cited 2024 Jul 29]

Song J, Bradeen JM, Naess SK, Raasch JA, Wielgus SM, Haberlach GT et al (2003) Gene RB cloned from Solanum bulbocastanum confers broad spectrum resistance to potato late blight. Proc Natl Acad Sci USA [Internet] 100(16):9128–9133. Available from: https://pubmed.ncbi.nlm.nih.gov/12872003/ [cited 2024 Jul 29]

Sparvoli F, Cominelli E (2015) Seed biofortification and phytic acid reduction: a conflict of interest for the plant? Plants [Internet] 4:728. Available from: /pmc/articles/PMC4844270/ [cited 2024 Jul 30]

Srivastava RK, Yadav OP, Kaliamoorthy S, Gupta SK, Serba DD, Choudhary S et al (2022) Breeding drought-tolerant pearl millet using conventional and genomic approaches: achievements and prospects. Front Plant Sci [Internet] 13. Available from: /pmc/articles/PMC9021881/ [cited 2024 Jul 31]

Stetter MG, Zeitler L, Steinhaus A, Kroener K, Biljecki M, Schmid KJ (2016) Crossing methods and cultivation conditions for rapid production of segregating populations in three grain amaranth species. Front Plant Sci [Internet] 7:205627. Available from: www.frontiersin.org [cited 2024 Jul 27]

Sudha SN, Jayakumar R, Sekar V (1999) Introduction and expression of the cry1Ac gene of Bacillus thuringiensis in a cereal-associated bacterium, Bacillus polymyxa. Curr Microbiol [Internet] 38(3):163–167. Available from: https://pubmed.ncbi.nlm.nih.gov/9922467/ [cited 2024 Jul 29]

Sugimoto T, Kato M, Yoshida S, Matsumoto I, Kobayashi T, Kaga A et al (2012) Pathogenic diversity of Phytophthora sojae and breeding strategies to develop Phytophthora-resistant soybeans. Breed Sci [Internet] 61(5):511. Available from: /pmc/articles/PMC3406798/ [cited 2024 Jul 29]

Sun X, Zhang H, Yang Z, Xing X, Fu Z, Li X et al (2024) Overexpression of GmPAP4 enhances symbiotic nitrogen fixation and seed yield in soybean under phosphorus-deficient condition. Int J Mol Sci 25(7)

Swarnapriya R, Swarnapriya R (2021) Breeding approaches for biotic stress resistance in vegetables. Plant Breed Curr Futur Views [Internet]. Available from: https://www.intechopen.com/chapters/74317 [cited 2024 Jul 25]

Tao Y, Xie Z, Chen W, Glazebrook J, Chang HS, Han B et al (2003) Quantitative nature of Arabidopsis responses during compatible and incompatible interactions with the bacterial pathogen Pseudomonas syringae. Plant Cell [Internet]. 15(2):317–330. Available from: https://pubmed.ncbi.nlm.nih.gov/12566575/ [cited 2024 Jul 26]

Thakur P, Kumar S, Malik JA, Berger JD, Nayyar H (2010) Cold stress effects on reproductive development in grain crops: an overview. Environ Exp Bot 67(3):429–443

The Development and Regulation of Bt Brinjal in India. ISAAA Brief 38-2009. ISAAA.org [Internet]. Available from: https://www.isaaa.org/resources/Publications/briefs/38/default.asp [cited 2024 Jul 29]

(2006) Transgenic crops engineered with insecticidal activity. ABCs Gene Cloning [Internet]:143–145. Available from: https://link.springer.com/chapter/10.1007/0-387-28679-9_13 [cited 2024 Jul 29]

Tricoli DM, Carney KJ, Russell PF, McMaster JR, Graff DW, Hadden KC et al (1995) Field evaluation of transgenic squash containing single or multiple virus coat protein gene constructs for resistance to cucumber mosaic virus, watermelon mosaic virus 2, and zucchini yellow mosaic virus. Bio/Technology [Internet] 13(12):1458–1465. Available from: https://www.nature.com/articles/nbt1295-1458 [cited 2024 Jul 29]

Tuskan GA, DiFazio S, Jansson S, Bohlmann J, Grigoriev I, Hellsten U et al (2006) The genome of black cottonwood, Populus trichocarpa (Torr. & Gray). Science [Internet] 313(5793): 1596–1604. Available from: https://pubmed.ncbi.nlm.nih.gov/16973872/ [cited 2024 Jul 31]

Umar OB, Ranti LA, Abdulbaki AS, Bola AL, Abdulhamid AK, Biola MR et al (2021) Stresses in plants: biotic and abiotic. Curr Trends Wheat Res [Internet]. Available from: https://www.intechopen.com/chapters/79012 [cited 2024 Jul 25]

Wahid A, Gelani S, Ashraf M, Foolad MR (2007) Heat tolerance in plants: an overview. Environ Exp Bot 61(3):199–223

Walford SA, Wu Y, Llewellyn DJ, Dennis ES (2011). GhMYB25-like: a key factor in early cotton fibre development. Plant J [Internet] 65(5):785–797. Available from: https://pubmed.ncbi.nlm.nih.gov/21235650/ [cited 2024 Jul 30];

Wang Y, Cheng X, Shan Q, Zhang Y, Liu J, Gao C et al (2014) Simultaneous editing of three homoeoalleles in hexaploid bread wheat confers heritable resistance to powdery mildew. Nat Biotechnol [Internet] 32(9):947–951. Available from: https://www.researchgate.net/publication/264054741_Simultaneous_editing_of_three_homoeoalleles_in_hexaploid_bread_wheat_confers_heritable_resistance_to_powdery_mildew [cited 2024 Jul 30]

Wang F, Wang C, Liu P, Lei C, Hao W, Gao Y et al (2016) Enhanced rice blast resistance by CRISPR/ Cas9-targeted mutagenesis of the ERF transcription factor gene OsERF922. PLoS One 11(4)

Wang Z, Ma LY, Cao J, Li YL, Ding LN, Zhu KM et al (2019) Recent advances in mechanisms of plant defense to Sclerotinia sclerotiorum. Front Plant Sci [Internet] 10:472930. Available from: www.frontiersin.org [cited 2024 Jul 26]

Wani SH, Samantara K, Razzaq A, Kakani G, Kumar P (2022) Back to the wild: mining maize (Zea mays L.) disease resistance using advanced breeding tools. Mol Biol Rep 49(6):5787–5803

Williams DJ, Denno BD (2014) Scale insect genus-group names and their families (Hemiptera: Sternorrhyncha: Coccoidea). Zootaxa [Internet] 3813(1):1–84. Available from: https://www.biotaxa.org/Zootaxa/article/view/zootaxa.3813.1.1 [cited 2024 Jul 29]

Wilson RA, Talbot NJ (2009) Under pressure: investigating the biology of plant infection by Magnaporthe oryzae. Nat Rev Microbiol 7(3):185–195. Available from: https://www.nature.com/articles/nrmicro2032 [cited 2024 Jul 25]

Wilson IW, Moncuquet P, Ellis M, White RG, Zhu QH, Stiller W et al (2021) Characterization and genetic mapping of black root rot resistance in Gossypium arboreum L. Int J Mol Sci [Internet] 22(5):1–17. Available from: /pmc/articles/PMC7961528/ [cited 2024 Jul 29]

Xu K, Xu X, Fukao T, Canlas P, Maghirang-Rodriguez R, Heuer S et al (2006) Sub1A is an ethylene-response-factor-like gene that confers submergence tolerance to rice. Nat [Internet] 442(7103):705–708. Available from: https://www.nature.com/articles/nature04920 [cited 2024 Jul 31]

Yang Y, Zhou Y, Sun J, Liang W, Chen X, Wang X et al (2022) Research progress on cloning and function of Xa genes against rice bacterial blight. Front Plant Sci [Internet] 3:847199. Available from: www.frontiersin.org [cited 2024 Jul 30]

Ye GY, Shu QY, Yao HW, Cui HR, Cheng XY, Hu C et al (2001) Field evaluation of resistance of transgenic rice containing a synthetic cry1Ab gene from Bacillus thuringiensis Berliner to two stem borers. J Econ Entomol [Internet] 94(1):271–276. Available from: https://pubmed.ncbi.nlm.nih.gov/11233125/ [cited 2024 Jul 29]

Zeng X, Luo Y, Vu NTQ, Shen S, Xia K, Zhang M (2020) CRISPR/Cas9-mediated mutation of OsSWEET14 in rice cv. Zhonghua11 confers resistance to Xanthomonas oryzae pv. oryzae without yield penalty. BMC Plant Biol [Internet] 20(1) Available from: /pmc/articles/PMC7333420/ [cited 2024 Jul 30]

Zhang Y, Massel K, Godwin ID, Gao C (2018a) Applications and potential of genome editing in crop improvement. Genome Biol [Internet] 19(1) Available from: https://pubmed.ncbi.nlm.nih.gov/30501614/ [cited 2024 Jul 31]

Zhang H, Lang Z, Zhu JK (2018b) Dynamics and function of DNA methylation in plants. Nat Rev Mol Cell Biol [Internet] 19(8):489–506. Available from: https://www.nature.com/articles/s41580-018-0016-z [cited 2024 Jul 31]

Zhao L, Lei J, Huang Y, Zhu S, Chen H, Huang R et al (2016) Mapping quantitative trait loci for heat tolerance at anthesis in rice using chromosomal segment substitution lines. Breed Sci [Internet] 66(3):358. Available from: /pmc/articles/PMC4902453/ [cited 2024 Jul 31]

Zhao Y-q, Tian Y-l, Wang L-m, Geng G-m, Zhao W-j, Hu B-s et al (2019) Fire blight disease, a fast-approaching threat to apple and pear production in China. J Integr Agric 18(4):815–820

Zhou X, Zhao Y, Ni P, Ni Z, Sun Q, Zong Y (2023) CRISPR-mediated acceleration of wheat improvement: advances and perspectives. J Genet Genomics 50(11):815–834

Marker-Assisted Selection and Genomics Tools for Improving Crops

5

Priti Upadhyay, Vineeta Dixit, P. K. Bhati, and Manish Kumar Vishwakarma

Abstract

Since the middle of the twentieth century, the yield of crops has been achieved through traditional breeding practices. With the advancement of molecular technology and statistical models, breeders now have enhanced the capacity to select desirable phenotypes with the aid of genotyping information. Whole-genome analysis and information also become valuable input in this regard. This innovation has resulted in more accurate and efficient crop growth using genome-based approaches, including molecular markers, genomic selection (GS), and genome editing. Molecular markers, particularly single-nucleotide polymorphisms (SNPs), are critical for identifying genomic regions linked to crucial characteristics, boosting breeding precision and efficiency.

Genomics, genetic markers, reference genomes, protein and sequence databases, transcriptomes, and even gene expression profiles have equipped plant breeding with essential resources. These resources assist with feature identification, broaden the understanding of genetic variation, accelerate genomic mapping, facilitate marker-assisted selection (MAS), and improve genetic mapping thereby speeding up breeding activities. The use of Clustered Regularly Interspaced Short Palindromic Repeats (CRISPR)/Cas9 techniques for gene editing advances the pace of the breeding process. These methods, particularly marker-assisted selection (MAS) and genomic selection (GS) enable accurate

P. Upadhyay
Division of Vegetable Sciences, Indian Agricultural Research Institute, New Delhi, India

V. Dixit
Nilamber-Pitamber University, Medininagar, Palamu, Jharkhand, India

P. K. Bhati · M. Kumar Vishwakarma (✉)
Borlaug Institute for South Asia (BISA) India, DPS Marg, NASC Complex, New Delhi, India
e-mail: M.Vishwakarma@cgiar.org

Y. Ahlawat et al. (eds.), *Next-Generation Strategies for Crop Improvement*,
https://doi.org/10.1007/978-981-95-0309-4_5

selection and outcome prediction, which in turn increases yield, disease tolerance, and stress resilience in plants. Such methods are vital when dealing with advanced traits influenced by numerous genes and environmental factors. This chapter discusses in detail the integration of molecular tools into modern crop engineering, underscoring their benefits, applications, and power to change the agricultural industry.

Keywords

Molecular marker · CRISPR/Cas9 · MAS · Genomic selection · SNPs

5.1 Breeding and Improvement of Crops

Crop breeding is a science-driven method of producing new, enhanced plant varieties with desired characteristics such as increased yield, better nutritional quality, disease resistance, and stress tolerance (Chapman et al. 2002). It is critical for addressing global food security, enhancing agricultural productivity, and ensuring sustainable farming practices all over the world. Traditional breeding methods that include the crossing of selected plants, despite their historical success, face several challenges at present to meet modern agricultural demands that have been caused by human actions, climate change, population growth, and evolving dietary needs (Xiong et al. 2022). Conventionally, breeding methods rely mainly on the principles of Mendelian genetics, including the selection of parental lines with desired traits and their subsequent hybridization (Acquaah 2015). The process involves three main steps: selection, hybridization, and segregation. Identifying and selecting parent plants with desirable traits, such as resistance to specific biotic or abiotic stress, is selection. Whereas hybridization is the process to cross two contrast parents to produce first filial generation with the recombination, segregation and selection involve the evaluation of the offspring across generations to choose the desirable individuals. The traditional breeding techniques include mass selection, pure-line selection, and hybrid breeding.

Traditional breeding has significantly contributed to agricultural advancements since prehistoric times. This began with domestication and the selection of lineages exhibiting desired characteristics. During the Green Revolution, high-yielding wheat and rice varieties were developed, significantly increasing food output in developing nations (Evenson and Gollin 2003). Several examples of conventional or traditional breeding have resulted in high-yielding and disease-resistant agricultural products, including maize, wheat, rice, and other crops. Conventional breeding methods also produced better abiotic stress tolerance and quality attributes.

Conventional breeding approaches have balanced the population and its need for food and provided many historical achievements. But as the global population burst, scientists worked on speeding up the procedure as there are many challenges in traditional breeding addressing contemporary agricultural issues. A few are listed below:

(a) Developing a new crop variety using traditional methods often requires 10–15 years due to the need for multiple generations of selection and evaluation.
(b) Relying on existing genetic resources limits the potential for introducing novel traits. Genetic bottlenecks, particularly in monoculture systems, exacerbate this problem (Tanksley and McCouch 1997).
(c) Traditional breeding methods struggle to quickly handle the numerous issues faced by climate change, such as rising temperatures, altering rainfall patterns, and the appearance of new pests and diseases (Ceccarelli 2012).
(d) Addressing combined stresses such as pest infestations and drought conditions requires complex trait integration, which is challenging with traditional methods.
(e) Modern consumers demand crops with improved nutritional profiles, flavour, and processing qualities, which require precise trait development beyond the capabilities of traditional approaches.
(f) Modern biotechnological technologies, such as marker-assisted selection (MAS) and gene editing, provide more precision and efficiency than traditional methods.

To resolve the confrontations of conventional breeding mentioned in the previous section, researchers and scientists worked harder, and as a result of their efforts, breeding methods became an integration of conventional and ever-growing innovative technologies. This integration of traditional knowledge (modern molecular and genomic tools) holds great promise for the development of resilient, high-performing crop varieties that can sustainably meet global food demands. Many breeding methods evolved with biotechnological inventions such as marker-assisted selection, genomic selection (GS), and gene editing. A small description of the various integrations is given below:

1. *Marker-assisted selection (MAS)*: MAS enables breeders to pick plants with desired features based on genetic markers, speeding up the breeding process (Beckmann and Soller 1986; Vishwakarma et al. 2022).
2. *Genomic selection*: Predictive modelling based on genomic data enables a more accurate selection of parental lines and offspring.
3. *Genome editing*: Tools such as CRISPR/Cas9 allow precise genetic modifications to introduce or enhance specific traits (Hickey et al. 2019).

5.1.1 Tools of Modern Crop Improvement

Innovation in agriculture is needed to feed the world's growing population under changing climate and shrinking arable land. Conventional breeding techniques are time taking, laborious and not precise to handle the critical issues adequately. Modern crop improvement has various refined tools and techniques, including molecular technologies, as potent tools that allow for increased accuracy, efficiency, and speed in the generation of crop varieties with improved attributes. Before discussing the integration of molecular tools into conventional ways of breeding, what is a molecular marker? It is important to know. Molecular markers are critical tools in genetics and biotechnology, offering unique insights into the genetic makeup of organisms (Vishwakarma et al. 2022; Bidyananda et al. 2024). Molecular markers are valuable tools for plant and animal breeding, biodiversity conservation, and genetic study. This chapter examines the fundamentals of molecular markers, including their types, uses, and importance in modern research. Molecular markers that are unique DNA sequences or fragments harbouring certain regions within an organism's genome (Amiteye 2021) act as genetic 'signposts', allowing researchers to track trait inheritance, examine genetic variation, and locate relevant genes.

5.1.2 Molecular Markers

Molecular markers are specific DNA sequences that can be used to identify genetic differences between individuals, populations, or species (Amiteye 2021). They serve as genetic landmarks and are widely utilized in fingerprinting, genetic mapping, plant and animal breeding, forensic studies, and evolutionary biology.

Molecular markers can be broadly categorized into three types based on their molecular basis and detection methods (Table 5.1). Following are some popular markers that have been extensively used in crop improvement:

1. *Hybridization-Based Markers*

(a) *Restriction Fragment Length Polymorphisms (RFLPs):*

Restriction fragment length polymorphisms (RFLPs) are a type of molecular marker that relies on variations in DNA sequences that are fragmented through restriction enzyme recognition sites (Lander and Botstein 1989). These variations lead to differences in the length of DNA fragments when the genome is digested with specific restriction enzymes. The resulting fragments are then separated using gel electrophoresis and detected through hybridization with a labelled probe in a process known as Southern blotting. Since RFLPs are co-dominant markers, they allow differentiation between homozygous and heterozygous individuals, making them valuable for genetic mapping, DNA fingerprinting, marker-assisted selection, and evolutionary studies. However, despite their usefulness, RFLPs have limitations, including the requirement for large amounts of high-quality DNA, the labour-intensive nature of the technique, and the relatively low throughput compared to more modern marker systems like single-nucleotide polymorphisms (SNPs) and simple sequence repeats (SSRs).

2. *Polymerase Chain Reaction (PCR)-Based Markers*
   (i) *Random Amplified Polymorphic DNA (RAPD):*
   Random amplified polymorphic DNA (RAPD) is based on the amplification of random DNA segments using short, arbitrary primers in a polymerase chain reaction (PCR) (Kumar and Gurusubramanian 2011). This

Table 5.1 Types of molecular markers and their advantages and limitations

Molecular marker	Brief	Advantages	Limitations
1. Hybridization-based marker			
RFLP	Uses restriction enzymes to detect DNA sequence variations, resulting in different fragment lengths	High reproducibility Co-dominant marker No prior sequence information is needed	Time-consuming and labour-intensive Requires high-quality DNA Limited genome coverage
2. PCR and hybridization-based marker			
AFLP	Combines restriction digestion and selective PCR amplification to detect DNA polymorphisms	High reproducibility High multiplexing capability No prior sequence information is needed	Technically demanding Dominant marker requires high-quality DNA
Cleaved amplified polymorphic sequences (CAPS)	PCR products are digested with restriction enzymes to reveal polymorphisms	Co-dominant marker Simple and reproducible	Requires prior sequence information Limited to specific regions
3. PCR-based marker			
RAPD	Amplifies DNA segments using random primers, revealing polymorphisms based on primer binding sites	Quick and inexpensive Requires small amounts of DNA No prior sequence information is needed	Low reproducibility Dominant marker Sensitive to reaction conditions
SSR (microsatellites)	Tandem repeats of 1–6 base pairs that vary in number among individuals, used as polymorphic markers	Co-dominant marker Highly polymorphic and reproducible requires small amounts of DNA	Development is expensive and labour-intensive
Inter simple sequence repeats (ISSR)	Amplifies regions between SSRs using single primers, revealing polymorphic bands	No prior sequence information is needed High reproducibility Inexpensive	Dominant marker May produce complex banding patterns
4. Sequencing-based markers			
Single-nucleotide polymorphism (SNP)	Single base pair changes in DNA sequences, widely distributed across the genome	High abundance in the genome Co-dominant marker Amenable to high-throughput genotyping	Requires prior sequence information Development costs can be high
Sequence-tagged sites (STS)	Short, unique DNA sequences amplified by PCR and used as genomic landmarks	Co-dominant marker Easy to interpret Highly reproducible	Requires prior sequence information Limited genome coverage
Expressed sequence tags (ESTs)	Short sequences derived from expressed genes, used to study functional genomics	Links to gene function Co-dominant marker Useful for gene discovery	Limited to expressed genes May miss non-coding regions

(continued)

Table 5.1 (continued)

Molecular marker	Brief	Advantages	Limitations
Diversity arrays technology (DArT)	Microarray-based technique that identifies polymorphisms by hybridization of DNA fragments	High-throughput and cost-effective No prior sequence information is needed	May miss rare polymorphisms Requires specialized equipment
InDels	Presence or absence of short insertions or deletions (1 to 50 base pairs) in the DNA sequence at specific genomic loci	High reproducibility, cost-effective, simplicity in detection, co-dominant inheritance, abundance in genomes, wide application	Limited polymorphism, size constraint for detection, allelic dropouts, difficulty in scoring

marker relies on the presence or absence of amplification products to detect genetic variation among individuals and does not require prior knowledge of the genome. RAPD markers are dominant in nature, as they do not distinguish homozygous and heterozygous alleles, that limit their usefulness in certain genetic studies. However, RAPD is widely used due to its simplicity, cost-effectiveness, and ability to generate a large number of polymorphic bands easily. It is commonly applied in genetic diversity studies, population genetics, plant and animal breeding, and species identification (Bidyananda et al. 2024). Despite its advantages, RAPD has limitations, including low reproducibility, sensitivity to experimental conditions, and potential issues with primer specificity, making it less reliable compared to other molecular marker systems like simple sequence repeats (SSRs) or single-nucleotide polymorphisms (SNPs).

(ii) *Amplified Fragment Length Polymorphism (AFLP):*
Amplified fragment length polymorphism (AFLP) marker is the combination of uses of restriction enzyme digestion and selective PCR amplification to detect genetic polymorphisms across a genome (Paun and Schönswetter 2012). In this method, genomic DNA is first digested with specific restriction enzymes, and short adapter sequences are ligated to the resulting fragments. Selective primers are then used to amplify a subset of these fragments via PCR, generating a unique fingerprint of DNA fragments that can be analysed using gel electrophoresis or capillary electrophoresis. AFLP is highly polymorphic, reproducible, and does not require prior knowledge of the genome, making it useful for genetic diversity studies, population genetics, phylogenetics, and marker-assisted breeding. Unlike random amplified polymorphic DNA (RAPD), AFLP offers greater reliability and resolution, but it is more labour-intensive and requires specialized equipment. Despite these challenges, AFLP

remains a powerful tool for analysing genetic variation, particularly in non-model organisms and species with complex genomes (Costa et al. 2016).

(iii) *Simple Sequence Repeats (SSR) or Microsatellites:*
Simple sequence repeats (SSR), well known as microsatellites, are short, tandemly repeated DNA sequences consisting of one to six base pair motifs that are distributed throughout the genome (Gebhardt 2007). This is highly polymorphic due to variations in the number of repeat units caused by replication slippage. SSR markers are co-dominant that distinguish homozygous and heterozygous alleles, making them highly informative for genetic studies. With the aid of their high reproducibility, abundance, and genome-wide distribution it is widely used in population genetics, genetic mapping, forensic analysis, and marker-assisted breeding (Diouf et al. 2024). It requires PCR amplification using primers flanking the repeat region, followed by fragment analysis through gel or capillary electrophoresis. It requires prior sequence information for primer design and is labour-intensive to develop. However, their high polymorphism and accuracy list them as favourite molecular markers in genetics and genomics (Amiteye 2021).

(iv) *Sequence-Tagged Sites (STS):*
Sequence-tagged sites (STS) are short, unique DNA sequences that have specific site in the genome and can be easily amplified using polymerase chain reaction (PCR) (Talbert et al. 1994). These sequences are typically 200 to 500 base pairs long and are defined by their known location and specific nucleotide sequence, making them useful for genome mapping and comparative genomics. STS markers are co-dominant, allowing for the differentiation of alleles in genetic studies. They are widely used in physical mapping, chromosome mapping, and the identification of genetic variations in different species (Mondini et al. 2009). STS markers provide site specific and reproducible as they are based on unique sequences. They can also serve as a foundation for other marker systems, such as microsatellites (SSRs) and expressed sequence tags (ESTs). Despite their usefulness, the development of STS markers requires prior sequence information, which can limit their application in species with un-sequenced genomes. However, they remain an important tool in genomics and molecular biology for identifying and mapping genetic loci.

5.1.3 Sequence-Based Markers

(a) *Expressed Sequence Tags (ESTs):*
Expressed sequence tags (ESTs) are short, sequenced fragments of complementary DNA (cDNA) that represent portions of expressed genes within an organism (Parkinson and Blaxter 2009). Typically ranging from 200 to 800 base pairs,

ESTs are generated by sequencing randomly selected cDNA clones from a library, providing a snapshot of actively transcribed genes. ESTs are important for gene discovery, genome annotation, and comparative genomics, helping researchers identify functional genes and their expression patterns across different tissues and developmental stages. ESTs are also used in the development of genetic linkage maps and in the identification of simple sequence repeats (SSRs) or single-nucleotide polymorphisms (SNPs) within coding regions. Since they originate from transcribed genes, ESTs focus on functionally relevant regions of the genome, making them valuable for studying gene expression and evolutionary relationships (Ueno et al. 2012). However, their utility is limited to expressed genes, and they may not fully represent the entire genome. Despite this limitation, ESTs have played a crucial role in functional genomics and in bridging the gap between molecular markers and gene function.

(b) *Single-Nucleotide Polymorphisms (SNPs):*
Single-nucleotide polymorphisms (SNPs) are the most common type of genetic variation, occurring when a single nucleotide in the DNA sequence is altered at a specific position within the genome (Rafalski 2002; Amiteye 2021). These variations, which involve the substitution of one base (adenine, thymine, cytosine, or guanine) for another, are widely distributed across both coding and non-coding regions of the genome. SNPs are highly stable and co-dominant markers, allowing researchers to distinguish between different alleles with high accuracy (Amiteye 2021). Due to their abundance and genome-wide distribution, SNPs are extensively used in genetic mapping, association studies, evolutionary biology, and personalized medicine. Advanced techniques such as DNA sequencing, microarrays, and genotyping assays are commonly used for SNP detection. While SNPs provide precise genetic information, their low polymorphism content per site means that large numbers of SNP markers are often required for comprehensive genetic analysis. Despite this, SNPs remain one of the most powerful and widely used molecular markers in modern genomics.

5.1.4 Comparison of Marker Systems

A comparative description of five molecular marker systems RFLP, RAPD, AFLP, SSR, and SNP based on key characteristics such as polymorphism level, reproducibility, ease of use, cost, genome coverage, and automation (Table 5.2). Restriction fragment length polymorphism (RFLP) exhibits a moderate level of polymorphism and high reproducibility but is labour-intensive, costly, and offers low genome coverage, making it less suitable for high-throughput studies. Random amplified polymorphic DNA (RAPD), on the other hand, is a simple and cost-effective technique with high polymorphism, but its low reproducibility limits its reliability. Amplified fragment length polymorphism (AFLP) provides high polymorphism and reproducibility with broader genome coverage than RFLP and RAPD, though it requires moderate effort and cost. Simple sequence repeats or microsatellites (SSR)

Table 5.2 Comparison of different molecular markers for different attributes

Feature	RFLP	RAPD	AFLP	SSR	SNP
Polymorphism level	Moderate	High	High	High	Very high
Reproducibility	High	Low	High	High	Very high
Ease of use	Labor-intensive	Simple	Moderate	Simple	Complex
Cost	High	Low	Moderate	Moderate	High
Genome coverage	Low	Moderate	High	Moderate	Very high
Automation	Low	High	Moderate	High	High

are highly polymorphic, reproducible, and easy to use, with moderate genome coverage and automation potential, making them a preferred choice for genetic mapping and population studies. Single-nucleotide polymorphism (SNP) markers offer the highest level of polymorphism, reproducibility, and genome coverage, making them ideal for high-throughput genotyping and precision breeding. However, SNP detection is complex and costly, requiring advanced sequencing technologies. Overall, each marker system has its advantages and limitations, with SNPs being the most precise but resource-intensive, while SSRs provide a balance of reliability and cost-effectiveness.

The term 'molecular tools' refers to a range of methods and instruments that use molecular biology to investigate the genetic composition of plants. The methods where molecular markers have been applied are marker-assisted selection (MAS), quantitative trait loci (QTL) mapping, genomic selection (GS), genome editing technologies (e.g. CRISPR/Cas9), transcriptomics, proteomics, and metabolomics.

5.1.5 Marker-Assisted Selection (MAS) and QTL Mapping

Marker-assisted selection involves the use of molecular markers linked to desirable traits to select superior individuals from different generations of breeding populations. Although the idea was initially conceived by Soller and Hazan in Soller and Plotkin-Hazan 1977, the phrase 'marker-assisted selection' was first used almost 30 years ago in reference to its possible application by Beckmann and Soller in Beckmann and Soller 1986. A small number of non-DNA-based assays were employed to fix advantageous variants at specific loci (Koebner and Summers 2003), which led to the popularity of marker-assisted selection (MAS) and mapping or tagging genes with molecular markers (Xu and Crouch 2008) (Fig. 5.1). Sax in 1923 reported the use of genetic markers as a reliable tool for plant breeders. Later, other researchers emphasised the development and application of marker-assisted selection (MAS) in plant breeding for virtually all targeted crops, highlighting four major advantages:

(a) MAS is especially useful for features with limited penetrance, complicated inheritance, or high measurement costs, which are difficult to regulate by traditional phenotypic selection.

Fig. 5.1 An overview of crop improvement through marker-assisted selection

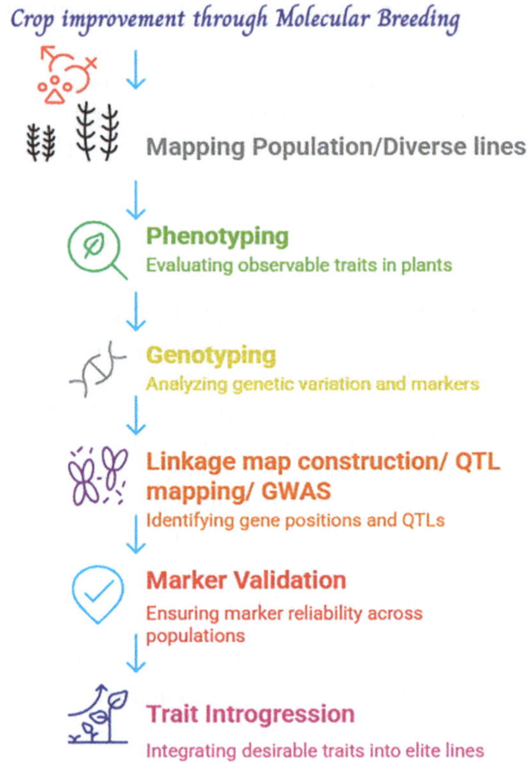

(b) It enables the selection of traits whose expression depends on specific environmental conditions or developmental stages, which might otherwise complicate phenotypic assessment.
(c) MAS accelerates backcross (BC) breeding and aids in the maintenance of recessive alleles during the backcrossing process.
(d) It enables the pyramiding of many quantitative trait loci (QTL) for a single complex feature, such as drought tolerance or other adaptive characteristics, and the combination of various monogenic traits, such as pest and disease resistance or quality attributes (Xu and Crouch 2008; Vishwakarma et al. 2014; Vishwakarma et al. 2016; Yadav et al. 2015).

This approach accelerates the breeding cycle by eliminating the need for extensive phenotypic evaluations (Huang et al. 1997). Vishwakarma et al. (2022) have described the marker-assisted selection in detail and depicted that molecular marker can speed up the breeding cycle of agronomic crops.

The following techniques are commonly employed in MAS breeding projects (Ribaut and Hoisington 1998):

1. Choosing plants that possess the target QTLs or genes associated with desired qualities is known as *'foreground selection'*.

The marker-assisted selection (MAS) includes a method, i.e. 'foreground selection' is used in crop improvement to find and choose plants with desired genes early on (Collard et al. 2005). This method guarantees that intended features are precisely incorporated into elite crop types while also speeding up the breeding process. In order to identify plants with desirable alleles prior to phenotypic expression, it uses molecular markers like SSRs, SNPs, or RFLPs that are linked to certain genes of interest. There are several phases in the foreground selection process. Primers unique to the target gene are used for PCR amplification after DNA is first collected from early seedlings. The targeted gene's existence or absence is then determined by genotyping or gel electrophoresis. Only plants with the target gene are kept for additional breeding based on this study.

There are a number of benefits to foreground selection, including shorter breeding cycle times, more selection precision, and less dependence on ambient factors for trait assessment. Foreground selection greatly increases the effectiveness of crop improvement programs when paired with background selection, which aids in recovering the recurrent parent genome, and phenotypic selection, which guarantees the expression of desired traits. In the end, this results in the creation of crop types with high yields that are resistant to disease and stress (Collard et al. 2005; Roychowdhury et al. 2014).

2. *Background selection* is the process of removing undesirable genomic areas while keeping the desired characteristic by using molecular markers.

In crop breeding, background selection is a marker-assisted selection (MAS) technique that speeds up the recurrent parent genome's recovery in backcross breeding initiatives (Hasan et al. 2015). By adding select genes of relevance from a donor parent while keeping the majority of the genetic background of an elite recurrent parent, it is especially helpful in creating superior agricultural varieties (Mohanavel et al. 2024).

It takes several generations of crossing in traditional backcross breeding to include the desired gene while restoring the genetic composition of the recurrent parent. However, background selection expedites this process by early detection and selection of plants that closely resemble the recurrent parent utilizing molecular markers dispersed across the genome. Background selection's main benefits include increasing selection efficiency, decreasing linkage drag-unwanted genetic material from the donor parent and speeding up the breeding process. Background selection increases the accuracy and efficacy of crop improvement initiatives when combined with phenotypic selection (Vishwakarma et al. 2014; Vishwakarma et al. 2016). In order to determine how much of the recurrent parent genome has been restored, backcross progeny is screened using DNA markers. In order to reduce the number of backcrosses, generations required to complete genome recovery, plants that are most similar to the recurrent parent are chosen for additional backcrossing (Chukwu et al. 2020).

3. *Gene Pyramiding for Crop Improvement*

By combining many beneficial genes into a single crop variety, a breeding approach known as 'gene pyramiding' can increase the crop's tolerance to environmental stress, diseases, and pests while simultaneously enhancing yield and quality features (Dormatey et al. 2020). This method aids in creating high-performing crop types with long-lasting resistance that can tolerate biotic and abiotic stresses. Marker-assisted selection (MAS), which enables breeders to monitor the presence of target genes at every breeding stage, is largely responsible for the accumulation of many genes controlling the same or distinct characteristics within a single genotype.

Numerous crops have benefited from the effective application of gene pyramiding to increase nutritional value, abiotic stress tolerance, disease resistance, and insect resistance. For instance, the pyramiding of blast resistance genes *Pi1*, *Pi2*, and *Pi9* in rice has produced long-lasting protection against *Magnaporthe oryzae* (Jiang et al. 2012; Singh et al. 2013; Hsu et al. 2020), whereas the combination of rust resistance genes *Sr2*, *Sr24*, and *Sr25* in wheat has enhanced resistance to stem rust (*Puccinia graminis*) (Liu et al. 2020; Gautam et al. 2020; Wang et al. 2023). In cotton, bollworm (*Helicoverpa armigera*) resistance has been improved due to the pyramiding of Bt genes *Cry1Ac* and *Cry2Ab* (Cui et al. 2011; Li et al. 2014). Rasheed et al. (2023) found that in maize, stacking the drought tolerance genes *DREB1A* and *ZmVPP1* has enhanced survival in water-deficit situations, but in cotton, pyramiding for heat tolerance genes has improved tolerance against high temperature (Saleem et al. 2024).

Gene pyramiding has also been beneficial for nutritional improvement. To increase the amount of provitamin A, for example, the genes for phytoene synthase and phytoene desaturase (PDS) were combined to create Golden Rice (Mallikarjuna-Swamy et al. 2021; Paine et al. 2005). Biofortified wheat cultivars with higher micronutrient content have been produced as a consequence of the pyramiding of genes linked to zinc and iron absorption (Bouis and Saltzman 2017).

Long-lasting resistance to diseases and pests, less reliance on chemical pesticides, increased stress tolerance for better climate change adaptation, and greater nutritional quality are just a few benefits of gene pyramiding. Nevertheless, this method has drawbacks, including the requirement for extremely accurate DNA markers, the possibility of unexpected phenotypes due to genetic interactions between stacked genes, and the lengthy breeding cycles needed for field testing and stability.

Gene pyramiding is still a potent technique for creating improved crop varieties that support sustainable agriculture and food security in spite of these obstacles. Breeders are now able to produce durable, high-yielding, and nutrient-rich crop types because of developments in molecular breeding and genomic technologies.

5.1.6 Advantages of MAS Over Conventional Breeding

MAS is superior to conventional breeding techniques in a number of ways (Collard et al. 2005).

- Enhanced selection efficiency: environmental impacts are eliminated since genotype, not phenotype, is the basis for selection.
  Shorter breeding time: new types are developed more quickly thanks to early-stage selection.
- Accurate trait selection: by facilitating precise selection of preferred alleles, MAS enhances breeding efficiency.

5.1.6.1 QTL Mapping: A tool for Crop Improvement

Any DNA region linked to a particular phenotype or trait that differs within a population is known as a QTL (Complex Trait Consortium et al. 2003). In contrast to discrete categories (seed colour), other features are usually quantitative, meaning they vary continuously (e.g. disease resistance or yield). Studying QTLs is challenging as they are frequently impacted by several genes and environmental variables. Quantitative trait loci (QTL) mapping is a statistical technique that uses the link between genetic markers and trait values to determine which parts of the genome affect a quantitative trait's variance (Broman and Sen 2009). It is proven that molecular marker-assisted selection (MAS) is very promising approach. The techniques enabled breeders to identify and utilize genetic loci associated with agronomic traits, such as yield, disease resistance, stress tolerance, and quality attributes. Conventional breeding relies on phenotypic selection, which is time-consuming, influenced by environmental factors, and inefficient for complex traits. In contrast, molecular breeding provides a precise, efficient, and high-throughput approach to developing superior crop varieties.

But when molecular markers were used in combination with traditional breeding via QTL mapping finds many genomic areas linked with complicated characteristics like drought tolerance or grain production (Manjunath et al. 2024), This allowed breeders to target specific loci for improvement. Furthermore, the development of various new sequence-based molecular biology technologies, as well as the advancement of quantitative genetics, provides the historical framework for increasing many quantitative features in crops. Early attempts to understand the inheritance of complex traits led to the concept of QTL mapping. Some traits, viz. yield, plant height, and disease resistance, are influenced by several loci and do not follow straightforward Mendelian inheritance. Ronald A. Fisher's 1918 work, which linked Mendelian genetics with biometric techniques and developed a statistical framework, laid the theoretical groundwork for the study of polygenic features (Visscher and Goddard 2019). The development of genetic linkage maps in the early twentieth century, particularly Thomas Hunt Morgan's work on Drosophila, enabled a practical method of relating traits to specific chromosomal locations. These concepts were immediately applied to plants, with maize (*Zea mays*) being one of the first crops to construct linkage maps (Semagn et al. 2006). However, a lack of reliable genetic markers impeded early attempts to evaluate quantitative traits.

The 1980s saw the advent of molecular markers, which revolutionized quantitative genetics research by their application for QTL mapping. According to Botstein et al. (1980), restriction fragment length polymorphisms (RFLPs) were one of the first markers that made high-resolution genetic mapping possible. The first systematic framework for QTL investigations was provided by Paterson et al. (1988), who showed the value of RFLP markers in mapping QTLs for yield-related parameters in tomatoes (Kimbara et al. 2018). This method demonstrated that intricate characteristics might be broken down into separate genetic elements, each contributing to phenotypic variance.

Advances in statistics accompanied these molecular technical innovations. Using maximum likelihood techniques and interval mapping (Lander and Botstein 1989), the accuracy of QTL localization was greatly increased. Subsequently, Zeng (1994) suggested composite interval mapping (CIM), which improved the strength and precision of QTL discovery by accounting for background genetic variation.

Microsatellite markers, or simple sequence repeats (SSRs), replaced RFLPs in the 1990s because of their codominance, quantity, and use. QTL mapping was significantly transformed in the 2000s with the emergence of single-nucleotide polymorphisms (SNPs), which provided high-throughput and affordable genotyping platforms (Rafalski 2002). At the same time, QTL mapping was integrated with physical genome maps by sequencing model plant genomes, including rice (Goff et al. 2002) and Arabidopsis (The Arabidopsis Genome Initiative 2000). This enabled the accurate identification of candidate genes.

5.1.6.2 Basic Principles of QTL Mapping

As discussed earlier, genomic areas known as quantitative trait loci (QTL) regulate intricate, quantitative features that are impacted by a variety of genes and environmental factors (Collard et al. 2005). A statistical method for determining the position and impact of these loci in a genome is called QTL mapping. The basic principles and procedure require:

1. Genetic variation: to identify QTLs, a segregating population with adequate genetic variation is necessary (Hill 2010). F2 populations, recombinant inbred lines (RILs), double haploids (DHs), and backcross populations are often utilized populations in QTL mapping (Singh et al. 2021).
2. Molecular markers: genetic maps are created using DNA markers including single-nucleotide polymorphisms (SNPs), amplified fragment length polymorphisms (AFLPs), simple sequence repeats (SSRs), and restriction fragment length polymorphisms (RFLPs) (Kushanov et al. 2021; Liu et al. 2024).
3. Phenotypic evaluation: to determine marker-trait relationships, precise plant height, yield, and disease resistance measurements are essential (Li et al. 2022).
4. Statistical analysis: major marker-trait relationships are found using a variety of statistical techniques, including multiple QTL mapping, interval mapping, and composite interval mapping.

5.1.7 QTL Mapping Techniques

Several approaches have been used for QTL mapping, and but the two popular techniques are explained below.

5.1.7.1 Linkage Mapping

Linkage mapping is a genctic analysis method used to identify regions of the genome associated with quantitative traits by examining the co-segregation of genetic markers with phenotypic variation in a population (Nantawan et al. 2019). This method depends on producing a mapping population from two genetically distinct parental lines, such as F_2, backcross (BC), recombinant inbred lines (RILs), or doubled haploids (DH). Researchers can ascertain the approximate chromosomal positions of genes regulating a characteristic of interest by using the linkage mapping principle, which states that loci governing a given trait are inherited jointly with neighbouring molecular markers.

The creation of a segregating population is the first step in the linkage QTL mapping process. This is followed by genotyping using molecular markers such as restriction fragment length polymorphisms (RFLPs), single-nucleotide polymorphisms (SNPs), and simple sequence repeats (SSRs). The population is simultaneously phenotyped for the attribute being studied, such as stress tolerance, disease resistance, or yield. After that, linkage analysis is carried out using statistical techniques that assess the position, impact, and interaction of QTLs within the genome, such as interval mapping, composite interval mapping (CIM), or multiple QTL mapping.

Many crops have benefited from the effective application of linkage QTL mapping to enhance significant agronomic features. This approach has found QTLs for blast resistance, drought tolerance, and crop production in rice (*Oryza sativa*) (Carrillo et al. 2021; Dwiningsih et al. 2023). Linkage mapping has aided in the identification of QTLs governing grain protein content in barley and wheat (Fan et al. 2017; Miao et al. 2022) and rust resistance in wheat (*Triticum aestivum*) (Rollar et al. 2021; Tong et al. 2024). Similarly, this method has been used to identify QTLs associated with kernel size and drought tolerance in maize (*Zea mays*) (Hu et al. 2021). In order to increase oil content and resistance to the soybean cyst nematode, the technique has also been widely applied to soybeans (Glycine max) (Shaibu et al. 2020).

Linkage QTL mapping has several drawbacks in spite of its benefits. In contrast to genome-wide association studies (GWAS), which make use of natural genetic variation in a variety of populations, the resolution is comparatively poor since it depends on recombination events within a particular group. Furthermore, QTL impacts might differ based on genetic backgrounds and environmental factors, and mapping populations require time to establish (Wang et al. 2014). For more accurate trait identification and breeding applications, researchers frequently combine linkage mapping with high-resolution methods like fine mapping and genomic selection to get over these restrictions.

All things considered, linkage QTL mapping is still a useful technique in genetics and plant breeding, assisting scientists in deciphering complicated features and creating better crop varieties with desired attributes. The effectiveness and precision of QTL mapping are anticipated to increase with the development of next-generation sequencing (NGS) and bioinformatics, making it a crucial strategy for contemporary crop improvement initiatives.

5.1.7.2 Genome-Wide Association Mapping

A potent method for QTL mapping, genome-wide association study (GWAS) finds correlations between phenotypic features and genetic markers in a variety of populations (Uffelmann et al. 2021). In contrast to linkage mapping, which depends on biparental populations, GWAS finds loci linked to complex characteristics by using natural genetic variation among a large number of unrelated people (Wu et al. 2020). By allowing for the high-resolution mapping of the genes governing important agronomic, physiological, and disease resistance features, this technique has completely changed the field of plant and animal genetics (Table 5.3).

Gathering a genetically diverse population is the first step in the GWAS process. High-throughput genotyping utilizing molecular markers like single-nucleotide polymorphisms (SNPs) comes next. In parallel, under field or controlled settings, the population is phenotyped for the desired characteristic (He et al. 2017). Then, using statistical models that take into consideration population structure and individual relatedness, marker-trait relationships are found using models like genome-wide efficient mixed models and mixed linear models. GWAS offers far greater resolution than conventional QTL mapping by using past recombination events within a population (He et al. 2023).

GWAS has been effectively used to enhance traits in a variety of crops. GWAS has discovered genes associated with disease resistance, grain quality, and drought tolerance in rice (*Oryza sativa*) (Huang et al. 2012). This method has been applied to map loci linked to protein content, rust resistance, value-added quality traits and yield in wheat (Tong et al. 2024; Khan et al. 2024; Vishwakarma et al. 2024). Genes governing kernel size, flowering period, and abiotic stress tolerance have been identified in maize (*Zea mays*) as a result of GWAS (Liu et al. 2021; Wang et al. 2016). Similarly, GWAS has been used to identify genes in soybean (*Glycine max*) that are linked to oil content, nodulation efficiency, and resistance to the soybean cyst nematode (Cao et al. 2017).

Small impact sizes of discovered loci, population structure confounding, and the need for large sample sizes to guarantee statistical power are some of the difficulties that GWAS encounters despite its benefits. Researchers combine GWAS with transcriptome analysis, genomic selection, and functional validation investigations to get around these restrictions and improve the precision and relevance of results. Modern crop breeding efforts that aim to create high-yielding, stress-resistant, and nutritionally improved varieties rely heavily on GWAS, which is becoming increasingly better because to advancements in sequencing technologies and analytics.

Table 5.3 QTL identification using GWAS approach in various crops

Crop	Trait	Molecular marker	Key findings	References
Melon	Yield-related traits	SNP	Identified QTLs for yield and fruit size; advanced by GWAS for precision mapping	Lian et al. (2021)
Mango	Fruit quality traits and yield	SNP	Identified QTLs for yield and fruit size; advanced by GWAS for precision mapping	Srivastav et al. (2023)
Wheat	Grain yield, drought, and heat tolerance	SNP	QTLs for yield and stress tolerance	Sukumaran et al. (2018)
Rice	Rice grain yield	SNP	QTLs for rice grain yield under reproductive stage drought stress	Yadav et al. (2019)
Wheat	Component traits of drought and heat tolerance	SNPs	A total of 22 stable marker-trait associations (MTAs) identified across the locations that have practical utilization in future wheat breeding programmes	Devate et al. (2022)
Maize	Yield-related ear traits	SNP	19 MQTL-detected yield-related ear trait reported	Qian et al. (2023)
Maize	Yield traits under water-deficit stress and well-watered conditions	SNPs	Identified 27 quantitative trait loci (QTLs) associated with various morphophysiological and yield-related traits, with 13 QTLs identified under well-watered conditions and 12 under water-deficit conditions	Sarkar et al. (2023)
Wheat	Disease resistance (rust, fusarium head blight)	SSR, SNP	52 significant marker-trait associations for FHB	Serajazari et al. (2023)
Soybean	Yield-related traits under well-watered and drought-stressed conditions	SNPs	39 SNPs were identified that have significant linkage with the yield-related traits	Li et al. (2023)
Soybean	Seed oil content and yield	SNP	3 loci located on chromosomes 5 and 20 influencing oil composition	Cao et al. (2017)

In order to address genotype-by-environment interactions that are crucial for agronomic characteristics, QTL mapping started to include multi-environment trials as plant genomic resources increased. Transcriptomics, metabolomics, and high-throughput phenotyping are further integrated into modern QTL mapping to reveal the molecular markers underpinnings of quantitative characteristics (Hayat et al.

2025). Next-generation sequencing (NGS) has transformed marker development by offering high-throughput, low-cost, and accurate methods for genomic research (Panahi et al. 2024). Unlike older approaches, NGS can swiftly sequence whole genomes or transcriptomes, providing a thorough understanding of genetic diversity. This feature is extremely useful for detecting molecular markers such as single-nucleotide polymorphisms (SNPs) and microsatellites, which are critical in breeding programs, evolutionary research, and disease diagnoses. Traditional QTL methods are now supplemented by technologies like genome-wide association studies (GWAS), which use dense SNP datasets and natural variation to analyse traits (Sukumaran et al. 2018). The continuous evolution of QTL mapping reflects its central role in plant genetics and breeding. NGS helps to construct functional markers by correlating genomic differences to phenotypic features. NGS uses RNA sequencing (RNA-seq) to identify differentially expressed genes under diverse environments, allowing the discovery of markers linked with certain features or disorders (Panahi et al. 2024). This integrated approach not only improves our understanding of genetic mechanisms but also speeds up the creation of tailored treatments in agriculture, health, and conservation biology. So, basically, by dissecting the genetic basis of complex traits, QTL mapping laced with NGS has provided insights that drive crop improvement efforts worldwide, demonstrating its enduring significance in the post-genomics era.

5.1.7.3 Genomic Selection (GS)

GS is another area where molecular markers have proved useful extensively for plant selection. Genomic selection uses genome-wide markers and predictive models to estimate the genetic potential of individuals without extensive field testing. Genomic selection is a transformative breeding approach that leverages genome-wide marker data to predict the performance of untested individuals in plant populations. By integrating more innovative and advanced statistical models and high-throughput genotyping, this technique allows breeders to make a precise decision about which individuals to select for further breeding cycles. This method has gained prominence in crop improvement programs due to its efficiency, accuracy, and potential to accelerate the development of high-performing crop varieties.

Unlike traditional MAS, which focuses on a limited number of significant markers associated with traits of interest, GS uses genome-wide markers to estimate the genomic estimated breeding value (GEBV) of individuals (Fig. 5.2). This approach accounts for the collective effects of all genetic loci, including both major and minor contributors, making it especially suitable for traits governed by complex inheritance patterns.

The genomic selection process typically involves:

1. Training population development: A subset of the breeding population is phenotyped and genotyped to establish a prediction model.
2. Model training: Statistical techniques such as ridge regression best linear unbiased prediction (RR-BLUP), genomic BLUP (GBLUP), and machine learning algorithms link genetic markers to phenotypic variables.

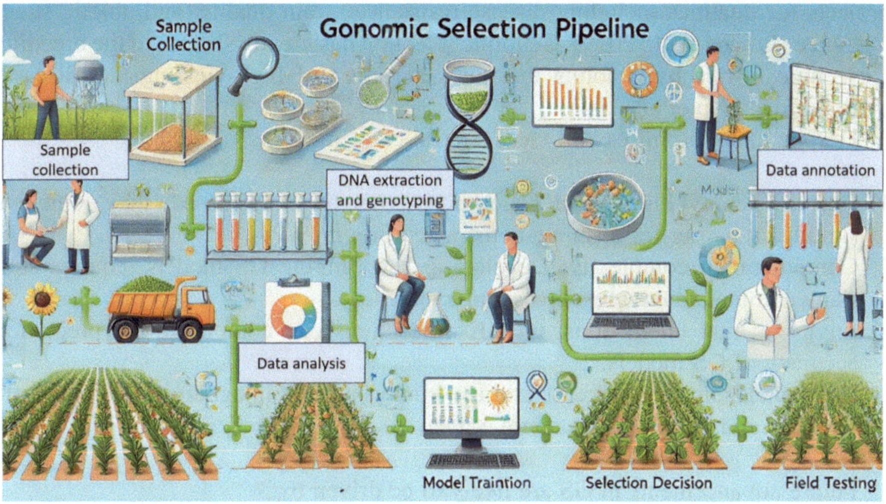

Fig. 5.2 Genomic selection pipeline for crop improvement

3. Prediction: The trained model is applied to genotyped but non-phenotyped individuals to predict their GEBVs.
4. Selection and validation: Individuals with high GEBVs are selected for subsequent breeding or field trials, and their performance is validated.

Genomic selection has been widely applied in crop improvement in the following ways:

(a) Accelerated breeding cycles: By predicting performance without extensive field trials, GS significantly reduces the time required to develop new varieties.
(b) Improved selection for complex traits: GS is incredibly successful for qualities like yield, drought tolerance, and disease resistance, which are controlled by a variety of genes and environmental conditions.
(c) Enhanced genetic gain: The comprehensive use of genome-wide data ensures a more accurate capture of genetic variance, leading to higher genetic gains per unit of time.
(d) Facilitating population management: GS aids in maintaining genetic diversity and optimizing parent selection in breeding populations.
(e) Adaptation to climate change: By identifying and selecting traits that enhance resilience to abiotic stresses, GS supports the development of climate-smart crop varieties.

Genomic selection has been used in maize breeding projects to increase drought resistance and grain production. Research by Beyene et al. (2019) showed how useful genomic selection is for choosing hybrids that perform better in water-limited environments. Genomic selection has been effectively used in wheat development projects for qualities including disease resistance, grain quality, and abiotic stress tolerance (Juliana et al., 2019). Genomic selection has greatly shortened the breeding cycle time in rice by enabling the quick selection of high-yielding and disease-resistant lines (Xu et al., 2021).

Despite many promises, GS faces several challenges:

1. Cost of genotyping: While genotyping costs have decreased, they remain a significant barrier for resource-constrained programs.
2. Computational complexity: Developing and running GS models requires substantial computational resources and expertise.
3. Environmental interactions: G × E interactions can complicate the prediction accuracy of GS models.
4. Data management: Efficient storage, analysis, and integration of large-scale genomic and phenotypic data are critical.

Future advancements in GS are likely to be driven by:

1. Integration with high-throughput phenotyping: Combining GS with advanced phenotyping platforms will enhance prediction accuracy and trait discovery.
2. Machine learning and AI: The use of advanced algorithms to improve model performance and capture complex trait architectures.
3. CRISPR and genomic editing: Synergizing GS with genomic editing tools to accelerate the incorporation of beneficial alleles.
4. Global collaboration: Sharing data, resources, and expertise among international research organizations to democratize access to GS technologies.

Genomic selection offers a paradigm shift in crop development, providing an unprecedented opportunity to address global issues, including food security, climate change, and sustainability. By allowing more efficient and precise selection procedures, GS is positioned to play a critical role in the production of robust and high-performing agricultural varieties.

5.1.7.4 Genome Editing for Positive Modifications in Crops

Over the last decade, breakthrough genome editing technologies have transformed the study of genetics by giving precise tools for changing DNA sequences in live creatures. CRISPR/Cas9 has emerged as the most well-known and commonly utilized technology because of its simplicity, efficiency, and adaptability. The steps involved in genome editing are briefly illustrated in Fig. 5.3.

5.1.8 CRISPR/Cas9 System

The CRISPR/Cas9 system is based on a natural bacterial defence mechanism that protects against viral infections. In genome editing, CRISPR/Cas9 uses a guide RNA (gRNA) to target the Cas9 nuclease to a specific DNA region and create double-

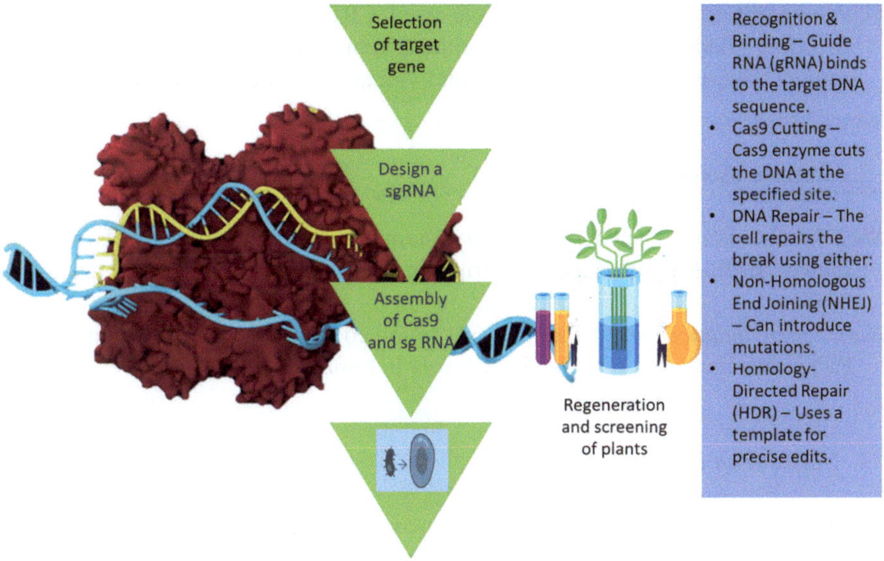

Fig. 5.3 Pictorial representation for procedure involved in CRISPR Cas9 gene-edited plant development

strand breaks. The cell's natural repair mechanisms subsequently patch the breaks, either by non-homologous end joining (NHEJ) or homology-directed repair (HDR), resulting in specific genetic alterations.

Genome editing technologies have been successfully applied in agriculture to enhance crop traits such as yield, disease resistance, and stress tolerance. Some notable examples include:

1. **Rice (*Oryza sativa*)**

Disease resistance: Using CRISPR/Cas9, scientists have knocked out the susceptibility gene *OsSWEET14*, conferring resistance to bacterial blight, a serious disease caused by *Xanthomonas oryzae* (Zeng et al. 2020).

Yield improvement: CRISPR/Cas9 has been employed to modify genes involved in grain size and number, such as Gn1a and DEP1, leading to increased rice yields (Thiruppathi et al. 2024).

2. **Wheat (*Triticum aestivum*)**

Drought tolerance: Genome editing has been used to change the expression of genes such as TaERF3, which improves drought tolerance in wheat. It maintained the expression of wheat dehydration-responsive element binding protein 2 (TaDREB2) and wheat ethylene-responsive factor 3 (TaERF3) (Kim et al. 2018).

Gluten modification: Researchers have used CRISPR/Cas9 to knock out genes encoding gliadins, a component of gluten, to produce wheat with reduced gluten content, catering to gluten-sensitive individuals (Sánchez-León et al. 2024).

3. **Tomato (*Solanum lycopersicum*)**

Shelf-life extension: The CRISPR/Cas9 system has been utilized to knock out genes such as SlMlo1, which delays fruit ripening and extends shelf life (Nekrasov et al. 2017).

Nutritional enhancement: Editing genes involved in carotenoid biosynthesis, like PSY1, has increased provitamin A levels in tomatoes (D'Ambrosio et al. 2018).

4. **Maize (*Zea mays*)**

Herbicide resistance: By editing genes such as ALS2, scientists have developed maize varieties resistant to herbicides, allowing for better weed management (Svitashev et al. 2016).

Nitrogen use efficiency: CRISPR/Cas9 has been applied to modify genes that regulate nitrogen uptake and assimilation, improving nitrogen use efficiency in maize (Zhao et al. 2018).

From the above examples, it is very clear that the potential of genome editing in agriculture is vast, promising improvements in crop resilience, nutritional quality, and environmental sustainability. However, challenges remain, including regulatory hurdles, public acceptance, and ethical considerations. Addressing these problems is critical for genome editing technologies' broad acceptance and effectiveness in agriculture.

5.1.8.1 Other Applications of Molecular Markers

With advancements in sequencing technologies and bioinformatics, molecular markers will continue to play a pivotal role in genomics research. Innovations like high-throughput marker discovery and integration with genome-editing tools promise exciting crop and livestock improvement opportunities.

Omics Technologies

- *Transcriptomics*: Studies gene expression under specific conditions, aiding in identifying genes responsible for stress responses
- *Proteomics*: Analyses protein profiles, offering insights into functional traits
- *Metabolomics*: Identifies metabolic pathways involved in traits like flavour or stress tolerance

5.1.8.2 Transcriptomics

Transcriptomics is the study of the entire collection of RNA transcripts generated by the genome under certain conditions or inside a given cell. This field sheds light on gene expression patterns, regulatory networks, and molecular processes underpinning crop attributes. Researchers may use high-throughput RNA sequencing to uncover genes that control critical qualities like yield, quality, and stress tolerance. This understanding facilitates the development of high-yielding, stress-tolerant, and nutritionally superior crop types. There are some examples of such studies that have been given in Table 5.4.

Transcriptomics with its integration of new advanced technologies has revolutionized plant research, and the following are the innovations:

- *High-throughput RNA sequencing*: Allows for the comprehensive analysis of gene expression.
- *Alternative splicing*: Helps in understanding the diversity of proteins produced by a single gene.
- *Gene discovery and regulatory networks*: Identifies new genes and their regulatory mechanisms.
- *Functional characterization*: Determines the roles of specific genes in crop traits.
- *Multi-omics integration*: Combines transcriptomics with other omics approaches for a holistic understanding.
- *Single-cell transcriptomics*: Provides insights into the heterogeneity of cell populations within a tissue.

5.1.8.3 Proteomics

Proteomics is the large-scale study of proteins, focussing on their structural and functional properties. Proteomics is useful in crop improvement because it gives insights into the molecular mechanisms underpinning plant growth, development, and responses to environmental challenges. By examining the whole collection of proteins produced in a plant, researchers identified several essential proteins involved in critical qualities like yield, disease resistance, and stress tolerance in number of crops (Table 5.5). This information may be utilized to create higher performance and resilience crop types.

Some prime advancements in proteomics studies include:

- *High-throughput technologies*: Techniques like mass spectrometry (MS) and two-dimensional (2D) gel electrophoresis allow for the comprehensive analysis of proteins.
- *Post-translational modifications (PTMs)*: Proteomics can reveal PTMs, which are crucial for understanding protein function and regulation.
- *Protein-protein interactions*: Identifying interactions between proteins helps in understanding complex biological processes.
- *Functional characterization*: Determines the roles of specific proteins in crop traits.

Table 5.4 Recent transcriptomic studies in various crops to understand and unravel the mechanism for different traits

Crop species	Trait studied	Objective of study	Methodology used	Key findings	Reference details
Maize	Nitrogen utilization	Investigate transcriptomic responses to nitrogen supplementation in maize hybrids	Plant BRB-seq	Identified differential expression of stress-responsive genes in response to varying nitrogen levels	Ying et al. (2024)
Rice	Polygenic traits	Apply transcriptome-based predictions for polygenic traits in rice using eQTLs	RNA-Seq, eQTL analysis	Improved predictive ability for polygenic traits using transcriptome data	Tanaka et al. (2024)
Cotton	Fibre development	Dissect cell-type-specific functions during fibre development	Single-cell transcriptomics	Revealed genes involved in fibre cell differentiation and elongation	Li et al. (2024)
Peanut	Peg development	Explore gene expression during peg development in peanuts	Spatial transcriptomics	Identified novel genes regulating peg geotropism and development	Liu et al. (2021)
Soybean	Stress responses	Analyse cell-type-specific responses to abiotic stresses	Single-cell transcriptomics	Discovered genes associated with high temperature and salt stress tolerance	Liu et al. (2023)
Cassava	Disease resistance	Investigate gene expression linked to disease resistance in cassava	Spatial transcriptomics	Uncovered candidate resistance genes with spatial expression patterns	Song et al. (2022)
Sorghum	Genes related to grain cell wall biosynthesis	Single-cell transcriptomes to understand trait evolution	Single-cell RNA sequencing	Identified the genes involved in the biosynthesis and remodelling of cell walls both in the endosperm and outer layers of sorghum grain during its development	Costes et al. (2024)
Pearl millet	Drought tolerance	mRNA library from stressed and control was studied to unravel the mechanism behind drought tolerance	RNA-Seq	Provided insights into genes controlling traits like drought tolerance	Dudhate et al. (2018)
Foxtail millet	Salt tolerance	Advanced understanding of gene functions for salt tolerance	RNA-seq	Transcription factor families bHLH, WRKY, AP2/ERF, and MYB-MYC were found to play critical roles in foxtail millet's response to salt stress	Han et al. (2022)
Chilli	Fruit-related traits	Gene expression of genes for fruit development and ripening	Transcriptomic profiling	Assessed the impact of genetic modification compared to environmental and varietal variations	Razo-Mendivil et al. (2021)

Table 5.5 Examples of some studies that have used proteomics in several crops

Crop species	Trait studied	Objective of study	Key findings	Reference details
Rice	Drought tolerance	Investigate the proteomic changes under drought stress conditions	Identified drought-responsive proteins and pathways involved in stress tolerance	Hamzelou et al. (2020)
Wheat	Salt and drought stress	Analyse the proteomic response to salt stress in wheat cultivars	Discovered proteins linked to ion transport and stress defence mechanisms	Peng et al. (2009)
Maize	Root development	Study the proteome of primary root development	Provided high-resolution tissue-specific proteome and phosphoproteome atlas of maize primary roots	Marcon et al. (2015)
Soybean	Nitrogen fixation	Explore the proteins involved in nitrogen fixation in soybean nodules	Key nodulin proteins and enzymes essential for efficient nitrogen fixation were identified	Yao et al. (2022)
Tomato	Fruit ripening	Characterize the proteome during tomato fruit ripening stages	Identified key enzymes involved in ethylene biosynthesis and cell wall modification	Pontiggia et al. (2019)
Barley	Leaf rust	Examine the proteomic response to fungal pathogen infection	Discovered defence-related proteins and pathways activated during infection and proteomic characterization of *Rph15* gene	Bernardo et al. (2012)
Cotton	Fibre development	Investigate proteins involved in fibre elongation and development	Key proteins linked to cell wall biosynthesis and elongation were identified	Zhou et al. (2019)
Potato	Cold stress tolerance	Study the proteomic changes in potatoes under cold stress conditions	Identified proteins related to antioxidant defence and osmoprotection	Li et al. (2021)
Chilli	Selenium stress	Analyse the proteomic response to selenium stresses	Identified common and unique stress-responsive proteins across different stress types	Zhang et al. (2019)
Oil palm	Oil quality	Investigate the proteome associated with oil biosynthesis in oil palm	Identified key enzymes in the fatty acid biosynthesis pathway contributing to oil content	Kok et al. (2021)

- *Integration with other omics approaches*: Combines proteomics with genomics, transcriptomics, and metabolomics for a systems biology approach.

It has wide application in crop improvement, as presented below:

- *Disease resistance*: Proteomics can identify proteins involved in plant defence mechanisms, leading to the development of disease-resistant crop varieties.
- *Stress tolerance*: By understanding the proteins involved in stress responses, researchers can develop crops more tolerant to abiotic stresses such as drought, salinity, and extreme temperatures.
- *Yield improvement*: Proteomics can identify proteins associated with high yield, enabling the development of high-yielding crop varieties.
- *Quality enhancement*: Proteomics can help improve crops' nutritional quality by identifying proteins involved in nutrient biosynthesis and accumulation.

5.1.8.4 Metabolomics

Metabolomics is the study of the entire collection of metabolites (small molecules) found in a biological sample. Plant metabolomics is a rapidly developing field within plant science and systems biology. This requires careful analysis of tiny chemicals (metabolites) found in plant cells and tissues. These metabolites include a wide range of chemicals such as sugars, amino acids, organic acids, secondary metabolites (such as flavonoids and alkaloids), lipids, and other things. Metabolomics can help us understand how individual metabolites affect plant physiological processes, growth, and responses to both biotic and abiotic stressors (Manickam et al. 2023). This can lead to the discovery of metabolites linked with certain features or functions. Metabolomics allows for a better knowledge of plant metabolic networks and pathways. Researchers can determine how plants respond to environmental stimuli or genetic changes by examining variations in metabolite profiles at various stages of growth (Kanagarajan 2023).

Metabolomics plays an important role in crop improvement and biotechnology. The combination of metabolomics and other omics data (genomics, transcriptomics, and proteomics) provides a more comprehensive understanding of plant life. This systems biology method allows researchers to understand the complicated connections inside organisms. This subject contributes to understanding the functional roles of particular metabolites in plant physiology, development, and responses to biotic and abiotic stressors.

Metabolomics has the following features:

- *High-throughput technologies*: Techniques like LC-MS and GC-MS allow for the comprehensive analysis of metabolites.
- *Integration with other omics approaches*: Combines metabolomics with genomics, transcriptomics, and proteomics for a systems biology approach.
- *Identification of metabolic pathways*: Helps understand how plants react to environmental cues or genetic modifications.

- *Trait-specific markers*: Identifies metabolites associated with desirable traits, aiding crop improvement.

Turner et al. (2016) employed metabolomics to investigate *Sorghum bicolor* (L.) in terms of its production of grain (food), stem sugar (food, fuel), and cellulosic biomass (forage, fuel). They examined phenotypically diverse genotypes for leaf metabolites as well as a morpho-physiological study. In a non-targeted metabolomics analysis of leaf tissue, it was found that 956 out of 1181 metabolites (81%) differed among the lines (Analysis of Variance (ANOVA), FDR-adjusted $p < 0.05$). Metabolomics was used to study associations between two or more morphophysiological features, demonstrating that chlorogenic and shikimic acid are associated with photosynthesis, early plant development, and final biomass metrics in sorghum. This work shows how metabolomics may be integrated with morphophysiological data to reveal links between plant metabolism, growth, and architecture. Kang et al. (2019) used metabolomics and physiological approaches to evaluate the various biochemical alterations and physiological modifications in the leaves and roots of two wheat genotypes in response to drought. The findings showed that drought caused a decrease in all physiological and yield-related traits. The findings further suggested that resource allocation was compromised for leaves in the sensitive genotype and was instead directed more towards roots. The outcomes showed that the metabolic composition of roots and shoots differs, with the shoot metabolome exhibiting greater variability than that of the root. This study discovered that shoots' metabolic response to dryness differs from that of roots, with certain growth metabolites (protein, sugar, etc.) showing a mirrored increase in both locations. Protein synthesis and the energy cycle were active in both organs, and the organs were metabolically stimulated to improve water intake and growth, thereby reducing drought consequences.

In the above study demonstrated that transcriptomics, proteomics, and metabolomics combined to give a complete knowledge of the molecular pathways controlling agricultural characteristics. These technologies may apply wisely to speed the creation of high-yielding, stress-tolerant, and nutritionally superior crop varieties, opening the path for sustainable agriculture and food security.

5.2 Conclusion

To meet the needs of a growing world population, crop yields must be increased under the climate change scenario predicted for the coming decades and the threat of the emergence of new pathogens. Various approaches to crop improvement, such as molecular marker-assisted selection and breeding, locating genes and QTL at accurate sites in the genome, and using molecular markers as a single marker at the genome level with GS, have become popular for achieving the goal of a food-secure world.

Prospects of these techniques, including next-generation sequencing, are very promising as these provide more accurate information about the genetic material, i.e. DNA and RNA, and through the given ways, it would come out as a beneficial tool for the pan-genome development: Expanding the genetic reference for crops will enhance breeding precision. Artificial intelligence (AI) integration: AI-driven analytics can optimize breeding strategies by leveraging large datasets from molecular studies. CRISPR advances: Improved genome-editing techniques, such as base editing and prime editing, will allow more refined modifications. Climate-resilient crops: Molecular tools will play a key role in developing crops that can withstand extreme environmental conditions. Genome sequencing and its low cost, the ongoing development of novel molecular markers, the application of high-throughput phenotyping technologies, and speed breeding procedures will enable us to meet these new challenges. These new systems incorporating molecular technology and gene editing tools will be critical to agriculture's future.

References

Acquaah G (2015) Conventional plant breeding principles and techniques. Advances in plant breeding strategies: breeding, biotechnology and molecular tools. Springer, pp 115–158

Amiteye S (2021) Basic concepts and methodologies of DNA marker systems in plant molecular breeding. Heliyon 7(10). https://doi.org/10.1016/j.heliyon.2021.e08093

Arabidopsis Genome Initiative genomeanalysis@ tgr. org genomeanalysis@ gsf. de (2000) Analysis of the genome sequence of the flowering plant *Arabidopsis thaliana*. Nature 408(6814): 796–815

Beckmann JS, Soller M (1986) Restriction fragment length polymorphism in genetic improvement. Oxf Surv Plant Mol Biol 3:197–250

Bernardo L, Prinsi B, Negri AS, Cattivelli L, Espen L, Valè G (2012) Proteomic characterization of the Rph15 barley resistance gene-mediated defence responses to leaf rust. BMC Genomics 13: 1–16

Beyene Y, Gowda M, Olsen M, Robbins KR, Pérez-Rodríguez P, Alvarado G, Dreher K, Gao SY, Mugo S, Prasanna BM, Crossa J. (2019) Empirical comparison of tropical maize hybrids selected through genomic and phenotypic selections. Frontiers in plant science. Nov 22;10:1502

Bidyananda N, Jamir I, NowakowskaK VV, Vendrame WA, Devi RS, Nongdam P (2024) Plant genetic diversity studies: insights from DNA marker analyses. Int J Plant Biol 15(3):607–640

Botstein D, White RL, Skolnick M, Davis RW (1980) Construction of a genetic linkage map in man using restriction fragment length polymorphisms. Am J Hum Genet 32(3):314

Bouis HE, Saltzman A (2017) Improving nutrition through biofortification: a review of evidence from HarvestPlus, 2003 through 2016. Glob Food Sec 12:49–58

Broman KW, Sen S (2009) A Guide to QTL Mapping with R/qtl, vol 46. Springer, New York

Cao Y, Li S, Wang Z, Chang F, Kong J, Gai J, Zhao T (2017) Identification of major quantitative trait loci for seed oil content in soybeans by combining linkage and genome-wide association mapping. Front Plant Sci 8:1222

Carrillo MGC, Martin F, Variar M, Bhatt JC, Perez-Quintero AL, Leung H, Leach JE, Vera Cruz CM (2021) Accumulating candidate genes for broad-spectrum resistance to rice blast in a drought-tolerant rice cultivar. Sci Rep 11(1):21502

Ceccarelli S (2012) Plant breeding with farmers. A technical manual. ICARDA, Aleppo

Chapman B, Mondragón-Jacobo C, Bunch RA, Paterson AH (2002) Breeding and biotechnology. Cacti: biology and uses, pp 255–271

Chukwu SC, Rafii MY, Ramlee SI, Ismail SI, Oladosu Y, Muhammad II, Musa I, Ahmed M, Jatto MI, Yusuf BR (2020) Recovery of recurrent parent genome in a marker-assisted backcrossing against rice blast and blight infections using functional markers and SSRs. Plants 9(11):1411

Collard BC, Jahufer MZZ, Brouwer JB, Pang ECK (2005) An introduction to markers, quantitative trait loci (QTL) mapping and marker-assisted selection for crop improvement: the basic concepts. Euphytica 142:169–196

Complex Trait Consortium (2003) The nature and identification of quantitative trait loci: a community's view. Nat Rev Genet 4(11):911–916

Costa R, Pereira G, Garrido I, Tavares-de-Sousa MM, Espinosa F (2016) Comparison of RAPD, ISSR, and AFLP molecular markers to reveal and classify orchardgrass (Dactylis glomerata L.) germplasm variations. PLoS One 11(4):e0152972

Costes C, Navarro Sanz S, Calatayud C, Soriano A, Mameri H, Terrier N, Francin-Allami M (2024) Transcriptomic analysis of developing sorghum grains to detect genes related to cell wall biosynthesis and remodelling. BMC Genomic Data 25(1):14

Cui J, Luo J, Van Der Werf W, Ma Y, Xia J (2011) Effect of pyramiding Bt and CpTI genes on resistance of cotton to *Helicoverpaarmigera* (Lepidoptera: Noctuidae) under laboratory and field conditions. J Econ Entomol 104(2):673–684

D'Ambrosio C, Stigliani AL, Giorio G (2018) CRISPR/Cas9 editing of carotenoid genes in tomato. Transgenic Res 27(4):367–378

Devate NB, Krishna H, Parmeshwarappa SKV, Manjunath KK, Chauhan D, Singh S, Singh JB, Kumar M, Patil R, Khan H, Jain N (2022) Genome-wide association mapping for component traits of drought and heat tolerance in wheat. Front Plant Sci *13*:943033

Diouf M, Zoclanclounon YAB, Mboup PA, Diouf D, Malédon E, Rivallan R, Chair H, Dossa K (2024) Genome-wide development of intra-and inter-specific transferable SSR markers and construction of a dynamic web resource for yam molecular breeding: Y2MD. Plant Genome *17*(2):e20428

Dormatey R, Sun C, Ali K, Coulter JA, Bi Z, Bai J (2020) Gene pyramiding for sustainable crop improvement against biotic and abiotic stresses. Agronomy 10(9):1255

Dudhate A, Shinde H, Tsugama D, Liu S, Takano T (2018) Transcriptomic analysis reveals the differentially expressed genes and pathways involved in drought tolerance in pearl millet [*Pennisetum glaucum* (L.) R. Br]. PLoS One 13(4):e0195908

Dwiningsih Y, Thomas J, Kumar A, Gupta C, Gill N, Ruiz C, Alkahtani J, Baisakh N, Pereira A (2023) QTLs and candidate loci associated with drought tolerance traits of kaybonnet × ZHE733 recombinant inbred lines rice population. Int J Mol Sci 24(20):15167

Evenson RE, Gollin D (2003) Assessing the impact of the Green Revolution, 1960 to 2000. Science 300(5620):758–762

Fan C, Zhai H, Wang H, Yue Y, Zhang M, Li J, Wen S, Guo G, Zeng Y, Ni Z, You M (2017) Identification of QTLs controlling grain protein concentration using a high-density SNP and SSR linkage map in barley (Hordeum vulgare L.). BMC Plant Biol 17:1–14

Gautam T, Dhillon GS, Saripalli G, Singh R, Prasad P, Kaur S, Chhuneja P, Sharma PK, Balyan HS, Gupta PK (2020) Marker-assisted pyramiding of genes/QTL for grain quality and rust resistance in wheat (Triticum aestivum L.). Mol Breed 40:1–14

Gebhardt C (2007) Molecular markers, maps and population genetics. In: Potato biology and biotechnology. Elsevier Science BV, pp 77–89

Goff SA, Ricke D, Lan TH, Presting G, Wang R, Dunn M, Glazebrook J, Sessions A, Oeller P, Varma H, Hadley D (2002) A draft sequence of the rice genome (*Oryza sativa* L. ssp. *japonica*). Science 296(5565):92–100

Hamzelou S, Pascovici D, Kamath KS, Amirkhani A, McKay M, Mirzaei M, Atwell BJ, Haynes PA (2020) Proteomic responses to drought vary widely among eight diverse genotypes of rice (*Oryza sativa*). Int J Mol Sci 21(1):363

Han F, Sun M, He W, Guo S, Feng J, Wang H, Yang Q, Pan H, Lou Y, Zhuge Y (2022) Transcriptome analysis reveals molecular mechanisms under salt stress in leaves of foxtail millet (*Setariaitalica* L.). Plants 11(14):1864

Hasan MM, Rafii MY, Ismail MR, Mahmood M, Rahim HA, Alam MA, Ashkani S, Malek MA, Latif MA (2015) Marker-assisted backcrossing: a useful method for rice improvement. Biotechnol Biotechnol Equip 29(2):237–254

Hayat U, Ke C, Wang L, Zhu G, Fang W, Wang X, Chen C, Li Y, Wu J (2025) Using quantitative trait locus mapping and genomic resources to improve breeding precision in peaches: current insights and future prospects. Plants 14(2):175

He J, Meng S, Zhao T, Xing G, Yang S, Li Y, Guan R, Lu J, Wang Y, Xia Q, Yang B (2017) An innovative procedure of genome-wide association analysis fits studies on germplasm population and plant breeding. Theor Appl Genet 130:2327–2343

He L, Sui Y, Che Y, Wang H, Rashid KY, Cloutier S, You FM (2023) Genome-wide association studies using multi-models and multi-SNP datasets provide new insights into pasmo resistance in flax. Front Plant Sci 14:1229457

Hickey LT, N. Hafeez A, Robinson H, Jackson SA, Leal-Bertioli SC, Tester M, Gao C, Godwin ID, Hayes BJ, Wulff BB (2019) Breeding crops to feed 10 billion. Nat Biotechnol 37:744–754. https://doi.org/10.1038/s41587-019-0152-9

Hill WG (2010) Understanding and using quantitative genetic variation. Philos Transac Roy Soc B Biol Sci *365*(1537):73–85

Hsu YC, Chiu CH, Yap R, Tseng YC, Wu YP (2020) Pyramiding bacterial blight resistance genes in Tainung82 for broad-spectrum resistance using marker-assisted selection. Int J Mol Sci 21(4): 1281

Hu X, Wang G, Du X, Zhang H, Xu Z, Wang J, Chen G, Wang B, Li X, Chen X, Fu J (2021) QTL analysis across multiple environments reveals promising chromosome regions associated with yield-related traits in maize under drought conditions. Crop J 9(4):759–766

Huang N, Angeles E.R, Domingo J, Magpantay G, Singh S, Zhang G, Kumaravadivel N, Bennett J, Khush G.S. (1997) Pyramiding of bacterial blight resistance genes in rice: marker-assisted selection using RFLP and PCR. Theor. Appl. Genet. 95:313–320. https://doi.org/10.1007/s001220050565

Huang X, Zhao Y, Wei X, Li C, Wang A, Zhao Q, Li W, Guo Y, Deng L, Zhu C, Fan D (2012) Genome-wide association study of flowering time and grain yield traits in a worldwide collection of rice germplasm. Nat Genet 44(1):32–39

Jiang, Haichao, Yutao Feng, Liang Bao, Xin Li, Guanjun Gao, Qinglu Zhang, Jinghua Xiao, Caiguo Xu, and Yuqing He. Improving blast resistance of Jin 23B and its hybrid rice by marker-assisted gene pyramiding Mol Breed 30 (2012): 1679–1688.

Juliana, P., Montesinos-López, O.A., Crossa, J. et al. Integrating genomic-enabled prediction and high-throughput phenotyping in breeding for climate-resilient bread wheat. Theor Appl Genet 132, 177–194 (2019). https://doi.org/10.1007/s00122-018-3206-3

Kanagarajan S (2023) Plant metabolomics: current initiatives and future prospects. https://agris.fao.org/search/en/providers/122495/records/65a94da04ddf7c0cd7c9dd11

Kang Z, Babar MA, Khan N, Guo J, Khan J, Islam S, Shrestha S, Shahi D (2019) Comparative metabolomic profiling in the roots and leaves in contrasting genotypes reveals complex mechanisms involved in post-anthesis drought tolerance in wheat. PLoS One 14(3):e0213502

Khan H, Krishnappa G, Kumar S, Devate NB, Rathan ND, Kumar S, Mishra CN, Ram S, Tiwari R, Parkash O, Ahlawat OP (2024) Genome-wide association study identifies novel loci and candidate genes for rust resistance in wheat (Triticum aestivum L.). BMC Plant Biol 24(1):411

Kim D, Alptekin B, Budak H (2018) CRISPR/Cas9 genome editing in wheat. Funct Integr Genom 18:31–41

Kimbara J, Ohyama A, Chikano H, Ito H, Hosoi K, Negoro S, Miyatake K, Yamaguchi H, Nunome T, Fukuoka H, Hayashi T (2018) QTL mapping of fruit nutritional and flavor components in tomato (*Solanum lycopersicum*) using genome-wide SSR markers and recombinant inbred lines (RILs) from an intra-specific cross. Euphytica 214:1–12

Koebner RM, Summers RW (2003) 21st century wheat breeding: plot selection or plate detection? Trends Biotechnol 21(2):59–63

Kok SY, Namasivayam P, Ee GCL, Ong-Abdullah M (2021) Comparative proteomic analysis of oil palm (*Elaeisguineensis* Jacq.) during early fruit development. J Proteome 232:104052

Kumar NS, Gurusubramanian G (2011) Random amplified polymorphic DNA (RAPD) markers and its applications. Sci Vis 11(3):116–124

Kushanov FN, Turaev OS, Ernazarova DK, Gapparov BM, Oripova BB, Kudratova MK, Rafieva FU, Khalikov KK, Erjigitov DS, Khidirov MT, Kholova MD (2021) Genetic diversity, QTL mapping, and marker-assisted selection technology in cotton (Gossypium spp.). Front Plant Sci 12:779386

Lander ES, Botstein S (1989) Mapping mendelian factors underlying quantitative traits using RFLP linkage maps. Genetics 121(1):185–199. https://doi.org/10.1093/genetics/121.1.185

Li H, Luo W, Ji R, Xu Y, Xu G, Qiu S, Tang H (2021) A comparative proteomic study of cold responses in potato leaves. Heliyon 7(2)

Li L, Zhu Y, Jin S, Zhang X (2014) Pyramiding Bt genes for increasing resistance of cotton to two major lepidopteran pests: Spodoptera litura and Heliothisarmigera. Acta Physiol Plant 36:2717–2727

Li PT, Liu QK, Wei YY, Xing CZ, Xu ZP, Ding F, Liu YL, Lu QW, Hu N, Wang T et al (2024) Transcriptional landscape of cotton roots in response to salt stress at single-cell resolution. Plant Commun 5(100):740

Li S, Cao Y, Wang C, Yan C, Sun X, Zhang L, Wang W, Song S (2023) Genome-wide association mapping for yield-related traits in soybean (Glycine max) under well-watered and drought-stressed conditions. Front Plant Sci 14:1265574

Li T, Li Q, Wang J, Yang Z, Tang Y, Su Y, Zhang J, Qiu X, Pu X, Pan Z, Zhang H (2022) High-resolution detection of quantitative trait loci for seven important yield-related traits in wheat (Triticum aestivum L.) using a high-density SLAF-seq genetic map. BMC Genomic Data 23(1): 37

Lian Q, Fu Q, Xu Y, Hu Z, Zheng J, Zhang A, He Y, Wang C, Xu C, Chen B, Garcia-Mas J (2021) QTLs and candidate genes analyses for fruit size under domestication and differentiation in melon (Cucumis melo L.) based on high resolution maps. BMC Plant Biol 21:1–13

Liu D, Ye Y, Tang R, Gong Y, Chen S, Zhang C, Mei P, Chen J, Chen L, Ma C (2024) High-density genetic map construction and QTL mapping of a zigzag-shaped stem trait in tea plant (*Camellia sinensis*). BMC Plant Biol 24(1):382

Liu H, Hu D, Du P, Wang L, Liang X, Li H, Lu Q, Li S, Liu H, Chen X, Varshney RK (2021) Single-cell RNA-seq describes the transcriptome landscape and identifies critical transcription factors in the leaf blade of the allotetraploid peanut (*Arachis hypogaea* L.). Plant Biotechnol J 19(11):2261–2276

Liu R, Lu J, Zhou M, Zheng S, Liu Z, Zhang C, Du M, Wang M, Li Y, Wu Y, Zhang L (2020) Developing stripe rust resistant wheat (*Triticum aestivum* L.) lines with gene pyramiding strategy and marker-assisted selection. Genet Resour Crop Evol 67:381–391

Liu Z, Kong X, Long Y, Liu S, Zhang H, Jia J, Cui W, Zhang Z, Song X, Qiu L, Zhai J (2023) Integrated single-nucleus and spatial transcriptomics captures transitional states in soybean nodule maturation. Nat Plant 9(4):515–524

Mallikarjuna-Swamy BP, Marundan S Jr, Samia M, Ordonio RL, Rebong DB, Miranda R, Alibuyog A, Rebong AT, Tabil MA, Suralta RR, Alfonso AA (2021) Development and characterization of GR2E Golden rice introgression lines. Sci Rep 11(1):2496. [online]

Manickam S, Rajagopalan VR, Kambale R, Rajasekaran R, Kanagarajan S, Muthurajan R (2023) Plant metabolomics: current initiatives and future prospects. Curr Iss Mol Biol 45(11): 8894–8906

Manjunath KK, Krishna H, Devate NB, Sunilkumar VP, Patil SP, Chauhan D, Singh S, Kumar S, Jain N, Singh GP, Singh PK (2024) QTL mapping: insights into genomic regions governing component traits of yield under combined heat and drought stress in wheat. Front Genet 14: 1282240

Marcon C, Malik WA, Walley JW, Shen Z, Paschold A, Smith LG, Piepho HP, Briggs SP, Hochholdinger F (2015) A high-resolution tissue-specific proteome and phosphoproteome atlas of maize primary roots reveals functional gradients along the root axes. Plant Physiol 168(1):233–246

Miao Y, Jing F, Ma J, Liu Y, Zhang P, Chen T, Che Z, Yang D (2022) Major genomic regions for wheat grain weight as revealed by QTL linkage mapping and meta-analysis. Front Plant Sci 13: 802310

Mohanavel V, Muthu V, Kambale R, Palaniswamy R, Seeli P, Ayyenar B, Rajagopalan V, Manickam S, Rajasekaran R, Rahman H, Nallathambi J (2024) Marker-assisted breeding accelerates the development of multiple-stress-tolerant rice genotypes adapted to wider environments. Front Plant Sci 15:1402368

Mondini L, Noorani A, Pagnotta MA (2009) Assessing plant genetic diversity by molecular tools. Diversity 1(1):19–35

Nantawan U, Kanchana-Udomkan C, Bar I, Ford R (2019) Linkage mapping and quantitative trait loci analysis of sweetness and other fruit quality traits in papaya. BMC Plant Biol 19:1–11

Nekrasov V, Wang C, Win J, Lanz C, Weigel D, Kamoun S (2017) Rapid generation of a transgene-free powdery mildew resistant tomato by genome deletion. Sci Rep 7(1):482

Paine JA, Shipton CA, Chaggar S, Howells RM, Kennedy MJ, Vernon G, Wright SY, Hinchliffe E, Adams JL, Silverstone AL, Drake R (2005) Improving the nutritional value of Golden Rice through increased pro-vitamin A content. Nat Biotechnol 23(4):482–487

Panahi B, Jalaly HM, Hamid R (2024) Using the next-generation sequencing approach for the discovery and characterization of molecular plant markers. Current. Plant Biol:100412

Parkinson J, Blaxter M (2009) Expressed sequence tags: an overview. Expressed sequence tags (ESTs) generation and analysis. Methods Mol Biol 533:1–12

Paterson AH, Lander ES, Hewitt JD, Peterson S, Lincoln SE, Tanksley SD (1988) Resolution of quantitative traits into Mendelian factors by using a complete linkage map of restriction fragment length polymorphisms. Nature 335(6192):721–726

Paun O, Schönswetter P (2012) Amplified fragment length polymorphism: an invaluable fingerprinting technique for genomic, transcriptomic, and epigenetic studies. Plant DNA fingerprinting and barcoding. Method Mol Biol 862:75–87

Peng Z, Wang M, Li F, Lv H, Li C, Xia G (2009) A proteomic study of the response to salinity and drought stress in an introgression strain of bread wheat. Mol Cell Proteomics 8(12):2676–2686

Pontiggia D, Spinelli F, Fabbri C, Licursi V, Negri R, De Lorenzo G, Mattei B (2019) Changes in the microsomal proteome of tomato fruit during ripening. Sci Rep 9(1):14350

Qian F, Jing J, Zhang Z, Chen S, Sang Z, Li W (2023) GWAS and Meta-QTL Analysis of yield-related ear traits in Maize. Plants 12(22):3806

Rafalski A (2002) Applications of single nucleotide polymorphisms in crop genetics. Curr Opin Plant Biol 5(2):94–100

Rasheed A, Jie H, Ali B, He P, Zhao L, Ma Y, Xing H, Qari SH, Hassan MU, Hamid MR, Jie Y (2023) Breeding drought-tolerant maize (*Zea mays*) using molecular breeding tools: recent advancements and future prospective. Agronomy 13(6):1459

Razo-Mendivil FG, Hernandez-Godínez F, Hayano-Kanashiro C, Martínez O (2021) Transcriptomic analysis of a wild and a cultivated varieties of Capsicum annuum over fruit development and ripening. PLoS One 16(8):e0256319

Ribaut JM, Hoisington D (1998) Marker-assisted selection: new tools and strategies. Trends Plant Sci 3(6):236–239

Rollar S, Serfling A, Geyer M, Hartl L, Mohler V, Ordon F (2021) QTL mapping of adult plant and seedling resistance to leaf rust (*Puccinia triticina* Eriks.) in a multiparent advanced generation intercross (MAGIC) wheat population. Theor Appl Genet 134:37–51

Roychowdhury R, Taoutaou A, Hakeem KR, Gawwad MRA, Tah J (2014) Molecular marker-assisted technologies for crop improvement. Crop improvement in the era of climate change. IK International Publishing House, New Delhi, pp 241–258

Saleem MA, Malik W, Ahmad MQ, Arshad SF, Baig MMA, Asif M, Nauman M, Anwar M (2024) Gene pyramiding improved cell membrane stability under heat stress in cotton (Gossypium hirsutum L.). BMC Plant Biol 24(1):886

Sánchez-León S, Marín-Sanz M, Guzmán-López MH, Gavilán-Camacho M, Simón E, Barro F (2024) CRISPR/Cas9-mediated multiplex gene editing of gamma and omega gliadins: paving the way for gliadin-free wheat. J Exp Bot *75*(22):7079–7095

Sarkar B, Varalaxmi Y, Vanaja M, RaviKumar N, Prabhakar M, Yadav SK, Maheswari M, Singh VK (2023) Mapping of QTLs for morphophysiological and yield traits under water-deficit stress and well-watered conditions in maize. Front Plant Sci 14:1124619

Sax K (1923) The association of size differences with seed-coat pattern and pigmentation in *Phaseolus vulgaris*. Genetics 8:552–560

Semagn K, Bjørnstad Å, Ndjiondjop MN (2006) Principles, requirements and prospects of genetic mapping in plants. Afr J Biotechnol 5(25)

Serajazari M, Torkamaneh D, Gordon E, Lee E, Booker H, Pauls KP, Navabi A (2023) Identification of fusarium head blight resistance markers in a genome-wide association study of CIMMYT spring synthetic hexaploid derived wheat lines. BMC Plant Biol 23(1):290

Shaibu AS, Li B, Zhang S, Sun J (2020) Soybean cyst nematode-resistance: gene identification and breeding strategies. Crop J *8*(6):892–904

Singh A, Dikshit HK, Ramawat N, Kumar S (2021) QTL mapping for abiotic stress in legumes. In: Abiotic stress and legumes. Academic Press, pp 337–370

Singh VK, Singh A, Singh SP, Ellur RK, Singh D, Gopala Krishnan S, Bhowmick PK, Nagarajan M, Vinod KK, Singh UD, Mohapatra T (2013) Marker-assisted simultaneous but stepwise backcross breeding for pyramiding blast resistance genes Piz5 and Pi54 into an elite Basmati rice restorer line 'PRR 78'. Plant Breed 132(5):486–495

Soller M, Plotkin-Hazan J (1977) The use marker alleles for the introgression of linked quantitative alleles. Theor Appl Genet 51:133–137

Song J, Fan B, Shao X, Zang Y, Wang D, Min Y (2022) Single-cell transcriptome sequencing atlas of cassava tuberous root. Front Plant Sci 13:1053669

Srivastav M, Radadiya N, Ramachandra S, Jayaswal PK, Singh N, Singh S, Mahato AK, Tandon G, Gupta A, Devi R, Subrayagowda SH (2023) High resolution mapping of QTLs for fruit color and firmness in Amrapali/Sensation mango hybrids. Front Plant Sci 14:1135285

Sukumaran S, Reynolds MP, Sansaloni C (2018) Genome-wide association analyses identify QTL hotspots for yield and component traits in durum wheat grown under yield potential, drought, and heat stress environments. Front Plant Sci 9:81

Svitashev S, Schwartz C, Lenderts B, Young JK, Mark Cigan A (2016) Genome editing in maize directed by CRISPR–Cas9 ribonucleoprotein complexes. Nat Commun 7(1):1–7

Talbert LE, Blake NK, Chee PA, Blake TK, Magyar GM (1994) Evaluation of "sequence-tagged-site" PCR products as molecular markers in wheat. Theor Appl Genet 87:789–794

Tanaka R, Kawai T, Kawakatsu T, Tanaka N, Shenton M, Yabe S, Uga Y (2024) Transcriptome-based prediction for polygenic traits in rice using different gene subsets. BMC Genomics 25(1): 915

Tanksley SD, McCouch SR (1997) Seed banks and molecular maps: unlocking genetic potential from the wild. Science 277(5329):1063–1066

Thiruppathi A, Salunkhe SR, Ramasamy SP, Palaniswamy R, Rajagopalan VR, Rathnasamy SA, Alagarswamy S, Swaminathan M, Manickam S, Muthurajan R (2024) Unleashing the Potential of CRISPR/Cas9 Genome Editing for Yield-Related Traits in Rice. Plants 13(21):2972

Tong J, Zhao C, Liu D, Jambuthenne DT, Sun M, Dinglasan E, Periyannan SK, Hickey LT, Hayes BJ (2024) Genome-wide atlas of rust resistance loci in wheat. Theor Appl Genet 137(8):179

Turner MF, Heuberger AL, Kirkwood JS, Collins CC, Wolfrum EJ, Broeckling CD, Prenni JE, Jahn CE (2016) Non-targeted metabolomics in diverse sorghum breeding lines indicates primary and secondary metabolite profiles are associated with plant biomass accumulation and photosynthesis. Front Plant Sci 7:953

Ueno S, Moriguchi Y, Uchiyama K, Ujino-Ihara T, Futamura N, Sakurai T, Shinohara K, Tsumura Y (2012) A second-generation framework for the analysis of microsatellites in expressed sequence tags and the development of EST-SSR markers for a conifer, Cryptomeria japonica. BMC Genomics 13:1–16

Uffelmann E, Huang QQ, Munung NS, De Vries J, Okada Y, Martin AR, Martin HC, Lappalainen T, Posthuma D (2021) Genome-wide association studies. Nat Rev Methods Primer 1(1):59

Vishwakarma MK, Arun B, Mishra VK et al (2016) Marker-assisted improvement of grain protein content and grain weight in Indian bread wheat. Euphytica 208:313–321. https://doi.org/10.1007/s10681-015-1598-6

Vishwakarma MK, Mishra VK, Gupta PK et al (2014) Introgression of the high grain protein gene Gpc-B1 in an elite wheat variety of Indo-Gangetic Plains through marker assisted backcross breeding. Curr Plant Biol 1:60–67

Vishwakarma, M.K., Yadav, P.S., Rai, V.P., Kumar, U. and Joshi, A.K., 2022. Molecular markers and genomics assisted breeding for improving crop plants. In *Relationship Between Microbes and the Environment for Sustainable Ecosystem Services, Volume 1* (pp. 303–334). Elsevier.

Vishwakarma MK, Bhati PK, Kumar U, Singh RP, Kumar S, Govindan V, Mavi GS, Thiyagarajan K, Dhar N, Joshi AK (2024) Genetic dissection of value-added quality traits and agronomic parameters through genome-wide association mapping in bread wheat (T. aestivum L.). Front Plant Sci 15:1419227. https://doi.org/10.3389/fpls.2024.1419227

Visscher PM, Goddard ME (2019) From RA Fisher's 1918 paper to GWAS a century later. Genetics 211(4):1125–1130

Wang F, Zhang M, Hu Y, Gan M, Jiang B, Hao M, Ning S, Yuan Z, Chen X, Chen X, Zhang L (2023) Pyramiding of adult-plant resistance genes enhances all-stage resistance to wheat stripe rust. Plant Dis 107(3):879–885

Wang X, Pang Y, Zhang J, Zhang Q, Tao Y, Feng B, Zheng T, Xu J, Li Z (2014) Genetic background effects on QTL and QTL× environment interaction for yield and its component traits as revealed by reciprocal introgression lines in rice. Crop J 2(6):345–357

Wang X, Wang H, Liu S, Ferjani A, Li J, Yan J, Yang X, Qin F (2016) Genetic variation in ZmVPP1 contributes to drought tolerance in maize seedlings. Nat Genet 48(10):1233–1241

Wu Y, Zhou Z, Dong C, Chen J, Ding J, Zhang X, Mu C, Chen Y, Li X, Li H, Han Y (2020) Linkage mapping and genome-wide association study reveals conservative QTL and candidate genes for Fusarium rot resistance in maize. BMC Genomics 21:1–11

Xiong W, Reynolds M, Xu Y (2022) Climate change challenges plant breeding. Curr Opin Plant Biol 70:102308

Xu Y, Crouch JH (2008) Marker-assisted selection in plant breeding: from publications to practice. Crop Sci 48(2):391–407

Xu Y, Ma K, Zhao Y, Wang X, Zhou K, Yu G, Li C, Li P, Yang Z, Xu C (2021) Genomic selection: a breakthrough technology in rice breeding. Crop J 9(3):669–677

Yadav PS, Mishra VK, Arun B, Chand R, Vishwakarma MK, Vasistha NK, Mishra AN, Kalappanavar IK, Joshi AK (2015) Enhanced resistance in wheat against stem rust achieved by marker assisted backcrossing involving three independent Sr genes. Curr Plant Biol 2:25–33

Yadav S, Sandhu N, Singh VK, Catolos M, Kumar A (2019) Genotyping-by-sequencing based QTL mapping for rice grain yield under reproductive stage drought stress tolerance. Sci Rep 9(1):14326

Yao Y, Yuan H, Wu G, Ma C, Gong Z (2022) Proteome analysis of the soybean nodule phosphorus response mechanism and characterization of stress-induced ribosome structural and protein expression changes. Front Plant Sci 13:908889

Ying S, Webster B, Gomez-Cano L, Shivaiah KK, Wang Q, Newton L, Grotewold E, Thompson A, Lundquist PK (2024) Multiscale physiological responses to nitrogen supplementation of maize hybrids. Plant Physiol 195(1):879–899

Zeng X, Luo Y, Vu NTQ, Shen S, Xia K, Zhang M (2020) CRISPR/Cas9-mediated mutation of OsSWEET14 in rice cv. Zhonghua11 confers resistance to *Xanthomonas oryzae* pv. oryzae without yield penalty. BMC Plant Biol 20:1–11

Zeng ZB (1994) Precision mapping of quantitative trait loci. Genetics 136(4):1457–1468

Zhang C, Xu B, Geng W, Shen Y, Xuan D, Lai Q, Shen C, Jin C, Yu C (2019) Comparative proteomic analysis of pepper (Capsicum annuum L.) seedlings under selenium stress. PeerJ 7: e8020

Zhao Y, Liu Z, Duan F, An X, Liu X, Hao D, Gu R, Wang Z, Chen F, Yuan L (2018) Overexpression of the maize ZmAMT1;1a gene enhances root ammonium uptake efficiency under low ammonium nutrition. Plant Biotechnol Rep 12:47–56. https://doi.org/10.1007/s11816-018-0471-1

Zhou X, Hu W, Li B, Yang Y, Zhang Y, Thow K, Fan L, Qu Y (2019) Proteomic profiling of cotton fiber developmental transition from cell elongation to secondary wall deposition. Acta Biochim Biophys Sin 51(11):1168–1177

CRISPR-Cas and RNAi in Crop Engineering: Silencing Target Genes for Enhanced Traits

6

Anu Naruka, Ashish H. Bankar, Walunjkar Babasaheb Changdeo, Narayan Pandit Gurav, Anurag Malik, Himani Punia, and Svetla Sofkova-Bobcheva

Abstract

The Clustered Regularly Interspaced Short Palindromic Repeats (CRISPR)-Cas system is a powerful tool for modifying crop genome that has recently attracted much attention due to the ability to modify crops' genomes with high precision and specificity. This system based on bacterial immune mechanisms enables meaningful editing of genomes of crop plants, which opens up new possibilities to increase yield, tolerance to stress and diseases. Additional to RNA interference (RNAi) to suppress unwanted target genes, CRISPR-Cas has significantly extended the field of molecular breeding and biotechnology. Instead, new solutions have been developed that expand the use of CRISPR: base and prime editing; these do not work with double-strand breaks, so there is no need for homologous recombination and less likelihood of escaping the site of action to harm a new location. Previously, Cas systems are used for DNA targeting only

A. Naruka · W. B. Changdeo
Shri Vaishnav Institute of Agriculture, SVVV (Shri Vaishnav Vidyapeeth Vishwavidyalaya), Indore, India

A. H. Bankar · N. P. Gurav
Department of Agricultural Botany, SVKMs College of Agriculture, Tardi, Shirpur, Dhule, Maharashtra (Affiliated to MPKV, Rahuri), India

A. Malik
Division of Research and Innovation, Uttaranchal University, Dehradun, Uttarakhand, India

H. Punia (✉)
Saraswati Group of Colleges, SGC Group, Gharuan, Mohali, Punjab, India
e-mail: puniahimani@hau.ac.in

S. Sofkova-Bobcheva
School of Agriculture and Environment, Massey University, Palmerston North, New Zealand
e-mail: s.sofkova-bobcheva@massey.ac.nz

Y. Ahlawat et al. (eds.), *Next-Generation Strategies for Crop Improvement*,
https://doi.org/10.1007/978-981-95-0309-4_6

while now for RNA targeting Cas12 and Cas13 are used for multiple used in pathogen defense. Furthermore, gene drives and epigenetic approaches using CRISPR open up new directions in work on stable durable even transgenerational crop improvement. In RNA interference process, small interfering RNAs (siRNAs) and microRNAs (miRNAs) have been used for maintaining gene expression controlling biotic and abiotic stresses. The combination of CRISPR and RNAi thus enhances multiple features of interest including silencing of genes responsible for susceptibility to diseases, and at the same time amplifying positive attributes including nutritional value and efficiency of growth. Use of CRISPR-Cas technology in crop improvement has broad implications for dealing with global issues as those of food shortage as well as climate change. Efficient if stacking systems are available through multiplexed editing, high-throughput screens and synthetic biology interfaces CRISPR, which is important for next-generation crops. As regulatory approaches change, it is apparent that CRISPR-edited crops will be a key part of feeding the world's population further as well as meeting the challenge of maintaining agricultural production in the light of climate variability.

Keywords

CRISPR-Cas · siRNAs · miRNAs · RNAi · Genetically modified organisms

6.1 Introduction

6.1.1 Overview of Genetic Modification in Agriculture

Similarly, in agriculture, genetic modification also referred to as genetic engineering is a great leap forward in human ability to breed crops with desirable attributes. The history of genetic manipulation has a breakthrough in the early twentieth century because of the discovery of the DNA structure by James Watson and Francis Crick (Abdul Aziz et al. 2022). But the point for major advancement of agricultural biotechnology was not marked until the late 1970s and early 1980s with the development of recombinant DNA technology (Khan et al. 2016). Prospection is now possible in terms or engineering an organism consciously changing its genetics by techniques making it genetically modified organisms (GMO) to possess certain traits (Health NRC (US) C on I and AUE of GEF on H 2004a).

One of the first plants, genetically engineered, was a tobacco plant which was created in 1983, and it contained antibiotic resistance (Raman 2017). The completion of this accomplishment paved the way for further growths in crop development as this provided concrete proof that it is possible and can be done to alter the DNA of plants with a view of adding new traits to it (Altpeter et al. 2016). The process of biotechnology in food crops was initiated in the middle of the 1990s with genetically modified insect- and herbicide-resistant maize and soybean (Lucht 2015). These crops produce more yield and require less chemical application since these crops were developed to be resistant to insects and to withstand certain herbicides (Ofosu et al. 2023). When the *Flavr Savr* tomato, the first commercial success GM crop, became available for sale in US supermarkets in 1994, it marked the start of the new generation of crops that can be genetically tailored to possess improved nutritional values and antibiotic qualities to diseases, pest and environmental adversities (National Academies of Sciences E and M, Studies D on E and L, Resources B on A and N, Prospects C on GECPE and F 2016). These brought about the new age in agriculture that enhanced agricultural improvement with special help of biotechnology.

Indeed, it will be pertinent to say that the technique of genetic modification has perhaps the greatest importance in the improvement of agriculture. Several food needs are expected to rise significantly with the population growth. It is estimated that by the year 2050, the world's population will be at about 10 billion (Oliver 2014). There is such a need that might not be satisfied with the conventional methods of agriculture applied in modern society as there are problems like soil erosion, lack of water, and climate change (Rhodes 2014). The solution comes in the form of genetic modification whereby one is able to produce more elastic and productive crop sorts. To address problems concerning portion control or nutritional value, certain plant varieties can be engineered to withstand drought, have natural insect repellents that are pesticide-free and be nutritionally enriched (Parmar et al. 2017). In addition, using this sort of breeding genetic manipulation, it has become possible to introduce features which would otherwise be difficult or even impossible to breed naturally (Health NRC (US) C on I and AUE of GEF on H 2004b). Pest-resistant crops, for example, have been developed genetically and assist farmers control weed incursion without harming the crop in question. This has also favored the use of better agricultural practices such as reduced use of tillage that is known to lead to soil erosion and degradation (Gage et al. 2019).

6.1.2 Introduction to CRISPR-Cas System

The CRISPR-Cas system whose full meaning is clustered regularly interspaced short palindromic repeats and CRISPR-associated proteins has altered the area of genetic engineering since its discovery (Shmakova et al. 2022). As for the history of CRISPR, it is necessary to mention that it may originate from the 1980s when Japanese scientists, including Yoshizumi Ishino, studied another phenomenon when they noticed some repeating DNA sequences in the genome of *Escherichia coli* (Ishino et al. 2018). These sequences were previously only known as mere genetic relics, or as junk DNA, until their vital potential was unveiled in the early period of the present century (Heather and Chain 2016). These repeating sequences were established to be a part of a bacterial immune system in 2005 by three teams, of which Mojica at the University of Alicante and Jansen at Utrecht University were a part. When they further examined the Cas protein, they found that it was coded by the genes associated with the CRISPR locus (Shabbir et al. 2019). Such outcomes demonstrated that by recognizing and cutting affiliated filaments of the incipient guests employing preserved scrapings of former invaders' genomes, the CRISPR-Cas systems safeguard bacteria as well as archaea against invasive viruses, primarily bacteriophages (Loureiro and Da Silva 2019).

CRISPR-Cas system consists of two main categories: guide RNAs including crRNA and tracrRNA as well as Cas proteins. The molecules involved in cutting target DNA or RNA are known as Cas proteins (Xu and Li 2020). By evolution, several Cas proteins have been able to identify specific protospacer adjacent motif (PAMs) for cleaving of the targets at exact places. For instance, a double-strand break is generated about three base pairs prior to an NGG PAM by the Cas9 protein that is regularly used and derived from *Streptococcus pyogenes* (Gleditzsch et al. 2019). Two other guide RNAs, crRNA and tracrRNA, are responsible for steering the Cas proteins to target regions. The crRNA has tracrRNA complementary repeats for binding to the crRNA and spacer region consisting of sequences complementary to the target DNA (Karvelis et al. 2013). This is done by the tracrRNA and Cas protein that forms one ribonucleoprotein (RNP) complex that scans the genome for PAM as well as the target sequence. The DNA sequence that needs to be targeted is bound to the Cas protein, which then nicks it leading to double-strand breakage, which can be repaired by the natural DNA repair machinery of the host cell (Xu and Li 2020).

With respect to other conventional techniques of genetic modification which include transcription activator-like effector nucleases (TALENs) and zinc finger nucleases (ZFNs), one may perceive a boost with the technology of CRISPR-Cas (Gaj et al. 2013). These older techniques were time consuming and required laborious protein engineering for the purpose of recognizing and synthesizing specific DNA chains. On the other hand, effect of CRISPR-Cas systems can merely use easily programmable guide RNAs to channel the Cas proteins to the intended targets by effecting quicker, cheaper and more effective process (Mengstie and Wondimu 2021). Moreover, CRISPR-Cas technique is preeminent to ZFNs and TALENs regarding specificity and the least off-target effect. Scientists have also gone further in the targeting range and lowered the risks of off-target effects through modifying Cas9 with greater PAM tolerance and high DNA specificity (Gupta and Musunuru 2014). Because of these attributes, CRISPR-Cas has surpassed previous systems of genetic engineering and has been employed on numerous organisms including plants, animal, and even human cells. So, in this chapter, we have described the role of CRISPR-Cas technology, advancement in the field of genetic engineering, and RNA interference (RNAi) in crop improvement (Xu and Li 2020).

6.2 CRISPR-Cas Technology in Crop Engineering

Advancements in the field of the CRISPR-Cas9 system of genome editing have continued to transform crop genome editing because of its precision (Rasheed et al. 2021). Adapted from bacteria and archaea's adaptive immune system, the CRISPR/Cas system, notably Cas9, lets scientists revise plant genomes with extraordinary precision. Cas proteins and guide RNAs (gRNAs) of this system, which target, cleave, and edit desired DNA sequences in the crop genome, open up fantastic possibilities to enhance the crop phenotypes including crop yield, resistance to diseases or stress (Kick et al. 2017).

6.2.1 The CRISPR-Cas System Operates Through a Series of Well-Defined Steps

- Adaptation: When a bacterium is infected by virus, it takes the part of the virus's DNA and integrates itself into the CRISPR array as spacer. This spacer is the quasi-memory of the infection that has occurred genetically (Loureiro and Da Silva 2019).
- Expression: Here, the CRISPR array is transcribed into a large RNA called the CRISPR RNA that is further cleaved into smaller RNA molecules that are called crRNAs. The different crRNA identify with different spacer sequences (Nidhi et al. 2021).
- Interference: Specifically, the crRNA directs the Cas protein or, most commonly Cas9, to the target DNA sequence within the genome. Cas protein cuts the DNA in a region of the interest by generating double-strand breaks in the DNA helix (Mengstie and Wondimu 2021).

6.2.2 The Function of Cas Proteins and Guide RNAs

The Cas proteins and gRNAs have a tremendous importance to the CRISPR-Cas system since they determine both the efficiency and specificity of the system.

- Cas9: This protein is essential to the CRISPR-Cas9 system. It binds to the target DNA sequence through the interaction of the gRNA through base pairing (Janik et al. 2020). Cas9 has two nuclease domains that cut the DNA into two, thereby generating a double-strand break. This break can potentially disrupt the gene, or allow the introduction of additional genetic material via cellular repair processes (Xue and Greene 2021).
- Guide RNAs (gRNAs): These are synthetic RNA molecules which comprise two parts, first an antisense 20-nucleotide sequence for the target DNA and second is Cas9 binding scaffold. The gRNA is crucial in guiding Cas9 to the right locus in the genome hence making the editing appropriately precise (Wilkinson et al. 2019).

6.2.3 Mechanism of Action

The action of the CRISPR-Cas9 system can be summarized in the following steps:

- Design of gRNA: Scientists develop a gRNA which goes for the target gene of their interest that they want to modify. This gRNA can be bought from the supermarket or can be synthesized in the lab (Zhang et al. 2021).
- Formation of the RNP complex: The gRNA is then mixed with Cas9 to form a ribonucleoprotein (RNP) complex, which is the functional entity that enters the cell. This complex is then introduced into the target cells such as crop cells as shown in the following equation (Farboud et al. 2018).
- Target recognition: After entering the cell, the RNP complex begins the search for a homologous sequence in the genome. Another important step is the binding of Cas9 to DNA at the site where there is heterologous sequence—protospacer adjacent motif (PAM) sequence. There is in most cases a short, conserved sequence also called PAM sequence (e.g., NGG for Cas9), which is important for Cas9 to be recognized and bound to the target DNA (Mengstie and Wondimu 2021).
- DNA cleavage: Upon binding, Cas9 induces a conformational change that activates its nuclease domains, resulting in a double-strand break in the DNA (Xue and Greene 2021).
- DNA repair: The cell's natural repair mechanisms then attempt to fix the break. This can occur through two primary pathways: non-homologous end joining (NHEJ)—this error-prone repair method can lead to insertions or deletions (indels) that disrupt the target gene (Rodgers and Mcvey 2016).
- Homology-directed repair (HDR): If a donor DNA template is provided, the cell can use this template to repair the break, allowing for precise modifications, such as the insertion of new genes (Xue and Greene 2021).

6.2.4 Applications in Crop Improvements

The development of the CRISPR-Cas system has created new opportunities to advance the improvement of crops; it permits quicker, more accurate, and flexible strategies to increase crop yields, stress tolerance, disease resistance, and produce crops with added nutritional value (Liu et al. 2021). These applications are fundamental in meeting the increasing global population's food needs in the face of climate change, shrinking arable land, and growing population. Below, we explore three key areas where CRISPR-Cas has made significant contributions to crop improvement: increasing crop productivity and abiotic and biotic stress tolerance, improving resistance to diseases, and developing functional crop qualities.

6.2.4.1 Improving Yield and Stress Tolerances

It is therefore important to work on improving crop yields and the crops' ability to withstand the effects of the various environmental stresses. CRISPR-Cas allows targeted changes to the plant genes involved in growth and metabolism, productivity, and response to abiotic stresses.

Enhancing Yield

- Scientists have also described genes for plant height, biomass production, and days to flowering that can be targeted to increase yields. For example, in an experiment to modify rice (*Oryza sativa*), the researchers employed CRISPR-Cas9 to target and erase the *GS3* gene, which is responsible for the control of grain size. This led to the increase in grain length, and therefore, yield without compromising other desirable characters of the plant (Usman et al. 2021).
- Another example can be seen in wheat where CRISPR-Cas has been used to edit the *TaGW2* gene, which in turn helped enhance the grain weight and its size. This modification was done in a way that did not diminish the plants ability to fight off diseases; thus, the modification has a double benefit (Achary and Reddy 2021).

Stress Resistance

- Drought resistance: CRISPR-Cas has been used to modify genes that offer cures toward drought. For example, in maize (*Zea mays*), through CRISPR Tech the *ARGOS8* gene overseeing ethylene signalization pathways which is vital in stress indication responses was attenuated, consequently increasing the drought persistence and productivity of the seeds/maize plants under water-limited environment (Shi et al. 2017).
- Seasonal changes which are an important factor influencing growth of crops have changed due to climate change leading to increase or decrease in temperature. Through the use of CRISPR, heat and cold tolerance has been improved by the change of certain genes essential in thermal regulation. For example, the downregulation of *SlMAPK3* gene in tomatoes led to the increase in heat tolerance by increasing the capability of plants to carry out the process of photosynthesis under high temperature stress (Raza et al. 2019).
- It is therefore central to enhance nutrient use efficiency in crops to help reduce on used chemical fertilizers. CRISPR-Cas has been used to edit genes involved with nitrogen and phosphorus uptake in crops including maize and wheat. These crops are able to produce higher yields for a given area while at the same time reducing the amount of fertilizer needed to help increase nutrient uptake (Kumar et al. 2022).

6.2.4.2 Disease Resistance Through Targeted Gene Edits

Possibly the most significant use of CRISPR-Cas technology in agriculture is the enhancement of genetic resistance to diseases mentioned in Table 6.1. Bacterial, fungal, and viral diseases affect the crops and since they infect the plants, yields may reduce greatly. Conventional breeding for disease resistance requires a lot of time and many a times is not successful due to new changes in pathogen. CRISPR-Cas makes it possible to modify genes involved in plant-pathogen interactions, thus creating crops that are disease resistant.

6.2.4.3 Developing Crops with Improved Nutritional Profiles

Enhancing the quality of the produced crops in terms of nutritional value is a very vital step towards eliminating the scourge of malnutrition and food insecurity around

Table 6.1 List of CRISPR-Cas applications for disease resistance in different crops

S. no.	Disease	Crop	Pathogen	Target gene(s)	Outcome	References
1.	Bacterial blight	Rice	*Xanthomonas oryzae*	Xa23, Xa13	Enhanced resistance to bacterial blight	Li et al. (2020)
2.	Powdery mildew	Wheat	*Blumeria graminis*	MLO	Strong resistance to powdery mildew	Ingvardsen et al. (2023)
3.	Cassava mosaic virus	Cassava	*Cassava mosaic virus*	Viral genome	Reduced viral replication and disease incidence	Gomez et al. (2019)
4.	Citrus canker	Citrus	*Xanthomonas citri*	CsLOB1	Improved resistance to citrus canker	Huang et al. (2022a)
5.	Tomato yellow leaf curl virus	Tomato	*Tomato yellow leaf curl virus*	Viral genome	Increased resistance to viral infection	Tashkandi et al. (2018)
6.	Bacterial canker	Tomato	*Clavibacter michiganensis*	SlNPR1	Improved resistance to bacterial canker	Erdoğan et al. (2023)
7.	Fusarium wilt	Banana	*Fusarium oxysporum*	Foc2	Enhanced resistance to fusarium wilt	Tripathi et al. (2020)
8.	Black rot	Cabbage	*Xanthomonas campestris*	CaPGIP	Improved resistance to black rot	Lee et al. (2023a)
9.	Bacterial spot	Tomato, pepper	*Xanthomonas* spp.	Bs5	Resistance to bacterial spot	Ortega et al. (2024)
10.	Late blight	Potato	*Phytophthora infestans*	StCDF1	Increased resistance to late blight	Hou et al. (2023)
11.	Grey mold	Tomato, strawberry	*Botrytis cinerea*	SlPMEI	Enhanced resistance to grey mold	Lee et al. (2023b)
12.	Rice blast	Rice	*Magnaporthe oryzae*	Pi21	Increased resistance to rice blast	Yang et al. (2023)
13.	Verticillium wilt	Cotton	*Verticillium dahliae*	GhERF105	Enhanced resistance to verticillium wilt	Gao et al. 2017)
14.	Soybean mosaic virus	Soybean	*Soybean mosaic virus*	Viral genome	Reduced viral load and improved resistance	Barone et al. (2020)

15.	Anthracnose	Maize	*Colletotrichum graminicola*	ZmLOX3	Resistance to anthracnose	Jianing et al. (2022)
16.	Sclerotinia stem rot	Soybean	*Sclerotinia sclerotiorum*	*GmVQ*	Improved resistance to stem rot	Khan et al. (2023)
17.	Tobacco mosaic virus	Tobacco	*Tobacco mosaic virus*	Viral genome	Reduced viral replication	Liu and Nelson (2013)
18.	Southern corn leaf blight	Maize	*Cochliobolus heterostrophus*	*ZmHCT*	Resistance to leaf blight	Chen et al. (2023)
19.	Barley yellow dwarf virus	Barley	*Barley yellow dwarf virus*	Viral genome	Increased viral resistance	Alquicer et al. (2023)

Table 6.2 List of CRISPR-Cas applications for nutritional enhancement in crops

S. no.	Nutrient	Crop	Target gene(s)	Outcome	References
1.	Beta-carotene (vitamin A)	Rice	PSY1, CRTISO	Increased beta-carotene content (Golden Rice)	Mallikarjuna Swamy et al. (2021)
2	Iron and zinc	Wheat, rice	IRT1, ZIP7, NAS2	Enhanced iron and zinc accumulation	Kumar et al. (2022)
3.	Folate (vitamin B9)	Rice	GCH1, ADCS	Increased folate levels	Caddell et al. (2023)
4.	Vitamin C (ascorbic acid)	Tomato	GME	Higher vitamin C content	Do et al. (2022)
5.	Omega-3 fatty acids	Camelina	FAD3, FAD7	Increased production of omega-3 fatty acids	Jiang et al. (2017)
6.	Essential amino acids	Maize, soybean	BADC1, BAT1	Enhanced essential amino acid profiles	Zhu et al. (2021)
7.	Tocopherols (vitamin E)	Soybean	GmVTE4	Increased vitamin E content	Zhang et al. (2023)
8.	Reduced gluten content	Wheat	Gliadin	Low-gluten wheat for gluten-sensitive diets	Gil-Humanes et al. (2014)
9.	Lysine	Rice, maize	BADH, LDH	Increased lysine content for improved protein quality	Yang et al. (2021)
10.	Low phytic acid	Soybean, wheat	GmFAD2, TaIPK1	Reduced antinutritional phytic acid content	Jha et al. (2022)
11.	Vitamin D	Tomato	SlD2	Enhanced vitamin D content (through biofortification)	Li et al. (2022)
12.	Anthocyanin	Tomato, rice	ANT1, DFR	Increased anthocyanin production (improved antioxidant capacity)	Massa et al. (2022)
13.	Resistant starch	Potato	GBSS	Increased resistant starch for enhanced dietary fiber content	Birt et al. (2013)
14	Beta-glucan (dietary fiber)	Barley	HvCslF6	Enhanced beta-glucan content	Singla et al. (2024)
15.	Flavonoids	Tomato, grapes	MYB12, FLS1	Increased flavonoid levels (antioxidant and health benefits)	Ullah et al. (2020)

the world. CRISPR-Cas technology enables researchers to increase the content of macro- and micronutrients in plant food staples by modifying target genes controlling biosynthesis of these nutrients few are mentioned in Table 6.2. This could result into the development of crops that have high content of essential vitamins, minerals, and other quality compounds.

6.3 RNA Interference (RNAi) in Crop Improvement

RNA interference (RNAi) has emerged as one of the most powerful and prominent tools in the field of molecular biology and genetics for crop improvement. RNAi is an endogenous process of gene regulation which acts significantly in the gene expression of plants, animals and other such multicellular eukaryotes. In plants, RNAi has been applied in biotechnology for the generation of transgenic crops, and these crops have increased disease and pest resistance, enhanced nutritional and stress tolerance, and higher yields (Koeppe et al. 2023).

6.3.1 Understanding RNAi

RNA interference or RNA silencing is a normal cellular process where RNA molecules can suppress either gene expression or translation by neutralizing the certain mRNA molecules. Originally identified in 1990s, RNA interference has been identified to transform genetic and molecular biology due to the capability in gene regulation, especially in knocking down certain gene assemblies.

6.3.1.1 Mechanism of RNAi in Gene Silencing

The RNAi process also entails the degradation of the mRNA after the transcription process, with a view of hindering translation and therefore downregulation of the target gene. The mechanism primarily includes two stages: of launching and implementation.

- Initiation: The initiation phase involves the formation of small RNA molecules which are either in form of microRNA or small interfering RNA. These RNA molecules which are derived from double-stranded RNA (dsRNA) or stem-loop structures are cut by an enzyme known as Dicer. SiRNA is produced from long dsRNA through the activity of Dicer enzyme which chops the long dsRNA into small duplexes of 21 to 25 nucleotides in size (Song and Rossi 2017).
- Execution: These small RNA molecules enter the RNA-induced silencing complex that is a multiprotein complex. The RISC stands for RNA-induced silencing complex, and it expels the other strand of the siRNA or miRNA duplex leaving only this particular guide strand that is complementary to the measured mRNA. The guide strand localizes RISC to the target mRNA through the process called base pairing. Once RISC is formed, it can either degrade the mRNA directly or block translation depending on the extent of the interaction (Leuschner et al. 2006).

6.3.1.2 Types of RNA Molecules Involved: siRNA and miRNA

These two basic RNA species involved in RNAi process are called small interfering RNA (siRNA) and microRNA (miRNA), but they differ in their formation and working.

- Small interfering RNA (siRNA): siRNA is normally of exogenous nature; it is derived from viruses or is used in experiments in laboratory conditions. Some of the functions include playing role in defending against viral infections and in the regulation of the endogenous genes. siRNA works through a base pairing with the target mRNA, achieving a 100% or almost 100% match to the targeted mRNA and hence degrades the latter. As they are known to silence genes to afford their functions to be studied, siRNAs are particularly useful in experimental biology (Dana et al. 2017).
- MicroRNA (miRNA): On the other hand, miRNAs are endogenous molecules which participate in the biological processes and functions that modulate gene expression at the posttranscriptional level miRNA is a small noncoding RNA that comes from hairpin-shape precursors generated from genes located within the nucleus, which undergoes Dicer cleavage to generate a mature miRNA (O'Brien et al. 2018). In contrast to siRNA, which usually targets the species at the 5′ end of the target mRNAs with extraordinary complementarity, the miRNA targets the 3′ untranslated region (3′ UTR) of the target mRNAs with partial complementarily; hence, translation is repressed rather than mRNA degradation (Bartel 2009).

6.3.1.3 Natural Occurrence and Discovery in Plants

RNA interference was first observed in plants in the early 1990s while experimenting on breeding petunias with better color pigmentation (Agrawal et al. 2003). Surprisingly, instead of enhancing the pigmentation of flowers, new genes that were introduced seemed to switch off the endogenous floral pigmentation genes, producing white flowers. This occurrence was referred to as 'co-suppression', which acts as the basis of RNAi mechanism.

The natural occurrence of RNAi in plants serves multiple roles:

- Defense against viruses: Viral RNA is a target in the plant's immune defense against viral pathogens through RNA silencing, more commonly known as RNA interference. For instance, healthy plants can develop siRNA when RNA viruses infect them, and the siRNA/RISC can degrade the viral RNA thus stopping the infection (Akbar et al. 2022).
- Developmental regulation: Several miRNAs have been reported to control a number of developmental processes in plants including leaf morphogenesis, flowering time and root development. For example, miR156 in *Arabidopsis thaliana* is involved in the process of community transition from the juvenile phase to the adult phase by modulating SPL genes (Lawrence et al. 2020).
- Stress response: Gene silencing through RNA interference assists plants to cope with what is known as abiotic stresses such as the absence of water, salt content, and cold. Stress-responsive miRNA plays a significant role in regulating those

key genes controlling the plants adaptive mechanisms in order to restore inner balance under stressed conditions (Khraiwesh et al. 2011).

6.3.2 RNAi Management of Crop Gene and Its Improvement

RNAi technology has also been effectively used for enhancing the crop traits, for example, resistance to pests and diseases, stress tolerance, and enhanced nutritional quality. Below are some of the key areas where RNAi has made a significant impact:

Below are some of the key areas where RNAi has made a significant impact:

- Pest and insect resistance: By far the most popular and one of the earliest use of RNAi in agriculture was in the creation of crops that are resistant to specific pests. Consequently, plants can be bioengineered to produce dsRNA that interferes with insect pests' essential genes or renders them unable to reproduce. These methods are natural, thereby reducing the use of chemical pesticides thus making the environment green (Mamta and Rajam 2017).
  Example: Western corn rootworm: The use of RNAi technology has been applied to shield corn against the menace of western corn rootworm that is deemed a major threat to the maize crops. RNAi has helped change corn plant genes to develop resistance against rootworm attack, hence minimizing the use of chemicals in the control of pests (Darlington et al. 2022).
- Viral resistance: RNAi is also applied against plant viruses that may lead to huge crop yield reduction. A plant can metabolize one viral RNA by synthesizing a dsRNA or siRNA that counteracts the RNA of the respective virus and therefore, obstruct the replication and spread of viruses (Akbar et al. 2022).
  Example: papaya ringspot virus (PRSV): This virus was the first to demonstrate great success of RNAi in agriculture and particularly in creating papaya varieties that are resistant to the papaya ringspot virus. RNAi technology was utilized to develop transgenic papayas that contain siRNAs, which are effective in neutralizing the viral RNA, thus making the plants resistant to infections by the virus (Jia et al. 2017).
- Fungal and bacterial resistance: RNAi is a process that can help upgrade a plant's tendency to adapt to fungal and bacterial pathogen. Reducing Christmas tree gene expression could cause diseases—for pathogens, or increasing plant defense genes could defeat diseases (Mann et al. 2023).
  Example: pest and disease resistance: RNAi has been used in transformation to produce bananas that are naturally resistant to fusarium wilt—a fungal disease which is a menace to banana production globally. Today, through RNAi-degraded essential genes in the banana feed fungi, scientists have come up with resistant bananas (Ghag and Ganapathi 2019).
- Enhanced nutritional content: The RNAi can also be used in the regulation of nutritional quality in crops, which facilitates the creation of biofortified foods. Using genetic engineering, the master genes that regulate metabolic specificity

can be downregulated and crops produced with higher levels of vitamins, minerals or any other nutrient required by the human body (Kleter 2020).
Example: high oleic acid soybeans: Through the use of RNAi technology, deseeding of soybean varieties has been achieved by mutation of the lineolic acid gene. This has led to the development of high-quality soybeans that are healthy for human consumption and with better oil characteristic to enhance the shelf life of the final products (Zhang et al. 2014).

- Abiotic stress tolerance: The organisms in the plants experience various types of abiotic stresses which include drought stress, salinity stress, and temperature stress among others. Through RNAi technology, the intensity of the various environmental stresses affecting crops can be managed adequately by altering the rate of transcription of stress-responsive genes (Ali and Mehmood 2021).
  Example: drought tolerance in wheat: Using RNAi technology, genetically modified wheat strains with enhanced drought tolerance have been developed through manipulation of genes that reduce the wheat plant stress response abilities. These wheat varieties should be able to continue producing under water-deficit conditions, and it should be achieved given that climate change is likely to reduce the availability of water in the future (Bapela et al. 2022).

6.3.3 Challenges and Future Directions in RNAi-Based Crop Improvement

Despite its success, RNAi technology in agriculture faces several challenges:

- Off-target effects: Another issue that may be associated with RNAi is genome toxicity beginning with the prospect of off-target effect, whereby small RNA molecules may target other genes than the intended ones. Such altering can cause phenotypic changes or have an impact on crop yield or negatively influence the well-being of the organism. Several off-target effects are thus being reduced due to continued development in the field of bioinformatics as well as the design of siRNA (Seok et al. 2017).
- Gene escape and ecological impact: The expansion of RNAi-modified crops in the biosphere may be potentially threatening to nontarget organisms and the environment as a whole. There has been concern that dsRNA or siRNA might be transmitted to another species and cause altering the gene expression of the species. There is effort being made in terms of the regulatory framework, which gives an attempt at trying to evaluate these risks (Roberts et al. 2015).
- Regulatory and public acceptance: Like any GMO, the plants developed by using RNAi can only avail the markets only after passing the approved governmental regulatory policies which are often time consuming and costly. Also, consumer appreciation of RNAi crops is limited due to opposition from people especially in the developed countries that have banned GMO crops (Lucht 2015).

6.4 Silencing Target Genes for Desired Traits

MicroRNA technology has now emerged as one of the most effective and promising approaches for modifying crop traits, increasing resistance to pests and diseases and improving tolerance to environmental adversities in agriculture (Macovei et al. 2012). The most widely used mechanisms for inhibiting gene transcription include RNA interfering and the use of CRISPR-Cas system. Both strategies are strong tools to regulate the gene expression although these both strategies can initially be used in similar manner, but they differ from each other in terms of working principle, effectiveness, target selection, and the duration of activity (Huang et al. 2020).

6.4.1 Overview of Gene Silencing Techniques

RNAi is a gene silencing process that takes place at the posttranscriptional level, and apparent in plants, animals and other eukaryotic organisms. It entails the destruction of particular RNA called posttranscriptional gene regulation, which means that certain mRNA molecules cannot be translated into proteins. RNAi occurs through small RNA such as mainly small siRNA and miRNA which help the RNA-induced silencing complex (RISC) in directly cleaving mRNAs or their repression (Castel and Martienssen 2013).

In agriculture, RNAi has been adopted to silence any genes that have effects that society considers undesirable, for instance, vulnerability to attacks by diseases or pests. The technology has been most successful in producing virus, insect, and fungi resistant crops.

- Mechanism: This works by using dsRNA molecules in the plant cells which are also referred to as RNA interference. These dsRNAs are further cleaved into siRNAs through a protein called Dicer enzymes. The siRNAs are thereafter incorporated into the RISC complex that targets mRNAs and brings about their degradation or translational repression (Agrawal et al. 2003).
- Applications in agriculture: Some of the achievements of the RNAi include the production of virus-resistant papayas, insect-resistant corns, as well as crops with improved nutritional value. It is also useful in functional genomic analyses as the technique aids in making the corresponding gene nonfunctional by using RNAi (Halder et al. 2022).

6.4.2 Comparing Efficiency, Specificity, and Stability

- To measure the efficiency of a specific technique regarding gene silencing, it refers to eliminating or modifying the specific gene of interest. RNAi is widely believed to be effective for short-term experiments since it may greatly inhibit the genes and mRNA molecules' activity through degradation (Agrawal et al. 2003). Nevertheless, RNAi efficiency can differ owing to the chosen gene and the level of homology between the siRNA and target mRNA. Sometimes RNAi results in other than complete processing, and this is called as partial knockdown. On the other hand, CRISPR-Cas9 is much more effective in the gene knockout or, at the very least, mutagenesis (Knudsen-Palmer et al. 2024). The Cas9 protein catalyzes the DNA break and the cell's repair machinery results in modifications which make it impossible for the gene to be expressed or the desired change to occur for the next generation. For this reason, the CRISPR-Cas makes it easier for stable and heritable modifications to be achieved (Zhang et al. 2021).

- RNAi and CRISPR-Cas mechanisms depend on the recognition of the target sequences and while they both are highly efficient CRISPR-Cas is usually reported to have higher efficiency for gene targeting. In many CRISPR-Cas systems, the guide RNA can be programmed and it is complementary to a specific element of the DNA sequence in the targeted gene, therefore, making the intervention very accurate (Javaid and Choi 2021). However, as with most tools, there are side effects; gRNA-Cas9 combines with DNA and cuts undesired locations in the genome. However, conventional CRISPR/Cas9 system is error prone and results in off-target effects, which have been minimized by the current improvements in gRNA design and generation of high-fidelity Cas9 variants (Jacinto et al. 2020). Due to the fact that RNAi specificity relies on the ability of the siRNA to target a specific mRNA, the complementarity between the two should be given extra attention. SiRNAs can be designed to target particular genes but due to the possibility of obtaining partially matched siRNA sequences required genes other than the target are also silenced. This can lower the accuracy of RNAi-based gene silencing (Agrawal et al. 2003).
- This is the ability if gene silencing to remain suppressed or altered for a certain period of time. Most of the time, RNAi process is temporary, in a way that it does not affect the DNA of an organism, but suppresses a specific gene at the mRNA level. This is an indication that RNAi-mediated gene silencing is reversible and that normal gene expression can be restored whenever the siRNA molecules that had been formed degrade. This ability can, however, be very useful for certain crop protection scenario where one wishes to silence bioacoustics signals for only a short period of time (Agrawal et al. 2003). While CRISPR-Cas provides assured and, relative to TALENs and ZFNs, long-lasting genetic changes over vast numbers of target genes, its use can be risky. After Cas9 has made the cut, a mutation is formed and integrated into the genome of a plant and this can be transmitted to future generations. This makes CRISPR-Cas very suitable in making sustainable long-term alteration to crop genotypes and therefore crop quality (Erdoğan et al. 2023).

6.4.3 Trait Development Through Gene Silencing

Techniques that involve silencing of genes for crops, including RNAi and CRISPR-Cas, have now been central in creating insect resistant, drought tolerant, and many other crops, therefore meeting the world's growing demand (Touzdjian Pinheiro

Kohlrausch Távora et al. 2022). These methods help in targeted knockdown of the genes and improve the resistance against diseases, nutritional value of the crop and reduction in allergens and toxins. This section delves into how these strategies are applied for crop improvement, focusing on three critical areas: one might include disease resistance, improved nutrition, and allergen or toxin reduction.

6.4.3.1 Silencing Genes for Disease Resistance

Utilizing gene silencing in this capacity is one of the most effective meanings in the field of agriculture such as increasing crop protection from viruses, bacteria, and fungi few are mentioned in Table 6.3. These methods like chemical treatment or crop cycling, somehow or other, do not seem to address the problem in a quite effective and long-lasting manner. Genetic silencing is a specific approach of improving the plants natural resistance against pests and diseases without using toxic chemicals and increasing yield.

6.4.3.2 Modulating Metabolic Pathways for Improved Nutrient Content

Problem statement malnutrition is still a critical factor among world's communities and predominantly affects the developing world where people depend on basic food staples. RNA interference with utilization of other plant metabolic genes has been used as an effective strategy for enhancing nutritional quality in crops through use of biofortification techniques (Table 6.4).

6.4.3.3 Reducing Allergenicity and Toxicity in Crops

Gene silencing technologies are also applied to downregulate the allergenicity and toxicity of crops such that the crops would be safe for human consumption few are mentioned in Table 6.5. Protein within some crops that are seen as allergens can produce severe instant reactions to people with sensitiveness despite the health benefits come with consumption of these crops. All crops contain natural toxins that results into health complications if taken in large amounts.

6.5 Integration of CRISPR-Cas and RNAi Technologies

6.5.1 Synergistic Use of CRISPR-Cas and RNAi

The use of CRISPR-Cas system and RNAi can be combined to offer improved targeted approach to genomic manipulation for the improvement of crops. The above two methods provide latex model for gene regulation and silencing using different but related molecular mechanisms. While CRISPR-Cas facilitates precise gene editing, it permits gene knockout, knockin, or base editing;, RNAi on the other hand permits posttranscriptional gene silencing, hence stopping gene expression after the synthesis of the mRNA (Núñez-Álvarez et al. 2024).

The use of CRISPR-Cas technology especially CRISPR-Cas 9 enable targeted and precise breaks on desired genomic locations with the help of sgRNA. These extremely specific approaches enable the loss-of-function mutations, or gene editing

Table 6.3 List of gene silencing approaches for disease resistance in different crops

S. no.	Crop	Pathogen type	Gene silencing approach	Silenced gene/target	Result	References
1.	Papaya	Viral	RNA interference (RNAi) targeting viral RNA	*Papaya ringspot virus* (PRSV) RNA	Resistance to PRSV	Fuchs and Gonsalves (2007)
2.	Tomato	Bacterial	RNAi targeting susceptibility genes in plant	*Pseudomonas syringae* susceptibility gene	Enhanced resistance to bacterial speck	Que et al. (2010)
3.	Banana	Fungal	RNAi targeting fungal virulence factors	*Fusarium oxysporum* virulence genes	Resistance to fusarium wilt	Broggini et al. (2014)
4.	Apple	Bacterial	RNAi targeting susceptibility genes in plant	*Erwinia amylovora* susceptibility genes	Fire blight resistance	Tyagi et al. (2008)
5.	Rice	Viral	RNAi targeting viral RNA	*Rice Tungro bacilliform virus* RNA	Resistance to rice tungro virus	Lin et al. (2022)
6.	Potato	Fungal	RNAi targeting fungal cell wall-degrading enzyme	*Phytophthora infestans* glucanase gene	Resistance to late blight	Panwar et al. (2013)
7.	Wheat	Fungal	Host-induced gene silencing (HIGS) targeting pathogen RNA	*Puccinia triticina* (leaf rust)	Reduced fungal infection	Sindhu et al. (2009)
8.	Soybean	Nematode	RNAi targeting nematode parasitism genes	*Heterodera glycines* parasitism gene	Resistance to soybean cyst nematode	Walsh et al. (2019)
9.	Cassava	Viral	RNAi targeting viral RNA	*Cassava mosaic virus* (CMV) RNA	Resistance to CMV	Qi et al. (2019)
10.	Barley	Fungal	HIGS targeting pathogen RNA	*Blumeria graminis* (powdery mildew)	Improved resistance to powdery mildew	Wang et al. (2024)
11.	Cotton	Fungal	RNAi targeting fungal pathogenicity genes	*Verticillium dahliae* pathogenicity genes	Resistance to verticillium wilt	Ravelonandro et al. (2021)
12.	Peach	Viral	RNAi targeting viral RNA	*Plum pox virus* (PPV) RNA	Resistance to PPV	Jiang et al. (2020)
13.	Tobacco	Bacterial	RNAi targeting bacterial pathogenicity genes	*Xanthomonas* pathogenicity genes	Resistance to bacterial blight	Koeppe et al. (2023)
14.	Grape	Fungal	RNAi targeting fungal virulence genes	*Botrytis cinerea* virulence genes	Reduced infection by botrytis	Wang et al. (2016)

Table 6.4 List of biofortification through gene silencing in different crops

S. no.	Crop	Target nutrient	Gene silencing approach	Silenced gene/target	Result	References
1.	Golden Rice	Provitamin A (beta-carotene)	RNA interference (RNAi) targeting enzymes in carotenoid biosynthesis pathway	Carotenoid desaturase genes	Enhanced beta-carotene content (provitamin A)	Paine et al. (2005)
2.	Tomato	Vitamin E (tocopherols)	RNAi targeting enzyme in tocopherol biosynthesis	γ-TMT gene	Increased vitamin E levels	Das et al. (2021)
3.	Maize	Amino acids (lysine, tryptophan)	RNAi targeting genes responsible for the synthesis of storage proteins	Opaque-2 and α-zeins genes	Improved lysine and tryptophan content	Wang and Larkins (2001)
4.	Soybean	Omega-3 fatty acids	RNAi targeting enzyme in fatty acid biosynthesis pathway	Δ12-desaturase genes	Increased omega-3 fatty acid content	Román et al. (2012)
5.	Potato	Resistant starch	RNAi targeting starch branching enzymes	Starch branching enzyme II (SBEII)	Increased resistant starch content	Chen et al. (2021)
6.	Wheat	Iron (Fe) and zinc (Zn)	RNAi targeting transporter genes	Ferritin and ZIP transporter genes	Enhanced iron and zinc levels	Tanin et al. (2024)
7.	Cassava	Provitamin A (beta-carotene)	RNAi targeting carotenoid degradation enzyme	Carotenoid cleavage dioxygenase (CCD) genes	Increased beta-carotene accumulation	Giuliano (2017)
8.	Barley	Folate (vitamin B9)	RNAi targeting folate degradation enzymes	Folylpolyglutamate synthetase (FPGS) gene	Increased folate levels	Strobbe and Van Der Straeten (2017)
9.	Rice	Iron	RNAi targeting negative regulators of iron absorption	Iron transporter and NAS genes	Enhanced iron content	Wairich et al. (2022)
10.	Banana	Provitamin A	RNAi targeting carotenoid biosynthesis pathway enzymes	Phytoene desaturase (PDS)	Enhanced provitamin A levels	Paul et al. (2017)

Table 6.5 List of reducing allergenicity and toxicity through gene silencing in different crops

S. no.	Crop	Allergen/toxin	Gene silencing approach	Silenced gene/target	Result	References
1.	Peanut	Allergen (Ara h 2, Ara h 3)	RNA interference (RNAi) targeting allergen genes	Ara h 2 and Ara h 3	Reduced peanut allergenicity	Singh and Bhalla (2008)
2.	Soybean	Allergen (Gly m 5, Gly m 6)	RNAi targeting soybean seed storage protein genes	Gly m 5 and Gly m 6	Reduced allergen levels in soybeans	Herman et al. (2003)
3.	Wheat	Gluten (Gliadins)	RNAi targeting gluten protein synthesis	Gliadin genes	Reduced gluten content in wheat	Marín-Sanz and Barro (2022)
4.	Rice	Allergen (Ory s 1)	RNAi targeting allergenic proteins in rice	Ory s 1 (major allergen)	Reduced allergenic protein in rice	Feng et al. (2021)
5.	Tomato	Glycoalkaloids (α-tomatine)	RNAi targeting glycoalkaloid biosynthesis	GAME9 gene	Reduced α-tomatine content, lower toxicity	Al Sinani and Eltayeb (2017)
6.	Potato	Glycoalkaloids (α-solanine, α-chaconine)	RNAi targeting glycoalkaloid biosynthesis enzymes	GAME9 and SGT genes	Reduced glycoalkaloid content in potatoes	Umemoto et al. (2016)
7.	Apple	Allergen (Mal d 1)	RNAi targeting apple allergen genes	Mal d 1 (major allergen)	Decreased allergenicity in apple	Yang et al. (2012)
8.	Strawberry	Allergen (Fra a 1)	RNAi targeting allergen-encoding genes	Fra a 1 gene	Reduced allergen levels in strawberries	Guo et al. (2016)
9.	Maize	Mycotoxins (aflatoxins)	Host-induced gene silencing (HIGS) targeting fungal toxin synthesis	Aflatoxin biosynthesis genes (in *Aspergillus*)	Reduced aflatoxin contamination in maize	Thakare et al. (2017)
10.	Cotton	Gossypol (toxin)	RNAi targeting gossypol biosynthesis pathway	δ-Cadinene synthase gene	Reduced gossypol content in cotton seeds	Gao et al. (2022)

and gene additions such as base editing, with minimal effect on other genomic locations. In contrast, RNAi uses small double-stranded RNAs such as siRNA or miRNA that bind to complementary mRNA, causing its degradation and thereby preventing translation (Zhang et al. 2021).

In agricultural biotechnology, both these technologies can be implemented in a better way to regulate the genes. For instance, small RNA-based CRISPR-Cas can be used to create null alleles of target genes potentially involved in disease propensities or metabolic relations, whereas RNA interference can be utilized in parallel to modulate the expression of other genes with related functions at the transcriptional level (Liu et al. 2021). This integration is especially essential where there is need for the simultaneous regulation of multiple genes, where RNAi can knockdown multiple gene expression through targeting several mRNAs, and on the other hand CRISPR-Cas can make permanent alteration on specific DNA sequence (Barrangou et al. 2015).

6.5.1.1 Combining CRISPR and RNAi for Complex Trait Development

Most desirable character evolution, as in the case of increasing tolerance to stress, increasing nutrient quality, or raising yield potential, typically requires the management of many genes. CRISPR-Cas has the advantage over gene knockouts and producing germline edits in crops while the RNAi methodology operates in a temporary and reversible way. The synergy of these approaches beneficial in situations when the multigenic control of the trait is present or both the coding and noncoding regions are to be regulated (Poller et al. 2023).

For instance, in breeding crops with resistance to both biotic and abiotic stresses such as pathogens and diseases, harsh conditions such as drought and heat; CRISPR-Cas can be used to erase out certain genes such as ethylene-responsive factors or even heat-shock proteins (Erdoğan et al. 2023). At the same time, the control over the stress tolerance might be achieved by using RNAi to knock down genes involved in stress-induced pathways to have more flexible control over stress response. This double strategy is particularly useful when it comes down to minimizing contrasts between resistance and growth drive, RNAi can be applied to balance gene narrative depending on the environment (Abdellatef et al. 2021).

In suitable dense networks which include the biosynthesis of nutrients that involves many enzymes and regulatory proteins, CRISPR-Cas targets and alters the regulatory genes while RNAi decreases the expression of the enzymes (Zhang et al. 2021). For instance, in the biofortification of crops with higher vitamin A (provitamin A), CRISPR can modify the gene coding for enzyme synthesized before a particular step in the carotenoid biosynthesis pathway whereas RNAi can interfere with the gene coding for enzyme synthesized after a particular step in the carotenoid biosynthesis pathway. The integrated approach for the change of the trait makes the process more efficient and precise in changing the desired trait (Liu et al. 2021).

6.5.1.2 Potential for Stacked Traits and Multigene Targeting

Another benefit with the use of CRISPR-Cas and RNAi is in the sense that multiple genes can be edited at the same time that is very important when developing crops that have improved traits in many fronts pyramided genes are multiple beneficial genes including disease resistance, high yields or stress tolerance and so on compacted into one product (Zhang et al. 2021). Compared to CRISPR-Cas, for example, which enables several target genes to be permanently modified at once, RNAi can also be adapted to fine-tune the expression level of specific genes in response to environmental factors, thus offering greater flexibility over complex phenotypes (Unniyampurath et al. 2016).

For instance, in maize and rice, two crops of high throughput and nutritive value, respectively, the CRIPSR-Cas system can be applied to disable some genes that are considered disadvantageous in a desired plant such as susceptibility to pests or inefficiency in nourishment utilization, and the RNAi system can be used to regulate the activity of genes that are influential for yield. Therefore, while CRISPR makes changes on more permanent changes in cells, RNAi gives temporary adaptive advantage to allow the plant to grow the fastest most suited for the environment like water reaching the roots or nutrient availability in the soil (Liu et al. 2021).

In addition, the ability of both CRISPR and RNAi in targeting multiple genes is the central strength in tackling those traits, which are under polygenic control. CRISPR-Cas multiplexing system that can target multiple genes at a time and RNAi which can suppress multiple transcripts has made it easier for the researchers to work on complex traits such as drought resistance which usually involve more than one signaling pathways and gene networks. It gets to the point where one can select crops that are suitable for climate-smart agriculture because it becomes difficult to produce a crop that will withstand the effects of climate change (Xu and Li 2020).

6.5.2 Future Prospects

6.5.2.1 Emerging Trends in Genome Editing and RNAi Technology

Genome editing is an expanding "technology", and so too is RNA interference or RNAi, in which the RNA molecule is becoming a more precise tool through advancements in substrate technologies and genetic control. They are revolutionizing the breeding of crops shielding agriculture from processes that diminish the quality of the crop and improving food security around the world (Rajput et al. 2021).

An innovative area pointed that may be associated with the existing methods is the emergence of next-generation CRISPR technologies. Although CRISPR-Cas9 system has been revolutionary, the other such as CRISPR-Cas12 and CRISPR-Cas13 have been developed to have more capabilities (Zeballos and Gaj 2021). For instance, Cas12 nuclease can bind to both DNA and RNA and Cas13 nuclease has specificity towards only RNA making it more useful for the RNA-targeted gene knockdown and modulation. These systems, in addition, present less substrate genericities, and hence, compared to previous methods of gene editing, possess higher selectivity a factor that has tremendous value when it comes to creating crops with desired characteristics, be it disease resistance, or increased nutritive value (Huang et al. 2022b).

Besides, fresh techniques of base editing and prime editing are transforming the agenda of genome editing since they make it possible to switch between one base pair to another without hampering the DNA's double-strand breaks (Rees and Liu 2018). This enables minor changes to be made in a targeted approach, for example, changing point mutations that are responsible for adverse characteristics in crops. These technologies are supposed to be useful in enhancing the genetic makeup of the plants for enhanced performance in suboptimal conditions (Zhang et al. 2018).

RNAi technology, on the other hand, has undergone improvements such as through host-induced gene silencing (HIGS) and spray-induced gene silencing (SIGS). HIGS enables the plant to produce molecules of RNA that inhibit the genes of pathogen or pests so that the plant is protected without the use of popular chemical pesticides (Zand Karimi and Innes 2022). SIGS is less intrusive and less disadvantageous to the environment in the sense that it allows an exact application of RNA molecules onto the surfaces of the plant, which leads to gene silencing in pests or pathogens. Biotic agents of this technology can be incorporated in IPM to reduce environmental factors affecting farming activities, thus promoting the use of sustainable farming (Hoang et al. 2022).

6.5.2.2 Role in Sustainable Agriculture and Global Food Security

It is noteworthy that genome editing and RNAi technologies are pivotal to meet the dual needs of food security for the growing global population and sustainable agriculture. Indeed, the use of these technologies can assist in reducing the impact of climate change, land degradation, and water deficits, on agriculture production systems to suit crop productivity and stress.

- Developing climate-resilient crops: They also assist in the creation of crops that are more resilient to restricted water supply, heat, and salt stress, which has been considered one of the most important advantages that these technologies brought into sustainable farming (Ijaz et al. 2023). CRISPR-Cas can remove genes linked to drought sensitivity or heat-inactivated cells and develop plants ready to grow in unfavorable conditions. RNAi can therefore be used to knock down the expression of genes that are linked with stress metabolic reactions so as to achieve quantitative control of stress resistance depending on the environment (Sami et al. 2021). This is more so considering the fact that there are areas that continue to experience natural disasters such as floods, thereby making climate-resistant crops very useful in mitigating hunger.
- Enhancing nutritional quality: The other important use of CRISPR and RNAi technologies is enhancing nutritional content of crops, also known as biofortification. CRISPR/Cas9 can be applied in enhancing the provitamin A, iron, and zinc contents in the commonly consumed foods including rice, maize, and wheat (Kumar et al. 2022). This is where RNAi can step in and help to knock down genes that lead to precursors being channeled in undesired directions, thus helping to maximize the nutritional value of crops. This has the potential of

helping in eradication of malnutrition especially in the developing countries whereby there is restricted dietary diversity (Kamthan et al. 2015).

- Improving disease resistance: These technologies can also enhance global food security through the application of gene silencing: CRISPR and RNAi can help create crops with increased pathogens' resistance (Touzdjian Pinheiro Kohlrausch Távora et al. 2022). For instance, CRISPR has been applied in the knockout of susceptibility genes in tomatoes and renders them immune to powdery mildew while RNAi has been applied in silencing the genes of pathogens directly including fusarium wilt in bananas. This cuts on the use of chemical fertilizers and pesticides to a certain extent enabling a greener agriculture (Pramanik et al. 2021).
- Reducing pesticide and herbicide usage: Another significant use is the generation of plants with incorporated protection from pests and herbicides or use of resistant herbicides (Tudi et al. 2021). As concluded earlier with SIGS, RNAi-based pest control is safe, nontoxic, biodegradable, and has a long-lasting pest controlling effect. Using CRISPR-Cas, it is also possible to engineer herbicide-tolerant crops, which would lead to better weed management (Touzdjian Pinheiro Kohlrausch Távora et al. 2022). These technologies can mean the reduction of synthetic chemicals to be used in farming something that can make agriculture to be more sustainable to the environment.

6.6 Conclusion

The advances in molecular breeding tools have completely transformed the crop breeding arena; among which CRISPR-Cas systems and RNA interfering technique or RNA interference (RNAi) are some of the most potent techniques for modification of crops. As has been noted throughout this review, these tools have offered the means to exercise unique control and make precise and swift changes to plant genomes that could soon help to fast track the development of traits deemed vital to food security, environmental conservation as well as boosting the overall agricultural yields.

CRISPR-Cas system, especially the most popular one—CRISPR-Cas9, revolutionized the field of genetic engineering. This technique indeed involves the use of technology that is able to induce site-specific double-strand breaks in the DNA, leading to gene repair mechanism, which results in gene edits. It is important to note that the power of CRISPR as a technology is unparalleled because it enables scientists to directly hit genes that express for desirable characteristic like yield, disease resistance, drought tolerant, or the likes.

Compared to the random mutagenesis, transgenic approaches, one of the major benefits of CRISPR-Cas is that it is possible to edit plants without any integration of foreign DNA, making gene-edited crops more palatable for the regulating authorities and consumers. CRISPR-Cas9 has also displayed adaptability in numerous crops the staple crops such as rice, wheat and maize, and horticultural crops of high economic value. With the help of such method, not only knockout but more gentle manipulations with the gene, for example, insertions, and applying base editing can be made. This level of control leads to possibilities to enhance the qualitative and nutritional and biotic and abiotic stress-tolerant characters of crops.

Another tool that has been found useful in gene regulation and crop enhancement is RNA interference commonly referred to as RNAi. RNAi works on the principles by which plants normally manage genes where it uses small RNA molecules like siRNA or miRNA to degrade or to suppress the particular mRNA molecule, and hence the targeted gene.

RNAi has been utilized extensively for crop improvement especially for resistance to biotic stresses, quality, and shelf life, among others. For instance, RNAi has been used to genetically engineer, for example, virus-resistant papaya, nematode-resistant soybean, and insect-resistant cotton. Furthermore, the RNAi technique is versatile in the sense that is capable of targeting not only genes of the plant but also that of pests and pathogens and thus important in the management of pests and diseases.

CRISPR-Cas and RNAi methods are methods of gene silencing meaning that scientists can selectively turn off genes that are preferably linked to unfavored characteristics or amplify characteristics more preferred. RNA interference can be used to manage pests and diseases or enhance desirable attributes in value addition such as nutritional content and/or stress resistance.

CRISPR-Cas and RNAi are members of the same family of tools in the field of gene silencing, with the former being highly effective when designing for permanent gene knockouts and the latter providing the option of the transient mode. Combining both technologies may result in a better regulation of gene expression in crops, thus improving their general performance and resistance levels on a wider scale. The future of crop engineering belongs to the development of the technologies such as CRISPR-Cas and RNAi. Better methods inclusive of multiplexed gene editing, refined delivery systems, and better gene targeting tools will further amplify the possibilities of utilization of these technologies. Because of their capacity to proactively respond to the current problems that include food insecurity, climate change, and sustainable agriculture systems, they can be very useful in defining the future agriculture systems.

Altogether, CRISPR-Cas and RNAi technologies are radicalizing the principles of crop engineering by giving more specific and effective mechanisms for genetic manipulation and gene suppression. Because of these factors, they are very useful in improving the crop traits, the ability to resist stresses and in the development of improved systems of production of crops. Thus, with further advancements, these technologies are to be the integral part of providing food security and sustainable agriculture in the future years.

Author Contributions Conceptualization, methodology, investigation, data curation, writing, original draft preparation, A.N., A.H.B., W.B.C., N.P.G., and A.M.; study design, supervision, validation, H.P, S.S.B. and A.M; literature search, data collection, writing review and editing, and all authors have read and agreed to the published version of the manuscript.

Conflicts of Interest The authors declare no conflict of interest.

Funding This research received no external funding.

References

Abdellatef E, Kamal NM, Tsujimoto H (2021) Tuning beforehand: a foresight on RNA interference (RNAi) and in vitro-derived dsRNAs to enhance crop resilience to biotic and abiotic stresses. Int J Mol Sci [Internet] 22(14):7687. Available from: https://pmc.ncbi.nlm.nih.gov/articles/PMC8306419/ [cited 2024 Oct 19]

Abdul Aziz M, Brini F, Rouached H, Masmoudi K (2022) Genetically engineered crops for sustainably enhanced food production systems. Front Plant Sci [Internet]. 13. Available from: /pmc/articles/PMC9680014/ [cited 2024 Sep 5]

Achary VMM, Reddy MK (2021) CRISPR-Cas9 mediated mutation in GRAIN WIDTH and WEIGHT2 (GW2) locus improves aleurone layer and grain nutritional quality in rice. Sci Rep [Internet] 11(1):21941. Available from: /pmc/articles/PMC8578329/ [cited 2024 Oct 2]

Agrawal N, Dasaradhi PVN, Mohmmed A, Malhotra P, Bhatnagar RK, Mukherjee SK (2003) RNA interference: biology, mechanism, and applications. Microbiol Mol Biol Rev [Internet] 67(4): 657. Available from: https://pmc.ncbi.nlm.nih.gov/articles/PMC309050/ [cited 2024 Oct 19]

Akbar S, Wei Y, Zhang MQ (2022) RNA interference: promising approach to combat plant viruses. Int J Mol Sci [Internet] 23(10):5312. Available from: https://pmc.ncbi.nlm.nih.gov/articles/PMC9142109/ [cited 2024 Oct 19]

Al Sinani SSS, Eltayeb EA (2017) The steroidal glycoalkaloids solamargine and solasonine in Solanum plants. South African J Bot 112:253–269

Ali S, Mehmood A, Khan N. Uptake, Translocation, and consequences of nanomaterials on plant growth and stress adaptation. J Nanomater 2021;2021, 1

Alquicer G, Ibrahim E, Maruthi MN, Kundu JK (2023) Identifying putative resistance genes for barley yellow dwarf virus-PAV in wheat and barley. Viruses [Internet] 15(3) Available from: /pmc/articles/PMC10053949/ [cited 2024 Oct 2]

Altpeter F, Springer NM, Bartley LE, Blechl AE, Brutnell TP, Citovsky V et al (2016) Advancing crop transformation in the era of genome editing. Plant Cell [Internet]. 28(7):1510. Available from: /pmc/articles/PMC4981132/ [cited 2024 Sep 5]

Bapela T, Shimelis H, Tsilo TJ, Mathew I (2022) Genetic improvement of wheat for drought tolerance: progress, challenges and opportunities. Plants [Internet] 11(10):1331. Available from: https://pmc.ncbi.nlm.nih.gov/articles/PMC9144332/ [cited 2024 Oct 19]

Barone P, Wu E, Lenderts B, Anand A, Gordon-Kamm W, Svitashev S et al (2020) Efficient gene targeting in maize using inducible CRISPR-Cas9 and marker-free donor template. Mol Plant [Internet] 13(8):1219–27. Available from: https://pubmed.ncbi.nlm.nih.gov/32574856/ [cited 2024 Sep 14]

Barrangou R, Birmingham A, Wiemann S, Beijersbergen RL, Hornung V, Smith AVB (2015) Advances in CRISPR-Cas9 genome engineering: lessons learned from RNA interference. Nucleic Acids Res [Internet] 43(7):3407. Available from: https://pmc.ncbi.nlm.nih.gov/articles/PMC4402539/ [cited 2024 Oct 19]

Bartel DP (2009) MicroRNA target recognition and regulatory functions. Cell [Internet] 136(2): 215. Available from: https://pmc.ncbi.nlm.nih.gov/articles/PMC3794896/ [cited 2024 Oct 19]

Birt DF, Boylston T, Hendrich S, Jane JL, Hollis J, Li L et al (2013) Resistant starch: promise for improving human health. Adv Nutr [Internet] 4(6):587. Available from: https://pmc.ncbi.nlm.nih.gov/articles/PMC3823506/ [cited 2024 Oct 19]

Broggini GAL, Wöhner T, Fahrentrapp J, Kost TD, Flachowsky H, Peil A et al (2014) Engineering fire blight resistance into the apple cultivar 'Gala' using the FB_MR5 CC-NBS-LRR resistance gene of Malus × robusta 5. Plant Biotechnol J [Internet] 12(6):728–733. Available from: https://pubmed.ncbi.nlm.nih.gov/24618178/ [cited 2024 Sep 12]

Caddell D, Langenfeld NJ, Eckels MJH, Zhen S, Klaras R, Mishra L et al (2023) Photosynthesis in rice is increased by CRISPR/Cas9-mediated transformation of two truncated light-harvesting antenna. Front Plant Sci 14:1050483

Castel SE, Martienssen RA (2013) RNA interference (RNAi) in the nucleus: roles for small RNA in transcription, epigenetics and beyond. Nat Rev Genet [Internet] 14(2):100. Available from: https://pmc.ncbi.nlm.nih.gov/articles/PMC4205957/ [cited 2024 Oct 19]

Chen J, Hawkins E, Seung D (2021) Towards targeted starch modification in plants. Curr Opin Plant Biol 60:102013

Chen G, Xiao Y, Dai S, Dai Z, Wang X, Li B et al (2023) Genetic basis of resistance to southern corn leaf blight in the maize multi-parent population and diversity panel. Plant Biotechnol J [Internet] 21(3):506. Available from: /pmc/articles/PMC9946143/ [cited 2024 Oct 2]

Dana H, Mahmoodi Chalbatani G, Mahmoodzadeh H, Karimloo R, Rezaiean O, Moradzadeh A et al (2017) Molecular mechanisms and biological functions of siRNA. Int J Biomed Sci [Internet] 13(2):48. Available from: https://pmc.ncbi.nlm.nih.gov/articles/PMC5542916/ [cited 2024 Oct 19]

Darlington M, Reinders JD, Sethi A, Lu AL, Ramaseshadri P, Fischer JR et al (2022) RNAi for western corn rootworm management: lessons learned, challenges, and future directions. Insects [Internet] 13(1):57. Available from: https://pmc.ncbi.nlm.nih.gov/articles/PMC8779393/ [cited 2024 Oct 19]

Das AK, Gowda MM, Muthusamy V, Zunjare RU, Chauhan HS, Baveja A et al (2021) Development of maize hybrids with enhanced vitamin-E, vitamin-a, lysine, and tryptophan through molecular breeding. Front Plant Sci [Internet]. 12:659381. Available from: www.frontiersin.org [cited 2024 Sep 12]

Do JH, Park SY, Park SH, Kim HM, Ma SH, Mai TD et al (2022) Development of a genome-edited tomato with high ascorbate content during later stage of fruit ripening through mutation of SlAPX4. Front Plant Sci [Internet]. 13:836916. Available from: /pmc/articles/PMC9039661/ [cited 2024 Sep 12]

Erdoğan İ, Cevher-Keskin B, Bilir Ö, Hong Y, Tör M (2023) Recent developments in CRISPR/Cas9 genome-editing technology related to plant disease resistance and abiotic stress tolerance. Biology (Basel) [Internet] 12(7):1037. Available from: /pmc/articles/PMC10376527/ [cited 2024 Jul 30]

Farboud B, Jarvis E, Roth TL, Shin J, Corn JE, Marson A, et al. Enhanced genome editing with Cas9 ribonucleoprotein in diverse cells and organisms. J Vis Exp [Internet] 2018 2018(135): 57350. Available from: /pmc/articles/PMC6101420/ [cited 2024 Sep 9]

Feng Z, Yuan M, Zou J, Wu LB, Wei L, Chen T et al (2021) Development of marker-free rice with stable and high resistance to rice black-streaked dwarf virus disease through RNA interference. Plant Biotechnol J [Internet] 19(2):212. Available from: /pmc/articles/PMC7868976/ [cited 2024 Sep 12]

Fuchs M, Gonsalves D (2007) Safety of virus-resistant transgenic plants two decades after their introduction: lessons from realistic field risk assessment studies. Annu Rev Phytopathol [Internet] 45:173–202. Available from: https://pubmed.ncbi.nlm.nih.gov/17408355/ [cited 2024 Sep 12]

Gage KL, Krausz RF, Walters SA (2019) Emerging challenges for weed management in herbicide-resistant crops. Agric [Internet] 9(8):180. Available from: https://www.mdpi.com/2077-0472/9/8/180/htm [cited 2024 Sep 5]

Gaj T, Gersbach CA, Barbas CF (2013) ZFN, TALEN and CRISPR/Cas-based methods for genome engineering. Trends Biotechnol [Internet] 31(7):397. Available from: /pmc/articles/PMC3694601/ [cited 2024 Sep 5]

Gao W, Long L, Tian X, Xu F, Liu J, Singh PK et al (2017) Genome editing in cotton with the CRISPR/Cas9 system. Front Plant Sci [Internet]. 8. Available from: /pmc/articles/PMC5541054/ [cited 2024 Sep 14]

Gao W, Zhu X, Ding L, Xu B, Gao Y, Cheng Y et al (2022) Development of the engineered "glanded plant and glandless seed" cotton. Food Chem Mol Sci 5:100130

Ghag SB, Ganapathi TR (2019) RNAi-mediated protection against banana diseases and pests. 3 Biotech [Internet] 9(3):112. Available from: https://pmc.ncbi.nlm.nih.gov/articles/PMC6395469/ [cited 2024 Oct 19]

Gil-Humanes J, Pistón F, Altamirano-Fortoul R, Real A, Comino I, Sousa C et al (2014) Reduced-gliadin wheat bread: an alternative to the gluten-free diet for consumers suffering gluten-related pathologies. PLoS One [Internet] 9(3):e90898. Available from: https://pmc.ncbi.nlm.nih.gov/articles/PMC3951262/ [cited 2024 Oct 19]

Giuliano G (2017) Provitamin a biofortification of crop plants: a gold rush with many miners. Curr Opin Biotechnol 44:169–180

Gleditzsch D, Pausch P, Müller-Esparza H, Özcan A, Guo X, Bange G et al (2019) PAM identification by CRISPR-Cas effector complexes: diversified mechanisms and structures. RNA Biol [Internet] 16(4):504. Available from: /pmc/articles/PMC6546366/ [cited 2024 Sep 5]

Gomez MA, Lin ZD, Moll T, Chauhan RD, Hayden L, Renninger K et al (2019) Simultaneous CRISPR/Cas9-mediated editing of cassava eIF4E isoforms nCBP-1 and nCBP-2 reduces cassava brown streak disease symptom severity and incidence. Plant Biotechnol J [Internet] 17(2): 421. Available from: /pmc/articles/PMC6335076/ [cited 2024 Sep 10]

Guo Q, Liu Q, Smith NA, Liang G, Wang M-B (2016) RNA silencing in plants: mechanisms, technologies and applications in horticultural crops. Curr Genomics [Internet] 17(6):476. Available from: /pmc/articles/PMC5108043/ [cited 2024 Sep 12]

Gupta RM, Musunuru K (2014) Expanding the genetic editing tool kit: ZFNs, TALENs, and CRISPR-Cas9. J Clin Invest [Internet] 124(10):4154. Available from: /pmc/articles/PMC4191047/ [cited 2024 Sep 5]

Halder K, Chaudhuri A, Abdin MZ, Majee M, Datta A (2022) RNA interference for improving disease resistance in plants and its relevance in this clustered regularly interspaced short palindromic repeats-dominated era in terms of dsRNA-based biopesticides. Front Plant Sci [Internet] 13:885128. Available from: https://pmc.ncbi.nlm.nih.gov/articles/PMC9141053/ [cited 2024 Oct 19]

Health NRC (US) C on I and AUE of GEF on H (2004a) Methods and mechanisms for genetic manipulation of plants, animals, and microorganisms. Available from: https://www.ncbi.nlm.nih.gov/books/NBK215771/ [cited 2024 Jul 27]

Health NRC (US) C on I and AUE of GEF on H (2004b) Unintended effects from breeding. Available from: https://www.ncbi.nlm.nih.gov/books/NBK215778/ [cited 2024 Sep 5]

Heather JM, Chain B (2016) The sequence of sequencers: the history of sequencing DNA. Genomics [Internet] 107(1):1. Available from: /pmc/articles/PMC4727787/ [cited 2024 Sep 5]

Herman EM, Helm RM, Jung R, Kinney AJ (2003) Genetic modification removes an immunodominant allergen from soybean. Plant Physiol [Internet] 132(1):36–43. Available from: https://pubmed.ncbi.nlm.nih.gov/12746509/ [cited 2024 Sep 12]

Hoang BTL, Fletcher SJ, Brosnan CA, Ghodke AB, Manzie N, Mitter N (2022) RNAi as a foliar spray: efficiency and challenges to field applications. Int J Mol Sci [Internet] 23-(12) Available from: /pmc/articles/PMC9224206/ [cited 2024 Oct 2]

Hou X, Guo X, Zhang Y, Zhang Q (2023) CRISPR/Cas genome editing system and its application in potato. Front Genet [Internet] 14. Available from: /pmc/articles/PMC9968925/ [cited 2024 Sep 14]

Huang X, Chen Z, Liu Y (2020) RNAi-mediated control of CRISPR functions. Theranostics [Internet] 10(15):6661. Available from: https://pmc.ncbi.nlm.nih.gov/articles/PMC7295050/ [cited 2024 Oct 19]

Huang X, Wang Y, Wang N. Highly efficient generation of canker-resistant sweet Orange enabled by an improved CRISPR/Cas9 system. Front Plant Sci [Internet]. 2022a 12:769907. Available from: www.frontiersin.org [cited 2024 Sep 13]

Huang Z, Fang J, Zhou M, Gong Z, Xiang T (2022b) CRISPR-Cas13: a new technology for the rapid detection of pathogenic microorganisms. Front Microbiol [Internet] 13. Available from: /pmc/articles/PMC9650447/ [cited 2024 Oct 2]

Ijaz M, Khan F, Ahmed T, Noman M, Zulfiqar F, Rizwan M et al (2023) Nanobiotechnology to advance stress resilience in plants: current opportunities and challenges. Mater Today Bio 22: 100759

Ingvardsen CR, Massange-Sánchez JA, Borum F, Füchtbauer WS, Bagge M, Knudsen S et al (2023) Highly effective mlo-based powdery mildew resistance in hexaploid wheat without pleiotropic effects. Plant Sci 335:111785

Ishino Y, Krupovic M, Forterre P (2018) History of CRISPR-Cas from Encounter with a Mysterious Repeated Sequence to Genome Editing Technology. J Bacteriol [Internet] 200(7). Available from: /pmc/articles/PMC5847661/ [cited 2024 Sep 5]

Jacinto F V., Link W, Ferreira BI (2020) CRISPR/Cas9-mediated genome editing: From basic research to translational medicine. J Cell Mol Med [Internet] 24(7):3766. Available from: https://pmc.ncbi.nlm.nih.gov/articles/PMC7171402/ [cited 2024 Oct 19]

Janik E, Niemcewicz M, Ceremuga M, Krzowski L, Saluk-Bijak J, Bijak M (2020) Various aspects of a gene editing system – CRISPR–Cas9. Int J Mol Sci [Internet]. 21(24):1–20. Available from: /pmc/articles/PMC7767219. [cited 2024 Sep 5]

Javaid N, Choi S (2021) CRISPR/Cas system and factors affecting its precision and efficiency. Front Cell Dev Biol [Internet] 9:761709. Available from: https://pmc.ncbi.nlm.nih.gov/articles/PMC8652214/ [cited 2024 Oct 19]

Jha R, Yadav HK, Raiya R, Singh RK, Jha UC, Sathee L et al (2022) Integrated breeding approaches to enhance the nutritional quality of food legumes. Front Plant Sci 13:984700

Jia R, Zhao H, Huang J, Kong H, Zhang Y, Guo J et al (2017) Use of RNAi technology to develop a PRSV-resistant transgenic papaya. Sci Rep [Internet] 7(1):12636. Available from: https://pmc.ncbi.nlm.nih.gov/articles/PMC5626737/ [cited 2024 Oct 19]

Jiang WZ, Henry IM, Lynagh PG, Comai L, Cahoon EB, Weeks DP (2017) Significant enhancement of fatty acid composition in seeds of the allohexaploid, Camelina sativa, using CRISPR/Cas9 gene editing. Plant Biotechnol J [Internet]. 15(5):648. Available from: /pmc/articles/PMC5399004/ [cited 2024 Sep 12]

Jiang N, Yan J, Liang Y, Shi Y, He Z, Wu Y et al (2020) Resistance genes and their interactions with bacterial blight/leaf streak pathogens (Xanthomonas oryzae) in rice (Oryza sativa L.) – an updated review. Rice [Internet] 13(1):1–12. Available from: https://thericejournal.springeropen.com/articles/10.1186/s12284-019-0358-y [cited 2024 Sep 12]

Jianing G, Zhiming X, Rasheed A, Tiancong W, Qian Z, Zhuo Z et al (2022) CRISPR/Cas9 applications for improvement of soybeans, current scenarios, and future perspectives. Not Bot Horti Agrobot Cluj-Napoca [Internet] 50(2):12678–12678. Available from: https://www.notulaebotanicae.ro/index.php/nbha/article/view/12678 [cited 2024 Sep 14]

Kamthan A, Chaudhuri A, Kamthan M, Datta A. Small RNAs in plants: recent development and application for crop improvement. Front Plant Sci [Internet] 2015;6(APR):124569. Available from: www.frontiersin.org [cited 2024 Oct 2]

Karvelis T, Gasiunas G, Miksys A, Barrangou R, Horvath P, Siksnys V (2013) crRNA and tracrRNA guide Cas9-mediated DNA interference in Streptococcus thermophilus. RNA Biol [Internet] 10(5):841. Available from: /pmc/articles/PMC3737341/ [cited 2024 Sep 5]

Khan S, Ullah MW, Siddique R, Nabi G, Manan S, Yousaf M et al (2016) Role of recombinant DNA technology to improve life. Int J Genomics [Internet] 2016. Available from: /pmc/articles/PMC5178364/ [cited 2024 Sep 5]

Khan MA, Cowling WA, Banga SS, Barbetti MJ, Cantila AY, Amas JC et al (2023) Genetic and molecular analysis of stem rot (Sclerotinia sclerotiorum) resistance in Brassica napus (canola type). Heliyon [Internet] 9(9):e19237. Available from: /pmc/articles/PMC10477455/ [cited 2024 Oct 2]

Khraiwesh B, Zhu JK, Zhu J (2011) Role of miRNAs and siRNAs in biotic and abiotic stress responses of plants. Biochim Biophys Acta [Internet] 1819(2):137. Available from: https://pmc.ncbi.nlm.nih.gov/articles/PMC3175014/ [cited 2024 Oct 19]

Kick L, Kirchner M, Schneider S (2017) CRISPR-Cas9: from a bacterial immune system to genome-edited human cells in clinical trials. Bioengineered [Internet] 8(3):280. Available from: /pmc/articles/PMC5470512/ [cited 2024 Sep 5]

Kleter GA (2020) Food safety assessment of crops engineered with RNA interference and other methods to modulate expression of endogenous and plant pest genes. Pest Manag Sci [Internet] 76(10):3333. Available from: https://pmc.ncbi.nlm.nih.gov/articles/PMC7540038/ [cited 2024 Oct 19]

Knudsen-Palmer DR, Raman P, Ettefa F, De Ravin L, Jose AM (2024). Target-specific requirements for RNA interference can arise through restricted RNA amplification despite the lack of specialized pathways. bioRxiv [Internet]:527351. Available from: https://pmc.ncbi.nlm.nih.gov/articles/PMC9934570/ [cited 2024 Oct 19]

Koeppe S, Kawchuk L, Kalischuk M (2023) RNA interference past and future applications in plants. Int J Mol Sci [Internet] 24(11) Available from: /pmc/articles/PMC10253773/ [cited 2024 Sep 12]

Kumar D, Yadav A, Ahmad R, Dwivedi UN, Yadav K (2022) CRISPR-based genome editing for nutrient enrichment in crops: a promising approach toward global food security. Front Genet [Internet] 13. Available from: /pmc/articles/PMC9329789/ [cited 2024 Sep 12]

Lawrence EH, Springer CJ, Helliker BR, Poethig RS (2020) MicroRNA156-mediated changes in leaf composition lead to altered photosynthetic traits during vegetative phase change. New Phytol [Internet] 231(3):1008. Available from: https://pmc.ncbi.nlm.nih.gov/articles/PMC8299463/ [cited 2024 Oct 19]

Lee YR, Siddique MI, Kim DS, Lee ES, Han K, Kim SG et al (2023a) CRISPR/Cas9-mediated gene editing to confer turnip mosaic virus (TuMV) resistance in Chinese cabbage (Brassica rapa). Hortic Res [Internet] 10(6) Available from: /pmc/articles/PMC10261878/ [cited 2024 Sep 14]

Lee MB, Han H, Lee S (2023b) The role of WRKY transcription factors, FaWRKY29 and FaWRKY64, for regulating Botrytis fruit rot resistance in strawberry (Fragaria × ananassa Duch.). BMC Plant Biol [Internet] 23(1) Available from: /pmc/articles/PMC10494375/ [cited 2024 Sep 14]

Leuschner PJF, Ameres SL, Kueng S, Martinez J (2006) Cleavage of the siRNA passenger strand during RISC assembly in human cells. EMBO Rep [Internet] 7(3):314–20. Available from: https://www.embopress.org/doi/10.1038/sj.embor.7400637 [cited 2024 Oct 19]

Li C, Li W, Zhou Z, Chen H, Xie C, Lin Y (2020) A new rice breeding method: CRISPR/Cas9 system editing of the Xa13 promoter to cultivate transgene-free bacterial blight-resistant rice. Plant Biotechnol J [Internet] 18(2):313. Available from: /pmc/articles/PMC6953186/ [cited 2024 Sep 9]

Li J, Scarano A, Gonzalez NM, D'Orso F, Yue Y, Nemeth K et al (2022) Biofortified tomatoes provide a new route to vitamin D sufficiency. Nat Plants [Internet] 8(6):611. Available from: https://pmc.ncbi.nlm.nih.gov/articles/PMC9213236/ [cited 2024 Oct 19]

Lin X, Olave-Achury A, Heal R, Pais M, Witek K, Ahn HK et al (2022) A potato late blight resistance gene protects against multiple Phytophthora species by recognizing a broadly conserved RXLR-WY effector. Mol Plant 15(9):1457–1469

Liu C, Nelson RS (2013) The cell biology of tobacco mosaic virus replication and movement. Front Plant Sci [Internet] 4(FEB):40765. Available from: www.frontiersin.org [cited 2024 Oct 2]

Liu Q, Yang F, Zhang J, Liu H, Rahman S, Islam S et al (2021) Application of CRISPR/Cas9 in crop quality improvement. Int J Mol Sci [Internet]. 22(8) Available from: /pmc/articles/PMC8073294/ [cited 2024 Sep 11]

Loureiro A, Da Silva GJ (2019) CRISPR-Cas: converting a bacterial Defence mechanism into a state-of-the-art genetic manipulation tool. Antibiotics [Internet] 8(1) Available from: /pmc/articles/PMC6466564/ [cited 2024 Sep 5]

Lucht JM (2015) Public acceptance of plant biotechnology and GM crops. Viruses [Internet] 7(8): 4254. Available from: /pmc/articles/PMC4576180/ [cited 2024 Sep 5]

Macovei A, Gill SS, Tuteja N (2012) microRNAs as promising tools for improving stress tolerance in rice. Plant Signal Behav [Internet] 7(10):1296. Available from: https://pmc.ncbi.nlm.nih.gov/articles/PMC3493417/ [cited 2024 Oct 19]

Mallikarjuna Swamy BP, Marundan S, Samia M, Ordonio RL, Rebong DB, Miranda R et al (2021) Development and characterization of GR2E Golden rice introgression lines. Sci Rep [Internet] 11(1):1–12. Available from: https://www.nature.com/articles/s41598-021-82001-0 [cited 2024 Sep 12]

Mamta B, Rajam MV (2017) RNAi technology: a new platform for crop pest control. Physiol Mol Biol Plants [Internet] 23(3):487. Available from: https://pmc.ncbi.nlm.nih.gov/articles/PMC5567704/ [cited 2024 Oct 19]

Mann CWG, Sawyer A, Gardiner DM, Mitter N, Carroll BJ, Eamens AL (2023) RNA-based control of fungal pathogens in plants. Int J Mol Sci [Internet] 24(15):12391. Available from: https://pmc.ncbi.nlm.nih.gov/articles/PMC10418863/ [cited 2024 Oct 19]

Marín-Sanz M, Barro F (2022) RNAi silencing of wheat gliadins alters the network of transcription factors that regulate the synthesis of seed storage proteins toward maintaining grain protein levels. Front Plant Sci [Internet] 13. Available from: /pmc/articles/PMC9395171/ [cited 2024 Sep 12]

Massa S, Pagliarello R, Cemmi A, Di Sarcina I, Bombarely A, Demurtas OC et al (2022) Modifying anthocyanins biosynthesis in tomato hairy roots: a test bed for plant resistance to ionizing radiation and antioxidant properties in space. Front Plant Sci [Internet] 13:830931. Available from: https://pmc.ncbi.nlm.nih.gov/articles/PMC8909381/ [cited 2024 Oct 19]

Mengstie MA, Wondimu BZ (2021) Mechanism and applications of CRISPR/Cas-9-mediated genome editing. Biol Int 15:353. Available from: /pmc/articles/PMC8388126/. [cited 2024 Jul 29]

National Academies of Sciences E and M, Studies D on E and L, Resources B on A and N, Prospects C on GECPE and F (2016) Genetically engineered crops through 2015. Available from: https://www.ncbi.nlm.nih.gov/books/NBK424540/ [cited 2024 Sep 5]

Nidhi S, Anand U, Oleksak P, Tripathi P, Lal JA, Thomas G et al (2021) Novel CRISPR–Cas systems: an updated review of the current achievements, applications, and future research perspectives. Int J Mol Sci [Internet]. 22(7) Available from: /pmc/articles/PMC8036902/ [cited 2024 Sep 5]

Núñez-Álvarez Y, Espie-Caullet T, Buhagiar G, Rubio-Zulaika A, Alonso-Marañón J, Luna-Pérez E et al (2024) A CRISPR-dCas13 RNA-editing tool to study alternative splicing. Nucleic Acids Res [Internet] 2024:1–14. Available from: https://doi.org/10.1093/nar/gkae682 [cited 2024 Oct 19]

O'Brien J, Hayder H, Zayed Y, Peng C (2018) Overview of microRNA biogenesis, mechanisms of actions, and circulation. Front Endocrinol (Lausanne) [Internet] 9(AUG):402. Available from: https://pmc.ncbi.nlm.nih.gov/articles/PMC6085463/ [cited 2024 Oct 19]

Ofosu R, Agyemang ED, Márton A, Pásztor G, Taller J, Kazinczi G (2023) Herbicide resistance: managing weeds in a changing world. Agron [Internet] 13(6):1595. Available from: https://www.mdpi.com/2073-4395/13/6/1595/htm [cited 2024 Sep 5]

Oliver MJ (2014) Why we need GMO crops in agriculture. Mo Med [Internet] 111(6):492. Available from: /pmc/articles/PMC6173531/ [cited 2024 Sep 5]

Ortega A, Seong K, Schultink A, de Toledo Thomazella DP, Seo E, Zhang E et al (2024) CRISPR/Cas9-mediated editing of Bs5 and Bs5L in tomato leads to resistance against Xanthomonas. Plant Biotechnol J [Internet]. Available from: https://onlinelibrary.wiley.com/doi/full/10.1111/pbi.14404 [cited 2024 Sep 14]

Paine JA, Shipton CA, Chaggar S, Howells RM, Kennedy MJ, Vernon G et al (2005) Improving the nutritional value of Golden Rice through increased pro-vitamin A content. Nat Biotechnol [Internet] 23(4):482–487. Available from: https://pubmed.ncbi.nlm.nih.gov/15793573/ [cited 2024 Aug 1]

Panwar V, McCallum B, Bakkeren G, Host-induced gene silencing of wheat leaf rust fungus Puccinia triticina pathogenicity genes mediated by the Barley stripe mosaic virus (2013) Plant Mol Biol [Internet] 81(6):595–608. Available from: https://pubmed.ncbi.nlm.nih.gov/23417582/ [cited 2024 Sep 12]

Parmar N, Singh KH, Sharma D, Singh L, Kumar P, Nanjundan J et al (2017) Genetic engineering strategies for biotic and abiotic stress tolerance and quality enhancement in horticultural crops: a comprehensive review. 3 Biotech [Internet] 7(4). Available from: /pmc/articles/PMC5507805/ [cited 2024 Jul 27]

Paul JY, Khanna H, Kleidon J, Hoang P, Geijskes J, Daniells J et al (2017) Golden bananas in the field: elevated fruit pro-vitamin a from the expression of a single banana transgene. Plant Biotechnol J [Internet]. 15(4):520. Available from: /pmc/articles/PMC5362681/ [cited 2024 Sep 12]

Poller W, Sahoo S, Hajjar R, Landmesser U, Krichevsky AM (2023) Exploration of the noncoding genome for human-specific therapeutic targets – recent insights at molecular and cellular level. Cells [Internet] 12(22):2660. Available from: https://pmc.ncbi.nlm.nih.gov/articles/PMC10670380/ [cited 2024 Oct 19]

Pramanik D, Shelake RM, Park J, Kim MJ, Hwang I, Park Y et al (2021) CRISPR/Cas9-mediated generation of pathogen-resistant tomato against tomato yellow leaf curl virus and powdery mildew. Int J Mol Sci [Internet]. 22(4):1–18. Available from: /pmc/articles/PMC7917697/ [cited 2024 Oct 2]

Qi T, Guo J, Peng H, Liu P, Kang Z, Guo J (2019) Host-induced gene silencing: a powerful strategy to control diseases of wheat and barley. Int J Mol Sci [Internet]. 20(1) Available from: /pmc/articles/PMC6337638/ [cited 2024 Sep 12]

Que Q, Chilton MDM, de Fontes CM, He C, Nuccio M, Zhu T et al (2010) Trait stacking in transgenic crops: challenges and opportunities. GM Crops [Internet] 1(4):220–229. Available from: https://pubmed.ncbi.nlm.nih.gov/21844677/ [cited 2024 Sep 12]

Rajput M, Choudhary K, Kumar M, Vivekanand V, Chawade A, Ortiz R et al (2021) RNA interference and CRISPR/Cas gene editing for crop improvement: paradigm shift towards sustainable agriculture. Plants [Internet] 10(9) Available from: /pmc/articles/PMC8467553/ [cited 2024 Oct 2]

Raman R (2017) The impact of genetically modified (GM) crops in modern agriculture: a review. GM Crops Food [Internet] 8(4):195. Available from: /pmc/articles/PMC5790416/ [cited 2024 Sep 5]

Rasheed A, Gill RA, Hassan MU, Mahmood A, Qari S, Zaman QU et al (2021) A critical review: recent advancements in the use of CRISPR/Cas9 technology to enhance crops and alleviate global food crises. Curr Issues Mol Biol [Internet] 43(3):1950. Available from: /pmc/articles/PMC8929161/ [cited 2024 Sep 5]

Ravelonandro M, Briard P, Scorza R, Callahan A, Zagrai I, Kundu JK et al (2021) Robust Response to Plum pox virus Infection via Plant Biotechnology. Genes (Basel) [Internet] 12-(6) Available from: /pmc/articles/PMC8227089/ [cited 2024 Sep 12]

Raza A, Razzaq A, Mehmood SS, Zou X, Zhang X, Lv Y et al (2019) Impact of Climate Change on Crops Adaptation and Strategies to Tackle Its Outcome: A Review. Plants [Internet] 8(2). Available from: /pmc/articles/PMC6409995/ [cited 2024 Oct 2]

Rees HA, Liu DR (2018) Base editing: precision chemistry on the genome and transcriptome of living cells. Nat Rev Genet [Internet] 19(12):770. Available from: /pmc/articles/PMC6535181/ [cited 2024 Oct 2]

Rhodes CJ (2014) Soil erosion, climate change and global food security: challenges and strategies. Sci Prog 97(2):97–153

Roberts AF, Devos Y, Lemgo GNY, Zhou X (2015) Biosafety research for non-target organism risk assessment of RNAi-based GE plants. Front Plant Sci [Internet] 6(NOVEMBER):958. Available from: https://pmc.ncbi.nlm.nih.gov/articles/PMC4635219/ [cited 2024 Oct 19]

Rodgers K, Mcvey M (2016) Error-prone repair of DNA double-strand breaks. J Cell Physiol [Internet] 231(1):15. Available from: /pmc/articles/PMC4586358/ [cited 2024 Sep 9]

Román Á, Andreu V, Hernández ML, Lagunas B, Picorel R, Martínez-Rivas JM et al (2012) Contribution of the different omega-3 fatty acid desaturase genes to the cold response in soybean. J Exp Bot [Internet]. 63(13):4973. Available from: /pmc/articles/PMC3427996/ [cited 2024 Sep 12]

Sami A, Xue Z, Tazein S, Arshad A, He Zhu Z, Ping Chen Y et al (2021) CRISPR–Cas9-based genetic engineering for crop improvement under drought stress. Bioengineered [Internet] 12(1): 5814. Available from: /pmc/articles/PMC8808358/ [cited 2024 Oct 2]

Seok H, Lee H, Jang ES, Chi SW (2017) Evaluation and control of miRNA-like off-target repression for RNA interference. Cell Mol Life Sci C [Internet] 75(5):797. Available from: https://pmc.ncbi.nlm.nih.gov/articles/PMC11105550/ [cited 2024 Oct 19]

Shabbir MAB, Shabbir MZ, Wu Q, Mahmood S, Sajid A, Maan MK et al (2019) CRISPR-cas system: biological function in microbes and its use to treat antimicrobial resistant pathogens. Ann Clin Microbiol Antimicrob [Internet] 18(1):21. Available from: /pmc/articles/PMC6611046/ [cited 2024 Sep 5]

Shi J, Gao H, Wang H, Lafitte HR, Archibald RL, Yang M et al (2017) ARGOS8 variants generated by CRISPR-Cas9 improve maize grain yield under field drought stress conditions. Plant Biotechnol J [Internet] 15(2):207. Available from: /pmc/articles/PMC5258859/ [cited 2024 Oct 2]

Shmakova AA, Shmakova OP, Karpukhina AA, Vassetzky YS (2022) CRISPR/Cas: history and perspectives. Russ J Dev Biol 2022 534 [Internet] 53(4):272–282. Available from: https://link.springer.com/article/10.1134/S1062360422040075 [cited 2024 Sep 5]

Sindhu AS, Maier TR, Mitchum MG, Hussey RS, Davis EL, Baum TJ (2009) Effective and specific in planta RNAi in cyst nematodes: expression interference of four parasitism genes reduces parasitic success. J Exp Bot [Internet] 60(1):315. Available from: /pmc/articles/PMC3071771/ [cited 2024 Sep 12]

Singh MB, Bhalla PL (2008) Genetic engineering for removing food allergens from plants. Trends Plant Sci 13(6):257–260

Singla A, Gupta OP, Sagwal V, Kumar A, Patwa N, Mohan N et al (2024) Beta-glucan as a soluble dietary fiber source: origins, biosynthesis, extraction, purification, structural characteristics, bioavailability, biofunctional attributes, industrial utilization, and global trade. Nutrients [Internet] 16(6):900. Available from: https://pmc.ncbi.nlm.nih.gov/articles/PMC10975496/ [cited 2024 Oct 19]

Song MS, Rossi JJ (2017).Molecular mechanisms of Dicer: endonuclease and enzymatic activity. Biochem J [Internet] 474(10):1603. Available from: https://pmc.ncbi.nlm.nih.gov/articles/PMC5415849/ [cited 2024 Oct 19]

Strobbe S, Van Der Straeten D (2017) Folate biofortification in food crops. Curr Opin Biotechnol 44:202–211

Tanin MJ, Saini DK, Kumar P, Gudi S, Sharma H, Kaur JP et al (2024) Iron biofortification in wheat: past, present, and future. Curr Plant Biol 38:100328

Tashkandi M, Ali Z, Aljedaani F, Shami A, Mahfouz MM (2018). Engineering resistance against Tomato yellow leaf curl virus via the CRISPR/Cas9 system in tomato. Plant Signal Behav [Internet] 3(10). Available from: https://pubmed.ncbi.nlm.nih.gov/30289378/ [cited 2024 Sep 13]

Thakare D, Zhang J, Wing RA, Cotty PJ, Schmidt MA (2017) Aflatoxin-free transgenic maize using host-induced gene silencing. Sci Adv [Internet] 3(3). Available from: https://pubmed.ncbi.nlm.nih.gov/28345051/ [cited 2024 Sep 12]

Touzdjian Pinheiro Kohlrausch Távora F, de Assis dos Santos Diniz F, de Moraes Rêgo-Machado C, Chagas Freitas N, Barbosa Monteiro Arraes F, Chumbinho de Andrade E et al (2022) CRISPR/Cas- and topical RNAi-based technologies for crop management and improvement: reviewing the risk assessment and challenges towards a more sustainable agriculture. Front Bioeng Biotechnol [Internet] 10:28. Available from: /pmc/articles/PMC9274005/ [cited 2024 Oct 2]

Tripathi L, Ntui VO, Tripathi JN (2020).CRISPR/Cas9-based genome editing of banana for disease resistance. Curr Opin Plant Biol [Internet] 56:118–26. Available from: https://pubmed.ncbi.nlm.nih.gov/32604025/ [cited 2024 Sep 14]

Tudi M, Ruan HD, Wang L, Lyu J, Sadler R, Connell D et al (2021) Agriculture development, pesticide application and its impact on the environment. Int J Environ Res Public Health [Internet] 18(3):1–24. Available from: /pmc/articles/PMC7908628/ [cited 2024 Oct 2]

Tyagi H, Rajasubramaniam S, Rajam MV, Dasgupta I (2008) RNA-interference in rice against Rice tungro bacilliform virus results in its decreased accumulation in inoculated rice plants. Transgenic Res [Internet] 17(5):897. Available from: /pmc/articles/PMC2522301/ [cited 2024 Sep 12]

Ullah A, Munir S, Badshah SL, Khan N, Ghani L, Poulson BG et al (2020) Important flavonoids and their role as a therapeutic agent. Molecules [Internet] 25(22) Available from: /pmc/articles/PMC7697716/ [cited 2024 Jul 5]

Umemoto N, Nakayasu M, Ohyama K, Yotsu-Yamashita M, Mizutani M, Seki H, et al. Two cytochrome P450 monooxygenases catalyze early hydroxylation steps in the potato steroid Glycoalkaloid biosynthetic pathway. Plant Physiol [Internet] 2016;171(4):2458–2467. Available from: https://doi.org/10.1104/pp.16.00137 [cited 2024 Sep 12]

Unniyampurath U, Pilankatta R, Krishnan MN (2016) RNA interference in the age of CRISPR: will CRISPR interfere with RNAi? Int J Mol Sci [Internet]. 17(3) Available from: /pmc/articles/PMC4813155/ [cited 2024 Oct 2]

Usman B, Nawaz G, Zhao N, Liao S, Qin B, Liu F et al (2021) Programmed editing of Rice (Oryza sativa L.) OsSPL16 gene using CRISPR/Cas9 improves grain yield by modulating the expression of pyruvate enzymes and cell cycle proteins. Int J Mol Sci [Internet]. 22((1):1–19. Available from: /pmc/articles/PMC7795130/ [cited 2024 Oct 2]

Wairich A, Ricachenevsky FK, Lee S (2022) A tale of two metals: biofortification of rice grains with iron and zinc. Front Plant Sci 13:944624

Walsh HA, Vanderschuren H, Taylor S, Rey MEC (2019) RNA silencing of south African cassava mosaic virus in transgenic cassava expressing AC1/AC4 hp- RNA induces tolerance. Biotechnol Rep [Internet] 24:e00383. Available from: /pmc/articles/PMC6864324/ [cited 2024 Sep 12]

Wang X, Larkins BA (2001) Genetic analysis of amino acid accumulation in opaque-2 maize endosperm. Plant Physiol [Internet]. 125(4):1766. Available from: /pmc/articles/PMC88833/ [cited 2024 Sep 12]

Wang M, Weiberg A, Lin FM, Thomma BPHJ, Da HH, Jin H (2016) Bidirectional cross-kingdom RNAi and fungal uptake of external RNAs confer plant protection. Nat plants [Internet] 2(10): 16151. Available from: /pmc/articles/PMC5040644/ [cited 2024 Sep 12]

Wang Q, Pan G, Wang X, Sun Z, Guo H, Su X et al (2024) Host-induced gene silencing of the Verticillium dahliae thiamine transporter protein gene (VdThit) confers resistance to Verticillium wilt in cotton. J Integr Agric 23:3358

Wilkinson RA, Martin C, Nemudryi AA, Wiedenheft B (2019) CRISPR RNA-guided autonomous delivery of Cas9. Nat Struct Mol Biol [Internet] 26(1):14. Available from: /pmc/articles/PMC7703833/ [cited 2024 Sep 5]

Xu Y, Li Z (2020) CRISPR-Cas systems: overview, innovations and applications in human disease research and gene therapy. Comput Struct Biotechnol J [Internet] (18):2401. Available from: /pmc/articles/PMC7508700/ [cited 2024 Sep 5]

Xue C, Greene EC (2021) DNA repair pathway choices in CRISPR-Cas9 mediated genome editing. Trend Genet [Internet] 37(7):639. Available from: /pmc/articles/PMC8187289/ [cited 2024 Sep 5]

Yang XT, Song J, Campbell-Palmer L, Walker B, Zhang Z (2012) Allergen related gene expression in apple fruit is differentially controlled by ethylene during ripening. Postharvest Biol Technol 63(1):40–49

Yang QQ, Yu WH, Wu HY, Zhang CQ, Sun SSM, Liu QQ (2021) Lysine biofortification in rice by modulating feedback inhibition of aspartate kinase and dihydrodipicolinate synthase. Plant Biotechnol J 19(3):490–501

Yang J, Fang Y, Wu H, Zhao N, Guo X, Mackon E et al (2023) Improvement of resistance to rice blast and bacterial leaf streak by CRISPR/Cas9-mediated mutagenesis of Pi21 and OsSULTR3;6 in rice (Oryza sativa L.). Front Plant Sci [Internet] 14:1209384. Available from: http://skl.scau.edu.cn/home [cited 2024 Sep 14]

Zand Karimi H, Innes RW (2022) Molecular mechanisms underlying host-induced gene silencing. Plant Cell [Internet] 34(9):3183. Available from: /pmc/articles/PMC9421479/ [cited 2024 Oct 2]

Zeballos CMA, Gaj T (2021) Next-generation CRISPR technologies and their applications in gene and Cell therapy. Trends Biotechnol [Internet]. 39(7):692. Available from: /pmc/articles/PMC8166939/ [cited 2024 Oct 2]

Zhang L, Yang XD, Zhang YY, Yang J, Qi GX, Guo DQ et al (2014) Changes in oleic acid content of transgenic soybeans by antisense rna mediated posttranscriptional gene silencing. Int J Genomic [Internet] 2014:921950. Available from: https://pmc.ncbi.nlm.nih.gov/articles/PMC4147191/ [cited 2024 Oct 19]

Zhang Y, Massel K, Godwin ID, Gao C (2018) Applications and potential of genome editing in crop improvement. Genome Biol [Internet] 19(1). Available from: https://pubmed.ncbi.nlm.nih.gov/30501614/ [cited 2024 Jul 31]

Zhang D, Zhang Z, Unver T, Zhang B (2021) CRISPR/Cas: a powerful tool for gene function study and crop improvement. J Adv Res [Internet] 29:207. Available from: /pmc/articles/PMC8020163/ [cited 2024 Sep 9]

Zhang S, Agyenim-Boateng KG, Zhang S, Gu Y, Qi J, Azam M et al (2023) QTL mapping for seed tocopherol content in soybean. Agronomy [Internet] 13(5):1188. Available from: https://www.mdpi.com/2073-4395/13/5/1188/htm [cited 2024 Oct 19]

Zhu M, Tong L, Xu M, Zhong T (2021) Genetic dissection of maize disease resistance and its applications in molecular breeding. Mol Breed [Internet] 41(5). Available from: /pmc/articles/PMC10236108/ [cited 2024 Sep 14]

RNA Interference (RNAi): A Powerful Tool for Crop Improvement and Stress Resilience

7

Ankita Ahlawat, Shikha Nara, Ankit Kumari, Yogesh Ahlawat, Ranjana Jaiwal, Pawan K. Jaiwal, and Darshna Chaudhary

Abstract

RNA interference (RNAi) has emerged as a powerful and precise technique for crop improvement that introduces small RNA molecules into cells, where they bind to complementary mRNA sequences and cause sequence-specific mRNA degradation, downregulating undesirable genes associated with susceptibility towards biotic and abiotic stress. RNAi-based gene silencing has gained recognition as an effective tool for elucidating gene functions and improving the agronomic traits of crops. This technology represents a naturally occurring cellular defence mechanism that suppresses target gene expression species specifically, as each species possesses a unique genetic sequence. RNAi enables the manipulation of desired traits by silencing genes in pathogens, pests, or plants. The conserved nature of the RNAi mechanism across diverse organisms further broadens its pertinence in both basic research and practical applications. RNAi technology has been used to develop resistant plants against biotic stresses like insects, fungi, viruses, nematodes, and aphids and abiotic stresses like drought, salinity, and heat. These applications not only contribute to higher crop yields and better quality but also reduce the need for chemical pesticides and fertilizers. It has also been used to develop male-sterile lines, plant architecture modification,

A. Ahlawat · S. Nara · A. Kumari · P. K. Jaiwal (✉) · D. Chaudhary (✉)
Centre for Biotechnology, Maharshi Dayanand University, Rohtak, Haryana, India
e-mail: darshnarajan.cbt@mdurohtak.ac.in

Y. Ahlawat
University Centre for Research and Development, Chandigarh University, Mohali, Punjab, India

Allied Health Sciences, Datta Meghe Institute of Higher Education and Research, Wardha, Maharashtra, India

R. Jaiwal
Department of Zoology, Maharshi Dayanand University, Rohtak, Haryana, India

Y. Ahlawat et al. (eds.), *Next-Generation Strategies for Crop Improvement*,
https://doi.org/10.1007/978-981-95-0309-4_7

resistance against parasitic plants, and better nutritional quality. Advancing these RNAi applications is essential for developing crops that are more resilient, efficient, and better prepared to meet the challenges of sustainable farming and global food security.

Keywords

RNA interference (RNAi) · Crop improvement · Gene silencing · Disease resistance · Stress tolerance

7.1 Introduction

Humans depend entirely on plants and their products for basic requirements, yet the demand for food supply is significantly rising because of the increasing population. Global warming also worsens food security challenges (Seto and Ramankutty 2016). Inadequate plant growth and extensive crop loss have been caused by adverse environmental conditions in the last 10 years (Mickelbart et al. 2015). Biotic and abiotic stresses have critically influenced crop productivity at the global level and significantly affected several countries' food security. Biotic stresses include bacteria, insects, fungi, nematodes, and viruses, whereas abiotic stress mainly comprises drought, heat, salinity, cold, and heavy metals (Kopecka et al. 2023). Conventional plant breeding techniques have undoubtedly enhanced agricultural production to a greater extent by producing disease-resistant cultivars. Still, it is a slow practice involving the transfer of undesired traits. New strains of pathogens have emerged that attack conventionally produced resistant cultivars. Therefore, it is hypercritical to develop new techniques to tackle the new virulent strains of pathogens (Abdellatef et al. 2021). RNAi is a powerful tool of genetic engineering for precise gene manipulations to produce quality traits in plants. RNAi technology is gaining popularity for its sequence-dependent mode of action, a unique feature of its efficacy over traditional methods.

7.2 RNAi: History and Mechanism of Action

In the late 1980s, researchers discovered unexpected gene silencing when extra copies of a gene were introduced into plants. In petunias, attempts to over-express pigmentation genes led to reduced pigmentation, a phenomenon later called post-transcriptional gene silencing (Napoli et al. 1990). Similarly, in *Neurospora crassa*, introducing a gene triggered the silencing of both the introduced and endogenous copies, and the process was called quelling (Romano and Macino 1992). In 1998, Andrew Fire and his co-workers published a groundbreaking study in *Nature*. They showed that introducing double-stranded RNA (dsRNA) into the nematode *Caenorhabditis elegans* specifically silenced the corresponding complementary mRNA. This experiment established RNAi as a natural, sequence-specific mechanism for silencing genes (Fire et al. 1998).

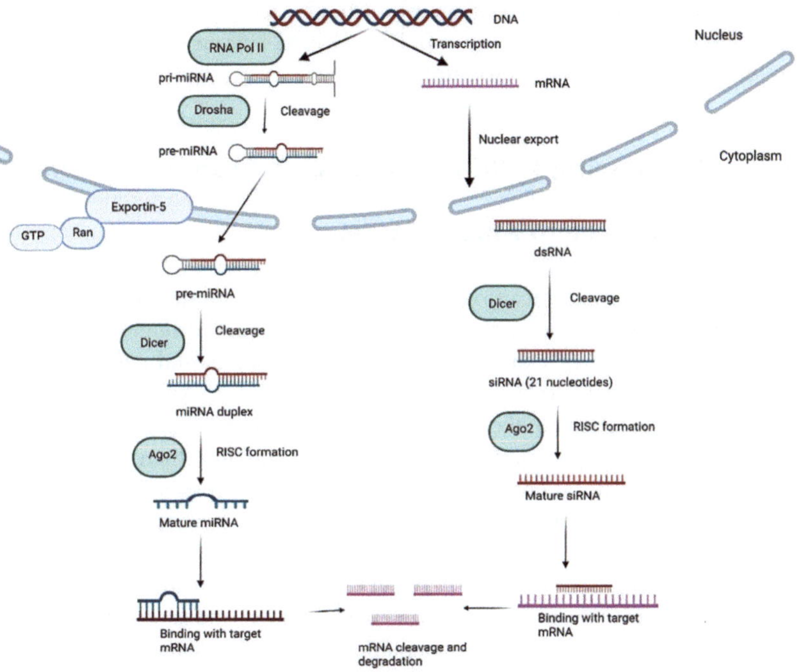

Fig. 7.1 The mechanism of RNA interference (RNAi) involves small interfering RNAs (siRNAs) or microRNAs (miRNAs) that mediate the suppression of gene expression

RNAi is a conserved gene silencing mechanism in eukaryotes that protects cells from viral infections. There are different mechanisms of RNAi based on the involvement of small RNA molecules such as siRNA, which is produced from exogenous dsRNA by the Dicer enzyme and works as an antiviral defence and transposon silencing. The second is miRNA—an endogenously transcribed non-coding RNA—which helps in fine-tuning gene expression during development, differentiation, and stress response. The RNAi process involves four key steps: (1) the Dicer, an RNase III endoribonuclease enzyme, splits double-stranded RNA (dsRNA) into siRNAs and miRNAs, (2) the RNA-induced silencing complex (RISC) assembles, (3) the guide strand of siRNA is incorporated into RISC, and (4) RISC cleaves the complementary mRNA, leading to its degradation (Fig. 7.1). The Argonaute protein within RISC is crucial in mRNA cleavage and passenger strand removal, ensuring specificity. The process is ATP-dependent, enabling multiple rounds of gene silencing. RNAi is fundamental to gene regulation and has extensive applications in functional genomics, biotechnology, and therapeutic development.

This technology harnesses a natural cellular mechanism where *double-stranded RNA (dsRNA)* triggers the degradation of *complementary messenger RNA (mRNA)*, thereby preventing the production of target proteins. RNAi can target essential genes in pests or pathogens, reducing their ability to infect or damage the crop. It can downregulate genes responsible for unwanted traits or improve nutritional content.

RNAi helps in managing tolerance to drought, salinity, and temperature stress by regulating stress-responsive genes. Thus, RNAi acts as a powerful, versatile, and targeted molecular tool in crop improvement.

7.3 Harnessing RNAi for Advancing Crop Improvement

7.3.1 Resistance to Abiotic Stress

Abiotic stress is a significant issue contributing to both qualitative and quantitative agricultural losses. Abiotic stresses such as cold, heat, drought, salinity, and metal toxicity hinder the growth and development of plants and cause significant yield losses (Zhang et al. 2023a, b; Kopecka et al. 2023). Advanced techniques have significantly accelerated the process of crop improvement by enabling targeted modifications and enhancing required traits, making them invaluable tools in addressing agricultural challenges posed by abiotic stresses (Varshney et al. 2011). RNAi has recently been the most exploited technique for coping with various abiotic stresses, as shown in Fig. 7.2. Plants produce numerous non-coding RNAs that regulate gene expression at multiple levels in response to abiotic stress. Studies have shown that miRNA expression fluctuates under stress conditions, influencing nutrient homeostasis and plant adaptation. In *Arabidopsis thaliana*, miR393 was unregulated under drought, salinity, and cold stress (Sunkar and Zhu 2004). RNAi-mediated RACK1silencing in *Oryza sativa* increased SOD accumulation, improving drought and ROS tolerance (Da-Hong et al. 2009). Additionally, targeting squalene

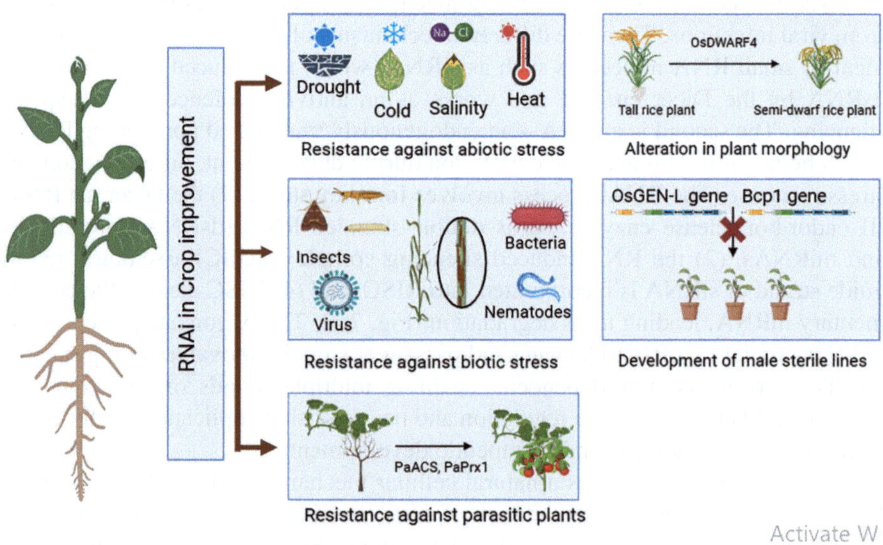

Fig. 7.2 Role of RNA interference in crop improvement

Table 7.1 Beneficial effects of RNAi-mediated gene silencing on abiotic stress resistance

Host plant	Gene(s) targeted	Result	References
A. thaliana	AtMYB74 (transcription factor)	Resistance to salt stress	Xu et al. (2015)
C. arietinum	DREB (transcription factor)	Improved resistance to drought stress	Hajyzadeh et al. (2015)
A. thaliana	*LCR*	Higher resistance to cold stress	Song et al. (2016)
C. limon	*ACO*	Growth improved in cold stress	Zhang et al. (2016)
H. vulgare	NAC transcription factor, *ARF*	Better growth in drought conditions	Fard et al. (2017)
A. thaliana	*ARF10*, *ARF16*, and *ARF17*	Resistance to heat stress	Lin et al. (2018)
N. tabacum	*BrDST71*	Enhanced resistance in drought stress	Park et al. (2018)
N. tabacum	*NtNCED3–2*	Resistance to high ABA	Yang et al. (2018)
O. sativa	*CKX 2*	Improved resistance to salt stress	Joshi et al. (2018)
O. sativa	*OsSPL3*	Higher resistance to cold stress	Zhou and Tang (2019)
A. thaliana	*AtPIMT2*	Enhanced resistance to salt stress	Ghosh et al. (2020)
M. sativa	*SPL9*	Improved resistance to drought stress	Hanly et al. (2020)
M. domestica	*MdSIZ1*	Resistance to drought stress	Chu et al. (2021)
G. hirsutum	*GhADF1*	Resistance to drought stress	Qin et al. (2022)
M. domestica	*MdGH3*	Higher resistance to drought stress	Jiang et al. (2022)
O. sativa	*OsABA8ox1*	Improved growth under salinity stress	Liu et al. (2022)

synthase (SQS) via RNAi enhanced drought resistance in rice (Manavalan et al. 2012), demonstrating RNAi's potential in developing climate-resilient crops. In addition to the model plant *Arabidopsis*, various other plant species have been targeted to create stress-resistant transgenic lines, as shown in Table 7.1.

7.3.2 Resistance to Biotic Stress

Pathogens are the organisms that cause diseases, such as bacteria, viruses, and fungi, along with insects and parasitic nematodes, which are key biotic stressors that impede crop plant species' growth, development, and productivity, ultimately affecting both their quality and quantity. Globally, these biotic stressors contribute to an estimated 40% loss in major staple and commercial crops (Oerke 2006; Savary et al. 2019; Chaudhary et al. 2022). RNAi has been extensively used to develop resistance against various biotic stressors (Fig. 7.2).

7.3.2.1 Nematode Resistance

Plant-parasitic nematodes (PPNs) are highly specialized and evolved pests that threaten global agriculture by damaging more than 2000 plant species. This broad host range makes them one of the most destructive groups of plant pests (Barker and Koenning 1998). PPNs feed on plant roots and other tissues, cause significant damage, and often create entry points for other possible pathogens of plants, such as bacteria and fungi. The global impact of PPNs is ambiguous, with billions of dollars in crop losses reported annually (Elling 2013). Root-knot nematodes (*Meloidogyne* spp.) and cyst nematodes (*Globodera* spp.) are the most destructive and widespread among them. Many conventional strategies have been tried to control them, such as nematicides, soil sterilization, land fallowing, crop rotation, and removing and burning diseased plants (Chitwood 2003). All these conventional approaches have some drawbacks and damage the environment by using chemicals (Tamilarasan and Rajam 2013). In these scenarios, RNA interference emerged as the most promising strategy to control the nematodes by simply targeting the effector protein genes (Table 7.2).

7.3.2.2 Resistance to Virus

Plant-infecting viruses are microscopic pathogens specifically targeting plants, causing various diseases that can lead to significant agricultural losses. Farmers use virus-resistant crop varieties to manage plant viral diseases, control vector populations and crop rotation, and maintain strict sanitation in farming operations. However, these conventional methods cannot control viral infections efficiently. RNA interference techniques have become widely used in producing transgenic lines with improved virus resistance (Pooggin 2017), as shown in Table 7.3. The pathogen-derived resistance (PDR) strategy involves introducing functional or non-functional pathogen sequences into a host genome or delivering them transiently into host cells. This approach effectively protects plants from viral infections. The first indication of an RNA-mediated mechanism for PDR was observed in transgenic lines expressing untranslatable sense or antisense versions of coat protein gene sequences of viruses (Lindbo and Dougherty 2005).

7.3.2.3 Resistance to Bacteria

Bacterial pathogens are a significant challenge in agriculture, mainly in tomatoes, soybeans, and bananas, causing devastating diseases that significantly constrain crop production and result in substantial economic losses worldwide each year. The impact extends to indirect losses, increasing costs for disease management through chemical treatments, resistant crop varieties, and advanced agricultural practices. RNAi has shown results in enhancing bacterial disease resistance, with significant success in *Arabidopsis thaliana* (Saurabh et al. 2014), though studies on other plant species remain limited. Some prokaryotes, such as bacteria and phytoplasmas, lack conserved RNAi machinery, making it impossible to induce RNAi in these plant

Table 7.2 RNAi-mediated resistance to plant-parasitic nematodes

Host species	Pathogen	Gene	Result	References
S. lycopersicum	*M. incognita*	*Mi-col-1* and *Lemmi-5*	Reduction in the number of females and eggs	Banerjee et al. (2018)
S. melongena	*M. incognita*	*Mi-msp-1*	Enhanced resistance to nematodes	Chaudhary et al. (2019)
G. max	*H. glycines*	*HgPrp17*and*HgY25*	Resistance to soybean cyst nematodes	Tian et al. (2019)
A. thaliana	*M. incognita*	*Mi-msp-2*	Decrease in the number of galls, egg masses, and females	Joshi et al. (2019)
A. thaliana	*M. incognita*	*Mi-msp3, Mi-msp5, Mi-msp18* and *Mi-msp24*	The number of galls reduced	Joshi et al. (2020)
N. benthamiana	*M. incognita*	Chitin synthase, glucose-6-phosphate isomerase, and trehalase	Decrease in the number of eggs and root knots	Mani et al. (2020)
O. sativa	*Meloidogyne graminicola*	*Flp1 and flp2*	Decrease in the number of galls and eggs	Hada et al. (2020)
S. lycopersicum	*M. incognita*	20 genes	Enhanced resistance	Iqbal and Jones (2020)
S. lycopersicum	*M. incognita*	Integrase	Reduction in the number of galls	Koulagi et al. (2020)
S. lycopersicum and V. angularis	*M. incognita*	*Mi-flp1, Mi-flp12*, and *Mi-flp18*	Reduced penetration ability	Banakar et al. (2020)
T. aestivum	*H. avenae*	RanBPM, serpin protease inhibitor, galectin, chitinase, cathepsin-L-like cysteine proteinase, FMRF-amide-like peptide, and venom allergen protein	Reduction in development and reproduction	Dutta et al. (2020)
A. thaliana	*M. incognita*	*Mi-msp10* and *Mi-msp23*	Reduction in gall formation and egg production	Kumar et al. (2022)
A. thaliana	*M. incognita*	*Minc16803*	number of galls and egg masses reduced	Moreira et al. (2024)

Table 7.3 Role of RNAi in conferring virus resistance

Host plant	Pathogen	Target gene (s)	Result	References
Oryza sativa	Rice black streak dwarf virus	*S7–2*	Higher resistance	Ahmed et al. (2017)
Solanum tuberosum	Potato virus X, Y, and S	*CP*	Nearly 100% resistance	Hameed et al. (2017)
Glycine max	Mungbean yellow mosaic India virus	*AC2*	Enhanced resistance	Ramesh et al. (2019)
Manihot esculenta	South African cassava mosaic virus	*AC1/AC4*	Decreased symptoms and viral loads	Walsh et al. (2019)
Nicotiana benthamiana	Cucumber green mottle mosaic virus	*CP, Rep, MP*	Enhanced resistance	Liang et al. (2019)
Vigna unguiculata	Bean common mosaic virus	*Nib, CP*	Protection against virus	Worrall et al. (2019)
Solanum lycopersicum	Tomato leaf curl New Delhi virus	*AC1*	Enhanced resistance	Sharma and Prasad (2020)
Solanum tuberosum	Potato virus Y	*elF4E*	Improved resistance	Miroshnichenko et al. (2020)
Glycine max	Soybean mosaic virus	*GmVma12*	Higher resistance	Luan et al. (2020)
Musa sp.	Banana bract mosaic virus	*BBrMV* replicase gene	Resistance to virus	Rs et al. (2021)
Saccharum officinarum	Sugarcane mosaic virus	*CP*	Higher resistance	Widyaningrum et al. (2021)
Solanum lycopersicum	Tomato spotted wilt ortho-tospovirus	*NSs*	Antiviral protection	ul ain Sajid and Elci (2024)

pathogens. *Agrobacterium tumefaciens* causes crown gall disease, which induces tumour-like growths. Resistance to crown gall was achieved in *A. thaliana* and *L. esculentum* by RNAi-mediated silencing of the *iaaM* (tryptophan monooxygenase) and *ipt* (isopentenyl transferase) oncogenes. Wild-type controls had an average tumorigenesis rate of 97.5%, while transgenic *Arabidopsis* lines had a 0.0%–1.5% tumorigenesis rate (Escobar et al. 2001). In 2002, *iaaM* and *ipt* were targeted in walnut (*Juglans regia* L.), which enhanced resistance to crown gall disease (Escobar et al. 2002). Citrus canker, caused by *Xanthomonas citri*, is a common disease in citrus fruits. Citrus *PDS* (phytoene desaturase) and *CalS1* (callose synthase) genes were targeted, with dsRNA vectors through *Agrobacterium*-mediated transformation. The effectiveness of gene silencing was validated by phenotypic and genetic analyses in lemon trees and other Rutaceae species (Enrique et al. 2011).

7.3.2.4 Resistance Against Insects

Insects are the largest group of animals on Earth, causes diseases in plants by feeding, chewing, sucking cell sap and act as carrier in transmitting pathogens.

Chemical pesticides, though effective against insects, have led to resistance and pose risks to human health and the environment (Sparks and Nauen 2015). As a safer, species-specific alternative, RNAi offers eco-friendly pest control by silencing essential genes, causing insect mortality without affecting non-target organisms (Whyard et al. 2009). Insect susceptibility to ingested dsRNA varies by species; coleopterans show strong RNAi responses, while lepidopterans are less responsive (Baum and Roberts 2014). For field application of RNAi, *in planta* transformation is generally raised by generating genetically modified (GM) crops expressing dsRNA against targeted insects (Table 7.4). Several studies have proved the efficacy of this technology in controlling insect pests (Mao et al. 2007; Baum et al. 2007; Pitino et al. 2011). Technical challenges, high costs, and public concerns limit the commercial use of transgenic RNAi, prompting the adoption of alternative delivery methods like exogenous dsRNA application through foliar spray, feeding, root drenching, or injection. These approaches have effectively silenced target genes in various studies (Table 7.4). Microinjection of dsRNA led to 91.4% gene downregulation in *Myllocerus undecimpustulatus* (Pinheiro et al. 2020), significant larval mortality in *Frankliniella occidentalis* (Han and Rotenberg 2024), and 74% downregulation of the hunchback gene in *Acyrthosiphon pisum* (Ye et al. 2019). The foliar spray of dsRNA in *Leptinotarsa decemlineata* (Máximo et al. 2020), *Bemisia tabaci* (Chakraborty and Ghosh 2022), and *Agrilus planipennis* (Pampolini and Rieske 2023) effectively silences the target gene expression. Efficient RNAi requires proper dsRNA uptake, stability in the gut, uptake and processing by epithelial cells, and effective systemic spread to silence the target gene (Cooper et al. 2019). However, the limiting efficacy of RNAi occurs due to refractory target gene selection, insufficient dsRNA uptake, dsRNA instability, improper dsRNA spreading, core RNAi machinery deficiency, etc.

7.3.2.5 Resistance Against Parasitic Plants

Parasitic plants obtain nutrients from host plants, often harming them. Plants have developed resistance mechanisms to defend against these threats, including physical barriers like thicker cell walls, specialized structures, and chemical responses that hinder parasitic growth. Some plants also use immune-like systems to counteract these attacks. Understanding these defences is key to improving agriculture and managing parasitism's ecological impacts (Gheysen and Vanholme 2007; Banerjee et al. 2017; Dutta et al. 2021). Parasitic plants are a group of fascinating plants that uptake some or all the nutrients from other host plants and are known as hemiparasites or holoparasites, respectively. Among the diverse parasitic plants, the Orobanchaceae family is the most prominent parasitic family, comprising around 2000 species within 90 genera (Li et al. 2019). It is a unique family with a broad spectrum of parasitic lifestyles, ranging from facultative hemiparasites to obligate holoparasites (Westwood et al. 2010). Parasitic plants from this family cause substantial damage to a wide range of crops across several plant families and also affect certain woody species, such as olive trees (Parker 2013). The RNAi approach is used to silence the viral genes of parasitic plants to control their growth in host plants, as shown in Fig. 7.2. *Striga asiatica* is a parasitic plant that attacks the roots of host

Table 7.4 Delivery methods and biological effects of RNAi-mediated gene silencing in insect pests

Target insect	Target gene(s)	Mode of delivery	Observed biological consequences in insects	References
Lepidoptera				
Helicoverpa armigera	Acetylcholinesterase *(AChE)*, ecdysone receptor *(EcR)*, v-ATPase-A *(vAA)*	Transgenic plant expressing dsRNAs	More than 68% mortality was observed by silencing *AChE*, while *EcR* downregulation showed molting defects	Sharif et al. (2022)
	Lipase and chitinase	CNP-dsRNA via feeding	100% mortality with reduced pupation and impaired moth emergence was observed	Kolge et al. (2023)
	Chitin biosynthesis pathway gene (*AGK, AGM,* and *UAP*)	Injection	Significant mortality (up to 69%) with developmental retardation and impaired chitin synthesis was observed	Das et al. (2023)
	Acetylcholinesterase (*Ace1*)	Transgenic plant expressing dsRNA	Significantly higher mortality of 55.5% was observed in lines 4 and 10	Jaiswal et al. 2024
	E93 (ecdysone-induced transcription factor)	Injection	A 60% mortality rate was observed at the larval-pupal stage	Wang et al. (2024)
Spodoptera litura	Insulin receptor	Microinjection	Impair ovarian development and decrease fecundity	Pan et al. (2022)
	Clock period (Per)	Microinjection	The mating rate significantly decreased by 33.33% due to disrupted sex pheromone communication behaviour	Yang et al. (2023)
	Gut receptor genes *CAD, ABCC2, ALP1,* and *APN*	Bacterial-mediated RNAi	Pre-treatment with CAD and ABCC2 dsRNA significantly reduced larvae susceptibility after Cry1ACF ingestion	Dutta et al. 2023
		Oral delivery		

(continued)

Table 7.4 (continued)

Target insect	Target gene(s)	Mode of delivery	Observed biological consequences in insects	References
	Juvenile hormone epoxide hydrolase		20% larval-pupal intermolt mortality and malformed pupae were recorded	Sarkar and Kalia 2023
	Trypsin-like serine protease (*TLSP*)	Microinjection	Significantly reduced TLSP expression and reduced male fertility	Yadav et al. (2024)
Coleoptera				
Diabrotica virgifera virgifera	*ABCB 1*	Artificial diet	Larvae became insensitive to Cry 3A toxin	Niu et al. (2020)
	Lethal giant larvae gene (*Lgl*)	Oral delivery	70% gene suppression was observed with no phenotypic abnormalities	Cooper et al. (2021)
	Spt 5, *Spt 6*, and *Rpll 33*	Artificial diet	90% mortality was observed	Knorr et al. (2021)
Leptinotarsa decemlineata	Proteasome subunit beta type-5	Foliar spray	90% mortality was observed with Ledprona	Rodrigues et al. (2021)
	Myosuppressin	Injection	Reduced oviposition (80%) and oocyte size (69%) were observed	Gui et al. (2021)
	Ecdysone receptor (*EcR*), ultraspiracle	Injection	Significantly inhibited oocyte development in female adult	Zhang et al. (2023a, b)
	Acetylcholinesterase (*ACE1*)	Foliar application	670 bp dsRNA showed the highest mortality with reduced weight compared to other dsRNAs	Julian-Chávez et al. (2024)
	Ago1	Injection	Significantly reduced the survival of insects when exposed to chlorantraniliprole	Bastarache et al. (2024)
Hemiptera				
Bemisia tabaci	Target of rapamycin (*TOR*)	Oral delivery	42.5% adult mortality was	Puri and Jindal (2021)

(continued)

Table 7.4 (continued)

Target insect	Target gene(s)	Mode of delivery	Observed biological consequences in insects	References
			recorded with reduced fecundity	
	Trehalose-6-phosphate synthase (*TPS*)	Transgenic plant-mediated RNAi	90% mortality was observed with retarded growth and decreased fecundity	Gong et al. (2022)
	v-ATPase	Transgenic plant expressing siRNA	A 57.1% mortality rate was observed	Pizetta et al. (2022)
	CAPAr	Oral delivery	A 30.74% mortality rate was recorded, with reduced survival and fecundity	Thakur and Jindal (2022)
	Alpha (α)-tubulin, ecdysone receptor (*ECR*), sucrase, trehalase (*TRE-1*), and trehalose transporter (*TRET-1*)	Oral delivery	Mortality rate was increased with increasing concentration of dsRNA with maximum mortality of $\geq 60\%$	Ramasamy et al. (2023)
Rhodnius prolixus	Insulin receptor	Injection	Significantly affected lipid synthesis and reproductive function	Silva-Oliveira et al. (2021)
	CYP4EM1 and *CYP4EM2*	Injection	Significantly reduced survival and increased susceptibility to deltamethrin	Paim et al. (2022)
	Eukaryotic translation initiation factor 3 subunit m (*eIF3m*)	–	Significantly affected ecdysis, ovary development, oviposition, and caused premature death	Ameijeiras et al. (2023)

plants. In 2007, de Framond and co-workers engineered transgenic maize for resistance against *S. asiatica* through the RNAi approach, and the results indicated slower growth of the parasite than complete resistance. The transgenic tomato was engineered to have a virus-induced gene-silencing vector against *PaACS, PaM6PR*, and *PaPrx1* genes of *Phelipanche aegyptiaca,* and its growth was significantly reduced (Dubey et al. 2017). Three genes (*PaM6PR, PaCWI*, and *PaSUS1)* of

Phelipanche aegyptiaca were silenced, and transgenic resistance in tomatoes against *P. aegyptiaca* was achieved (Farrokhi et al. 2019). *Orobanchecumana* is a holo parasite that causes serious damage to the yield of sunflowers; *EXPA6, QR1, CKX5, WRI1, PME4, PGLR5, JAR1, BXL6, PYL8,* and *BIG* were silenced by virus-induced gene silencing (VIGS) to express RNAi inducers and resulted into phenotypic abnormalities and significant resistance against *O. cumana* (Jiang et al. 2021).

Another major plant parasite globally distributed is *Cuscuta campestris*, which is commonly known as field dodder. It belongs to the Convolvulaceae family and is well known for its thin, yellow, and thread-like stems twisted around its host plants (Dawson et al. 1994). RNAi approach has been used to control Cuscuta by targeting the enzymes of critical biosynthetic pathways for its life cycle. Phytoene desaturase (*PDS*) is an enzyme involved in the biosynthesis of carotenoids. *PDS* in *C. campestris* was knocked down by VIGS using tobacco rattle virus in *Arabidopsis thaliana,* and it resulted in decreased accumulation of beta-carotene, synthesized from phytoene, and caused a significant reduction in the growth of *C. campestris* (Dyer et al. 2021).

7.3.3 Developing Male-Sterile Lines

One important characteristic deliberately chosen to increase crop output through hybridization significantly is the establishment of male sterility. This trait is essential for producing hybrid crops because it permits regulated cross-pollination, which eventually results in higher yields and higher-quality crops. Advances in genetic engineering have made it possible to create male-sterile plant varieties by manipulating specific genes. For example, technologies like RNA interference (RNAi) have successfully generated male-sterile varieties by silencing particular genes, as shown in Fig. 7.2. Transgenic rice plants were developed targeting the *OsGEN-L* gene via RNAi, resulting in low fertility of most transgenic lines, with some male-sterile lines. Transgenic plants displayed immature pollens that culminated in the development of defective microspores (Moritoh et al. 2005). *Bcp1* is a male-specific gene that actively functions in the tapetum and microspores. In a study, *Bcp1* was targeted in *Arabidopsis thaliana* using RNAi. Out of 58, around 49 *Arabidopsis* lines were transformed, resulting in sterile male lines. Transgenic plants were not different from non-transgenic plants in their morphology (Tehseen et al. 2010). S-adenosylmethionine decarboxylase (*SAMDC*) is an enzyme crucial for the biosynthesis of polyamines. In a study, *SAMDC* was silenced in the tapetal tissues of tomatoes using the RNAi approach to check its outcome on the development of tapetal tissue. Transgenic lines expressed siRNAs in the anther, producing sterile pollen grains with distorted structures. Using the RNAi approach, Bae and his co-workers targeted rice's allene oxide synthase (OsAOS) gene. This gene plays a crucial role in synthesizing jasmonic acid and developing anthers, resulting in transgenic plants exhibiting severe male sterility (Bae et al. 2010). *SAMDC* was targeted by RNAi in tomatoes, resulting in the development of sterile pollens, eventually developing the male-sterile lines, and the female fertility remained unaffected (Sinha and Rajam 2013). In a study, the ascorbic acid oxidase gene, *OsPTD1*, was downregulated by RNAi and resulted in retarded growth of the pollen tube that eventually resulted in male sterility (Chen et al. 2021a, b).

7.3.4 Changes in Plant Morphology Due to Climatic Conditions

Plant architecture, specifically the aerial portion of a higher plant, is primarily shaped by factors influencing shoot branching, plant height, and inflorescence morphology (Wang and Li 2006). Plant architecture can be strategically modified to reduce the adverse impacts of climate change on crop productivity by silencing genes via RNAi, as shown in Fig. 7.2. Optimizing traits like shoot branching, plant height, and canopy structure improves resilience by enhancing light interception, water-use efficiency, and abiotic stress tolerance. For instance, to create lodging-resistant cultivars with higher grain yields, tall rice, and wheat lines were switched to semi-dwarf lines during the Green Revolution (Peng et al. 1999). By optimizing the root architecture to maximize the absorption of water and nutrients, plants can be grown in soils prone to drought and nutrient deficiencies (de Dorlodot et al. 2007). miRNAs have been shown to regulate plant architecture, directly or indirectly manipulating plant structure, biomass accumulation, and yield (Rubinelli et al. 2013). Wang and Li have reviewed the key genes involved in plant architecture. Desirable traits in the architecture of plants can be achieved by manipulating those key genes using RNA interference techniques (Wang and Li 2006). Silencing the *OsDWARF4* gene in rice by RNAi, plants of small height were produced with erect leaf structures (Feldmann 2006). Using RNAi technology, the GA 20-oxidase gene (responsible for synthesizing gibberellic acid) was suppressed to transform the taller rice variety into a semi-dwarf variety. The resulting transgenic rice showed increased yield, attributed to a notable enhancement in panicle length, seed count per panicle, and the weight of individual seeds (Qiao et al. 2007). In *Populus* plants, overexpression of *Cg1* miRNA led to a significant reduction in internode length, enhanced growth of axillary meristems, and approximately a 30% decrease in stem lignin content compared to untransformed controls (Rubinelli et al. 2013). Brassinosteroid insensitive1 is a BR receptor, and its homolog Zmbri1 was knocked out in maize using RNAi. The results showed that transgenic plants showed dwarf heights with short internodes. Leaves were darker and twisted with reduced auricle formation (Kir et al. 2015). The *GA20*-oxidase gene was suppressed in apples by RNA interference; in transgenic plants with low height, an increase in the number of nodes with short internodal areas was observed (Zhao et al. 2016). Squamosa promoter binding protein-like7 *(SPL7)* regulates optimal plant architecture. Downregulation of *SPL7* in rice by RNAi produced dwarf plants with increased tillers, producing higher yields (Dai et al. 2018). In a study, *OsMADS18* was silenced in rice by RNAi, and transgenic lines showed delayed seed germination and young seedling growth (Yin et al. 2019). Squamosa promoter binding protein-like 14 (*OsSPL14*) binds to the promoters of (*OsPIN1b*) and (PILS6b). Downregulation of *OsPIN1b* or *PILS6b* increased the tiller count in rice (Li et al. 2022).

7.3.5 Improvement in Nutritional Qualities of Crop Plants

To improve health and protect rural populations of developing countries from chronic diseases, RNAi technology has been used over the past 20 years to create biofortified crops enriched with certain macro-, micro-, and phytonutrients (biofortification) and/or reduced levels of harmful compoundsin edible parts of the plants sustainably and economically without compromising yield or consumer preferences (Jaiwal et al. 2019) (Table 7.5).

7.3.6 Therapeutics

The natural flora of the world has proven to be an invaluable resource for enhancing health and treating a wide range of illnesses. Numerous plant species effectively remedy various medical conditions, particularly in regions such as Africa, Asia, and South America, where traditional medicine remains deeply rooted in cultural practices (Khalid et al. 2012). Plants have been utilized for centuries, offering therapeutic benefits derived from their bioactive compounds. Modern research is increasingly validating the medicinal properties of these plants, bridging the gap between traditional knowledge and contemporary science. This highlights the enduring importance of botanical resources in global healthcare systems. The plant genome comprises an estimated 20,000–60,000 genes, reflecting the complexity and diversity of plant genetic makeup. Among these, a significant subset—approximately 15%–25% is dedicated to synthesizing secondary metabolites (Rastegari et al. 2019). So, RNA interference can be used to monitor the quality and quantity of manipulation of the secondary metabolites. Manipulate these genes. Table 7.6 presents RNAi-mediated gene-silencing efforts in medicinal plants and their observed effects.

7.3.7 Development of Seedless Fruits

Parthenocarpy is a natural process of producing fruits and vegetables directly from the ovary without fertilization. This is especially important for plants that struggle with pollination or fertilization due to extreme temperatures or other challenging environmental conditions (Gorguet et al. 2005). As it ensures the production of fruits, Parthenocarpy is a desirable trait in plants. Recent research has revealed that fruits with no seeds can have better flavour and texture. Seedlessness can increase the fruit yield as it is independent of the probability criterion of pollination. Seedless fruits have a long shelf life because seeds are a source of ethylene, which is responsible for fruit ripening. Seedless fruits can be artificially produced by interfering with the genes responsible for seed formation. A balanced level of phytohormones is the key requirement of seed synthesis (Wang et al. 2005). We know that most plant developmental processes depend on auxin signalling, which is regulated by two families of transcription factors. One is the auxin response factor

Table 7.5 RNAi-mediated improvement in nutritionally essential traits in crop plants

Crop	Gene targeted	Trait improved with the advantage	Reference
Cotton	Stearoyl-acyl-carrier protein Δ9 desaturase and oleoyl-phosphatidylcholine ω6-desaturase	Greatly elevated stearic (from the normal 2%–3% to 40%) and oleic fatty acids (from 15% to 70%)	Liu et al. (2002)
Soybean	Omega-3 fatty acid desaturase (FAD3)	Reduced α-linoleic acid content	Flores et al. (2008)
Brassica	*BnFAE1*	Reduced erucic acid	Shi et al. (2015)
Tomato	Photomorphogenesis regulatory gene, *DET 1*	Increased both carotenoid and flavonoid contents	Davuluri et al. (2005)
Potato	Beta-carotene hydroxylase gene, *BCH*	Increased β-carotene and lutein contents	Van Eck et al. (2007)
Brassica napus	*ε-CYC*, lycopene *ε*-cyclase	Increased total carotenoid content and reduction in overall fatty acid content	Yu et al. (2008)
Brassica napus	Regulatory gene DE-ETIOLATED1, *DET 1*	Increased carotenoid levels and decreased levels of sinapate esters simultaneously	Wei et al. (2009)
Tomato	*SlNCED1* gene encodes 9-cis-epoxycarotenoid dioxygenase (NCED), a crucial enzyme involved in ABA biosynthesis	Decreased ABA resulted in ethylene synthesis, increased lycopene and β-carotene, and deep red fruits	Sun et al. (2012)
Maize	Glucan, water dikinase (and phosphoglucan, water dikinase), *AtGWD*	20-fold higher starch in leaves with no increase in total plant biomass	Weise et al. (2012)
Arabidopsis	*AtGWD* and phosphoglucan phosphatase (*AtSEX4)*	Sevenfold higher starch in leaves	Weise et al. (2012)
Wheat	Starch branching enzyme II (SBAIIa and SBEIIb)	Silencing both SBEIIa and SBEIIb led to starch containing up to 70% amylose, improving human health through its resistant starch content	Regina et al. (2005)
Barley	SBE IIa and SBEIIb	Repression of both increased high amylose > 70%	Regina et al. (2010)
Rice	*GluB*	Low glutelin content (named LGC-1)	Kusaba et al. (2003)
Wheat	γ-Gliadins	Reduced glutenin	Gil-Humanes et al. (2008)
-Do-	α-, γ-, and ω-Gliadin	Reduced gliadin content by up to 90% and decreased the formation of celiac disease-related epitopes	Gil-Humanes et al. (2010)
Maize	22 kDa	Enhanced lysin	Song et al. (2001)

(continued)

Table 7.5 (continued)

Crop	Gene targeted	Trait improved with the advantage	Reference
Wheat	Inositol pentakisphosphate kinase (IPK1)	Increase in Fe and Zn	Aggarwal et al. (2018)
Rice	Myo-inositol-3-phosphate synthase *(MIPS), inositol* 1,3,4,5,6-pentakisphosphate 2-kinase gene *(IPK1)*	Lowered phytic acid, an increase in Fe content in endosperm	Ali et al. (2013a, b)
Soybean	Gm MIPS1 (myoinositol 3-phosphate synthase)	Reduced phytate/increase in Fe, Zn, and Ca	Kumar et al. (2019)
Cotton	Delta-cadinene synthase	Reduced toxic terpenoid gossypol	SunilKumar et al. (2006)
Cotton	Delta-cadinene synthase	Ultra-low gossypol	Rathore et al. (2012)
Coffee	Theobromine synthase (*CaMXMT1*)	Reduced caffeine content in seed (decaffeinated coffee) by 70%	Ogita et al. (2004) and Pathak and Gogoi (2016)
Opium poppy	Codeine reductase (COR)	Blocked morphine formation and accumulated a pathway intermediate, reticuline	Allen et al. (2004)
Tobacco	Nicotine demethylase	Lowered nicotine	Lewis et al. (2008)
Cassava	Cytochrome P450 genes, CYP79D1/D2	Removed cyanogen, linamarin	Meena et al. (2017)
Potato	Sterol side chain reductase 2	Reduced steroidal glycoalkaloids	Sawai et al. (2014)
Grass pea (*Lathyrus sativus*)	BOAA (beta-N-oxalyl amino alanine)	Removed a neurotoxin amino acid, BOAA	Angaji et al. (2010)
-do-	CoA synthase	Removed a neurotoxin amino acid, BOAA	Shiv Kumar et al. (2011)
Onion	Lachrymatory factor synthase (LFS)	Tearless onion	Eady et al. (2008)
Apple	Mal d1 allergen gene	Reduced in vivo allergenicity to apple pollens in patients, production of hypoallergenic apple cultivars	Gilissen et al. (2005)
Tomato	Lyc e 3 tomato fruit allergen	Reduced allergenicity to tomato fruit allergens	Le et al. 2006
Peanut	Ara h 2 Peanut seed allergen	Alleviated peanut seed allergy	Dodo et al. (2008)
Potato	*StANS1*, potato asparagine synthetase 1	Reduced acrylamide-forming potential of potato tubers	Chawla et al. (2012)

Table 7.6 Role of RNAi in therapeutics

Plant species	Gene targeted	Result	References
Salvia miltiorrhiza	Chalcone synthase	Increase in the production of phenolic acids	Zhang et al. (2015)
Withania somnifera	Sterol glycosyltransferases (WsSGTL1)	Decrease in the level of glycosylated sterols	Saema et al. (2015)
Lithospermum erythrorhizon	*L. erythrorhizo*n EIN3-like protein gene 1 (LeEIL-1)	Reduced shikonin production	Fang et al. (2016)
Papaver somniferum	Codeinone reductase	Reduction in transcription of the COR gene was about 89 per cent	Khajvand et al. (2016)
Withania somnifera	Cycloartenol synthase (CAS)	Reduction in withanolide content	Mishra et al. (2016)
Artemisia annua	Pleiotropic drug resistance (PDR) transporter	The level of β-caryophyllene was decreased in transgenic *A. annua* plants expressing AaPDR3-RNAi	Fu et al. (2017)
Panax quinquefolius	UDP-glycosyltransferase gene Pq3-O-UGT2	The levels of ginsenoside Rd., protopanaxadiol-type, and total ginsenoside were reduced	Lu et al. (2017)
Artemisia annua	1-Deoxy-D-xylulose-5-phosphate reductoisomerase (DXR)	Artemisinin contents in leaves declined	Wang et al. (2018)
Withania somnifera	S-adenosyl L-methionine-dependent sterol-C24-methyltransferase type 1 (SMT1)	Reduced levels of campesterol, sitosterol, and stigmasterol and an increase in cholesterol content	Pal et al. (2019)
Betula platyphylla	Cycloartenol synthase (CAS) and β-amyrin synthase (βAS)	β-AS inhibition promotes betulinic acid synthesis. CAS inhibition promotes the conversion of 2,3-oxidosqualene to betulinic acid and oleanolic acid	Yin et al. (2020)
Citrus	CiMYB42 transcription factor	Downregulation of CiMYB42 by RNAi resulted in a dwarf phenotype and less nomilin accumulation	Zhang et al. (2020)
Ocimum basilicum	Coniferyl alcohol acyltransferase 1 (CAAT1)	Decreased levels of eugenol and accumulation of coniferyl alcohol and its derivatives	Dhar et al. (2020)
Panax quinquefolius	UDP-glycosyltransferase gene Pq-PPT-6,20-O-UGT1	Expression of dammarenediol synthase, protopanaxadiol synthase, and protopanaxatriol increased	Feng et al. (2021)

(continued)

Table 7.6 (continued)

Plant species	Gene targeted	Result	References
Trigonellafoenum-graecum	Sterol-specific glucosyltransferase (TfS3GT2)	Reduced levels of diosgenin or yamogenin-derived steroidal saponins	Gao et al. (2021)
Taxus chinensis	WRKY transcription factor	Positively regulates taxol biosynthesis genes directly and through the activation of other activators	Chen et al. (2021a, b)
Phalaenopsis bellina	PbABCG1 and PbABCG2	Decrease in monoterpene emission	Chang et al. (2023)

(ARF) family that binds to the promoter of auxin response genes to either activate or repress that particular gene. *Arabidopsis thaliana* consists of a large ARF family with 23 members, of which five are activators (ARF5-8 and 19), and the rest are repressors. A high rise in the expression of ARF8 in *Arabidopsis thaliana* and ARF7 in tomatoes was observed in non-pollinated flowers. It started decreasing after the pollination. They were transcription activators for the genes responsible for seed formation. Therefore, RNAi can suppress these transcription factors to achieve Parthenocarpy.

RNA interference suppressed the *Chalcone synthase* gene, producing parthenocarpic tomatoes (Schijlen et al. 2007). Decreased expression of SlARF7 in tomatoes leads to the production of seedless tomatoes (De Jong et al. 2009). Silencing of flavanol synthase by RNAi in tobacco causes a reduction in seeds (Mahajan et al. 2011). Apples develop from the thalamus, not from the ovary. It is possible to create seedless and coreless apples by inhibiting AGAMOUS-like genes responsible for carpel identity and using hormones to induce fruit formation. Suppressing AGAMOUS genes increases petal whorls and eliminates carpel development (Ireland et al. 2021). Silencing of the SlMPK4 gene produced seedless tomato fruits (Wang et al. 2022).

7.3.8 Increasing Fruit Shelf Life

Fruits and vegetables are more prone to spoilage than cereals due to their inherent nature and composition. Ethylene, a ripening hormone, is crucial in triggering, regulating, and coordinating the expression of various genes associated with the ripening process. In climacteric fruits, a surge in ethylene production is a critical event that marks the beginning of ripening. As reported in the Agricultural Research Data Book 2004, India faces significant postharvest losses in its fruit and vegetable sector, with an estimated 30%–40% of the total produce wasted due to spoilage. Therefore, it is necessary to increase the shelf life of fruits and vegetables using modern biotechnological techniques like RNAi. α-mannosidase (α-Man) and β-D-N-acetylhexosaminidase (β-Hex) are two ripening-specific N-glycoprotein modifying enzymes. These enzymes were suppressed by RNA interference, and transgenic lines resulted in enhanced fruit shelf life (Meli et al. 2010). In a study, aminocyclopropane-1-carboxylate synthase (ACS) gene homologs were targeted using RNAi to control the ethylene production in tomato fruits. Fruits derived from transgenic lines demonstrated a significant delay in ripening and an extended shelf life of approximately 45 days, along with enhanced juice quality (Gupta et al. 2013). In 2017, Yang and co-workers targeted the pectate lyase gene SlPL in tomatoes by RNA interference. The transgenic lines showed positive results, with a significant increase in the firmness of the walls of tomato fruits. Compared to wild-type fruits, SlPL-RNAi fruits exhibited increased levels of cellulose and hemicellulose, while the concentration of water-soluble pectin was reduced. The SlFSR (fruit shelf-life regulator) gene was repressed in tomatoes using RNAi, resulting in the fruit's prolonged shelf life. Its overexpression reduces shelf life (Zhang et al. 2018). Recently, pectate lyase SlPL16 (Solyc06g083580) was silenced by RNAi in tomatoes, decreasing the activity of pectate lyase enzyme and eventually increasing the sturdiness and shelf life of tomato and significantly decreasing the postharvest losses (Ren et al. 2023).

7.3.9 Flower Colour

In modern plant biotechnology, the artificial alteration of flower colours has become a significant subject of interest. RNAi approach can also be used to manipulate pigments of flowers to change their colour accordingly (Waterhouse et al. 1998). When mRNA levels of six genes responsible for anthocyanin pigments were examined in the unpigmented sector of Red-starflower of *Petunia hybrida*, a reduction in mRNA of only chalcone synthase (CHS) was discovered (Koseki et al. 2005). CHS siRNAs were found in those unpigmented areas, indicating that sequence-specific degradation of CHS mRNAs was the cause of the pigmentation. The flower colour of Torenia hybrid, a garden plant, was modulated by targeting CHS through RNA interference. The authentic blue colour of flowers was changed to white and other pastel shades (Fukusaki et al. 2004). Carotenoid cleavage dioxygenases (CCDs) are the enzymes responsible for the cleavage of carotenoids. In Chrysanthemums, CCD4 was targeted by RNAi, which changed the white-coloured petals to yellow (Ohmiya et al. 2009). Anthocyanin 5,3′-aromatic acyltransferase (5/3′AT) and flavonoid 3′,5′-hydroxylase (F3′5′H) are two key enzymes that are responsible for the vivid-blue colour of gentian flowers. Using the RNAi approach, genes coding these two enzymes were downregulated, and as a result, lilac and light-blue colour flowers were observed (Nakatsuka et al. 2010). CHS was suppressed in a monocotyledonous ornamental plant, *Tricyrtis* sp., by introducing an RNAi construct through Agrobacterium-mediated transformation to modulate the flower colour (Kamiishi et al. 2012). Flavanone 3-hydroxylase (F3H) is the enzyme that catalyses flavanones' biosynthesis. In 2017, Chaipanya and co-workers targeted F3H by RNAi using a Pja8F3H vector in water lily petals, and a change in flower colour was observed. Chalcone isomerase (chi) RNAi construct cloned in pBI121 vector was delivered in *Petunia hybrida* by using agroinfiltration (*Agrobacterium*-mediated transient assay), resulting in variegated petals of Petunia flowers (Keykha et al. 2016). Phytoene synthase (PSY) is a key enzyme that catalyses the rate-limiting step of the carotenoid biosynthesis pathway. RNAi silenced the *PSY* gene in the *Oncidium* orchid plant, and white flowers were obtained instead of colourful flowers (Liu et al. 2019).

7.3.10 Conclusions and Future Prospects

RNA interference (RNAi) has emerged as a revolutionary tool in molecular biology, offering a precise and efficient gene-silencing mechanism with vast applications in crop improvement. Over the years, RNAi technology has facilitated the development of genetically modified crops with enhanced resistance to pests, pathogens, and environmental stressors while improving nutritional quality, yield, and postharvest longevity. The ability of RNAi to target specific gene sequences in a species-specific manner makes it an invaluable asset in modern agriculture, enabling sustainable crop production with minimal environmental impact. Several commercialized RNAi-based crops have demonstrated the potential to reduce dependency on chemical pesticides and herbicides, promoting eco-friendly farming practices.

Despite these advancements, the widespread adoption of RNAi technology in agriculture faces several challenges and limitations. Regulatory concerns, biosafety issues, and public perception regarding genetically modified crops remain significant barriers to commercialization. Additionally, the durability and stability of RNAi-induced gene silencing under field conditions require further investigation to ensure long-term efficacy. Advancements in RNA delivery mechanisms, such as spray-induced gene silencing (SIGS) and nanotechnology-based RNAi applications, hold promise for overcoming some of these challenges and making RNAi a more adaptable and non-transgenic solution for crop protection.

Ensuring food security for the growing global population remains one of the most pressing challenges of twenty-first-century agriculture. According to the Global Hunger Index, more than one-third of countries worldwide face critical food shortages, struggling to meet nutritional demands (Kumar 2022). The increasing resistance to conventional agricultural practices and rapid environmental changes has exacerbated the pressure on natural resources. Over the last 15 years, RNAi has emerged as a revolutionary tool in molecular biology, providing a precise and efficient approach to gene silencing. This technology has shown remarkable potential in transforming agriculture by improving crop productivity and enabling the development of genotypes resistant to pests, diseases, and environmental stresses. Several key quality traits have already been successfully modified using RNAi, including caffeine-free coffee (Ogita et al. 2004), allergen-free grains (Kalunke et al. 2020), reduced antinutrient content in crops (Tian et al. 2011), low-nicotine tobacco (Deboer et al. 2011), and fruits with extended shelf life (Gupta et al. 2013). Like any technology, RNAi also has its limitations. Although it targets precise sequences and genes of interest, there is a chance that off-target effects could occur, leading to undesirable characteristics (Qiu et al. 2005). These developments may be essential for improving and protecting crops and solving the urgent global issue of food security. Effective planning is necessary to reduce or eliminate pleiotropic and off-target effects to optimize the efficacy of RNAi-based transgenic methods. For example, a comprehensive in silico analysis must be conducted to choose the best target gene sequences before creating the hairpin RNAi constructs.

One of the main barriers to RNAi adoption has been its high production cost and limited stability. However, advances in technology are reducing costs, while encapsulation methods are addressing stability issues. Over the past decade, encapsulating dsRNA in carriers such as liposomes, polyplex nanoparticles, virus-like particles, and bioclay has proven effective in protecting it from environmental degradation and enzymatic breakdown. Naked dsRNA is highly susceptible to degradation due to pH and enzymatic activity, whereas encapsulation not only enhances stability but also improves cellular uptake (Hernández-Soto and Chacón-Cerdas 2021).

Furthermore, RNA interference technology must advance from lab studies to real-world field implementations, moving from model plants to economically significant crop species (Kamthan et al. 2015). This strategic approach will ensure the broader impact of RNAi in advancing agricultural innovation.

Integrating RNAi with other cutting-edge technologies, including CRISPR-Cas gene editing, synthetic biology, and precision agriculture, could further enhance its effectiveness in crop improvement. Research should focus on developing cost-effective, scalable, and regulatory-compliant RNAi-based solutions that can be seamlessly integrated into traditional breeding programmes. Additionally, extensive field trials and multi-location studies are needed to validate RNAi-based traits' long-term stability and performance across diverse agro-climatic conditions. With continued advancements and collaborative efforts between researchers, policymakers, and agricultural industries, RNAi has the potential to revolutionize global food security by creating resilient, high-yielding, and nutritionally enhanced crops, ultimately contributing to sustainable and climate-smart agriculture for the future.

References

Abdellatef E, Kamal NM, Tsujimoto H (2021) Tuning beforehand: a foresight on RNA interference (RNAi) and *in vitro*-derived dsRNAs to enhance crop resilience to biotic and abiotic stresses. Int J Mol Sci 22(14):7687

Aggarwal S, Kumar A, Bhati KK, Kaur G, Shukla V, Tiwari S, Pandey AK (2018) RNAi-mediated downregulation of inositol pentakisphosphate kinase (IPK1) in wheat grains decreases phytic acid levels and increases Fe and Zn accumulation. Front Plant Sci 9:259

Ahmed MM, Bian S, Wang M, Zhao J, Zhang B, Liu Q, Yu H (2017) RNAi-mediated resistance to rice black-streaked dwarf virus in transgenic rice. Transgenic Res 26:197–207

Allen RS, Millgate AG, Chitty JA, Thisleton J, Miller JA, Fist AJ, Gerlach WL, Larkin PJ (2004) RNAi-mediated replacement of morphine with the nonnarcotic alkaloid reticuline in opium poppy. Nat Biotechnol 22:1559–1566. https://doi.org/10.1038/nbt1033

Ali N, Paul S, Gayen D, Sarkar SN, Datta SK, Datta K (2013a) RNAi mediated down regulation of myo-inositol-3-phosphate synthase to generate low phytate rice. Rice 6(1):12.

Ali N, Paul S, Gayen D, Sarkar SN, Datta K, Datta SK (2013b) Development of low phytate rice by RNAi mediated seed-specific silencing of inositol 1, 3, 4, 5, 6-pentakisphosphate 2-kinase gene (IPK1). PloS one 8(7):e68161

Ameijeiras P, Capriotti N, Ons S, Oliveira PL, Sterkel M (2023) eIF3 subunit M regulates blood meal digestion in *Rhodniusprolixus,* affecting ecdysis, reproduction, and survival. Insect Sci 30(5):1282–1292

Angaji S, Hedayati SS, Hosein R, Samad S, Shiravi S, Madani S (2010) Application of RNA interference in plants. Plant Omics 3(3):77–84

Bae HK, Kang HG, Kim GJ, Eu HJ, Oh SA, Song JT, Chung IK, Eun MY, Park SK (2010) Transgenic rice plants carrying RNA interference constructs of AOS (allene oxide synthase) genes show severe male sterility. Plant Breed 129(6):647–651

Banakar P, Hada A, Papolu PK, Rao U (2020) Simultaneous RNAi knockdown of three FMRFamide-like peptide genes, Mi-flp1, Mi-flp12, and Mi-flp18 provides resistance to root-knot nematode, Meloidogyne incognita. Front Microbiol 11:573916.

Banerjee S, Banerjee A, Gill SS, Gupta OP, Dahuja A, Jain PK, Sirohi A (2017) RNA interference: a novel source of resistance to combat plant parasitic nematodes. Front Plant Sci 8:834

Banerjee S, Gill SS, Gawade BH, Jain PK, Subramaniam K, Sirohi A (2018) Host-delivered RNAi of two cuticle collagen genes, Mi-col-1 and Lemmi-5, hampers structure and fecundity in *Meloidogyne incognita.* Front Plant Sci 8:2266

Barker KR, Koenning SR (1998) Developing sustainable systems for nematode management. Annu Rev Phytopathol 36(1):165–205

Bastarache P, Timani K, Ben Youssef M, Omakele E, Vickruck JL, Morin PJ (2024) Cyantraniliprole and thiamethoxam exposure changes the expression of transcripts associated with small non-coding RNA processing in the Colorado Potato Beetle. Insects 15(3):147

Baum JA, Roberts JK (2014) Progress towards RNAi-mediated insect pest management. Adv Insect Physiol Acad Press 47:249–295

Baum JA, Bogaert T, Clinton W, Heck GR, Feldmann P, Ilagan O, Johnson S, Plaetinck G, Munyikwa T, Pleau M, Vaughn T, Roberts J (2007) Control of coleopteran insect pests through RNA interference. Nat Biotechnol 25:1322–1326

Chakraborty P, Ghosh A (2022) Topical spray of dsRNA induces mortality and inhibits chilli leaf curl virus transmission by *Bemisiatabaci* Asia II. Cells 11(5):833

Chang L, Huang M, Kuo Z, Chen Y, Lin T, Jeng F, Yeh H, Tsai C, Chen H (2023) PbABCG1 and PbABCG2 transporters are required for the emission of floral monoterpenes in *Phalaenopsis bellina.* Plant J 114(2):279–292

Chaudhary S, Dutta TK, Tyagi N, Shivakumara TN, Papolu PK, Chobhe KA, Rao U (2019) Host-induced silencing of Mi-msp-1 confers resistance to root-knot nematode *Meloidogyne incognita* in eggplant. Transgenic Res 28:327–340

Chaudhary P, Agri U, Chaudhary A, Kumar A, Kumar G (2022) Endophytes and their potential in biotic stress management and crop production. Front Microbiol 13:933017

Chawla R, Shakya R, Rommens CM (2012) Tuber-specific silencing of asparagine synthetase-1 reduces the acrylamide-forming potential of potatoes grown in the field without affecting tuber shape and yield. Plant Biotechnol J 10:913–924. https://doi.org/10.1111/j.1467-7652.2012.00720.x

Chen Y, Zhang H, Zhang M, Zhang W, Ou Z, Peng Z, Yu L (2021a) Salicylic acid-responsive factor TcWRKY33 positively regulates taxol biosynthesis in *Taxus chinensis* in direct and indirect ways. Front Plant Sci 12:697476

Chen Y, Zhu W, Shi S, Wu L, Du S, Jin L, Yang K, Zhao W, Yang J (2021b) Use of RNAi with OsMYB76R as a reporter for candidate genes can efficiently create and verify gametophytic male sterility in rice. Front Plant Sci 12:728193

Chitwood DJ (2003) Nematicides. Encyclopedia of agrochemicals. https://doi.org/10.1002/047126363X.agr171

Chu B, Sun J, Dang H, Ma Z, Zhao S, Guan Q, Li X (2021) Apple SUMO E3 LIGASE MDSIZ1 negatively regulates drought tolerance. Immunity 9(25):26

Cooper AM, Silver K, Zhang J, Park Y, Zhu KY (2019) Molecular mechanisms influencing the efficiency of RNA interference in insects. Pest Manag Sci 75(1):18–28

Cooper AM, Song H, Shi X, Yu Z, Kim YH, Silver K, Zhang J, Zhu KY (2021) Molecular characterization and RNA interference responses of the lethal giant larvae gene in *Diabrotica virgifera* adults. Arch Insect Biochem Physiol 107(2):e21787

Da-Hong L, Hui L, Yan-li Y, Zhen PP, Li JS (2009) Down-regulated expression of the RACK1 gene by RNA interference enhances drought tolerance in rice. Rice Sci 16:14–20

Dai Z, Wang J, Yang X, Lu H, Miao X, Shi Z (2018) Modulation of plant architecture by the miR156f–OsSPL7–OsGH3. 8 pathways in rice. J Exp Bot 69(21):5117–5130

Das J, Kumar R, Shah V, Sharma AK (2023) Simple, cost-effective larval injection method for dsRNA delivery to induce RNAi response in *Helicoverpa armigera* (Hübner). J Appl Entomol 147(4):289–298

Davuluri GR, van Tuinen A, Fraser PD, Manfredonia A, Newman R, Burgess D, Brummell DA, King SR, Palys J, Uhlig J, Bramley PM, Pennings HM, Bowler C (2005) Fruit-specific RNAi-mediated suppression of DET1 enhances carotenoid and flavonoid content in tomatoes. Nat Biotechnol 23:890–895. https://doi.org/10.1038/nbt1108

Dawson JH, Musselman LJ, Wolswinkel PI, Dörr IN (1994) Biology and control of Cuscuta. Rev Weed Sci 6:265–317

de Dorlodot S, Forster B, Pagès L, Price A, Tuberosa R, Draye X (2007) Root system architecture: opportunities and constraints for genetic improvement of crops. Trends Plant Sci 12(10): 474–481

De Jong M, Wolters-Arts M, Feron R, Mariani C, Vriezen WH (2009) The *Solanum lycopersicum* auxin response factor 7 (SlARF7) regulates auxin signaling during tomato fruit set and development. Plant J 57(1):160–170

DeBoer KD, Dalton HL, Edward FJ, Hamill JD (2011) RNAi-mediated down-regulation of ornithine decarboxylase (ODC) leads to reduced nicotine and increased anatabine levels in transgenic *Nicotiana tabacum* L. Phytochemistry 72(4–5):344–355

Dhar N, Sarangapani S, Reddy VA, Kumar N, Panicker D, Jin J, Sarojam R (2020) Characterization of a sweet basil acyltransferase involved in eugenol biosynthesis. J Exp Bot 71(12):3638–3652

Dodo HW, Konan KN, Chen FC, Egnin M, Viquez OM (2008) Alleviating peanut allergy using genetic engineering: the silencing of the immunodominant allergen Ara h 2 leads to its significant reduction and a decrease in peanut allergenicity. Plant Biotechnol J 6:135–145. https://doi.org/10.1111/j.1467-7652.2007.00292.x

Dubey NK, Eizenberg H, Leibman D, Wolf D, Edelstein M, Abu-Nassar J, Marzouk S, Gal-On A, Aly R (2017) Enhanced host-parasite resistance based on down-regulation of *Phelipancheaegyptiaca* target genes is likely by mobile small RNA. Front Plant Sci 8:1574

Dutta TK, Papolu PK, Singh D, Sreevathsa R, Rao U (2020) Expression interference of a number of *Heteroderaavenae* conserved genes perturbs nematode parasitic success in *Triticum aestivum*. Plant Sci 301:110670

Dutta SD, Ganguly K, Lim KT (2021) RNAi-based gene silencing in plant-parasitic nematodes: a road toward crop improvements. In: CRISPR and RNAi systems, nanobiotechnology approaches to plant breeding and protection nanobiotechnology for plant protection. Elsevier, pp 579–596

Dutta TK, Santhoshkumar K, Veeresh A, Waghmare C, Mathur C, Sreevathsa R (2023) RNAi-based knockdown of candidate gut receptor genes altered the susceptibility of *Spodoptera frugiperda* and *Spodoptera litura* larvae to a chimeric toxin Cry1AcF. Peer J 11:e14716

Dyer S, Weir R, Manesiotis P, Dalzell JJ (2021) Knockdown of *phytoene desaturase*in the field dodder, *Cuscuta campestris*, by virus-induced gene silencing. BioRxiv 2021:04

Eady CC, Kamoi T, Kato M, Porter NG, Davis S, Shaw M, Kamoi A, Imai S (2008) Silencing onion lachrymatory factor synthase causes a significant change in the sulfur secondary metabolite profile. Plant Physiol 147:2096–2106. https://doi.org/10.1104/pp.108.123273

Elling AA (2013) Major emerging problems with minor *Meloidogyne species*. Phytopathology 103(11):1092–1102

Enrique R, Siciliano F, Favaro MA (2011) Novel demonstration of RNAi in citrus reveals importance of citrus callose synthase in defence against *Xanthomonas citri* subsp. *citri*. Plant Biotechnol J 9(3):394–407

Escobar MA, Civerolo EL, Summerfelt KR, Dandekar AM (2001) RNAi-mediated oncogene silencing confers resistance to crown gall tumorigenesis. PNAS 98(23):13437–13442

Escobar MA, Leslie CA, McGranahan GH, Dandekar AM (2002) Silencing crown gall disease in walnut (*Juglans regia* L.). Plant Sci 163(3):591–597

Fang R, Zou A, Zhao H, Wu F, Zhu Y, Zhao H, Yang Y (2016) Transgenic studies reveal the positive role of LeEIL-1 in regulating shikonin biosynthesis in *Lithospermum erythrorhizon* hairy roots. BMC Plant Biol 16:1–12

Fard EM, Bakhshi B, Keshavarznia R, Nikpay N, Shahbazi M, Salekdeh GH (2017) Drought-responsive microRNAs in two barley cultivars differ in their level of sensitivity to drought stress. Plant Physiol Biochem 118:121–129

Farrokhi Z, Alizadeh H, Alizadeh H Mehrizi FA (2019). Host-induced silencing of some important genes involved in osmoregulation of parasitic plant *Phelipanche aegyptiaca*. Mol Biotechnol 61:929–937

Feldmann KA (2006) Steroid regulation improves crop yield. Nat Biotechnol 24(1):46–47

Feng P, Li G, Wang X, Sun Y, Cui Y, Zhao S (2021) Identification and RNAi-based gene silencing of a novel UDP-glycosyltransferase from *Panax quinquefolius*. PCTOC 144:567–576

Fire A, Xu S, Montgomery MK, Kostas SA, Driver SE, Mello CC (1998) Potent and specific genetic interference by double-stranded RNA in *Caenorhabditis elegans*. Nature 391(6669): 806–811

Flores T, Karpova O, Su X, Zeng P, Bilyeu K, Sleper DA, Nguyen HT, Zhang ZJ (2008) Silencing of Gm FAD3 gene by siRNA leads to low α-linolenic acids (18:3) of fad3-mutant phenotype in soybean [Glycine max (Merr.)]. Transgenic Res 17:839–850

Fu X, Shi P, He Q, Shen Q, Tang Y, Pan Q, Tang K (2017) AaPDR3, a PDR transporter 3, is involved in sesquiterpene β-caryophyllene transport in *Artemisia annua*. Front Plant Sci 8:723

Fukusaki EI, Kawasaki K, Kajiyama SI, An CI, Suzuki K, Tanaka Y, Kobayashi A (2004) Flower color modulations of Toreniahybrida by downregulation of chalcone synthase genes with RNA interference. J Biotechnol 111(3):229–240

Gao J, Xu Y, Hua C, Li C, Zhang Y (2021) Molecular cloning and functional characterization of a sterol 3-O-Glucosyltransferase involved in the biosynthesis of steroidal saponins in *Trigonellafoenum-graecum*. Front Plant Sci 12:809579

Gheysen G, Vanholme B (2007) RNAi from plants to nematodes. Trends Biotechnol 25(3):89–92

Ghosh S, Kamble NU, Verma P, Salvi P, Petla BP, Roy S, Rao V, Hazra A, Varshney V, Kaur H, Majee M (2020) Arabidopsis protein l-isoaspartyl methyltransferase repairs isoaspartyl damage to antioxidant enzymes and increases heat and oxidative stress tolerance. J Biol Chem 295(3): 783–799

Gil-Humanes J, Pistón F, Hernando A, Alvarez JB, Shewry PR, Barro F (2008) Silencing of γ-gliadins by RNA interference (RNAi) in bread wheat. J Cereal Sci 48(3):565–568

Gil-Humanes J, Pistón F, Tollefsen S, Sollid LM, Barro F (2010) Effective shutdown in the expression of celiac disease-related wheat gliadin T-cell epitopes by RNA interference. Proc Natl Acad Sci USA 107:17023–17028. https://doi.org/10.1073/pnas.1007773107

Gilissen LJ, Bolhaar ST, Matos CI, Rouwendal GJ, Boone MJ, Krens FA, Zuidmeer L, Van Leeuwen A, Akkerdaas J, Hoffmann-Sommergruber K, Knulst AC, Bosch D, Van de Weg WE, Van Ree R (2005) Silencing the major apple allergen Mal d 1 by using the RNA interference approach. J Allergy Clin Immunol 115:364–369. https://doi.org/10.1016/j.jaci.2004.10.014

Gong C, Yang Z, Hu Y, Wu Q, Wang S, Guo Z, Zhang Y (2022) Silencing of the BtTPS genes by transgenic plant-mediated RNAi to control *Bemisia tabaci* MED. Pest Manag Sci 78(3): 1128–1137

Gorguet B, Van Heusden AW, Lindhout P (2005) Parthenocarpic fruit development in tomato. Plant Biol 7(02):131–139

Gui SH, Taning CNT, Smagghe G (2021) Myosuppressin influences fecundity in the Colorado potato beetle, *Leptinotarsa decemlineata*. Insect Sci 28(4):1191–1201

Gupta A, Pal RK, Rajam MV (2013) Delayed ripening and improved fruit processing quality in tomato by RNAi-mediated silencing of three homologs of 1-aminopropane-1-carboxylate synthase gene. J Plant Physiol 170(11):987–995

Hada A, Kumari C, Phani V, Singh D, Chinnusamy V, Rao U (2020) Host-induced silencing of FMRFamide-like peptide genes, flp-1 and flp-12, in rice impairs reproductive fitness of the root-knot nematode *Meloidogyne graminicola*. Front Plant Sci 11:894

Hajyzadeh M, Turktas M, Khawar KM, Unver T (2015) miR408 overexpression causes increased drought tolerance in chickpeas. Gene 555(2):186–193

Hameed A, Tahir MN, Asad S, Bilal R, Van Eck J, Jander G, Mansoor S (2017) RNAi-mediated simultaneous resistance against three RNA viruses in potatoes. Mol Biotechnol 59:73–83

Han J, Rotenberg D (2024) Microinjection-enabled gene silencing in first instar larvae of western flower thrips, *Frankliniella occidentalis*, reveals vital genes for larval survival. Insect Sci 0:1–13

Hanly A, Karagiannis J, Lu QS, Tian L, Hannoufa A (2020) Characterization of the role of SPL9 in drought stress tolerance in *Medicago sativa*. Int J Mol Sci 21(17):6003

Hernández-Soto A, Chacón-Cerdas R (2021) RNAi crop protection advances. Int J Mol Sci 22(22): 12148

Iqbal S, Jones MG (2020) Attempts to silence genes of the RNAi pathways of the root-knot nematode, *Meloidogyne incognita*, result in diverse responses, including an increase and no change in the expression of some genes. Front Plant Sci 11:501032

Ireland HS, Tomes S, Hallett IC, Karunairetnam S, David KM, Yao JL, Schaffer RJ (2021) Coreless apples generated by the suppression of carpel genes and hormone-induced fruit set. Fruit Res 1(1):1–9

Jaiswal S, Raman KV, Saini RP, Tilgam J, Saakre M, Dhandapani G, Sreevathsa R, Pattanayak D (2024) Host-delivered RNAi-mediated silencing of HaAce1, encoding the major isoform of acetylcholinesterase, imparts resistance against the polyphagous insect-pest, *Helicoverpaarmigera*. J Plant Biochem Biotechnol:1–12. https://doi.org/10.1007/s13562-024-00938-5

Jaiwal PK, Chhillar A, Chaudhary D, Jaiwal A (eds) (2019) Nutritional quality improvement in plants. Springer, Switzerland, pp 1–485

Jiang Z, Zhao Q, Bai R, Yu R, Diao P, Yan T, Duan H, Ma X, Zhou Z, Fan Y, Wuriyanghan H (2021) Host sunflower-induced silencing of parasitism-related genes confers resistance to invading *Orobanchecumana*. Plant Physiol 185(2):424–440

Jiang L, Zhang D, Liu C, Shen W, He J, Yue Q, Niu C, Yang F, Li X, Shen X, Hou N (2022) MdGH3. 6 is targeted by MdMYB94 and plays a negative role in apple water-deficit stress tolerance. Plant J 109(5):1271–1289

Joshi R, Sahoo KK, Tripathi AK, Kumar R, Gupta BK, Pareek A, Singla-Pareek SL (2018) Knockdown of an inflorescence meristem-specific cytokinin oxidase–OsCKX2 in rice reduces yield penalty under salinity stress conditions. Plant Cell Environ 41(5):936–946

Joshi I, Kumar A, Singh AK, Kohli D, Raman KV, Sirohi A, Chaudhury A, Jain PK (2019) Development of nematode resistance in Arabidopsis by HD-RNAi-mediated silencing of the effector gene et Mi-msp2. Sci Rep 9(1):17404

Joshi I, Kumar A, Kohli D, Singh AK, Sirohi A, Subramaniam K, Chaudhury A, Jain PK (2020) Conferring root-knot nematode resistance via host-delivered RNAi-mediated silencing of four Mi-msp genes in Arabidopsis. Plant Sci 298:110592

Julian-Chávez B, Siqueiros-Cendón TS, Torres-Castillo JA, Sinagawa-García SR, Abraham-Juárez MJ, González-Barriga CD, Rascón-Cruz Q, Siañez-Estrada LI, Arévalo-Gallegos S, Espinoza-Sánchez EA (2024) Silencing ACE1 genes with dsRNA of different lengths impairs larval development in *Leptinotarsa decemlineata*. Insects 15(12):1000

Kalunke RM, Tundo S, Sestili F, Camerlengo F, Lafiandra D, Lupi R, Larré C, Denery-Papini S, Islam S, Ma W, D'amico S (2020) Reduction of allergenic potential in bread wheat RNAi transgenic lines silenced for CM3, CM16 and 0.28 ATI genes. Int J Mol Sci 21(16):5817

Kamiishi Y, Otani M, Takagi H, Han DS, Mori S, Tatsuzawa F, Nakano M (2012) Flower color alteration in the liliaceous ornamental *Tricyrtis* sp. by RNA interference-mediated suppression of the chalcone synthase gene. Mol Breed 30:671–680

Kamthan A, Chaudhuri A, Kamthan M, Datta A (2015) Small RNAs in plants: recent development and application for crop improvement. Front Plant Sci 6:124569

Keykha F, Bagheri A, Moshtaghi N, Bahrami AR, Sharifi A (2016) RNAi-induced silencing in floral tissues of *Petunia hybrida* by agroinfiltration: a rapid assay for chalcone isomerase gene function analysis. Cell Mol Biol 62(10):26–31

Khajvand R, Ismaili A, Nazarian Firouz-Abadi F, Latifi AM (2016) Gene silencing of Codeinone Reductase in *Papaver somniferum* L., using virus-induced gene silencing technique. Cell Tissue J 7(4):407–415

Khalid H, Abdalla WE, Abdelgadir H, Opatz T, Efferth T (2012) Gems from traditional North-African medicine: medicinal and aromatic plants from Sudan. Nat Prod Bioprospect 2:92–103

Kir G, Ye H, Nelissen H, Neelakandan AK, Kusnandar AS, Luo A, Inzé D, Sylvester AW, Yin Y, Becraft PW (2015) RNA interference knockdown of BRASSINOSTEROID INSENSITIVE1 in maize reveals novel functions for brassinosteroid signalling in controlling plant architecture. Plant Physiol 169(1):826–839

Knorr E, Billion A, Fishilevich E, Tenbusch L, Frey ML, Rangasamy M, Gandra P, Arora K, Lo W, Geng C, Vilcinskas A (2021) Knockdown of genes involved in transcription and splicing reveals novel RNAi targets for pest control. Front Agron 3:715823

Kolge H, Kadam K, Ghormade V (2023) Chitosan nanocarriers mediated dsRNA delivery in gene silencing for *Helicoverpa armigera* biocontrol. PesticBiochemPhysiol 189:105292

Kopecka R, Kameniarova M, Cerny M, Brzobohaty B, Novak J (2023) Abiotic stress in crop production. Int J Mol Sci 24(7):6603

Koseki M, Goto K, Masuta C, Kanazawa A (2005) The star-type color pattern in *Petunia hybrida* 'Red Star' flowers is induced by sequence-specific degradation of chalcone synthase RNA. Plant Cell Physiol 46(11):879–1883

Koulagi R, Banerjee S, Gawade BH, Singh AK, Jain PK, Praveen S, Subramaniam K, Sirohi A (2020) Host-delivered RNA interference in tomato for mediating resistance against *Meloidogyne incognita* and Tomato leaf curl virus. PCTOC 143:345–361

Kumar A (2022) Agriculture for combating global starvation. In: Ansari SA et al (eds) Augmenting crop productivity in stress environment. Springer, pp 345–365

Kumar S, Bejiga G, Ahmed S, Nakkoul H, Sarker A (2011) Genetic improvement of grass pea for low neurotoxin (β-ODAP) content. Food Chem Toxicol 49:589–600

Kumar A, Kumar V, Krishnan V, Hada A, Marathe A, Jolly M, Sachdev A (2019) Seed-targeted RNAi-mediated silencing of GmMIPS1 limits phytate accumulation and improves mineral bioavailability in soybean. Sci Rep 9(1):7744

Kumar A, Joshi I, Changwal C, Sirohi A, Jain PK (2022) Host-delivered RNAi-mediated silencing of the root-knot nematode (*Meloidogyne incognita*) effector genes, Mi-msp10 and Mi-msp23, confers resistance in Arabidopsis and impairs the reproductive ability of the root-knot nematode. Planta 256(4):74

Kusaba M, Miyahara K, Iida S, Fukuoka H, Takano T, Sassa H, Nishimura M, Nishio T (2003) Low glutelin content1: a dominant mutation that suppresses the glutelin multigene family via RNA silencing in rice. Plant Cell 15:1455–1467. https://doi.org/10.1105/tpc.011452

Le LQ, Lorenz Y, Scheurer S, Fötisch K, Enrique E, Bartra J, Biemelt S, Vieths S, Sonnewald U (2006) Design of tomato fruits with reduced allergenicity by dsRNAi-mediated inhibition of ns-LTP (Lyc e 3) expression. Plant Biotechnol J 4:231–242

Lewis RS, Jack AM, Morris JW, Robert VJ, Gavilano LB, Siminszky B, Bush LP, Hayes AJ, Dewey RE (2008) RNA interference (RNAi)-induced suppression of nicotine demethylase activity reduces levels of a key carcinogen in cured tobacco leaves. Plant Biotechnol J 6:346–354

Li X, Feng T, Randle C, Schneeweiss GM (2019) Phylogenetic relationships in Orobanchaceae inferred from low-copy nuclear genes: consolidation of major clades and identification of a

novel position of the non-photosynthetic *Orobancheclade* sister to all other parasitic Orobanchaceae. Front Plant Sci 10:902
Li Y, He Y, Liu Z, Qin T, Wang L, Chen Z, Zhang B, Zhang H, Li H, Liu L, Zhang J (2022) OsSPL14 acts upstream of OsPIN1b and PILS6b to modulate axillary bud outgrowth by fine-tuning auxin transport in rice. Plant J 111(4):1167–1182
Liang C, Hao J, Li J, Baker B, Luo L (2019) Artificial microRNA-mediated resistance to cucumber green mottle mosaic virus in *Nicotiana benthamiana*. Planta 250:1591–1601
Lin JS, Kuo CC, Yang IC, Tsai WA, Shen YH, Lin CC, Liang YC, Li YC, Kuo YW, King YC, Lai HM (2018) MicroRNA160 modulates plant development and heat shock protein gene expression to mediate heat tolerance in Arabidopsis. Front Plant Sci 9:68
Lindbo JA, Dougherty WG (2005) Plant pathology and RNAi: a brief history. Annu Rev Phytopathol 43(1):191–204
Liu Q, Singh SP, Green AG (2002) High-stearic and high-oleic cottonseed oils produced by hairpin RNA-mediated post-transcriptional gene silencing. Plant Physiol 129:1732–1743. https://doi.org/10.1104/pp.001933
Liu YC, Yeh CW, Chung JD, Tsai CY, Chiou CY, Yeh KW (2019) Petal-specific RNAi-mediated silencing of the phytoene synthase gene reduces xanthophyll levels to generate new Oncidium orchid varieties with white-colour blooms. Plant Biotechnol J 17(11):2035
Liu X, Xie X, Zheng C (2022) RNAi-mediated suppression of the abscisic acid catabolism gene OsABA8ox1 increases abscisic acid content and tolerance to saline-alkaline stress in rice (*Oryza sativa* L.). Crop J 10(2):354–367
Lu C, Zhao S, Wei G, Zhao H, Qu Q (2017) Functional regulation of ginsenoside biosynthesis by RNA interferences of a UDP-glycosyltransferase gene in Panax ginseng and *Panax quinquefolius*. Plant Physiol Biochem 111:67–76
Luan H, Liao W, Song Y, Niu H, Hu T, Zhi H (2020) Transgenic plant generated by RNAi-mediated knocking down of soybean Vma12 and soybean mosaic virus resistance evaluation. AMB Express 10:1–10
Mahajan M, Ahuja PS, Yadav SK (2011) Post-transcriptional silencing of flavonol synthase mRNA in tobacco leads to fruits with arrested seed set. PLoS One 6(12):e28315
Manavalan LP, Chen X, Clarke J, Salmeron J, Nguyen HT (2012) RNAi-mediated disruption of squalene synthase improves drought tolerance and yield in rice. J Exp Bot 63:163–175
Mani V, Reddy CS, Lee SK, Park S, Ko HR, Kim DG, Hahn BS (2020) Chitin biosynthesis inhibition of *Meloidogyne incognita* by RNAi-mediated gene silencing increases resistance to transgenic tobacco plants. Int J Mol Sci 21(18):6626
Mao YB, Cai WJ, Wang JW, Hong GJ, Tao XY, Wang LJ, Huang YP, Chen XY (2007) Silencing a cotton bollworm P450 monooxygenase gene by plant-mediated RNAi impairs larval tolerance of gossypol. Nat Biotechnol 25:1307–1313
Máximo WP, Howell JL, Mogilicherla K, Basij M, Chereddy SC, Palli SR (2020) Inhibitor of apoptosis is an effective target gene for RNAi-mediated control of the Colorado potato beetle, *Leptinotarsa decemlineata*. Arch Insect Biochem Physiol 104(4):e21685
Meena AK, Verma LK, Kumhar BL (2017) RNAi, it's mechanism and potential use in crop improvement: a review. Int J Pure Appl Biosci 5:294–311. https://doi.org/10.18782/2320-7051.2890
Meli VS, Ghosh S, Prabha TN, Chakraborty N, Chakraborty S, Datta A (2010) Enhancement of fruit shelf life by suppressing N-glycan processing enzymes. PNAS 107(6):2413–2418
Mickelbart MV, Hasegawa PM, Bailey-Serres J (2015) Genetic mechanisms of abiotic stress tolerance that translate to crop yield stability. Nat Rev Genet 16(4):237–251
Miroshnichenko D, Timerbaev V, Okuneva A, Klementyeva A, Sidorova T, Pushin A, Dolgov S (2020) Enhancement of resistance to PVY in intragenic marker-free potato plants by RNAi-mediated silencing of eIF4E translation initiation factors. PCTOC 140:691–705
Mishra S, Bansal S, Mishra B, Sangwan RS, Asha Jadaun JS, Sangwan NS (2016) RNAi and homologous over-expression-based functional approaches reveal triterpenoid synthase

gene-cycloartenol synthase is involved in downstream withanolide biosynthesis in *Withaniasomnifera*. PLoS One 11(2):e0149691

Moreira VJ, Pinheiro DH, Lourenço-Tessutti IT, Basso MF, Lisei-de-Sa ME, Silva MC, Danchin EG, Guimarães PM, Grynberg P, Brasileiro AC, Macedo LL (2024) In planta RNAi targeting *Meloidogyne incognita, the* Minc16803 gene perturbs nematode parasitism and reduces plant susceptibility. J Pest Sci 97(1):411–427

Moritoh S, Miki D, Akiyama M, Kawahara M, Izawa T, Maki H, Shimamoto K (2005) RNAi-mediated silencing of OsGEN-L (OsGEN-like), a new member of the RAD2/XPG nuclease family, causes male sterility by defect of microspore development in rice. Plant Cell Physiol 46(5):699–715

Nakatsuka T, Mishiba KI, Kubota A, Abe Y, Yamamura S, Nakamura N, Nishihara M (2010) Genetic engineering of novel flower colour by suppression of anthocyanin modification genes in gentian. J Plant Physiol 167(3):231–237

Napoli C, Lemieux C, Jorgensen R (1990) Introduction of a chimeric chalcone synthase gene into petunia results in reversible co-suppression of homologous genes in trans. Plant Cell 2(4): 279–289

Niu X, Kassa A, Hasler J, Griffin S, Perez-Ortega C, Procyk L, Zhang J, Kapka-Kitzman DM, Nelson ME, Lu A (2020) Functional validation of DvABCB1 as a receptor of Cry3 toxins in western corn rootworm, *Diabrotica virgifera*. Sci Rep 10(1):15830

Oerke EC (2006) Crop losses to pests. J Agric Sci 144:31–43

Ogita S, Uefuji H, Morimoto M (2004) Application of RNAi to confirm theobromine as the major intermediate for caffeine biosynthesis in coffee plants with potential for construction of decaffeinated varieties. Plant Mol Biol 54:931–941

Ohmiya A, Sumitomo K, Aida R (2009) "Yellow Jimba": suppression of carotenoid cleavage dioxygenase (CmCCD4a) expression turns white chrysanthemum petals yellow. Hort J 78(4): 450–455

Paim RM, Pessoa GC, Nascimento BW, Nascimento AM, Pinheiro LC, Koerich LB, Diotaiuti L, Araujo RN, SantAnna MR, Gontijo NF, Pereira MH (2022) Effect of salivary CYP4EM1 and CYP4EM2 gene silencing on the life span of Chagas disease vector *Rhodniusprolixus (Hemiptera, Reduviidae)* exposed to a sublethal dose of deltamethrin. Insect Mol Biol 31(1): 49–59

Pal S, Rastogi S, Nagegowda DA, Gupta MM, Shasany AK, Chanotiya CS (2019) RNAi of sterol methyl transferase1 reveals its direct role in diverting intermediates towards withanolide/ phytosterol biosynthesis in *Withaniasomnifera*. Plant Cell Physiol 60(3):672–686

Pampolini F, Rieske LK (2023) Foliar application of dsRNA to induce gene silencing in emerald ash borer: systemic distribution, persistence, and bioactivity. Forests 14(9):1853

Pan X, Pei Y, Zhang C, Huang Y, Chen L, Wei L, Li C, Dong X, Chen X (2022) Effect of insulin receptor on juvenile hormone signal and fecundity in *Spodoptera litura* (F.). Insects 13(8):701

Park JS, Yu JG, Lee GH (2018) Drought tolerance induction in transgenic tobacco through RNA interference of BrDST71, a drought-responsive gene from Chinese cabbage. Hortic Environ Biotechnol 59:749–757

Parker C (2013) The parasitic weeds of the Orobanchaceae. In: Joel D, Gressel J, Musselman L (eds) Parasitic Orobanchaceae: parasitic mechanisms and control strategies. Springer, pp 313–344

Pathak K, Gogoi B (2016) RNA interference (RNAi): application in crop improvement: a review. Agric Rev 37:245–249

Peng J, Richards DE, Hartley NM (1999) 'Green revolution' genes encode mutant gibberellin response modulators. Nature 400(6741):256–261

Pinheiro DH, Taylor CE, Wu K, Siegfried BD (2020) Delivery of gene-specific dsRNA by microinjection and feeding induces RNAi response in Sri Lanka weevil, *Myllocerus undecimpustulatus undatus* Marshall. Pest Manag Sci 76(3):936–943

Pitino M, Coleman AD, Maffei ME, Ridout CJ, Hogenhout SA (2011) Silencing of aphid genes by dsRNA feeding from plants. PLoS ONE 6:e25709

Pizetta CS, Ribeiro WR, Ferreira AL, da Costa MM, Bonfim K, Pinheiro PV, Cabral GB, Aragão FJ (2022) RNA interference-mediated tolerance to whitefly (*Bemisiatabaci*) in genetically engineered tomato. PCTOC 148(2):281–291

Pooggin MM (2017) RNAi-mediated resistance to viruses: a critical assessment of methodologies. COVIRCO 26:28–35

Puri H, Jindal V (2021) Target of rapamycin (TOR) gene is vital for whitefly survival and reproduction. J Bio Sci 46(3):60

Qiao F, Yang Q, Wang CL, Fan YL, Wu XF, Zhao KJ (2007) Modification of plant height via RNAi suppression of OsGA20ox2 gene in rice. Euphytica 158:35–45

Qin L, Zhang H, Li J, Zhu Y, Jiao G, Wang C, Wu S (2022) Down-regulation of GhADF1 in cotton (*Gossypium hirsutum*) improves plant drought tolerance and increases fibre yield. Crop J 10(4): 1037–1048

Qiu S, Adema CM, Lane T (2005) A computational study of off-target effects of RNA interference. NAR 33(6):1834–1847

Ramasamy A, Keppanana R, Karuppannasamy A, Bhargava CN, Karakatti P, Srivastava S, Thiruvengadam V, Kesavan S, Ramasamy GG, Muthugounder M, Sagar D (2023) Oral delivery of cognate double-stranded RNAs for the effective suppression of key functional genes in Asia-I genetic group of whiteflies, Bemisiatabaci. SSRN Electron J. https://doi.org/10.2139/ssrn.4572182

Ramesh SV, Shivakumar M, Praveen S, Chouhan BS, Chand S (2019) Expression of short hairpin RNA (shRNA) targeting AC2 gene of Mungbean yellow mosaic India virus (MYMIV) reduces the viral titre in soybean. 3 Biotech 9:1–9

Rastegari AA, Yadav AN, Yadav N, Sarshari NT (2019) Bioengineering of secondary metabolites. In: New and future developments in microbial biotechnology and bioengineering. Elsevier, pp 55–68

Rathore KS, Sundaram S, Sunilkumar G, Campbell LM, Puckhaber L, Marcel S, Palle SR, Stipanovic RD, Wedegaertner TC (2012) Ultra-low gossypol cottonseed: generational stability of the seed-specific, RNAi-mediated phenotype and resumption of terpenoid profile following seed germination. Plant Biotechnol J 10:174–183

Regina A, Kosar-Hashemi B, Li Z et al (2005) Starch branching enzyme IIb in wheat is expressed at low levels in the endosperm compared to other cereals and encoded at a non-syntenic locus. Planta 222:899–909. https://doi.org/10.1007/s00425-005-0032-z

Regina A, Kosar-Hashemi B, Ling S, Li Z, Rahman S, Morell M (2010) Control of starch branching in barley defined through differential RNAi suppression of starch branching enzyme IIa and IIb. J Experimen Bot 1(61):1469–1482

Ren H, Zhao Q, Feng Y, Tang P, Wang Y, Jiang J, Hu C, Wang Y, Cui B, Xie X, Li Y, Zhao X, Gu H, Huang J, Zhang Y (2023) Gene-specific silencing of SlPL16, a pectate lyase coding gene, extends the shelf life of tomato fruit. Postharvest Biol Technol 201:112368

Rodrigues TB, Mishra SK, Sridharan K, Barnes ER, Alyokhin A, Tuttle R, Kokulapalan W, Garby D, Skizim NJ, Tang YW, Manley B (2021) The first sprayable double-stranded RNA-based biopesticide product targets proteasome subunit beta type-5 in the Colorado potato beetle (*Leptinotarsa decemlineata*). Front Plant Sci 12:728652

Romano N, Macino G (1992) Quelling: transient inactivation of gene expression in Neurospora crassa by transformation with homologous sequences. Mol Microbiol 6(22):3343–3353

Rs L, Ck H, Kb S, Alex S (2021) Transgenic banana plants carrying ihpRNA cassette targeting viral replicase gene show resistance against Banana bract mosaic virus. J Hortic Sci Biotechnol 96(3): 324–329

Rubinelli PM, Chuck G, Li X, Meilan R (2013) Constitutive expression of the Corngrass1 microRNA in poplar affects plant architecture and stem lignin content and composition. BMSBEO 54:312–321

Saema S, Rahman LU, Niranjan A, Ahmad IZ, Misra P (2015) RNAi-mediated gene silencing of WsSGTL1 in *W. somnifera* affects growth and glycosylation patterns. Plant Signal Behav 10(12):e1078064

Sarkar S, Kalia VK (2023) Silencing of juvenile hormone epoxide hydrolase gene in *Spodoptera litura* (Lepidoptera: Noctuidae) by oral delivery of double-stranded RNA. Biologia 78(2): 429–442

Saurabh S, Vidyarthi AS, Prasad D (2014) RNA interference: concept to reality in crop improvement. Planta 239:543–564

Savary S, Willocquet L, Pethybridge SJ, Esker P, McRoberts N, Nelson A (2019) The global burden of pathogens and pests on major food crops. Nat Ecol Evol 3(3):430–439

Sawai S, Ohyama K, Yasumoto S, Seki H, Sakuma T, Yamamoto T, Umemoto N (2014) Sterol side chain reductase 2 is a key enzyme in the biosynthesis of cholesterol, the common precursor of toxic steroidal glycoalkaloids in potato. The Plant Cell 26(9):3763–3774

Schijlen EG, De Vos CH, Martens S, Jonker HH, Rosin FM, Molthoff JW, Tikunov YM, Angenent GC, Van Tunen AJ, Bovy AG (2007) RNA interference silencing of chalcone synthase, the first step in the flavonoid biosynthesis pathway, leads to parthenocarpic tomato fruits. Plant Physiol 144(3):1520–1530

Seto KC, Ramankutty N (2016) Hidden linkages between urbanization and food systems. Science 352:943–945

Sharif MN, Iqbal MS, Alam R, Awan MF, Tariq M, Ali Q, Nasir IA (2022) Silencing of multiple target genes via ingestion of dsRNA and PMRi affects development and survival in *Helicoverpa armigera*. Sci Rep 12(1):10405

Sharma N, Prasad M (2020) Silencing AC1 of tomato leaf curl virus using artificial microRNA confers resistance to leaf curl disease in transgenic tomato. Plant Cell Rep 39:1565–1579

Shi J, Lang C, Wu X, Liu R, Zheng T, Zhang D, Chen J, Wu G (2015) RNAi knockdown of fatty acid elongase1 alters fatty acid composition in Brassica napus. Biochem Biophys Res Commun 466:518–522

Silva-Oliveira G, De Paula IF, Medina JM, Alves-Bezerra M, Gondim KC (2021) Insulin receptor deficiency reduces lipid synthesis and reproductive function in the insect *Rhodnius prolixus*. Biochim Biophys Acta Mol Cell Biol Lipid 1866(2):158851

Sinha R, Rajam MV (2013) RNAi silencing of three homologues of the S-adenosylmethionine decarboxylase gene in the tapetal tissue of tomato results in male sterility. Plant Mol Biol 82: 169–180

Song R, Llaca V, Linton E, Messing J (2001) Sequence, regulation, and evolution of the maize 22-kD α zein gene family. Genome Res 11:1817–1825

Song JB, Gao S, Wang Y, Li BW, Zhang YL, Yang ZM (2016) miR394 and its target gene LCR are involved in cold stress response in Arabidopsis. Plant Gene 5:56–64

Sparks TC, Nauen R (2015) IRAC: mode of action classification and insecticide resistance management. Pestic Biochem Physiol 121:122–128

Sun L, Yuan B, Zhang M, Wang L, Cui M, Wang Q, Leng P (2012) Fruit-specific RNAi-mediated suppression of SlNCED1 increases both lycopene and β-carotene contents in tomato fruit. J Exp Bot 63:3097–3108. https://doi.org/10.1093/jxb/ers026

Sunilkumar G, Campbell LM, Puckhaber L, Stipanovic RD, Rathore KS (2006) Engineering cottonseed for use in human nutrition by tissue-specific reduction of toxic gossypol. Proc Natl Acad Sci USA 103:18054–18059. https://doi.org/10.1073/pnas.0605389103

Sunkar R, Zhu JK (2004) Novel and stress-regulated microRNAs and other small RNAs from Arabidopsis. Plant Cell 16(8):2001–2019

Tamilarasan S, Rajam MV (2013) Engineering crop plants for nematode resistance through host-derived RNA interference. Cell Dev Biol 2(2):114

Tehseen M, Imran M, Hussain M, Irum S, Ali S, Mansoor S, Zafar Y (2010) Development of male sterility by silencing Bcp1 gene of Arabidopsis through RNA interference. Afr J Biotechnol 9(19):2736–2741

Thakur S, Jindal V (2022) RNAi knockdown of the CAPAr gene affects the survival and fecundity of *Bemisiatabaci* (Gennadius). J Asia Pac Entomol 25(3):101940

Tian B, Wei F, Shu H, Zhang Q, Zang X, Lian Y (2011) Decreasing erucic acid level by RNAi-mediated silencing of fatty acid elongase 1 (BnFAE1. 1) in rapeseeds (*Brassica napus* L.). Afr J Biotechnol 10(61):13194–13201

Tian B, Li J, Vodkin LO, Todd TC, Finer JJ, Trick HN (2019) Host-derived gene silencing of parasite fitness genes improves resistance to soybean cyst nematodes in stable transgenic soybean. Theor Appl Genet 132:2651–2662

ul ain Sajid Q, Elci E (2024) Knockdown of orthotospovirus-derived silencing suppressor gene by plant-mediated RNAi approach induces viral resistance in tomato. Physiol Mol Plant Pathol 131:102264

Van Eck JO, Conlin BR, Garvin DF, Mason H, Navarre DA, Brown CR (2007) Enhancing beta-carotene content in potato by RNAi-mediated silencing of the beta-carotene hydroxylase gene. Am J Potato Res 84:331–342

Varshney RK, Bansal KC, Aggarwal PK, Datta SK, Craufurd PQ (2011) Agricultural biotechnology for crop improvement in a variable climate: hope or hype? Trends Plant Sci 16:363–371

Walsh HA, Vanderschuren H, Taylor S, Rey ME (2019) RNA silencing of South African cassava mosaic virus in transgenic cassava expressing AC1/AC4 hp-RNA induces tolerance. Biotechnol Rep 24:e00383

Wang Y, Li J (2006) Genes controlling plant architecture. Curr Opin Biotechnol 17(2):123–129

Wang H, Jones B, Li Z, Frasse P, Delalande C, Regad F, Bouzayen M (2005) The tomato Aux/IAA transcription factor IAA9 is involved in fruit development and leaf morphogenesis. Plant Cell 17(10):2676–2692

Wang CH, Lei XY, Xia J, Wang JW (2018) Effect of down-regulating 1-deoxy-D-xylulose-5-phosphate reductoisomerase by RNAi on growth and artemisinin biosynthesis in *Artemisia annua* L. Plant Growth Regul 84:549–559

Wang J, Li M, Zhuo S, Liu Y, Yu X, Mukhtar S, Lu G (2022) Mitogen-activated protein kinase 4 is obligatory for late pollen and early fruit development in tomatoes. Hort Res 9:uhac048

Wang K, Sun L, Zhang M, Chen S, Xie G, An S, Chen W, Zhao X (2024) E93 is involved in regulating larval-pupal metamorphosis and the development of the wing and ovary in *Helicoverpaarmigera*, and it shows great potential for RNAi-based pest control. J Integr Agric 24(5):1916–1929

Waterhouse PM, Graham MW, Wang MB (1998) Virus resistance and gene silencing in plants can be induced by simultaneous expression of sense and antisense RNA. PNAS 95(23): 13959–13964

Wei S, Li X, Gruber MY, Li R, Zhou R, Zebarjadi A, Hannoufa A (2009) RNAi-mediated suppression of DET1 alters the levels of carotenoids and sinapate esters in seeds of Brassica napus. J Agric Food Chem 57:5326–5333. https://doi.org/10.1021/jf803983w

Weise SE, Aung K, Jarou ZJ, Mehrshahi P, Li Z, Hardy AC, Carr DJ, Sharkey TD (2012) Engineering starch accumulation by manipulation of phosphate metabolism of starch. Plant Biotechnol J 10:545–554. https://doi.org/10.1111/j.1467-7652.2012.00684.x

Westwood JH, Yoder JI, Timko MP, Depamphilis CW (2010) The evolution of parasitism in plants. Trends Plant Sci 15:227–235

Whyard S, Singh AD, Wong S (2009) Ingested double-stranded RNAs can act as species-specific insecticides. Insect Biochem Mol Biol 39(11):824–832

Widyaningrum S, Pujiasih DR, Sholeha W, Harmoko R, Sugiharto B (2021) Induction of resistance to sugarcane mosaic virus by RNA interference targeting coat protein gene silencing in transgenic sugarcane. Mol Biol Rep 48(3):3047–3054

Worrall EA, Bravo-Cazar A, Nilon AT, Fletcher SJ, Robinson KE, Carr JP, Mitter N (2019) Exogenous application of RNAi-inducing double-stranded RNA inhibits aphid-mediated transmission of a plant virus. Front Plant Sci 10:265

Xu R, Wang Y, Zheng H, Lu W, Wu C, Huang J, Yan K, Yang G, Zheng C (2015) Salt-induced transcription factor MYB74 is regulated by the RNA-directed DNA methylation pathway in Arabidopsis. J Exp Bot 66(19):5997–6008

Yadav P, Seth RK, Reynolds SE (2024) A sperm-activating trypsin-like protease from the male reproductive tract of *Spodoptera litura*: proteomic identification, sequence characterization, gene expression profile, RNAi and the effects of ionizing radiation. J Insect Physiol 156:104664

Yang Y, Zhou Q, Xu J, Li Q, Zhang S (2018) RNA interference of NtNCED3 reduces drought tolerance and impairs plant growth through feedback regulation of isoprenoids in *Nicotiana tabacum*. Environ Exp Bot 155:332–344

Yang HH, Li JQ, Ma S, Yao WC, Chen YW, El Wakil A, Dewer Y, Zhu XY, Sun L, Zhang YN (2023) RNAi-mediated silencing of SlitPer disrupts sex pheromone communication behaviour in *Spodoptera litura*. Pest Manag Sci 79(10):3993–3998

Ye C, An X, Jiang YD, Ding BY, Shang F, Christiaens O, Taning CN, Smagghe G, Niu J, Wang JJ (2019) Induction of RNAi core machinery's gene expression by exogenous dsRNA and the effects of pre-exposure to dsRNA on the gene silencing efficiency in the pea aphid (*Acyrthosiphonpisum*). Front Physiol 9:1906

Yin X, Liu X, Xu B, Lu P, Dong T, Yang D, Ye T, Feng YQ, Wu Y (2019) OsMADS18, a membrane-bound MADS-box transcription factor, modulates plant architecture and the abscisic acid response in rice. J Exp Bot 70(15):3895–3909

Yin J, Yang J, Ma H, Liang T, Li Y, Xiao J, Zhan Y (2020) Expression characteristics and function of CAS and a new beta-amyrin synthase in triterpenoid synthesis in birch (*Betula platyphylla* Suk.). Plant Sci 294:110433

Yu B, Lydiate DJ, Young LW, Schäfer UA, Hannoufa A (2008) Enhancing the carotenoid content of Brassica napus seeds by downregulating lycopene epsilon cyclase. Transgenic Res 17:573–585

Zhang S, Li H, Liang X, Yan Y, Xia P, Jia Y, Liang Z (2015) Enhanced production of phenolic acids in *Salvia miltiorrhiza* hairy root cultures by combining the RNAi-mediated silencing of chalcone synthase gene with salicylic acid treatment. Biochem Eng J 103:185–192

Zhang X, Wang W, Wang M, Zhang HY, Liu JH (2016) The miR396b of *Poncirus trifoliata* functions in cold tolerance by regulating ACC oxidase gene expression and modulating ethylene–polyamine homeostasis. Plant Cell Physiol 57(9):1865–1878

Zhang L, Zhu M, Ren L, Li A, Chen G, Hu Z (2018) The SlFSR gene controls fruit shelf-life in tomato. J Exp Bot 69(12):2897–2909

Zhang P, Liu X, Yu X, Wang F, Long J, Shen W, Zhao X (2020) The MYB transcription factor CiMYB42 regulates limonoid biosynthesis in citrus. BMC Plant Biol 20:1–12

Zhang X, Jin L, Li G (2023a) RNAi-mediated functional analysis reveals the regulation of oocyte vitellogenesis by ecdysone signalling in two Coleoptera species. Biology 12(10):1284

Zhang Y, Xu J, Li R, Ge Y, Li Y, Li R (2023b) Plants' response to abiotic stress: mechanisms and strategies. Int J Mol Sci 24(13):10915

Zhao K, Zhang F, Yang Y, Ma Y, Liu Y, Li H, Dai H, Zhang Z (2016) Modification of plant height via RNAi suppression of MdGA20-ox gene expression in apple. JASHS 141(3):242–248

Zhou M, Tang W (2019) MicroRNA156 amplifies transcription factor-associated cold stress tolerance in plant cells. Mol Gen Genomics 294:379–393

Omics Approaches in Crop Science: Transcriptomics, Proteomics, and Metabolomics for Enhanced Crop Traits

8

Walunjkar Babasaheb Changdeo, Anu Naruka, Narayan Pandit Gurav, Ashish H. Bankar, Anurag Malik, Himani Punia, and Natalia Repkina

Abstract

Evolving the science of omics transformed crop science by offering three methods, namely transcriptomics, proteomics, and metabolomics, to provide systemic knowledge about the mechanisms of biological regulations influencing plant development besides stress adaptation capacity. With research in transcriptomics, scientists could assess the mechanisms of how genes express themselves for basic mechanisms governing significant agronomic characters like yield formation, management, and nutrient use. The proteomic approach strengthens transcriptomics by estimating functional proteins, besides providing data regarding the dynamics of protein interaction and patterns of posttranslational modification. The metabolic profiles from metabolomics analysis display existing plant conditions of physiology, but define basic metabolic pathways that need enhancement for better traits. These omics platforms form a systems biology

W. B. Changdeo · A. Naruka
Shri Vaishnav Institute of Agriculture, SVVV (Shri Vaishnav Vidyapeeth Vishwavidyalaya), Indore, India

N. P. Gurav · A. H. Bankar
Department of Agricultural Botany, SVKMs College of Agriculture, Tardi, Shirpur, Dhule, Maharashtra (Affiliated to MPKV, Rahuri), India

A. Malik
Division of Research and Innovation, Uttaranchal University, Dehradun, Uttarakhand, India

H. Punia
Saraswati Group of Colleges, SGC Group, Gharuan, Mohali, Punjab, India

N. Repkina (✉)
Institute of Biology of Karelian Research Centre, Russian Academy of Sciences, Petrozavodsk, Russian Federation Pushkinskaya
e-mail: nrt9@ya.ru

Y. Ahlawat et al. (eds.), *Next-Generation Strategies for Crop Improvement*,
https://doi.org/10.1007/978-981-95-0309-4_8

framework that establishes connections between genetic information and observable characteristics. By integrating different platforms, scientists discover those targeted crop-enhancement genes together with biomarkers and pathways that the genetic breeding methods, together with genome-editing techniques, harness through CRISPR-Cas. Omics technologies are part of key tools in developing crop varieties resistant to climate stresses including drought together with salinity and heat stress exposure. This chapter shows how omics methods transform breeding practices to ensure world food security and maintain environmentally sustainable farming methods.

Keywords

Omics approaches · Transcriptomics · Proteomics · Metabolomics · Crop improvement · Sustainable agriculture · Stress tolerance · Systems biology

8.1 Introduction

The growing world population along with climate change effects makes sustainable farming practices more essential while demanding better crop production methods. Overcoming these challenges relies on a better understanding of molecular, biochemical, and physiological procedures that govern plant growth, development, and stress tolerance (Benitez-Alfonso et al. 2023). Omics approaches, which are transcriptomics and proteomics combined with metabolomics and powerful analytical capabilities, are now emerging as fundamental tools for crop science to look into complex regulatory pathways that may affect agronomic features. Through research in transcriptomics, scientists analyze RNA expressions for the detection of regulatory networks that influence plant characteristics such as yield and stress tolerance and nutrient management abilities (Vasupalli et al. 2024). Through proteomic analysis, researchers obtain information regarding the distributions of protein populations and protein-protein networks, as well as about structural changes in proteins. Metabolomics analyzes small molecule metabolites developed in the cellular activity of plants that show both the physiological condition of plants and the reactions of plants toward environmental elements (Al-Amrani et al. 2021).

Each standing alone contributes distinct forms of investigation. By integrating their analysis we obtain comprehensive systems-level knowledge of plant biology enabling scientists to bridge genetic data with observable phenotypic manifestations. An integrated approach has uncovered potential genes and pathways with biomarkers that researches use for enhancing crop traits (Cembrowska-Lech et al. 2023). Such integration serves modern breeding methods by making them more efficient with genome-editing tools, particularly CRISPR-Cas, to speed up the development of adaptable and productive crops.

The paper investigates contemporary research involving transcriptomics and proteomics together with metabolomics within crop science applications. The discussion demonstrates how omics technologies enhance agronomic traits while solving food security issues and climate change while displaying their future role in shaping sustainable agriculture systems.

8.2 Transcriptomics: Deciphering Gene Expression in Crop Plants

8.2.1 Importance of Transcriptomics for Understanding Gene Expression

Transcriptomics is the analysis of the whole RNA transcriptome, mRNA, rRNA, tRNA, non-coding RNA, etc., in a single cell or multiple cells. In other words, transcriptomics is centered on the "transcriptome," which in its simplified meaning represents the entire inventory of RNA transcripts derived from transcription (Gawronski and Kim 2017). Developing knowledge of this branch of molecular biology is crucial in providing insights into gene expression and regulation and how they respond to distinctive environmental or developmental signals (Hill et al. 2020). Transcriptomics embraces processes such as RNA sequencing, microarray analysis, and any other techniques that make it possible to quantify, identify, and characterize RNA transcripts in a given sample (Lowe et al. 2017). These techniques allow the investigator to monitor gene expression in different types of tissues, at different developmental ages, or under particular physiological or environmental conditions (Singh et al. 2018). With transcriptomics, it is possible to quantify and identify the RNA molecules present in the cells, and this will in one way help scientists identify which among the genes are in active use in certain cells under given conditions (Wang et al. 2008).

Gene Expression Profiling Transcriptomics offers a view of the level of gene activity at any point in time; it tells you which genes are active or inactive in a given circumstance, stress, disease, or other factors of the environment (Breschi et al. 2017). This data is important in developing the profiles of gene expression in different states or conditions within biological organisms. For instance, in cancer biology transcriptomics support identification of differential gene expression profiles that defines various types of cancer aiding in its diagnosis and prognosis (Roberts et al. 2022).

Regulatory Mechanisms of Genes Transcriptomics also helps in the gene regulation because it studies the relations between different RNA types. Micro RNAs are among the noncoding RNAs which can post-transcriptionally control how genes are regulated for the proteins that are finally produced (Statello et al. 2020). Through the study of the transcriptome, one can understand these regulatory circuits and impact factors as well as effects on cell functions, differentiation of tissues, as well as development of the whole organism (Wang et al. 2024).

Identifying Biomarkers and Therapeutic Targets Transcriptomics plays an important role in the discovery of molecular biomarkers that are a sign of a certain biological condition. For instance, deep transcriptome analyses have highlighted certain mRNA and noncoding RNAs as diagnostic markers of cardiac and metabolic disorders, neurodegenerative diseases, etc. (Supplitt et al. 2021). These biomarkers are essential in the diagnostics of the diseases at their early stages and for evaluation or the effectiveness of the treatment processes (Wang et al. 2008). Moreover, transcriptomics helps predict potential therapeutic agents since it identifies which genes are abnormal in disease conditions.

Functional Genomics and Systems Biology Basically, transcriptomics is important in functional genomics, which focuses on the relationship of all the expressed genes to different physical and developmental processes (Khodadadian et al. 2020). When they combine transcriptomic information with other "omics" data including proteomics and metabolomics, scientists are able to develop system biology models that accurately and comprehensively depict the interactions that occur within cells as well as tissues (Manzoni et al. 2016). These models offer an integrated approach to depict cellular functions and thereby show how alterations in gene expression impact the molecular processes and how diseases interfere with standard functions (Hill et al. 2020).

Personalized Medicine Transcriptomics has immense possibilities in the area of personalized medicine because it creates the opportunity for the doctor to decide about the exact treatment depending on the gene expression level in an organism (Wang and Wang 2023). For instance, through the transcriptome it may be possible to establish how the patient's gene expression could influence the metabolism of the drug, so as to improve the prescription and to reduce side effects (Khodadadian et al. 2020). The method is especially useful in oncology, as treatment strategies for cancer are now being created based on the transcriptomic characteristics of tumors.

In general, transcriptomics is a key tool for understanding the global patterns of gene expression for decoding regulatory processes and cellular roles (Tyagi et al. 2022). In recent years, transcriptomics has become crucial in between and by expanding the knowledge with regards to gene expression and control in different fields that include molecular biology to medicine (Koks et al. 2020). The outcomes derived from transcriptomic research are not only enriching basic science knowledge but are also setting the stage for major innovations in diagnostics, therapeutics, and individualized medicine (Tsakiroglou et al. 2023).

8.2.2 Applications of Transcriptomics in Crop Improvement

Transcriptomics has also proved to be a powerful technique for improving crop resistance and yield, nutritional quality and quality control in plant biotechnology applications (Yang et al. 2021). The analysis of RNA transcriptomes ultimately allows for effective identification of how plants react to environmental changes, increasing yield and biomass, and increasing nutritional profile through the process of biofortification. To help the reader understand these key applications in detail, the following subtopics have been provided (Dueñas et al. 2021).

8.2.2.1 Studying Stress Response at the Transcript Level

Other environmental factors such as drought, salinity, diseases, and pests have a direct influence on the yield and the quality of the produce (Muhammad et al. 2024). Transcriptomic profiling reveals the identities of genes that are upregulated or downregulated in plants under such stresses. This is because stress-induced genes enable researchers to readily observe the molecular changes and pathways that underpin increased stress tolerance (Yuan et al. 2021). For example, transcriptomic analysis of drought-tolerant rice and maize varieties shows that particular genes, for instance, dehydration-responsive element-binding (DREB), and NAM, ATAF, and CUC (NAC) transcription factors, are involved in the stress response. Through knowledge and understanding of these kinds of regulatory mechanisms, researchers and breeders can produce crop varieties that will be more suitable for environments characterized by such conditions (He et al. 2024).

8.2.2.2 Transcriptomic Insights into Yield and Biomass Enhancement

Increasing yield and biomass production is highly recommended to support food demand around the world. Since yield and biomass are quantitatively spirited phenotypes, genes regulating these traits and auxiliary genes have been discovered through transcriptomic studies like those related to photosynthesis efficiency, growth rate regulation, and flowering time (Ain et al. 2022). For instance, transcriptomics approach in wheat and maize has grouped genes in the Photosystem II and RUBISCO complexes, which are crucial for optimizing photosynthesis and hence biomass (Caruana et al. 2022). Also, transcriptomics enables gene expression analysis to be made when exposed to plant growth hormones, which influence cell division and elongation processes that affect biomass and yield (Friero et al. 2023).

8.2.2.3 Biofortification and Nutritional Enhancement

Biofortification seeks to introduce micronutrients including vitamins and minerals as well as essential amino acids into the crops genetically (Yadava et al. 2018). Transcriptomics is useful in biofortification since it helps to establish the genes responsible for nutrient production and deposition. In rice, for example, transcriptomic studies have identified key genes such as *OsNAS1* and *OsNAS2* involved in iron and zinc accumulation (Dueñas et al. 2021). These genes can be targeted to develop biofortified rice varieties that provide higher nutritional value.

Beyond minerals, transcriptomics also supports the enhancement of essential vitamins like provitamin A in crops. This is evident in the case of Golden Rice, where transcriptomics identified and subsequently enhanced the expression of genes involved in beta-carotene synthesis, leading to increased vitamin A content.

8.2.3 Case Studies in Crop Transcriptomics

In crop research, transcriptomics has provided solutions for evolving breeding programs, identifying stress tolerance, yield, and quality, particularly in staple diets like rice and maize. These studies include massive parallel sequencing of RNA transcripts whereby specific genes and a regulatory network that controls these favorable phenotypes can be pinpointed (Abdul Aziz and Masmoudi 2025). The following section describes examples of how transcriptomics has been applied to certain crops, some of the main findings of transcriptomics, and their uses in contemporary breeding practices.

8.2.3.1 Transcriptomics in Rice for Stress Tolerance and Quality Enhancement

Rice (*Oryza sativa*) is one of the world's most important staple food crops, and enhancing the germplasm of the crop to stress factors such as drought and salinity is imperative to feed the growing population in the world. Stress-related genes and pathways have been discovered by transcriptomic analysis on rice (Hoang et al. 2016). For instance, the protein transcription factors, which include DREB (dehydration-responsive element-binding) and NAC (NAM, ATAF, and CUC), have been named as positive regulators of drought tolerance under drought-stressful conditions (Sahid et al. 2023). These genes are rated as robust stress signaling, water, and osmotic regulation genes, thus being ideal candidates for selection in rice breeding for drought-resistant types. Besides stress tolerance, transcriptomics has enhanced the rice grain quality, and strain analysis of the OsSWEET gene family provided understanding about the carbohydrate movement and grain filling in rice grain quality formation and yield formation (Sahebi et al. 2018). Differences in the regulation of these genes have been utilized in developing varieties of rice with enhanced grain dimension, mass, and value.

8.2.3.2 Transcriptomics in Maize for Yield and Nutritional Quality Improvement

Maize is popular for food, feed, and industrial uses crops grown in both developed and developing countries. More specifically, studies based on the maize transcriptome have helped to identify genes that can enhance yield as well as those that help the plant tolerate stress. For example, it was found that under salt stress, genes like *HKT1* and *SOS1* are involved in controlling sodium ion homeostasis, thereby promoting salt tolerance (Li et al. 2017). Furthermore, research on the identification of genes for nitrogen use efficiency in maize at the transcriptome level has also provided understanding of genes involved in nitrogen acquisition and remobilization in maize, which are essential for genetic enhancement under solution nitrogen-limited environments. Transcriptomics has also been used in increasing the nutritional value of the maize grains. For instance, research carried out to study the *ZmPSY1* gene that is responsible for carotenoid biosynthesis aims at producing biofortified maize with high provitamin A carotenoids (Maqbool et al.

2018). This is particularly relevant in a bid to reduce the incidence of vitamin A deficiency in groups who rely on maize as their staple food.

8.3 Proteomics: Understanding Protein Function in Crop Development

8.3.1 Overview of Proteomics and Its Significance in Studying Protein Functions in Crops

In the recent past, proteomics, the study of proteins at a large scale, has turned out to be crucial in agricultural research to study the functions and interactions of proteins in crops. Whereas genomics offers information on sequences of DNA, proteomics is interested in proteins that are products of Gene expression and are directly involved in cellular activities (Al-Amrani et al. 2021). They are essential for virtually all aspects of plant life cycle, growth and development, defense, and stress responses. Hence, proteomics intervention on crops gives information on other intricate bio systems and helps in realizing the aims of crop development, particularly concentrating on stress tolerance, yield, and quality (Liu et al. 2019).

In crop research, proteomics makes it possible for researchers to study the impact of environmental stress whereby stress like water limitation, high salt levels, and diseases can be studied in view of the change in protein expression patterns. For instance, stress by drought affects certain proteins in maize relative to water stress, allowing breeders to focus on certain pathways when developing new crops. Proteomics plays a vital role in nutrition biosynthesis and the level of metabolic activity, which would be advantageous in enhancing the quality of plants such as rice and wheat (Ahmad et al. 2016).

8.3.2 Applications of Proteomics in Crop Trait Improvement

System-level functional genomics, particularly proteomics, is widely used in crop research to pursue important agronomic traits including stress tolerance, growth, development, and reproduction. In addition to offering information on the functional proteins and regulatory networks that are at work within plants, proteomics allows for the determination of target proteins and pathways that can improve the yield, quality, and robustness of crops (Yan et al. 2022). This feature makes the use of this approach particularly effective in explaining polygenic characteristics that include several genes and regulatory pathways.

8.3.2.1 Proteomic Insights into Stress Tolerance and Regulatory Networks

For crop plants, abiotic factors such as drought, salinity, temperature stress, and oxidative stress are quite common, along with biotic stresses from disease and insects. Proteomics gives the picture of the proteins and the molecular signal transduction pathways that are engaged in response to these stimuli (Ahmad et al. 2016). For instance, in response to drought, proteomics in crops like maize and rice unveiled proteins involved in osmotic stress and reactive oxygen species (ROS) detoxification to support efforts required in energy production, according to the proteome maps of immersed plants. These proteins include heat shock proteins, aquaporines, and antioxidant enzymes, all of which are precious in stress acclimation (Hamzelou et al. 2020).

Another application of proteomics is to discover proteins and molecular mechanisms that mediate various stress responses by using signaling pathways. Proteomic analysis also highlights that stress-responsive TFs such as WRKY, MYB, and the NAC families are major signal mediators (Bawa et al. 2022). Such mechanisms can be manipulated by increasing or decreasing the level of these regulatory proteins with a view of creating crop varieties with the ability to withstand different stress conditions. It is within such insights that new marker-assisted breeding and engineering continue the development of stress-tolerant crop varieties (Gill et al. 2024).

8.3.2.2 Studying Proteins Related to Growth, Development, and Reproductive Success

Apart from stress tolerance, proteomics is widely employed for analysis of proteins related to growth, development as well as reproductive success of organisms. That is why these fundamental processes dictate crop yield and quality, which must be considered to improve crop production (Villalobos-López et al. 2022). Research on proteomics of crops is based on proteins associated with cell wall synthesis, photosynthesis, and nutrient as well as hormonal control of plant growth and development. For instance, proteomic analysis in rice aimed at finding proteins that are related to biomass accumulation or yield for this plant identified proteins related to photosynthesis, carbon metabolism, and nutrient transport (Sah et al. 2022). Besides, abacus, gibberellin, and cytokinin signal transduction-related proteins have also been investigated to elucidate the mechanism of root and shoot development. With the knowledge of proteins that coordinate developmental periods, it is possible for breeders to focus on specific growth phases for improvement of the production rates (Weiss and Ori 2007).

Proteomics also plays a useful role in explaining the reproductive success in crops through the identification of proteins that play a role in pollen and seed formation and fertilization. In wheat and in maize, proteins that affect pollen quality and seed filling, which are important for yield, have been identified (Ji et al. 2024).

The integration of proteomics into crop improvement strategies is a necessity because it is the only technique that can link gene expression with functional consequences. As proteomic technologies evolve, the extent of high-throughput analysis accessible to researchers will be increased, therefore allowing greater identification of protein-protein interactions, posttranslational modifications, and capture of dynamic responses to stresses. When combined with additional omics technologies, proteomics will improve the efficiency of crop trait identification and create a high-yield crop that can withstand adverse environmental conditions.

8.3.3 Case Studies in Crop Proteomics

Several traits, including pest and disease resistance, response to environmental challenge, and yield, have been improved using proteomics in crops. Commonly, proteomics is the study of protein expression and its regulation, knowledge that assists a plant in coping with adverse conditions (Ahmad et al. 2016). This approach has resulted in several successful applications including the discovery of stress-related proteins and increased key metabolic pathways, resulting in improved crop performance and quality. Real-life examples of wheat, soybean, and tomato reveal the role of proteomics in strengthening the plant resistance and increasing the yield potential, proving the great advances of agricultural science.

8.3.3.1 Wheat: Enhancing Drought and Salinity Tolerance

Wheat is one of the most widely cultivated cereal crops; nevertheless, drought and salinity are major constraints to yield and quality. Potential target proteins selected for wheat proteomic research were those associated with stress response, including osmotic adjustment, ROS, and ion homeostasis. For instance, in proteomics, the definitive appearance of heat shock proteins (HSPs), aquaporin, and antioxidant enzymes like superoxide dismutase comes into operation during drought stress conditions. They assist in the protection of the cell compartment from stress by nipping oxidation and regulating water content, and since wheat is used in environments that are dry and saline, it would also allow wheat to survive such harsh conditions (Singh et al. 2022a).

Thus, proteomic data have been applied to identify the key proteins that can be targeted for improving stress tolerance by applying selective breeding programs. Therefore, the crop varieties that are resistant to stress factors such as droughts and salinity have been produced, which support the yield stability (Alrajeh et al. 2024).

8.3.3.2 Soybean: Improving Yield and Disease Resistance

Soybean is both a protein and an oil, and its yields are under threat from both biotic factors like pests and diseases and pressure from the need to provide very high yields. In yield improvement and disease resistance, proteomics has been applied to define a wide range of proteins. For instance, pathogenesis-related (PR) proteins and enzymes in soybean have been found to be upregulated during fungal attacks and bacterial blight according to Wang et al. (2020). PR proteins improve the rigidity of the cell walls as well as slow the rate of pathogen growth, making it a natural barrier (Du et al. 2023).

The proteomic analysis has also shown that storage proteins like glycinin and beta-conglycinin are important in determining seed quality and yield. Owing to understanding how these proteins are regulated, farmers have been able to genetically modify soybeans with high protein content and disease repellent, bringing about a change in crop yield and quality (Guo et al. 2022).

8.3.3.3 Tomato: Enhancing Heat Tolerance and Fruit Quality

Tomato, an extensively grown horticultural crop, is sensitive to heat stress that reduces its fruit yield and quality. Studies of proteomics in tomato have targeted heat shock proteins and the proteins involved in the heat-responsive antioxidants. These proteins prevent the detrimental consequences of heat stress through the stabilization of proteins and by combating free radical injury in cells. Further, proteomics approaches have been used to map proteins related to lycopene and carotenoids, which are highly important in the coloration and nutritional quality of fruits (Li et al. 2021).

To date, these proteomic findings have been applied by breeding programs targeting tomatoes with better heat tolerance and other desirable fruit quality characteristics. Such advancement has enabled the producers to grow tomato varieties that are high yielding and of high quality in warmer areas, producing benevolence both to the producers and the consumers (Hoshikawa et al. 2021).

These cases illustrate the revolution that proteomics has brought about in crop science and offer concrete ideas for producing higher-yielding, stress-free crops. Through the association of protein function to certain traits, proteomics provides the prospect for a more targeted illustration of the genetic makeup of crop plants than what has been made available by breeding practices. The future perspectives in crop proteomics are the extension of proteomics together with genomics and metabolomics, which is expected to provide more in-depth knowledge of plant biology. Moreover, as more proteomic studies are developed, it will be possible to explore protein temporal changes that will assist crop resistance and production as relation to environmental conditions.

8.4 Metabolomics: Profiling Metabolic Pathways for Enhanced Crop Traits

Metabolomics is defined as the ambitious discipline that focuses on the identification of metabolites with reference to small molecules involved in various metabolic processes in organisms. Metabolomics is a potent strategy for characterizing plant metabolome and mapping plant metabolic networks, growth, development, stress tolerance, and nutrient accumulation. As put forward by the Metabolomics Society, metabolomics is the large-scale study of metabolites within a biological system at an arbitrary time point (Manickam et al. 2023). In crops, metabolomics has also been useful in offering an understanding of the outstanding metabolic processes that dictate development, growth, and adaptability to the outdoor environment. Through these pathways, the researchers are able to find out the molecular aspects of characteristics such as taste, nutritional quality, disease susceptibility, and stress resistance like a drought or high-salt tolerance (Marchev et al. 2021).

Among the major strengths of metabolomics is its capacity to provide a "momentary picture" of a plant metabolic system as influenced by environment, genetic makeup, or disease influence. For example, stress conditions such as drought or pathogen infection change a plant's metabolism to release certain metabolites that can act as a biomarker of stress tolerance. Knowing these biomarkers assists in growing stress-tolerant plants by selecting breeds genetically inclined to stress tolerance. Furthermore, metabolomics contributes to biofortification because it charts compartments for nutrient accumulation to breed better crops nutritionally (Marchev et al. 2021).

8.4.1 Applications of Metabolomics in Crop Development

Metabolomics is a young discipline that is trying to unravel the entire profile of cell metabolites through the identification and quantification of cellular metabolites, hence reflecting an integral approach toward understanding biochemical pathways in living organisms. Primary and secondary metabolites play pivotal roles in the world of plant physiology and biochemistry (Shen et al. 2023). Crop science has witnessed revolutions, which have also offered new scopes to enhance productivity, nutritional quality, resilience, and marketability. This section discusses three significant areas of metabolomics application in crop improvement, namely adaptation and resilience under stress, nutritional quality and biofortification, and flavor, aroma, and quality improvement (Marchev et al. 2021).

8.4.1.1 Metabolomic Insights into Stress Adaptation and Resilience

Plants face constant exposure to a variety of abiotic stresses such as drought, salinity, temperature stress, oxidative stress, and biotic stresses including pathogens and herbivory. These stresses have the potential to negatively impact crop yield and quality. Metabolomics enables an overall view on how plants cope with these challenges, thereby facilitating the elucidation of metabolic pathways and the identification of biomarkers unique to stress conditions (Shiade et al. 2024).

Metabolic plant response under stress conditions often leads to dramatic changes of the plant metabolome, such as the biosynthesis of osmo-protectants, anti-oxidative compounds, and stress-related secondary metabolites. In drought stress, for example, plants often accumulate proline, which is one of the main osmolytes maintaining osmotic homeostasis in cells (Kumar et al. 2021). Flavonoids and phenolic compounds are also upregulated in oxidative stress as scavengers of reactive oxygen species (ROS). Generalized by metabolomics studies, those stress-specific upregulated metabolites that stabilized membranes and proteins, such as glycine betaine and trehalose, increased the resistance to stress.

Salinity stress is the other principal challenge, triggering the accumulation of metabolites that mitigate ionic toxicity and oxidative damage, which include quaternary ammonium compounds and polyamines (Singh et al. 2022b). Metabolomic profiling under salinity stresses in rice reported a higher spermidine content, which is well correlated with greater ion homeostasis and lowered lipid peroxidation. Heat shock similarly induces secondary metabolites that include carotenoids, capable of protecting the chloroplast against thermal damage in addition to eliciting the induction of heat shock proteins (Chen et al. 2021).

8.4.1.2 Identification of Metabolites Related to Nutritional Quality and Biofortification

Global health faces two critical challenges because of malnutrition together with the lack of essential micronutrients. The field of metabolomics serves as a critical method to improve staple crops' nutritional characteristics by analyzing essential nutrients that appear as metabolites. Analyzing both primary and secondary metabolites including carbohydrates, proteins, lipids, together with vitamins, flavonoids, and polyphenols, allows researchers to improve crop nutritional content (Marchev et al. 2021).

Metabolomic scans of rice resulted in the discovery of carotenoids including beta-carotene, as promising candidates for biofortification research. Scientists created Golden Rice through genetic engineering to produce a provitamin A-rich variety. Biofortification research using metabolomics investigated how to increase lysine and tryptophan levels in maize to make cereal-based diets more nutritious. Table 8.1 shows the health benefits of biofortified crops (Li et al. 2022a).

Multidimensional metabolomics methods have helped to mitigate widespread deficiencies in iron and zinc, which affect billions of people across the world. Metabolome analysis of wheat and pearl millet showed pathways that control Phytosiderophore biosynthesis and accumulation, allowing increased iron and zinc uptake. Different national programs are utilizing biofortified versions of crops that contain increased levels of essential nutrients (Yogendra et al. 2024).

Through metabolomics, researchers were able to improve nutritional phytochemicals while achieving micronutrient deficiency solutions. Researchers used metabolic techniques to boost lycopene and flavonoid production in tomatoes, thus increasing these proven antioxidant and anti-inflammatory compounds. Research on cruciferous vegetables including broccoli and kale has utilized metabolomics approaches to study glycosylates because they show cancer prevention properties (Mukherjee et al. 2024).

Metabolomic approaches contribute to enhancing food products with dietary fiber content. Using metabolite profiling for cellulose and hemicellulose and lignin biosynthetic pathways researchers have achieved high-fiber wheat and barley strains. These new developments create solutions for nutrient deficiencies while supporting

Table 8.1 Health benefits of biofortified crops

Key metabolites identified	Examples of biofortified crops	Nutritional/health benefits
Carotenoids (e.g., beta-carotene)	Golden rice	Enriched with provitamin A
Lysine and tryptophan	Biofortified maize	Enhanced essential amino acid content
Phytosiderophores	Wheat, pearl millet	Improved iron and zinc uptake
Lycopene, flavonoids	Tomatoes	Antioxidant and anti-inflammatory properties
Glycosylates	Broccoli, kale	Linked to cancer prevention
Cellulose, hemicellulose	Wheat, barley	Increased dietary fiber content

environmentally friendly farming by decreasing dependency on chemical fertilization and synthetic supplementation.

8.4.1.3 Flavor, Aroma, and Crop Quality Improvement Through Metabolic Profiling

The sensory qualities of flavor together with aroma and texture determine both consumer interest and agricultural product sales potential. Through metabolomics, researchers gain access to powerful tools that analyze the underlying metabolites affecting food taste and smell as well as product quality attributes (Shu et al. 2023).

Mainly, the flavor composition of fruits and vegetables exists from a combination of sugars, organic acids, and volatile organic compounds (VOCs). Metabolomic analysis of tomatoes detects three specific VOCs: hexanal, cis-3-hexenol, and beta-ionone, which impact its characteristic fragrance. The analysis of strawberry metabolites identified furanones and esters as fundamental compounds that create their sweetness and their fruity aroma. Strawberry and tomato cultivar development through metabolite selection has produced varieties with improved flavor characteristics (Zárate-Carbajal et al. 2024).

The aroma characteristics of rice alongside coffee are leading factors that determine crop quality. Analysis of fragrant rice varieties like basmati and jasmine detected 2-acetyl-1-pyrroline (2AP) through metabolomics as the major compound behind their dominant aroma. The identification of genetic and environmental elements affecting 2AP production enables purposeful plant breeding for aromatic rice varieties. The selection of high-quality coffee beans for top-grade quality depends on identifying furans, pyrazines, and aldehydes as VOCs that produce coffee's distinct aroma (Varatharajan et al. 2021).

Metabolites greatly influence both texture profiles and quality in the markets of products. The combination of pectins and hemicelluloses allows scientists to estimate the firmness of fruits, including apples and pears, since these metabolites control the processes of synthesis and breakdown of cell walls. Laboratory studies with a focus on metabolites have helped researchers to create new varieties of plants that show larger storage capabilities and reduced waste after harvesting (Ozyigit et al. 2023).

Metabolomic analysis drives improvement in both visual quality and nutritional value assessment. Antioxidant defenses of fruits and vegetables are identified in unique compounds that cause red, purple, and orange pigmentations. Profiling and enhancing these metabolites have led breeders to develop visually appealing as well as nutritionally enhanced crops such as purple carrots and golden kiwifruits (Mattioli et al. 2020).

8.5 Integrative Multi-omics Approaches for Comprehensive Crop Improvement

Complete information about the plant traits at biological processes is provided through crop improvement based on the alignment of multi-omics approaches, such as transcriptomics, proteomics, and metabolomics. Altogether, the technologies give a researcher all details about molecular processes and unveil the complex mechanisms for traits while also enhancing crop productivity and nutritional aspects.

8.5.1 Importance of Integrating Transcriptomics, Proteomics, and Metabolomics

Multiple omics layers, when combined, provide crucial systemic views that one would require to study the interlinked activities of genes, proteins, and metabolites. The analysis of transcriptomics indicates which genes are being actively expressed, but proteomics sheds light on the functional proteins that result from activated genes. Metabolic profiling fills in the gaps with existing datasets by generating identification and numerical assessment of biochemical compounds that are produced because of the action of protein enzymes (Hasin et al. 2017).

The benefits of cogenetic analysis to researchers are that they break down the complex traits, such as drought resistance combined with disease immunity and nutrient processing capability, since these traits incorporate simultaneous gene networks and regulatory elements. In studies on drought-tolerant crops, transcriptional analysis reveals stress-responsive genes, and proteomic assessment of them reveals stress-related proteins, whereas metabolomic data exposes osmo-protectants and antioxidants (Li et al. 2022b).

The synthesis of such database types can be greatly supported through computational tools combined with bioinformatics methods. By integrating WGCNA with machine learning algorithms, scientists are able to align the data pathways for both transcriptomic and proteomic as well as metabolomic to help discover critical regulatory hubs and identify biomarkers.

8.5.2 Multi-omics Case Studies in Crop Research

These case reports show that multi-omics techniques elucidate trait-related molecular mechanisms while providing information that directly benefits plant breeding initiatives (Table 8.2).

8.5.3 Challenges and Opportunities in Multi-omics Integration

Despite its transformative potential, multi-omics integration faces several challenges:

Table 8.2 Case studies demonstrating the strength of integrated omics approaches in crop research

Case study	Omics integration	Key insights
Drought tolerance in rice	Transcriptomics, proteomics, and metabolomics	Identified stress-responsive transcription factors, heat shock proteins, and osmo-protectants
Nutritional enhancement in maize	Genomics, transcriptomics, and metabolomics	Enhanced provitamin A and amino acid content through targeted metabolic pathway engineering
Disease resistance in tomato	Transcriptomics, metabolomics, and proteomics	Revealed upregulation of defense-related genes and accumulation of antifungal secondary metabolites
Salt tolerance in barley	Multi-omics (genomics, transcriptomics, proteomics, and metabolomics)	Identified ion transporters, compatible solutes, and stress-responsive proteins
Flavor improvement in tea	Metabolomics and transcriptomics integration	Enhanced biosynthesis of catechins and volatile organic compounds for better flavor profiles

- Technical Challenges: Generation of good-quality omics data requires standardization of preparatory methods of sample preparation and data acquisition technologies and normalization standards.
- Computational Challenges: Such a manipulation of multi-omics large-scale datasets calls for a strong performing bioinformatics infrastructure and vast computational capacity.
- Reproducibility: The reproduction and consistency of results between experiments are seriously hampered because quite crucial variations arise from the differences in the experiment setups and data handling.

New breakthrough technologies combined with new methodologies provide promising options for a solution. AI combined with machine learning methods is now processing multi-omics data to detect patterns and establish predictive models for exploring complex characteristics. Precision agriculture is now more reachable due to integration frameworks of data along with deep learning networks, which continue to evolve multi-omics analysis methods (Sheikh et al. 2024).

8.6 Future Directions and Prospects for Omics Technologies in Crop Science

The scientific discipline of omics technology that includes genomics, transcriptomics, proteomics, and metabolomics has vast potential to revolutionize the practices in plant science. Researchers predict improvements based on the integration of current technologies with next-generation AI and ML, along with rapid phenotyping. The combined use of all these technologies will uncover complex biological structures responsible for crop traits, hence leading to a fully mature understanding of the growth and developmental cycles of plants, along with mechanisms related to environmental responses (Sheikh et al. 2024).

Omics methods, in the sense of studying cellular heterogeneity at single-cell resolution, now seem to manifest as a potential application avenue in crop science. This method now makes very specific cellular stress responses in plant tissue visible so researchers can better deduce crop improvement targets. The burden for spatially resolved omics technologies remains ever on the increase since better views into the molecular component movement of plant tissues begin to show up (Liu et al. 2024).

The field of synthetic biology stands to gain many benefits from future progress in omics research. The analysis of omics data allows researchers to develop new pathways that produce valuable compounds while building crop resilience. Synthetic biology enables researchers to produce biofortified crops through the combination of metabolomics with optimized contents of essential nutrients and health-promoting compounds (Barnum et al. 2021).

Interaction between high-throughput phenotyping technologies and omics methods allows faster discovery of genotype-phenotype connections. The platforms enable simultaneous observation of broad plant populations in controlled settings to relate core molecular analyses to agricultural traits. The method proves essential in resolving global problems between climate change and food security (Wani et al. 2022).

Big omics datasets will be managed and shared through the combination of cloud computing and blockchain technology. Computing technologies, together with blockchain systems, guarantee the protection and duplication of information access that makes it possible for researchers across industries to partner together. Precise agricultural practices will benefit from omics integration, which would allow for constant monitoring of crop health and development in making data-driven decisions on farms.

Future development of omics technologies depends significantly on ethical guidelines accompanied by regulatory regulations. Established policies should guide the use of genetically modified crops along with efforts to safeguard smallholder farmers in the developing world who need equal access to omics-driven innovation.

8.7 Conclusion

Crop science includes an omics approach to acquire detailed information about molecular and biochemical pathways that control traits in plants. The tools with environmental stress response facilitate scientists' capacity to understand how complex biological networks regulate growth and development. Research in the area of biology by applying transcriptomics enables scientists to find expression patterns of genes while revealing control mechanisms over crucial traits in agriculture. Proteomic tools extend previous studies by quantifying protein levels but defining operational molecules and their assembly properties. Using metabolite profiling, metabolomics captures snapshots that highlight the physiological status of plants as it detects cellular processes end products. These technologies of omics thus show their capacity to develop a whole biological system model that connects DNA sequences to observable characteristics and physical properties. The combined omics approach helps scientists detect genes along with biological indicators of change and designate metabolic pathways that lead to yield enhancement and stress resistance and nutrient elements management. Omics-based insights provide precise functionality to breeding techniques and CRISPR-Cas genome-editing methods which creates climate-proof and high-yielding crops. Through deep insights into plant science and precise molecular-phenotypic data connections these analytical tools drive novel innovations which enable crop development for enhanced environmental resilience.

References

Abdul Aziz M, Masmoudi K. Molecular breakthroughs in modern plant breeding techniques. Hortic Plant J 2025 11(1):15–41

Ahmad P, Abdel Latef AAH, Rasool S, Akram NA, Ashraf M, Gucel S (2016) Role of proteomics in crop stress tolerance. Front Plant Sci [Internet] 7(September):1336. Available from: https://pmc.ncbi.nlm.nih.gov/articles/PMC5014855/ [cited 2025 Jan 28]

Ain NU, Haider FU, Fatima M, Habiba, Zhou Y, Ming R (2022) Genetic determinants of biomass in C4 crops: molecular and agronomic approaches to increase biomass for biofuels. Front Plant Sci [Internet] 13:839588. Available from: https://pmc.ncbi.nlm.nih.gov/articles/PMC9260593/ [cited 2024 Dec 18]

Al-Amrani S, Al-Jabri Z, Al-Zaabi A, Alshekaili J, Al-Khabori M (2021) Proteomics: concepts and applications in human medicine. World J Biol Chem [Internet] 12(5):57. Available from: https://pmc.ncbi.nlm.nih.gov/articles/PMC8473418/ [cited 2025 Jan 28]

Alrajeh S, Naveed Khan M, Irhash Putra A, Al-ugaili DN, Alobaidi KH, Al Dossary O et al (2024) Mapping proteomic response to salinity stress tolerance in oil crops: towards enhanced plant resilience. J Genet Eng Biotechnol 22(4):100432

Barnum CR, Endelman BJ, Shih PM (2021) Utilizing plant synthetic biology to improve human health and wellness. Front Plant Sci [Internet] 12:691462. Available from: www.frontiersin.org [cited 2025 Jan 29]

Bawa G, Liu Z, Zhou Y, Fan S, Ma Q, Tissue DT et al 2022 Cotton proteomics: dissecting the stress response mechanisms in cotton. Front Plant Sci [Internet] 13:1035801. Available from: https://pmc.ncbi.nlm.nih.gov/articles/PMC9714328/ 17 [cited 2025 Jan 28]

Benitez-Alfonso Y, Soanes BK, Zimba S, Sinanaj B, German L, Sharma V et al (2023) Enhancing climate change resilience in agricultural crops. Curr Biol 33(23):R1246–R1261

Breschi A, Gingeras TR, Guigó R (2017) Comparative transcriptomics in human and mouse. Nat Rev Genet [Internet] 18(7):425. Available from: https://pmc.ncbi.nlm.nih.gov/articles/PMC6413734/ [cited 2024 Dec 18]

Caruana L, Orr DJ, Carmo-Silva E (2022). Rubiscosome gene expression is balanced across the hexaploid wheat genome. Photosynth Res [Internet] 152(1):1–11. Available from: https://link.springer.com/article/10.1007/s11120-022-00897-9 [cited 2024 Dec 18]

Cembrowska-Lech D, Krzemińska A, Miller T, Nowakowska A, Adamski C, Radaczyńska M et al (2023) An integrated multi-omics and artificial intelligence framework for advance plant phenotyping in horticulture. Biology (Basel) [Internet] 12(10):1298. Available from: https://pmc.ncbi.nlm.nih.gov/articles/PMC10603917/ [cited 2025 Jan 29]

Chen Y, Huang W, Zhang F, Luo X, Hu B, Xie J (2021) Metabolomic profiling of Dongxiang wild Rice under salinity demonstrates the significant role of amino acids in Rice salt stress. Front Plant Sci [Internet] 12:729004. Available from: www.frontiersin.org [cited 2025 Jan 29]

Du Y, Amin N, Ahmad N, Zhang H, Zhang Y, Song Y et al (2023) Identification of the function of the pathogenesis-related protein GmPR1L in the resistance of soybean to Cercospora sojina hara. Genes (Basel) [Internet] 14(4):920. Available from: https://pmc.ncbi.nlm.nih.gov/articles/PMC10137329/ [cited 2025 Jan 28]

Dueñas CJ, Slamet-Loedin I, Macovei A (2021) Transcriptomics View over the Germination Landscape in Biofortified Rice. Genes (Basel) [Internet] 12(12):2013. Available from: https://pmc.ncbi.nlm.nih.gov/articles/PMC8700799/ [cited 2024 Dec 18]

Friero I, Larriba E, Martínez-Melgarejo PA, Justamante MS, Alarcón MV, Albacete A et al (2023) Transcriptomic and hormonal analysis of the roots of maize seedlings grown hydroponically at low temperature. Plant Sci 326:111525

Gawronski KAB, Kim J (2017) Single cell transcriptomics of non-coding RNAs and their cell-specificity. Wiley Interdiscip Rev RNA [Internet] 8(6). https://doi.org/10.1002/wrna.1433. Available from: https://pmc.ncbi.nlm.nih.gov/articles/PMC5675046/ [cited 2024 Dec 18]

Gill SS, Khan NA, Agarwala N, Singh K, Sunkar R, Tuteja N (2024) ncRNAs in plant development and stress responses. Plant Physiol Biochem (214):108950

Guo B, Sun L, Jiang S, Ren H, Sun R, Wei Z et al (2022) Soybean genetic resources contributing to sustainable protein production. Theor Appl Genet [Internet] 135(11):4095–4121. Available from: https://link.springer.com/article/10.1007/s00122-022-04222-9 [cited 2025 Jan 28]

Hamzelou S, Pascovici D, Kamath KS, Amirkhani A, McKay M, Mirzaei M et al (2020) Proteomic responses to drought vary widely among eight diverse genotypes of rice (Oryza sativa). Int J Mol Sci [Internet] 21(1):363. Available from: https://pmc.ncbi.nlm.nih.gov/articles/PMC6982093/ [cited 2025 Jan 28]

Hasin Y, Seldin M, Lusis A (2017) Multi-omics approaches to disease. Genome Biol [Internet] 18(1):1–15. Available from: https://genomebiology.biomedcentral.com/articles/10.1186/s13059-017-1215-1 [cited 2025 Jan 29]

He Z, Zhang P, Jia H, Zhang S, Nishawy E, Sun X et al (2024) Regulatory mechanisms and breeding strategies for crop drought resistance. New Crop 1:100029

Hill MS, Vande Zande P, Wittkopp PJ (2020) Molecular and evolutionary processes generating variation in gene expression. Nat Rev Genet [Internet] 22(4):203. Available from: https://pmc.ncbi.nlm.nih.gov/articles/PMC7981258/ [cited 2024 Dec 18]

Hoang TML, Tran TN, Nguyen TKT, Williams B, Wurm P, Bellairs S et al (2016) Improvement of salinity stress tolerance in rice: challenges and opportunities. Agron [Internet] 6(4):54. Available from: https://www.mdpi.com/2073-4395/6/4/54/htm [cited 2025 Jan 28]

Hoshikawa K, Pham D, Ezura H, Schafleitner R, Nakashima K (2021) Genetic and molecular mechanisms conferring heat stress tolerance in tomato plants. Front Plant Sci [Internet] 12:786688. Available from: https://pmc.ncbi.nlm.nih.gov/articles/PMC8739973/ [cited 2025 Jan 28]

Ji Y, Hewavithana T, Sharpe AG, Jin L (2024) Understanding grain development in the Poaceae family by comparing conserved and distinctive pathways through omics studies in wheat and maize. Front Plant Sci [Internet] 15:1393140. Available from: https://pmc.ncbi.nlm.nih.gov/articles/PMC11295249/ [cited 2025 Jan 28]

Khodadadian A, Darzi S, Haghi-Daredeh S, Eshaghi FS, Babakhanzadeh E, Mirabutalebi SH et al (2020) Genomics and transcriptomics: the powerful technologies in precision medicine. Int J Gen Med [Internet] 13:627. Available from: https://pmc.ncbi.nlm.nih.gov/articles/PMC7509479/ [cited 2024 Dec 18]

Koks G, Pfaff AL, Bubb VJ, Quinn JP, Koks S (2020) At the dawn of the transcriptomic medicine. Exp Biol Med [Internet] 246(3):286. Available from: https://pmc.ncbi.nlm.nih.gov/articles/PMC7876640/ [cited 2024 Dec 18]

Kumar M, Patel MK, Kumar N, Bajpai AB, Siddique KHM (2021) Metabolomics and molecular approaches reveal drought stress tolerance in plants. Int J Mol Sci [Internet] 22(17):9108. Available from: https://pmc.ncbi.nlm.nih.gov/articles/PMC8431676/ [cited 2025 Jan 29]

Li P, Cao W, Fang H, Xu S, Yin S, Zhang Y et al (2017) Stranscriptomic profiling of the maize (Zea mays L.) leaf response to abiotic stresses at the seedling stage. Front. Plant Sci 8:252782
Li Y, Li X, Zhang J, Li D, Yan L, You M et al (2021) Physiological and proteomic responses of contrasting alfalfa (Medicago sativa L.) varieties to high temperature stress. Front Plant Sci [Internet]. 12:753011. Available from: www.frontiersin.org [cited 2025 Jan 28]
Li Z, Gao J, Wang B, Xu J, Fu X, Han H et al (2022a) Rice carotenoid biofortification and yield improvement conferred by endosperm-specific overexpression of OsGLK1. Front Plant Sci [Internet] 13:951605. Available from: https://pmc.ncbi.nlm.nih.gov/articles/PMC9335051/ [cited 2025 Jan 29]
Li M, Li H, Sun A, Wang L, Ren C, Liu J et al (2022b) Transcriptome analysis reveals key drought-stress-responsive genes in soybean. Front Genet [Internet] 13:1060529. Available from: https://pmc.ncbi.nlm.nih.gov/articles/PMC9742610/ [cited 2025 Jan 29]
Liu Y, Lu S, Liu K, Wang S, Huang L, Guo L (2019) Proteomics: a powerful tool to study plant responses to biotic stress. Plant Methods [Internet] 15(1):1–20. Available from: https://plantmethods.biomedcentral.com/articles/10.1186/s13007-019-0515-8 [cited 2025 Jan 28]
Liu X, Peng T, Xu M, Lin S, Hu B, Chu T et al (2024) Spatial multi-omics: deciphering technological landscape of integration of multi-omics and its applications. J Hematol Oncol [Internet] 17(1):1–24. Available from: https://jhoonline.biomedcentral.com/articles/10.1186/s13045-024-01596-9 [cited 2025 Jan 29]
Lowe R, Shirley N, Bleackley M, Dolan S, Shafee T (2017) Transcriptomics technologies. PLoS Comput Biol [Internet] 13(5):e1005457. Available from: https://pmc.ncbi.nlm.nih.gov/articles/PMC5436640/ [cited 2024 Dec 18]
Manickam S, Rajagopalan VR, Kambale R, Rajasekaran R, Kanagarajan S, Muthurajan R (2023) Plant metabolomics: current initiatives and future prospects. Curr Issues Mol Biol [Internet] 45(11):8894. Available from: https://pmc.ncbi.nlm.nih.gov/articles/PMC10670879/ [cited 2025 Jan 28]
Manzoni C, Kia DA, Vandrovcova J, Hardy J, Wood NW, Lewis PA et al (2016) Genome, transcriptome and proteome: the rise of omics data and their integration in biomedical sciences. Brief Bioinform [Internet] 19(2):286. Available from: https://pmc.ncbi.nlm.nih.gov/articles/PMC6018996/ [cited 2024 Dec 18]
Maqbool MA, Aslam M, Beshir A, Khan MS (2018) Breeding for provitamin a biofortification of maize (Zea mays L.). Plant Breed [Internet] 137(4):451–469. Available from: https://onlinelibrary.wiley.com/doi/full/10.1111/pbr.12618 [cited 2025 Jan 28]
Marchev AS, Vasileva LV, Amirova KM, Savova MS, Balcheva-Sivenova ZP, Georgiev MI (2021) Metabolomics and health: from nutritional crops and plant-based pharmaceuticals to profiling of human biofluids. Cell Mol Life Sci C [Internet] 78(19–20):6487. Available from: https://pmc.ncbi.nlm.nih.gov/articles/PMC8558153/ [cited 2025 Jan 28]
Mattioli R, Francioso A, Mosca L, Silva P (2020) Anthocyanins: a comprehensive review of their chemical properties and health effects on cardiovascular and neurodegenerative diseases. Molecules [Internet] 25(17):3809. Available from: https://pmc.ncbi.nlm.nih.gov/articles/PMC7504512/ [cited 2025 Jan 29]
Muhammad M, Waheed A, Wahab A, Majeed M, Nazim M, Liu YH et al (2024) Soil salinity and drought tolerance: an evaluation of plant growth, productivity, microbial diversity, and amelioration strategies. Plant Stress 11:100319
Mukherjee S, Chopra H, Goyal R, Jin S, Dong Z, Das T et al (2024) Therapeutic effect of targeted antioxidant natural products. Discov Nano [Internet] 19(1):1–28. Available from: https://link.springer.com/article/10.1186/s11671-024-04100-x [cited 2025 Jan 29]
Ozyigit II, Dogan I, Hocaoglu-Ozyigit A, Yalcin B, Erdogan A, Yalcin IE et al (2023) Production of secondary metabolites using tissue culture-based biotechnological applications. Front Plant Sci [Internet] 14:1132555. Available from: https://pmc.ncbi.nlm.nih.gov/articles/PMC10339834/ [cited 2025 Jan 29]
Roberts AGK, Catchpoole DR, Kennedy PJ (2022) Identification of differentially distributed gene expression and distinct sets of cancer-related genes identified by changes in mean and

variability. NAR Genomics Bioinforma [Internet] 4(1):lqab124. Available from: https://pmc.ncbi.nlm.nih.gov/articles/PMC8759562/ [cited 2024 Dec 18]
Sah SK, Jumaa S, Li J, Reddy KR (2022) Proteomic analysis response of rice (Oryza sativa) leaves to ultraviolet-B radiation stress. Front Plant Sci [Internet] 13:871331. Available from: https://pmc.ncbi.nlm.nih.gov/articles/PMC9536139/ [cited 2025 Jan 28]
Sahebi M, Hanafi MM, Rafii MY, Mahmud TMM, Azizi P, Osman M et al (2018) Improvement of drought tolerance in rice (Oryza sativa L.): genetics, genomic tools, and the WRKY gene family. Biomed Res Int [Internet] 2018:3158474. Available from: https://pmc.ncbi.nlm.nih.gov/articles/PMC6106855/ [cited 2025 Jan 28]
Sahid S, Roy C, Shee D, Shee R, Datta R, Paul S (2023) ZFP37, C3H, NAC94, and bHLH148 transcription factors regulate cultivar-specific drought response by modulating r40C1 gene expression in rice. Environ Exp Bot 214:105480
Sheikh M, Iqra F, Ambreen H, Pravin KA, Ikra M, Chung YS (2024) Integrating artificial intelligence and high-throughput phenotyping for crop improvement. J Integr Agric 23(6): 1787–1802
Shen S, Zhan C, Yang C, Fernie AR, Luo J (2023) Metabolomics-centered mining of plant metabolic diversity and function: past decade and future perspectives. Mol Plant 16(1):43–63
Shiade SRG, Zand-Silakhoor A, Fathi A, Rahimi R, Minkina T, Rajput VD et al (2024) Plant metabolites and signaling pathways in response to biotic and abiotic stresses: exploring bio stimulant applications. Plant Stress 12:100454
Shu N, Chen X, Sun X, Cao X, Liu Y, Xu YJ (2023) Metabolomics identify landscape of food sensory properties. Crit Rev Food Sci Nutr 63(27):8478–8488
Singh KP, Miaskowski C, Dhruva AA, Flowers E, Kober KM (2018) Mechanisms and measurement of changes in gene expression. Biol Res Nurs [Internet] 20(4):369. Available from: https://pmc.ncbi.nlm.nih.gov/articles/PMC6346310/ [cited 2024 Dec 18]
Singh P, Kumar V, Sharma J, Saini S, Sharma P, Kumar S et al (2022a) Silicon supplementation alleviates the salinity stress in wheat plants by enhancing the plant water status, photosynthetic pigments, proline content and antioxidant enzyme activities. Plants [Internet] 11(19):2525. Available from: https://www.mdpi.com/2223-7747/11/19/2525/htm [cited 2025 Jan 28]
Singh P, Choudhary KK, Chaudhary N, Gupta S, Sahu M, Tejaswini B et al (2022b) Salt stress resilience in plants mediated through osmolyte accumulation and its crosstalk mechanism with phytohormones. Front Plant Sci [Internet]. 13:1006617. Available from: https://pmc.ncbi.nlm.nih.gov/articles/PMC9552866/ [cited 2025 Jan 29]
Statello L, Guo CJ, Chen LL, Huarte M (2020). Gene regulation by long non-coding RNAs and its biological functions. Nat Rev Mol Cell Biol [Internet] 22(2):96–118. Available from: https://www.nature.com/articles/s41580-020-00315-9 [cited 2024 Dec 18]
Supplitt S, Karpinski P, Sasiadek M, Laczmanska I (2021) Current achievements and applications of transcriptomics in personalized cancer medicine. Int J Mol Sci [Internet] 22(3):1422. Available from: https://pmc.ncbi.nlm.nih.gov/articles/PMC7866970/ [cited 2024 Dec 18]
Tsakiroglou M, Evans A, Pirmohamed M (2023) Leveraging transcriptomics for precision diagnosis: lessons learned from cancer and sepsis. Front Genet [Internet] 14:1100352. Available from: https://pmc.ncbi.nlm.nih.gov/articles/PMC10036914/ [cited 2024 Dec 18]
Tyagi P, Singh D, Mathur S, Singh A, Ranjan R (2022) Upcoming progress of transcriptomics studies on plants: An overview. Front Plant Sci [Internet] 13:1030890. Available from: https://pmc.ncbi.nlm.nih.gov/articles/PMC9798009/ [cited 2024 Dec 18]
Varatharajan N, Sekaran DC, Murugan K, Chockalingam V, Varatharajan N, Sekaran DC et al (2021) Rice aroma: biochemical, genetics and molecular aspects and its extraction and quantification methods. Integr Adv Rice Res [Internet]. Available from: https://www.intechopen.com/chapters/77895 [cited 2025 Jan 29]
Vasupalli N, Bhat JA, Jain P, Sri T, Islam MA, Shivaraj SM et al (2024) Omics big data for crop improvement: opportunities and challenges. Crop J 12(6):1517–1532

Villalobos-López MA, Arroyo-Becerra A, Quintero-Jiménez A, Iturriaga G (2022) Biotechnological advances to improve abiotic stress tolerance in crops. Int J Mol Sci [Internet] 23(19):12053. Available from: https://pmc.ncbi.nlm.nih.gov/articles/PMC9570234/ [cited 2025 Jan 28]

Wang RC, Wang Z (2023) Precision medicine: disease subtyping and tailored treatment. Cancers (Basel) [Internet] 15(15):3837. Available from: https://pmc.ncbi.nlm.nih.gov/articles/PMC10417651/ [cited 2024 Dec 18]

Wang Z, Gerstein M, Snyder M (2008) RNA-Seq: a revolutionary tool for transcriptomics. Nat Rev Genet [Internet] 10(1):57–63. Available from: https://www.nature.com/articles/nrg2484 [cited 2024 Nov 9]

Wang H, Xu Y, Zhang Z, Zhang G, Tan C, Ye L (2024) Development and application of transcriptomics technologies in plant science. Crop Des 3(2):100057

Wang X, Zhao Z, Guo N, Wang H, Zhao J, Xing H (2020) Comparative proteomics analysis reveals that lignin biosynthesis contributes to brassinosteroid-mediated response to Phytophthora sojae in soybeans. Journal of Agricultural and Food Chemistry, 68(19):5496–5506

Wani SH, Samantara K, Razzaq A, Kakani G, Kumar P (2022) Back to the wild: mining maize (Zea mays L.) disease resistance using advanced breeding tools. Mol Biol Rep 49(6):5787–5803

Weiss D, Ori N (2007) Mechanisms of cross talk between gibberellin and other hormones. Plant Physiol [Internet] 144(3):1240. Available from: https://pmc.ncbi.nlm.nih.gov/articles/PMC1914132/ [cited 2025 Jan 28];

Yadava DK, Hossain F, Mohapatra T (2018) Nutritional security through crop biofortification in India: status & future prospects. Indian J Med Res [Internet] 148(5):621. Available from: https://pmc.ncbi.nlm.nih.gov/articles/PMC6366255/ [cited 2024 Dec 18]

Yan S, Bhawal R, Yin Z, Thannhauser TW, Zhang S (2022) Recent advances in proteomics and metabolomics in plants. Mol Hortic [Internet] 2(1):17. Available from: https://pmc.ncbi.nlm.nih.gov/articles/PMC10514990/ [cited 2025 Jan 28]

Yang Y, Saand MA, Huang L, Abdelaal WB, Zhang J, Wu Y et al (2021) Applications of multi-omics technologies for crop improvement. Front Plant Sci [Internet] 12:563953. Available from: www.frontiersin.org [cited 2024 Dec 18]

Yogendra K, Sanivarapu H, Avuthu T, Gupta SK, Durgalla P, Banerjee R et al (2024) Comparative metabolomics to unravel the biochemical mechanism associated with rancidity in pearl millet (Pennisetum glaucum L.). Int J Mol Sci [Internet] 25(21):11583. Available from: https://www.mdpi.com/1422-0067/25/21/11583/htm [cited 2025 Jan 29]

Yuan L, Zheng Y, Nie L, Zhang L, Wu Y, Zhu S et al (2021) Transcriptional profiling reveals changes in gene regulation and signaling transduction pathways during temperature stress in wucai (Brassica campestris L.). BMC Genom [Internet] 22(1):687. Available from: https://pmc.ncbi.nlm.nih.gov/articles/PMC8456696/ [cited 2024 Dec 18]

Zárate-Carbajal A, Lagarda-Clark EA, Mikhaylin S, Duarte-Sierra A (2024) Untargeted metabolomic analysis of strawberries exposed to pulsed electric fields and cold plasma before postharvest storage. Futur Foods 9:100364

Integration of Omics Technologies in Crop Breeding Program: Data Analytics and Machine Learning in Crop Genomics

9

Sarahani Harun and Rabiatul-Adawiah Zainal-Abidin

Abstract

Recent advancements in genome sequencing, bioinformatics, and omics technologies, including genomics, transcriptomics, proteomics, and metabolomics, have revolutionized molecular studies and crop breeding by enabling unprecedented discoveries in gene functionality, molecular mechanisms, and crop protection strategies. Functional genomics and high-throughput omics approaches have significantly enhanced our understanding of crop genetics, facilitating the development of new and improved crop varieties. However, the complexity of big data and the heterogeneity of omics datasets necessitate advanced analytical tools such as machine learning (ML). Integrating ML with omics approaches has transformed data analysis, enabling efficient processing of massive datasets and accelerating genetic advancements in agricultural breeding. Machine learning algorithms, including supervised, unsupervised, and deep learning techniques, integrate diverse datasets, handle high-dimensional data, and classify biological functions with significant accuracy and efficiency. These methods uncover hidden patterns, predict gene functions, and prioritize candidate genes for experimental validation. This chapter explores recent advances at the intersection of machine learning and crop genomics, focusing on ML algorithms and models for omics data analysis, as well as the application

S. Harun (✉)
Centre for Bioinformatics Research, Institute of Systems Biology (INBIOSIS), Universiti Kebangsaan Malaysia, Bangi, Selangor, Malaysia
e-mail: sarahani@ukm.edu.my

R.-A. Zainal-Abidin
Biotechnology and Nanotechnology Research Centre, Malaysian Agricultural Research and Development Institute (MARDI), Serdang, Malaysia
e-mail: rabiatul@mardi.gov.my

Y. Ahlawat et al. (eds.), *Next-Generation Strategies for Crop Improvement*,
https://doi.org/10.1007/978-981-95-0309-4_9

of ML-based tools to address global challenges such as food security and climate resilience while significantly accelerating crop breeding in the future.

Keywords

Bioinformatics · Functional genomics · Machine learning · Data analytics · Crop improvement

9.1 Introduction

The rapid increase in the global population presents a critical challenge to the world's food supply. By 2050, the global population is projected to reach 9.9 billion, necessitating a 60% increase in agricultural productivity to meet the rising demand for food and essential commodities (Martin-Rios et al. 2020; Nath et al. 2024). Compounding this challenge are additional factors, such as urbanization, which is reducing the availability of agricultural land, and the intensifying effects of climate change, including extreme weather events, such as floods, prolonged droughts, and rising temperatures in both tropical and temperate regions. Over the past decade, there has been a marked increase in compound drought and heatwave events, severely affecting agriculture. For instance, He et al. (2022) reported that over 92% of wheat-growing regions have experienced at least one such event per season, with their frequency and duration increasing by over 28% since the 1980s (He et al. 2022). These challenges necessitate innovative adaptation strategies for global food production systems. The impact of climate variability on food security is evident, with changes in rainfall patterns, rising temperatures, and more frequent extreme weather events posing significant threats to agricultural productivity. Given the current trends in global consumer behavior and the mounting pressure on food systems, it is crucial to develop innovative technologies—such as genetic modification—that enhance the resilience and productivity of staple crops (Voss-Fels et al. 2019; Bowerman et al. 2023). Global food security will face severe risks without these advancements in the coming decades.

In recent years, genome sequencing and bioinformatics advancements have provided powerful tools for studying agricultural plants and model species at the molecular level. While sequencing data is essential, bioinformatics alone cannot fully understand gene functions. Therefore, functional genomics has become a crucial field, focusing on uncovering gene functions through high-throughput molecular approaches such as full-length cDNA libraries, gene expression microarrays, and next-generation sequencing (NGS) technologies, which profile gene expression patterns. Bioinformatics tools are essential for analyzing complex molecular datasets, allowing researchers to generate hypotheses about gene functions, protein interactions, and other molecular mechanisms (Pushparaj 2019). Integrating these methodologies accelerates our understanding of gene functionality, facilitating the discovery of biological pathways vital for improving crop traits and addressing agricultural challenges. Building on these advancements, crop improvement efforts have increasingly incorporated a range of omics approaches, such as genomics, transcriptomics, proteomics, metabolomics, and bioinformatics, into breeding programs (Chao et al. 2023). These techniques aim to comprehensively understand crop genetics and the complex molecular interactions among genes, proteins, and metabolites, particularly in complex traits. Such knowledge is essential for tackling challenges in breeding programs, particularly in improving traits such as yield, stress tolerance, and disease resistance.

Due to the large, complex, and diverse datasets generated by omics technologies, data integration and analysis have become increasingly crucial. Machine learning (ML) and data analytics are revolutionizing these omics analyses by enabling researchers to process and interpret these massive datasets efficiently (Cembrowska-Lech et al. 2023). Machine learning has been defined as a data-driven paradigm in which computational algorithms can learn from data themselves by identifying patterns, relationships, and structures without relying on statistical assumptions to discover new knowledge (Yan and Wang 2023). Machine learning is a subfield of artificial intelligence technology and has been categorized into three major categories: supervised, unsupervised, and semi-supervised.

The supervised learning algorithm learns patterns based on labeled examples of past data. The unsupervised learning algorithms seek to uncover patterns without the assistance of labeled data. Algorithms or models such as naïve Bayes (NB), random forest (RF), and support vector machine (SVM) are applied in supervised learning, while k-means clustering, hierarchical clustering, and principal component analysis (PCA) are suitable for unsupervised learning (Koh et al. 2024). Large language models (LLMs) are gaining significant traction in various areas, including their emerging application in crop genomics. LLM, such as generative pretrained transformers (GPT), can be described as self-supervised learning, where the model refines its understanding by generating insights from input data (Lam et al. 2024). Recent studies have demonstrated the versatility of LLMs in biological research, including extracting protein-protein and protein-chemical associations from textual data (Weber et al. 2023) and predicting functional regions across 48 plant genomes (Mendoza-Revilla et al. 2024). These examples highlight the immense potential of LLMs to drive advancements in biological and genomic studies.

This chapter summarizes the ML-assisted omics techniques and their application in crop breeding improvement. Although most ML methods and tools can be used in both animals and plants, this chapter focuses mainly on recent developments in the application of ML to crop research and breeding. We highlighted ML-assisted omics techniques and their application in crop breeding improvement. In addition, challenges and methods to apply ML and omics in improving crops to be resistant and highly tolerant to abiotic and biotic stresses are discussed. Integrating ML with the omics platform accelerates researchers to unravel valuable information and translational output to enhance crop resilience, especially against climate change and emerging crop diseases.

9.2 Current State of Omics in Crop Breeding

The development of advanced tools and methods, including genetic engineering, hybridization, and molecular marker-based breeding strategies, has led to enormous developments in plant breeding. These advancements have made it easier to produce improved plant varieties. Plant breeders now gather and analyze diverse datasets, including omics, to identify desired traits and effectively incorporate these methods into their programs. The application of omics in plant breeding has completely revolutionized the field of agricultural biotechnology, making omics technologies the preferred choice for plant breeders seeking to develop new crop varieties (Kim et al. 2020; Yoosefzadeh Najafabadi et al. 2023). Researchers can accelerate germplasm screening by leveraging omics tools and ultimately develop superior crop varieties (Sharma et al. 2024). However, the full potential of these technologies depends on effectively integrating vast and diverse datasets from multiple sources. This integration is crucial for uncovering the biological processes that govern desirable plant traits and their complex interactions (Chao et al. 2023).

Numerous omics databases have been established to support these efforts, providing access to large-scale datasets for crops, such as rice (Itoh et al. 2018), maize (Lawrence 2007), wheat (Ma et al. 2021), and soybean (Liu et al. 2023). These databases offer extensive information on genetic composition, epigenetic regulation, gene expression, protein functions, and metabolic pathways, serving as invaluable resources to enhance the precision and efficiency of modern breeding programs (Chao et al. 2023). By utilizing such comprehensive datasets, researchers can identify key genes and pathways involved in crop traits, enabling breeders to select and develop new plant varieties with desirable characteristics more efficiently. This has led to a significant increase in recent studies focused on applying omics technologies in crop breeding (Fig. 9.1).

Currently, genomic data is necessary for downstream research, whereby genome databases accessible to the public will enable researchers to address various biological questions in plant biology and agriculture. Many biological datasets and bioinformatics tools have been developed and made available to other researchers working on related topics (Raza et al. 2021; Marsh et al. 2021). The genome of the model organism, *Arabidopsis thaliana*, was published in 2000, marking a significant milestone in plant genomics. Since then, over 700 plant genomes have been sequenced and assembled, leading to groundbreaking discoveries in trait genetics and crop improvement (Song et al. 2023a). After reference genome assemblies were completed, efforts to sequence individual genomes were launched in order to investigate genetic variation among plant populations. Initially, these efforts faced difficulties because of the high cost of DNA sequencing. The first study in the 1001 Genomes Project, for example, sequenced only three ecotypes of *A. thaliana*, namely the reference (Col-0) and two others (Bur-0 and Tsu-1), utilizing short-read Illumina technology with coverages ranging from 15- to 20-fold (Ossowski et al. 2008).

The evolution of sequencing technologies has significantly transformed plant breeding and genomics. Early methods, such as Sanger sequencing, were known

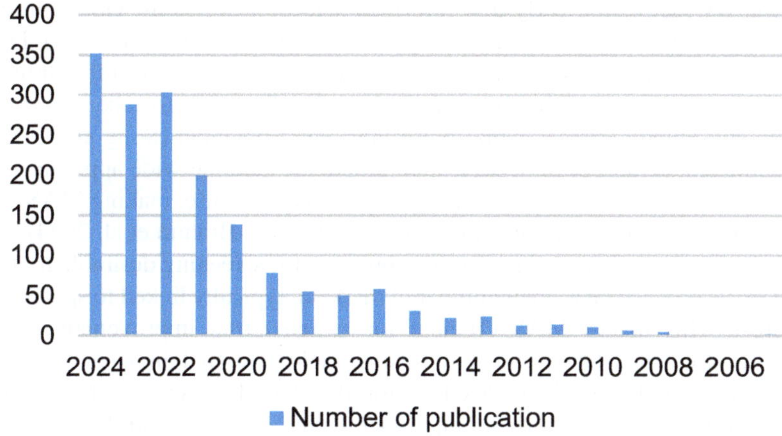

Fig. 9.1 Statistics on land plant omics data and related literature were analyzed by retrieving the number of studies on land plant omics research from the PubMed database using the search term "omics crop"

for their accuracy but were hindered by slow processing speed and high costs, limiting their application to small-scale studies. The introduction of next-generation sequencing (NGS) marked a significant breakthrough, offering high-throughput and cost-effective solutions for sequencing entire genomes. Platforms such as Illumina generate millions of short reads in parallel, enabling rapid sequencing crucial for large-scale initiatives such as genome-wide association studies (GWAS) and population genomics (Metzker 2009; Chun et al. 2014). More recently, third-generation sequencing (TGS) technologies, including PacBio and Oxford Nanopore, have further advanced genomics by providing long-read sequencing capabilities. These advancements enable detailed examination of complex genomes, particularly those with high levels of polyploidy or repetitive sequences—common features in many crop species. The longer reads contribute to more complete and accurate genome assemblies, which are essential for understanding structural variations and gene functions (Singh et al. 2022a; Ruperao et al. 2023; Scarano et al. 2024).

Genomic applications in crop breeding are extensive and transformative. For example, GWAS utilize sequencing data to identify genetic variants linked to significant phenotypic traits by analyzing the correlations between genetic markers and traits across large populations. This approach has played a crucial role in pinpointing genes associated with desirable traits, such as drought tolerance, pest resistance, and improved yield, thus providing breeders with valuable targets for selection (Sahito et al. 2024). Marker-assisted selection (MAS) uses molecular markers linked to specific traits to accelerate breeding. Marker-assisted selection (MAS) accelerates the breeding cycle by allowing breeders to identify plants with desirable genetic profiles before traits are fully expressed. Similarly, genomic selection (GS) relies on genome-wide marker data to predict breeding values, enabling accurate selection of complex, quantitative traits such as yield and stress tolerance. By facilitating early and informed selection decisions, GS significantly improves the efficiency of breeding programs (Chang-Brahim et al. 2024). Despite recent advancements, MAS remains widely utilized to screen desirable traits efficiently. For example, MAS has been applied in wheat breeding to enhance resistance to biotic and abiotic stresses while preserving yield potential (Song et al. 2023b). One significant advantage of MAS is that, unlike the genome-wide approach of GS, it requires only a small number of markers for breeders to use, making MAS a cost-effective option for practical breeding programs (Chang-Brahim et al. 2024). While MAS is a widely used and cost-effective approach for screening desirable traits, it is limited by its reliance on a small number of markers, typically linked to specific traits of interest. This makes MAS efficient for targeted trait improvement, such as enhancing stress resistance and maintaining yield potential in wheat. However, MAS cannot comprehensively account for the complex genetic architecture of quantitative traits influenced by multiple loci (Tong and Nikoloski 2021).

Transcriptomics has become a fundamental element in crop breeding, providing valuable insights into gene expression and the regulatory mechanisms that drive complex traits. Functional genomics, enhanced by transcriptomic methodologies, enables researchers to explore the intricate dynamics of gene function and regulation. High-throughput technologies, especially RNA sequencing (RNA-Seq), have revolutionized the field by enabling precise and comprehensive gene expression profiling across different developmental stages, tissues, and environmental conditions. Utilizing next-generation sequencing (NGS), RNA-Seq quantifies transcript levels, identifies differentially expressed genes, and detects alternative splicing events, offering an in-depth view of transcriptomes (Tyagi et al. 2022). This technology is instrumental in understanding plant responses to abiotic stresses, such as drought, salinity, and heat, as well as biotic stresses, including pathogen attacks and pest infestations. By mapping these responses, RNA-Seq uncovers critical genes and pathways involved in photosynthesis, nutrient assimilation, and defense mechanisms, offering targets for crop improvement (Zhang et al. 2019; Varshney et al. 2020; Pazhamala et al. 2021).

In addition to RNA-Seq, third-generation sequencing technologies such as PacBio and Oxford Nanopore facilitate full-length transcript sequencing, capturing isoform diversity and overcoming the limitations of short-read sequencing (Scarano et al. 2024). These methods enable the construction of complete transcriptomes, particularly in polyploid crops or species with highly repetitive genomes. Although less prevalent in recent years, technologies such as microarrays remain useful for specific gene expression studies due to their cost-effectiveness. Advanced bioinformatics tools are critical in analyzing transcriptomic data, employing algorithms for read alignment, differential gene expression analysis, and gene co-expression network construction. These analyses identify candidate genes linked to traits such as disease resistance, stress tolerance, and yield enhancement (Movahedi et al. 2012; Serin et al. 2016). Integrating transcriptomics with other omics datasets, such as proteomics and metabolomics, strengthens breeding strategies by providing a systems-level understanding of crop biology. Ultimately, transcriptomics empowers breeders to identify molecular markers, validate gene targets, and refine crop selection processes, driving the development of high-performing, resilient crop varieties tailored to global agricultural challenges (Mahmood et al. 2022; Chao et al. 2023).

9.3 Application of Machine Learning in Omics and Crop Breeding

Climate change poses significant threats to agricultural systems, highlighting the significance of understanding various defense mechanisms in crops, enabling them to be resistant or highly tolerant to both abiotic and biotic stress factors. Biotic stress occurs due to pathogenic bacteria, fungi, viruses, nematodes, pest infestations, and the invasion of parasitic plants (Vo et al. 2021). Abiotic stress arises from adverse environmental conditions, including drought, salinity, flooding, extreme temperatures, heavy metals, and radiation (Vo et al. 2021). Therefore, safeguarding crops from the ravages of diseases, pests, and adverse environmental conditions is essential to maintain agricultural productivity and secure the global food supply. A study by Savary et al. (2019) revealed that pathogen diseases and pest infestations have resulted in significant production losses in rice, with an average loss of 30.3%, maize at 22.6%, and potatoes at 17.2%. Additionally, rice production decreased by 3.2%, maize by 7.4%, and soybean by 3.1% due to climate change factors, including drought, flooding, extreme heat, and cold (Ullah et al. 2021).

Cutting-edge ML approaches have significantly enhanced genomics and transcriptomics analyses in plant-pathogen interactions (Yang et al. 2019), facilitated gene discovery and the prioritization of stress-responsive genes (Shaik and Ramakrishna 2014), inferred gene regulatory networks (Gupta et al. 2021), and improved genomic predictions of plant phenotypes (Xu et al. 2022). These advancements contribute to a deeper understanding of regulation mechanisms in traits of interest and the identification of target genes, paving the way for knowledge-driven molecular breeding programs.

Machine learning (ML) algorithms, such as random forest (RF) (Qi 2012), support vector machine (SVM) (Pisner and Schnyer 2020), and artificial neural network (ANN) (Zou et al. 2008) are widely used to predict essential crop traits such as yield, disease resistance, and drought tolerance. These models help breeders toidentify high-potential plants in early stage, thus shortening breeding cycles (Yoosefzadeh Najafabadi et al. 2023). Incorporating data analytics and ML into crop breeding programs, alongside omics approaches, has significantly enhanced the ability to process and interpret complex datasets (Yoosefzadeh Najafabadi et al. 2023). By identifying patterns, predicting trait performance, and uncovering hidden correlations that traditional analytical methods may overlook, ML algorithms lead to more accurate breeding outcome predictions and accelerate genetic advancements.

Proteomics and metabolomics play a crucial role in elucidating the changes in proteins and metabolites that occur during the defense responses of plants when facing abiotic and biotic stressors. Subsequently, both omics approaches provide valuable insights into plant resilience and stress adaptation mechanisms. A proteome is defined as the abundance of various proteoforms, encompassing isoforms and post-translational modifications of each protein within a specific sample (Neely et al. 2023). In proteomics, four key studies include sequence, structure, expression, and interaction network, offering a holistic view of the complex interactions within plant cells (Murmu et al. 2024). Metabolomics investigates metabolites in biological systems and is crucial for understanding the plant metabolome, especially in revealing pathways involved in stress conditions or diseases (Manickam et al. 2023). Despite these advancements, both fields face significant challenges that hinder their full potential.

In proteomics, a major obstacle lies in predicting protein structure and function, mainly when limited information is available (Huang et al. 2023). Other issues involve measuring protein expression levels using isobaric tags for relative and absolute quantitation (iTRAQ) methods and predicting peptide retention times and fragmentation spectra (Neely et al. 2023). Challenges in metabolomics include extraction and analyzing massive metabolite data generated during the metabolites profiling (Kisiel et al. 2023), integrating different data formats from multiple metabolomics platforms, such as gas chromatography-mass spectrometry (GC-MS), high-performance liquid chromatography (HPLC), and nuclear magnetic resonance spectroscopy (NMR), as well as choosing the best statistical analysis to observe the metabolites profiling especially for untargeted metabolites (Villate et al. 2020). Moreover, traditional statistical models face limitations in handling this complexity due to high dimensionality, uncertainty, and non-independence of variables in plant omics (Murmu et al. 2024).

However, the advent of machine learning has introduced innovative solutions to address and interpret the complexity of vast proteomics and metabolomics datasets. Significant applications of ML in proteomics and metabolomics include pattern recognition, feature selection, and predictive modeling. Pattern recognition employs clustering and classification techniques to interpret complex proteomic and metabolomics data patterns. This approach is crucial for uncovering protein expression, structure, and function relationships. Feature selection highlights the most significant proteins and metabolites from large datasets and enhances model performance by focusing on the most discriminative features (Sirocchi et al. 2024). This approach is particularly critical in untargeted metabolites, where thousands of proteins or metabolites may be detected. Predictive models are used to forecast outcomes or infer relationships in heterogeneous datasets including predicting metabolites and transcripts associated with specific plant traits (Sprenger et al. 2018). The following sections discuss the strategies used by ML for genomics, transcriptomics, proteomics, and metabolomics analyses. Table 9.1 summarizes ML-based tools for application in omics research.

Table 9.1 Summary of machine learning (ML)-based tools for application in omics research

Tools	Algorithm	Function	Link and references
ir-HSP	SVM	An online prediction tool to detect heat shock protein (HSP), their families, and DnaJ	http://cabgrid.res.in:8080/ir-hsp/ Meher et al. (2018)
InterSPPI	RF	A Web server to predict the plant-pathogen protein-protein interaction network in plant diseases	http://zzdlab.com/intersppi/Index.Php Yang et al. (2019)
ASRpro	SVM RF ADB XGBoost	An online prediction tool to predict proteins associated with multiple abiotic stress traits in plants	https://iasri-sg.icar.gov.in/asrpro/ Meher et al. (2024)
MetaClean	Decision tree (DT) logistic regression NB NN SVM RF	An R package to build a classifier for automatically assessing peak integration from untargeted metabolomic analysis	https://github.com/GauravPandeyLab/MetaClean Chetnik et al. (2020)
Retip (retention time prediction) for metabolomics	RF BRNN XGBoost lightGBM Keras H2O autoML	An R package to predict retention time for small molecules in high-pressure liquid chromatography (HPLC) MS data analysis workflows	https://www.retip.app/ Bonini et al. (2020)
strucToolbox	SVM	ML-based toolbox for preprocessing and normalizing metabolites data	https://www.bioconductor.org/packages/release/bioc/html/structToolbox.html Lloyd et al. (2020)
CANOPUS	DNN	ML-based tools to predict metabolite structural classes based on the MS/MS spectrum. Requires installation	https://bio.informatik.uni-jena.de/software/canopus/ Dührkop et al. (2021)

9.3.1 Machine Learning in Crop Genomics and Transcriptomics

Genomic selection (GS) is deeply intertwined with machine learning (ML) methodologies. In GS, predictive models are developed using a training population that is both genotyped and phenotyped. Using ML techniques, such as regression-based models, classification-based models, and multi-trait models, GS can predict phenotypic traits in unseen genotypes, enabling more accurate selection for breeding. Furthermore, ML-powered GS models have revolutionized breeding by dramatically accelerating the breeding cycle (Tong and Nikoloski 2021). By

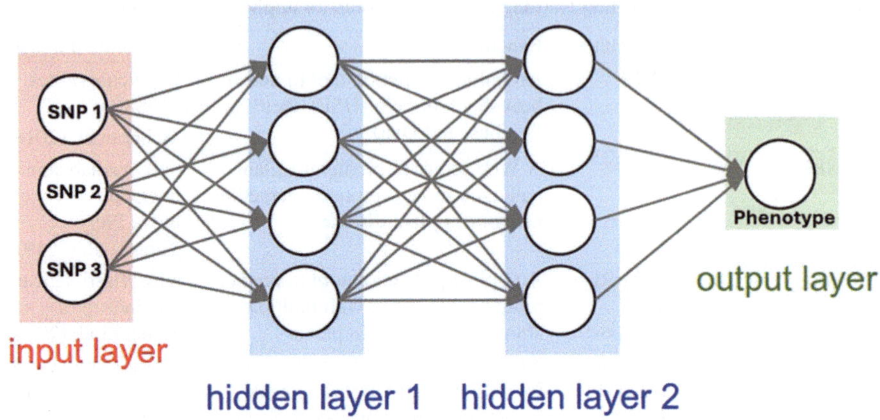

Fig. 9.2 Overview of deep learning approaches for genomic selection. Genomic selection models utilizing deep learning, with SNPs as the input and phenotypic traits as the output, are categorized based on specific architectures and algorithms, highlighting their ability to capture complex genotype-phenotype interactions (Manuel González-Camacho et al. 2018; Tong and Nikoloski 2021)

predicting phenotypes without the need for lengthy field trials, GS reduces the time to develop new cultivars, cutting breeding cycles to 6 years for soybean, 10 years for cassava, and 10 years for wheat (de Oliveira et al. 2012; Matei et al. 2018; Ahmar et al. 2020). These advancements align with the goals of modern plant breeding, in which ML plays a pivotal role in managing high-dimensional data and optimizing complex predictions. Building on this, advanced ML techniques such as neural network (NN) and deep learning have further enhanced the potential of genomic selection. Neural network, inspired by biological neural systems, consist of input, hidden, and output layers. In genomic selection, the input comprises genotypic data, such as SNPs, while the output corresponds to the predicted phenotypic traits (Fig. 9.2). The strength of deep learning lies in its ability to model complex, higher-order interactions within the data, often leading to improved prediction accuracy and better insights into genotype-phenotype relationships (Manuel González-Camacho et al. 2018; Tong and Nikoloski 2021).

The omics integration is particularly valuable for elucidating gene regulatory networks (GRNs), which govern gene expression and coordinate complex biological processes in plants. By leveraging ML and artificial intelligence advancements, researchers can analyze large datasets to infer GRNs, uncover hidden regulatory patterns, and predict gene interactions that influence crop traits (Singh et al. 2022b). As these fields converge, understanding GRNs through multi-omics data and predictive analytics will enable the development of high-yield, resilient crop varieties that can better address global food security needs (Yang et al. 2021). GRN inference in plants predominantly relies on expression data, using correlation-based methods and mutual information (MI) algorithms to predict potential regulatory relationships between genes (Haque et al. 2019). Among the earliest approaches in gene co-expression analysis were Pearson's correlation coefficient and Spearman's rank correlation coefficient.

9.3.2 Machine Learning in Crop Proteomics

Machine learning (ML) techniques have been extensively applied in omics analyses, such as data integration, classification of genes and proteins, and prioritization of metabolites and association with specific traits. In contrast, efforts to apply ML in crop proteomics have been relatively limited compared to genome and transcriptome datasets. Neely et al. (2023) have addressed several applications of ML in the proteomics field, such as predicting enzymatic digestion, protein and peptide fractionation, liquid chromatography, ion mobility, and simulating LC-MS/MS data. Additionally, ML has been used to predict peptide retention times and fragmentation spectra.

Previous work has utilized sequence-attributed features in ML to classify proteins and predict their function. For instance, Rakesh et al. (2023) presented an SVM classifier to differentiate between the plastid and non-plastid proteins and classify them into chloroplast, chromoplast, etioplast, and amyloplast based on their function. Among the protein sequence features were amino acid composition, dipeptide composition, and protein physicochemical properties. Similarly, Zhang et al. (2018) combined sequence-based features (i.e., amino acid composition, protein position weight matrix, and gene co-expression) by utilizing deep neural networks (DNN) and SVM to predict plant mitochondrial protein in *Arabidopsis* and potatoes (Zhang et al. 2018). This integrative approach paves the way for further developing proteins with a lack of targeting information. In addition, SVM has been extensively employed in analyzing biological sequences due to its capability to handle data noise and accommodate high-dimensional features.

Gupta and Zou (2019) have proposed an ML-driven approach using a generative adversarial network (GAN) to optimize protein function prediction by generating synthetic DNA sequences encoding proteins. Their method, which offers a reliable and efficient solution, reassures researchers about the progress in the field and empowers them with advanced tools to design better protein structures. Recent studies have underscored the pivotal role of DL in addressing specific protein challenges. For instance, Ahmed et al. (2023) introduced a novel activation function, Gaussian Error Linear Unit with Sigmoid (SIELU), in a deep learning model for classifying unknown abiotic stress protein sequences, outperforming other models with high accuracies ranging from 80.78% to 95.11%. This underscores the potential of DL in protein research. Recent attention has been given to combining different ML models for detecting and classifying proteins into various functions or categories. For example, a predictive model based on SVM, NB, and KNN was built to predict the allergenicity of proteins in plants. Using multiple models improves prediction accuracy and allows for selecting the best-performing model, ultimately leading to comprehensive and robust results in which researchers can have confidence (Nedyalkova et al. 2022).

Several prediction tools have been developed using ML techniques to detect protein families, classify proteins into abiotic stress traits, and predict plant and pathogen interactions. For instance, Meher et al. (2018) presented ir-HSP, an online prediction tool to detect heat shock protein (HSP), their families, and DnaJ. The authors utilized protein sequence features of HSP and DnaJ, including G-spaced dipeptide compositions (GPC) as pseudo amino acid composition (PAAC). They employed SVM as the ML model, particularly utilizing the radial basis function (RBF) kernel. Their computational methods offer a reliable and efficient solution to enhance HSP prediction, making it a valuable tool for proteomics research.

In addition, Yang et al. (2019) introduced the inter-species protein-protein interaction predictor (InterSPPI) web server to predict the plant-pathogen protein-protein interaction network in plant diseases. The authors employed the RF algorithm to train the classification model based on an integrative feature encoding design. The primary data sources used in this study are the predicted PPIs from genomic information-based methods.

Next, ASRpro, an online prediction tool to predict proteins associated with multiple abiotic stress traits in plants, including cold, drought, heat, light, oxidative, and salt, was introduced. They utilized four ML algorithms to develop the predictive model: SVM, RF, adaptive boosting, and XGBoost. The authors retrieved protein sequences of stress-responsive proteins from multiple databases and translated them into K-mer compositional and autocross covariance (ACC) to train the predictive model. This online prediction tool significantly supports wet lab experiments in identifying stress-responsive genes and proteins for combating abiotic stress traits in crops (Meher et al. 2024).

The insights gained from proteomic and ML approaches significantly contribute to unraveling the molecular mechanisms by which plants adapt to environmental challenges, ultimately leading to advancements in crop yield improvement and stress resilience.

9.3.3 Machine Learning in Crop Metabolomics

Integrating machine learning (ML) with bioinformatics in metabolomics analysis enables large and complex dataset analyses. This approach unveils patterns, relationships, and trends that are concealed in the raw data, from data preprocessing to biological interpretation. For instance, Matyushin et al. (2020) utilized deep learning (a convolutional neural network) to rank small molecules identified from low-resolution GC-MS data. This was a response to the limitations of the library search procedure, which often returns incorrect small molecules despite the correct compound being available. The DL model requires a comprehensive set of input features to ensure optimal results. This comprehensive dataset enhances model effectiveness and accuracy, especially for a search in a tandem MAS library.

Identifying unknown metabolites is challenging due to the lack of commercial standard compounds. However, the use of ML can be a game-changer in this regard, as it can be used to predict the functions of these hidden metabolites. In this context,

Mahood et al. (2023) presented an ML-based analysis to annotate fragmented peaks with limited information in the library. Using ML coupled with information theory and co-accumulation clustering, they significantly improved the annotation. The potential of ML in predicting metabolite functions is intriguing, and further validation of these ML-based annotations could open up new avenues for metabolome-wide analyses.

ML can be used to classify and predict the functions of different metabolites based on their structural properties and other features. In this regard, Liu et al. (2017) reported using SVM and network clustering algorithms to classify 216 plants into metabolite groups and their plant species. Using the ML approach, they predicted relationships in unknown plant species-metabolite. The findings from this study demonstrated that ML classification can complement traditional morphology-based plant taxonomy. A study by Moore et al. (2019) used SVM and RF models to build a predictive model to classify plant general metabolism and specialized metabolism genes from multiple features (i.e., gene duplication, evolutionary rate, and gene expression) with high prediction accuracy. This approach has facilitated the mapping of the predicted specialized metabolism and genes to pathways that have not yet been able to be mapped.

Fürtauer et al. (2018) have presented a predictive supervised ML model to classify metabolic responses under control or stress conditions from proteomics and metabolomics datasets. Three ML algorithms, SVM, DT, and nearest neighbor, were used to train the metabolite and protein data in stress conditions. This approach offers a method for classifying data based on patterns, especially by integrating two types of omics data: metabolomics and proteomics.

Model validation is a crucial aspect of ML research, as Bai et al. (2024) demonstrated. They proposed an automated ML approach, employing AutoGluon-Tabular integration with multi-omics data from *Arabidopsis* to predict genes synthesizing terpenoids, alkaloids, and phenolics. The study also highlighted the importance of model validation in maize and tomatoes, showing that higher prediction accuracy is achieved with more data used in the model validation. Future studies should focus on integrating pretrained protein language models for specific secondary metabolite classes, further emphasizing the importance of model validation in ML research.

ML algorithms, including XGBoost, Decision Tree (DT), Neural Network, and Random Forest, have shown substantial accuracy in predicting consumer preferences for fruit flavor attributes such as sweetness, sourness, umami, and flavor intensity. In this study, Colantonio et al. (2022) analyzed metabolomic data from tomato and blueberry accessions alongside consumer sensory ratings and tested 18 ML models to predict consumer perceptions of fruit flavor. This approach enables high-throughput, accurate flavor phenotyping and highlights important genes and metabolites that could be targeted for marker-assisted selection (MAS) or genome editing. Ultimately, these tools could help breeders develop fruit varieties with flavors that appeal more to consumers.

Several ML tools have been developed to accelerate tasks in metabolomics data analysis. For instance, MetaClean, an R package, was developed using eight ML algorithms to build a classifier for automatically assessing peak integration from untargeted metabolomic analysis (Chetnik et al. 2020). In another effort, metabolite annotation remains crucial in predicting untargeted metabolomics. Hence, Retip (retention time prediction for metabolomics) was developed to predict retention time for small molecules in high-pressure liquid chromatography (HPLC) MS data analysis workflows (Bonini et al. 2020). The authors used five ML algorithms, including Random Forest, Bayesian regularized neural network, XGBoost, lightGBM, and Keras, to build a stable, accurate, and fast RT prediction model coupled with biochemical structural databases, such as CheBI, DrugBank, and KNApSAcK. strucToolbox also provides ML methods instead of statistical models as a data analysis method for preprocessing and normalizing metabolite data (Lloyd et al. 2020). These tools significantly reduce the manual work involved in these tasks.

Dührkop et al. (2021) offered valuable insights into applying an ML-based tool, namely CANOPUS, by predicting metabolite structural classes based on the MS/MS spectrum. The CANOPUS (class assignment and ontology prediction using mass spectrometry) uses a deep neural network to predict 2497 compound classes from fragmentation spectra, targeting and predicting compounds with structures not recorded in any database or publication and lacking tandem mass spectrometry training data. CANOPUS has accelerated the annotation of novel or unique compounds as it can rapidly annotate them. Hence, it is interestingly used in chemotaxonomic plant investigations and to illuminate global changes in plant metabolome under different conditions (Mahood et al. 2023). These two applications show the importance of integrating ML models in metabolomics data as practical tools to analyze comparative metabolomics data. In conclusion, several ML applications have been widely used in metabolomics, including prioritizing and ranking compounds, predicting low-resolution compounds, integrating metabolomics and other omics data, and classifying small molecules into different categories.

9.4 Challenges and Future Directions

Despite the omics advancements, plant genomics faces several challenges, primarily due to the increasing volume of data generated and the complexity of most plant genomes. High-throughput sequencing generates massive datasets that require robust bioinformatics tools and computational infrastructure for data processing, storage, and analysis. Polyploid genomes and high levels of repetitive sequences also present unique challenges in genome assembly and interpretation. Future developments in plant genomics will focus on integrating multi-omics data, spanning genomics, transcriptomics, proteomics, and metabolomics, to create a comprehensive view of crop biology. Integrating machine learning (ML) and omics technologies in crop breeding presents numerous challenges that must be addressed to fully leverage their potential for enhancing agricultural resilience and productivity. One significant challenge is data integration. The rapid advancement of various omics technologies produces vast and diverse datasets, encompassing genomic, transcriptomic, proteomic, and metabolomic data. Combining these heterogeneous data sources is essential for a comprehensive understanding of plant biology. However, inconsistencies in data formats and methodologies can impede researchers' ability to integrate datasets meaningfully, thereby limiting the scope of analyses that could inform breeding strategies.

Additionally, the computational complexity of advanced ML models poses substantial barriers to widespread implementation. While ML algorithms are promising for processing large datasets, they often require considerable computational resources and expertise. This challenge is particularly pronounced for researchers and institutions in developing regions, where access to high-end computing facilities and technical knowledge may be limited. As a result, this resource disparity can exacerbate inequalities in agricultural innovation and research capabilities on a global scale. Another critical concern is model validation and reproducibility. It is vital to ensure that ML models generalize effectively to real-world scenarios for their practical application in crop breeding. Issues such as model overfitting, where models perform exceptionally well on training data but struggle to predict outcomes on unseen datasets, must be thoroughly addressed. Comprehensive validation against various datasets is necessary to establish reliability and confidence in the predictions made by these models. Without such validation, the practical utility of ML models in guiding breeding decisions remains uncertain.

References

Ahmar S, Gill RA, Jung KH et al (2020) Conventional and molecular techniques from simple breeding to speed breeding in crop plants: recent advances and future outlook. Int J Mol Sci 21: 2590. https://doi.org/10.3390/IJMS21072590

Ahmed B, Haque MA, Iquebal MA et al (2023) DeepAProt: deep learning based abiotic stress protein sequence classification and identification tool in cereals. Front Plant Sci 13:1–12. https://doi.org/10.3389/fpls.2022.1008756

Bai W, Li C, Li W et al (2024) Machine learning assists prediction of genes responsible for plant specialized metabolite biosynthesis by integrating multi-omics data. BMC Genomics 25:1–12. https://doi.org/10.1186/s12864-024-10258-6

Bonini P, Kind T, Tsugawa H et al (2020) Retip: retention time prediction for compound annotation in untargeted metabolomics. Anal Chem 92:7515–7522. https://doi.org/10.1021/acs.analchem.9b05765

Bowerman AF, Byrt CS, Roy SJ et al (2023) Potential abiotic stress targets for modern genetic manipulation. Plant Cell 35:139–161. https://doi.org/10.1093/PLCELL/KOAC327

Cembrowska-Lech D, Krzemińska A, Miller T et al (2023) An integrated multi-omics and artificial intelligence framework for advance plant phenotyping in horticulture. Biology 12:1298. https://doi.org/10.3390/BIOLOGY12101298

Chang-Brahim I, Koppensteiner LJ, Beltrame L et al (2024) Reviewing the essential roles of remote phenotyping, GWAS and explainable AI in practical marker-assisted selection for drought-tolerant winter wheat breeding. Front Plant Sci 15:1319938. https://doi.org/10.3389/FPLS.2024.1319938/BIBTEX

Chao H, Zhang S, Hu Y et al (2023) Integrating omics databases for enhanced crop breeding. J Integr Bioinform 20:20230012. https://doi.org/10.1515/JIB-2023-0012/ASSET/GRAPHIC/J_JIB-2023-0012_FIG_003.JPG

Chetnik K, Petrick L, Pandey G (2020) MetaClean: a machine learning-based classifier for reduced false positive peak detection in untargeted LC–MS metabolomics data. Metabolomics 16:117. https://doi.org/10.1007/s11306-020-01738-3

Chun HJE, Khattra J, Krzywinski M et al (2014) Second-generation sequencing for cancer genome analysis. Cancer Genom Bench Personal Med:13–30. https://doi.org/10.1016/B978-0-12-396967-5.00002-5

Colantonio V, Ferrão LFV, Tieman DM et al (2022) Metabolomic selection for enhanced fruit flavor. Proc Natl Acad Sci USA 119:1–11. https://doi.org/10.1073/pnas.2115865119

de Oliveira EJ, de Resende MDV, da Silva SV et al (2012) Genome-wide selection in cassava. Euphytica 187:263–276. https://doi.org/10.1007/S10681-012-0722-0/TABLES/5

Kim KD, Kang Y, Kim C (2020) Application of genomic big data in plant breeding: past, present, and future. Plants 9:1–25. https://doi.org/10.3390/plants9111454

Dührkop K, Nothias LF, Fleischauer M et al (2021) Systematic classification of unknown metabolites using high-resolution fragmentation mass spectra. Nat Biotechnol 39:462–471. https://doi.org/10.1038/s41587-020-0740-8

Fürtauer L, Pschenitschnigg A, Scharkosi H et al (2018) Combined multivariate analysis and machine learning reveals a predictive module of metabolic stress response in *Arabidopsis thaliana*. Mol Omics 14:437–449. https://doi.org/10.1039/c8mo00095f

Gupta A, Zou J (2019) Feedback GAN for DNA optimizes protein functions. Nat Mach Intell 1: 105–111. https://doi.org/10.1038/s42256-019-0017-4

Gupta C, Ramegowda V, Basu S, Pereira A (2021) Using network-based machine learning to predict transcription factors involved in drought resistance. Front Genet 12:1–24. https://doi.org/10.3389/fgene.2021.652189

Haque S, Ahmad JS, Clark NM et al (2019) Computational prediction of gene regulatory networks in plant growth and development. Curr Opin Plant Biol 47:96–105. https://doi.org/10.1016/J.PBI.2018.10.005

He Y, Fang J, Xu W, Shi P (2022) Substantial increase of compound droughts and heatwaves in wheat growing seasons worldwide. Int J Climatol 42:5038–5054. https://doi.org/10.1002/JOC.7518

Huang J, Lin Q, Fei H et al (2023) Discovery of deaminase functions by structure-based protein clustering. Cell 186:3182–3195. https://doi.org/10.1016/j.cell.2023.05.041

Itoh T, Kawahara Y, Tanaka T (2018) Databases for rice omics studies. Rice Genom Genet Breed:541–554. https://doi.org/10.1007/978-981-10-7461-5_28

Kisiel A, Krzemińska A, Cembrowska-Lech D, Miller T (2023) Data Sci Plant Metab Metab 13:1–19. https://doi.org/10.3390/metabo13030454

Koh E, Sunil RS, Lam HYI, Mutwil M (2024) Confronting the data deluge: how artificial intelligence can be used in the study of plant stress. Comput Struct Biotechnol J 23:3454–3466. https://doi.org/10.1016/j.csbj.2024.09.010

Lam HYI, Ong XE, Mutwil M (2024) Large language models in plant biology. Trends Plant Sci 29: 1145–1155. https://doi.org/10.1016/j.tplants.2024.04.013

Lawrence CJ (2007) MaizeGDB. Plant Bioinform:331–345. https://doi.org/10.1007/978-1-59745-535-0_16

Liu K, Abdullah AA, Huang M et al (2017) Novel approach to classify plants based on metabolite-content similarity. Biomed Res Int 2017:1–12. https://doi.org/10.1155/2017/5296729

Liu Y, Zhang Y, Liu X et al (2023) SoyOmics: a deeply integrated database on soybean multi-omics. Mol Plant 16:794–797. https://doi.org/10.1016/j.molp.2023.03.011

Lloyd GR, Jankevics A, Weber RJM (2020) Struct: an R/Bioconductor-based framework for standardized metabolomics data analysis and beyond. Bioinformatics 36:5551–5552. https://doi.org/10.1093/bioinformatics/btaa1031

Ma S, Wang M, Wu J et al (2021) WheatOmics: a platform combining multiple omics data to accelerate functional genomics studies in wheat. Mol Plant 14:1965–1968. https://doi.org/10.1016/j.molp.2021.10.006

Mahmood U, Li X, Fan Y et al (2022) Multi-omics revolution to promote plant breeding efficiency. Front Plant Sci 13:1062952. https://doi.org/10.3389/FPLS.2022.1062952/BIBTEX

Mahood EH, Bennett AA, Komatsu K et al (2023) Information theory and machine learning illuminate large-scale metabolomic responses of Brachypodium distachyon to environmental change. Plant J 114:463–481. https://doi.org/10.1111/tpj.16160

Manickam S, Rajagopalan VR, Kambale R et al (2023) Plant metabolomics: current initiatives and future prospects. Curr Issues Mol Biol 45:8894–8906. https://doi.org/10.3390/cimb45110558

Manuel González-Camacho J, Ornella L, Pérez-Rodríguez P et al (2018) Applications of machine learning methods to genomic selection in breeding wheat for rust resistance. Plant Genome 11: 170104. https://doi.org/10.3835/PLANTGENOME2017.11.0104

Marsh JI, Hu H, Gill M, Batley J, Edwards D (2021) Crop breeding for a changing climate: integrating phenomics and genomics with bioinformatics. Theor Appl Genet. 134(6):1677–1690. https://doi.org/10.1007/s00122-021-03820-3

Martin-Rios C, Hofmann A, Mackenzie N (2020) Sustainability-oriented innovations in food waste management technology. Sustainability 13:210. https://doi.org/10.3390/SU13010210

Matei G, Woyann LG, Milioli AS et al (2018) Genomic selection in soybean: accuracy and time gain in relation to phenotypic selection. Mol Breed 38:1–13. https://doi.org/10.1007/S11032-018-0872-4/METRICS

Matyushin DD, Sholokhova AY, Buryak AK (2020) Deep learning driven GC-MS library search and its application for metabolomics. Anal Chem 92:11818–11825. https://doi.org/10.1021/acs.analchem.0c02082

Meher PK, Sahu TK, Gahoi S, Rao AR (2018) ir-HSP: improved recognition of heat shock proteins, their families and sub-types based on g-spaced di-peptide features and support vector machine. Front Genet 8:1–17. https://doi.org/10.3389/fgene.2017.00235

Meher PK, Sahu TK, Gupta A et al (2024) ASRpro: a machine-learning computational model for identifying proteins associated with multiple abiotic stress in plants. Plant Genome 17:1–13. https://doi.org/10.1002/tpg2.20259

Mendoza-Revilla J, Trop E, Gonzalez L et al (2024) A foundational large language model for edible plant genomes. Commun Biol 7:835. https://doi.org/10.1038/s42003-024-06465-2

Metzker ML (2009) Sequencing technologies – the next generation. Nat Rev Genet 11:31–46. https://doi.org/10.1038/nrg2626

Moore BM, Wang P, Fan P et al (2019) Robust predictions of specialized metabolism genes through machine learning. Proc Natl Acad Sci USA 116:2344–2353. https://doi.org/10.1073/pnas.1817074116

Movahedi S, Movahedi S, Bel MVAN et al (2012) Comparative co-expression analysis in plant biology. Plant Cell Environ 35:1787–1798. https://doi.org/10.1111/j.1365-3040.2012.02517.x

Murmu S, Sinha D, Chaurasia H et al (2024) A review of artificial intelligence-assisted omics techniques in plant defense: current trends and future directions. Front Plant Sci 15:1–17. https://doi.org/10.3389/fpls.2024.1292054

Nath B, Chen G, O'Sullivan CM, Zare D (2024) Research and technologies to reduce grain postharvest losses: a review. Foods 13:1875. https://doi.org/10.3390/FOODS13121875

Nedyalkova M, Vasighi M, Azmoon A et al (2022) Sequence-based prediction of plant allergenic proteins: machine learning classification approach. ACS Omega 8:3698. https://doi.org/10.1021/acsomega.2c02842

Neely BA, Dorfer V, Martens L et al (2023) Toward an integrated machine learning model of a proteomics experiment. J Proteome Res 22:681–696. https://doi.org/10.1021/acs.jproteome.2c00711

Ossowski S, Schneeberger K, Clark RM et al (2008) Sequencing of natural strains of *Arabidopsis thaliana* with short reads. Genome Res 18:2024–2033. https://doi.org/10.1101/GR.080200.108

Pazhamala LT, Kudapa H, Weckwerth W et al (2021) Systems biology for crop improvement. Plant Genome 14:e20098. https://doi.org/10.1002/TPG2.20098

Pisner DA, Schnyer DM (2020) Support vector machine. Machin Learn Method Appl Brain Disord:101–121. https://doi.org/10.1016/B978-0-12-815739-8.00006-7

Pushparaj PN (2019) Introduction to functional bioinformatics. Essent Bioinform I:235–254. https://doi.org/10.1007/978-3-030-02634-9_11
Qi Y (2012) Random forest for bioinformatics. Ensemb Mach Learn:307–323. https://doi.org/10.1007/978-1-4419-9326-7_11
Rakesh K, Sitanshu SS, Ruchi V, Weirick T (2023) MU-LOC: a machine-learning method for predicting mitochondrially localized proteins in plants. Front Plant Sci 14:1–9. https://doi.org/10.1021/acsomega.2c02842
Raza A, Tabassum J, Kudapa H, Varshney RK (2021) Can omics deliver temperature resilient ready-to-grow crops? Crit Rev Biotechnol. 41(8):1209–1232. https://doi.org/10.1080/07388551.2021.1898332
Ruperao P, Rangan P, Shah T et al (2023) The progression in developing genomic resources for crop improvement. Life 13:1668. https://doi.org/10.3390/LIFE13081668/S1
Sahito JH, Zhang H, Gishkori ZGN et al (2024) Advancements and prospects of genome-wide association studies (GWAS) in maize. Int J Mol Sci 25:1918. https://doi.org/10.3390/IJMS25031918
Savary S, Willocquet L, Pethybridge SJ et al (2019) The global burden of pathogens and pests on major food crops. Nat Ecol Evol 3:430–439. https://doi.org/10.1038/s41559-018-0793-y
Scarano C, Veneruso I, De Simone RR et al (2024) The third-generation sequencing challenge: novel insights for the omic sciences. Biomolecules 14:568. https://doi.org/10.3390/BIOM14050568
Serin EAR, Nijveen H, Hilhorst HWM, Ligterink W (2016) Learning from co-expression networks: possibilities and challenges. Front Plant Sci 7:1–18. https://doi.org/10.3389/fpls.2016.00444
Shaik R, Ramakrishna W (2014) Machine learning approaches distinguish multiple stress conditions using stress-responsive genes and identify candidate genes for broad resistance in rice. Plant Physiol 164:481–495. https://doi.org/10.1104/pp.113.225862
Sharma L, Kumar S, Jaiswal VP, Shukla SK (2024) Advances and challenges in omics approaches for alleviating abiotic stresses and improving cane yield in sugarcane crop. In: Current omics advancement in plant abiotic stress biology: a volume in developments in applied microbiology and biotechnology, pp 269–276. https://doi.org/10.1016/B978-0-443-21625-1.00018-X
Singh A, Ramakrishna G, Kaila T et al (2022a) Next-generation sequencing technologies: approaches and applications for. Crop Improv:31–94. https://doi.org/10.1007/978-1-0716-2533-0_3
Singh KS, van der Hooft JJJ, van Wees SCM, Medema MH (2022b) Integrative omics approaches for biosynthetic pathway discovery in plants. Nat Prod Rep 39:1876–1896. https://doi.org/10.1039/D2NP00032F
Sirocchi C, Biancucci F, Donati M et al (2024) Exploring machine learning for untargeted metabolomics using molecular fingerprints. Comput Methods Prog Biomed 250:108163. https://doi.org/10.1016/j.cmpb.2024.108163
Song B, Ning W, Wei D et al (2023a) Plant genome resequencing and population genomics: current status and future prospects. Mol Plant 16:1252–1268. https://doi.org/10.1016/J.MOLP.2023.07.009/ASSET/49A7C524-4BC7-4EC9-821B-4A6677857016/MAIN.ASSETS/GR3.JPG
Song L, Wang R, Yang X et al (2023b) Molecular markers and their applications in marker-assisted selection (MAS) in bread wheat (Triticum aestivum L.). Agriculture 13:642. https://doi.org/10.3390/AGRICULTURE13030642
Sprenger H, Erban A, Seddig S et al (2018) Metabolite and transcript markers for the prediction of potato drought tolerance. Plant Biotechnol J 16:939–950. https://doi.org/10.1111/pbi.12840
Tong H, Nikoloski Z (2021) Machine learning approaches for crop improvement: leveraging phenotypic and genotypic big data. J Plant Physiol 257:153354. https://doi.org/10.1016/J.JPLPH.2020.153354
Tyagi P, Singh D, Mathur S et al (2022) Upcoming progress of transcriptomics studies on plants: an overview. Front Plant Sci 13:1030890. https://doi.org/10.3389/FPLS.2022.1030890/FULL

Ullah A, Bano A, Khan N (2021) Climate change and salinity effects on crops and chemical communication between plants and plant growth-promoting microorganisms under stress. Front Sustain Food Syst 5. https://doi.org/10.3389/fsufs.2021.618092

Varshney RK, Sinha P, Singh VK et al (2020) 5Gs for crop genetic improvement. Curr Opin Plant Biol 56:190–196. https://doi.org/10.1016/j.pbi.2019.12.004

Villate A, San Nicolas M, Gallastegi M et al (2020) Review: metabolomics as a prediction tool for plants performance under environmental stress. Plant Sci 303:1–12. https://doi.org/10.1016/j.plantsci.2020.110789

Vo KTX, Rahman MM, Rahman MM et al (2021) Proteomics and metabolomics studies on the biotic stress responses of rice: an update. Rice 14:1–16. https://doi.org/10.1186/s12284-021-00461-4

Voss-Fels KP, Stahl A, Hickey LT (2019) Q & A: modern crop breeding for future food security. BMC Biol 17:1–7. https://doi.org/10.1186/S12915-019-0638-4

Weber L, Barth F, Lorenz L et al (2023) PEDL1: protein-centered relation extraction from PubMed at your fingertip. Bioinformatics 39:1–4. https://doi.org/10.1093/bioinformatics/btad603

Xu Y, Zhang X, Li H et al (2022) Smart breeding driven by big data, artificial intelligence, and integrated genomic-enviromic prediction. Mol Plant 15:1664–1695. https://doi.org/10.1016/j.molp.2022.09.001

Yan J, Wang X (2023) Machine learning bridges omics sciences and plant breeding. Trends Plant Sci 28:199–210. https://doi.org/10.1016/j.tplants.2022.08.018

Yang S, Li H, He H et al (2019) Critical assessment and performance improvement of plant-pathogen protein-protein interaction prediction methods. Brief Bioinform 20:274–287. https://doi.org/10.1093/bib/bbx123

Yang Y, Saand MA, Huang L et al (2021) Applications of multi-omics technologies for crop improvement. Front Plant Sci 12:563953. https://doi.org/10.3389/FPLS.2021.563953/BIBTEX

Yoosefzadeh Najafabadi M, Hesami M, Eskandari M (2023) Machine learning-assisted approaches in modernized plant breeding programs. Genes 14:777. https://doi.org/10.3390/GENES14040777

Zhang N, Rao RSP, Salvato F et al (2018) MU-LOC: a machine-learning method for predicting mitochondrially localized proteins in plants. Front Plant Sci 9. https://doi.org/10.3389/fpls.2018.00634

Zhang W, Zuo C, Chen Z et al (2019) RNA sequencing reveals that both abiotic and biotic stress-responsive genes are induced during expression of steroidal Glycoalkaloid in potato tuber subjected to light exposure. Genes (Basel) 10. https://doi.org/10.3390/GENES10110920

Zou J, Han Y, So SS (2008) Overview of artificial neural networks. Methods Mol Biol 458:15–23. https://doi.org/10.1007/978-1-60327-101-1_2/FIGURES/6_2

Unraveling Solutions to the Complex Problems: Implementing Climate-Smart Crop Management Strategies

10

Mohd Anas, Nashra Waqar, P. M. S. Noor Mohamed Ibrahim, Minza Hasan, K. P. Jaseem, Yogesh Ahlawat, Pawan K. Jaiwal, and Darshna Chaudhary

Abstract

The agricultural industry is becoming increasingly challenged by climate change with fluctuating weather, degrading soil quality, water shortages, and pest invasions that each threatens farmer livelihoods, food security, and sustainability. This chapter explores Climate-Smart Crop Management Strategies (CSCMS) as a way to transform productivity, resilience, and sustainability. The chapter explains CSCMS by its key principles—productivity, resilience, and mitigation—and the key CSCMS strategies for each of the principles which include determining soil health, water resource management, crop diversification, integrated pest and nutrient management, and digital/new technologies, including precision farming and Internet-of-Things-based tools. The chapter showcases an implementation framework that stresses community engagement, human capacity development, supportive policy framework, and investing in infrastructure. The chapter also emphasizes potential agricultural economic benefits, environmental awareness and mitigation, and social acceptance, while discussing the barriers to adoption of CSCMS which encompass financial, knowledge, and cultural barriers. To

M. Anas · P. M. S. Noor Mohamed Ibrahim · M. Hasan · K. P. Jaseem
Department of Botany, Aligarh Muslim University, Aligarh, India

N. Waqar
Department of Biotechnology, Faculty of Science, Era University, Lucknow, India

Y. Ahlawat (✉)
University Centre for Research and Development, Chandigarh University, Mohali, Punjab, India

Allied Health Sciences, Datta Meghe Institute of Higher Education and Research, Wardha, Maharashtra, India
e-mail: ykahlawa@mtu.edu

P. K. Jaiwal · D. Chaudhary
Centre for Biotechnology, Maharshi Dayanand University, Rohtak, Haryana, India

Y. Ahlawat et al. (eds.), *Next-Generation Strategies for Crop Improvement*,
https://doi.org/10.1007/978-981-95-0309-4_10

illustrate practical examples and future approaches, the chapter provides case studies of successful projects such as drought-resilient farming and agroforestry, as well as emerging digital technologies such as artificial intelligence and remote sensing. By combining Agri-Tech and leveraging the benefits of collaboration across the agricultural landscape, CSCMS provides an effective pathway for securing future food systems and strengthening resilient landscapes for sustainable development.

Keywords

Climate-smart agriculture · Crop management · Sustainable agriculture · Climate change adaptation · Agricultural resilience · Crop yield improvement · Environmental sustainability · Precision farming

10.1 Introduction

The need for better agricultural practice is increasing day by day due to the catastrophic changes that are associated with the climate change, the weather has become unpredictable in almost every part of the world due to the global warming and emission of greenhouse gases (GHGs) in an enormous scale, farmers are greatly affected when abiotic and biotic factors both synergistically cause yield loss, by declining soil fertility, water scarcity, and crop loss due to pathogen infestation (Richard et al. 2022). To overcome these, Climate-Smart Crop Management Strategies (CSCMS) have come along as a revolutionary approach to supporting agricultural resilience, enhancing productivity, and promoting environmental sustainability (Fig. 10.1). Such strategies involve adjusting farm practice to respond to climate change so that productivity is sustained or even enhanced with reduced impact on the environment. CSCMS consider optimizing water use and nutrient use, integrating pest and disease management, as well as soil health through sustainable

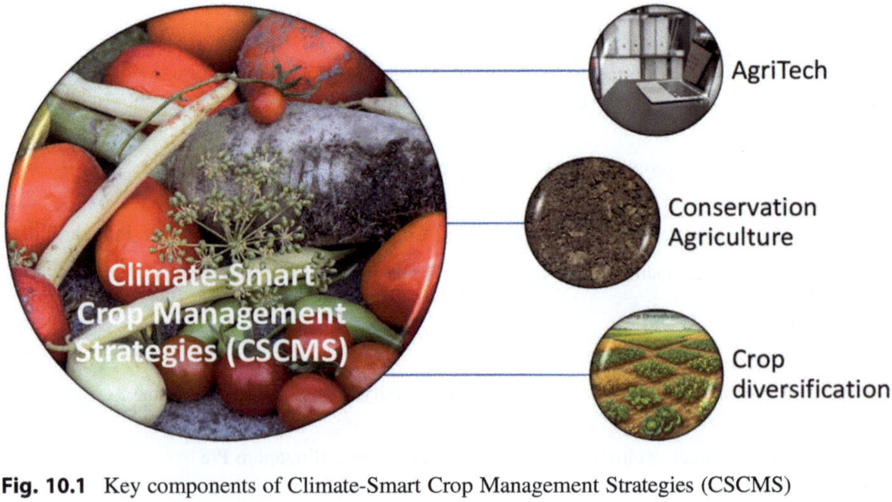

Fig. 10.1 Key components of Climate-Smart Crop Management Strategies (CSCMS)

land use. Through such strategies, farmers are in a position to adapt to climatic stresses, decrease their use of chemical inputs, and mitigate GHG emission (Bhattacharyya et al. 2020). Additionally, CSCMS utilizes emerging technologies, including climate forecasting software and precision agriculture (PA), to help farmers make informed choices using live data. Such practices not only contribute to yield stability but also promote long-term ecological balance.

By doing so, CSCMS are multifunctional in their role to improve agricultural productivity as well as promote environmental sustainability. Hence, use of Climate-Smart Crop Management Strategies is imperative in order to protect the agriculture of tomorrow in an age of accelerated climatic change in such a way that world food demands can be fulfilled without harming our earth's health (Gani et al. 2022). An integral part of these agricultural reforms is the use of technologies which would make the work more smarter and things easier for farmers, enhancing productivity. Denoted in the form of precision agriculture, IoT-based decision support, AI analytics, and advisory using digital technology, the revolutionary role that IoT and AI are playing to transform agriculture. Combined, they are at the core of Precision Agriculture—an intelligent, efficient, and sustainable means to farm. By making smart choices at all levels of the farm cycle, such technologies hold out promise of increased yields, plus an improved, more responsive food system (Senoo et al. 2024). Precision agriculture comprises a set of Geospatial tools and technologies such as Remote sensing, Geographic Information Systems (GIS), and Variable Rate Technology (VRT) which will play a crucial role in improving the understanding and approach of crop management (Fig. 10.2). These tools not only help mitigate the impact of climate catastrophe but if used intellectually, they could help prevent yield losses by early planning and implementation (Krishnababu et al. 2024).

The use of AI is also seeing a boom in the field of agriculture; machine learning and predictive analytics have a tremendous impact on modern-day farming practices, these tools have made the task of data analysis much easier providing reliable insights to farmers regarding their field conditions, nutrients, and soil status. For instance, IIT Kharagpur's Gramin Krishi Mausam Sewa is forecasting climatic risks and advising farmers for better agricultural decision-making. This model helps rural farmers sow their seeds in favorable weather conditions which will protect their crops from a wide range of abiotic and biotic stressors. Using predictive analytics, we are able to get informed regarding the uncertainties of weather much earlier which is a boon for farmers. Since AI has a multidimensional approach, it can analyze and predict various biotic and abiotic stress factors simultaneously. Biotic factors are still a great threat to crop production, and various conventional methods are used currently to control phytopathogens, such as bacteria, fungi, and nematodes, remain to pollute the soil and environment. One of the positive impacts of AI and IoT in the field of plant protection is the early detection of pest and disease outbreaks through high-resolution images taken by drones and satellites. For instance, Berlin-based AI startup Plantix and a Tel Aviv company Taranis are involved in early detection of diseases using artificial intelligence tools (Azenzem and Eddomairi 2025). Besides crop protection, the demand for AI-based strategies is also in other forms of agricultural practices like animal husbandry, aquaculture, sericulture,

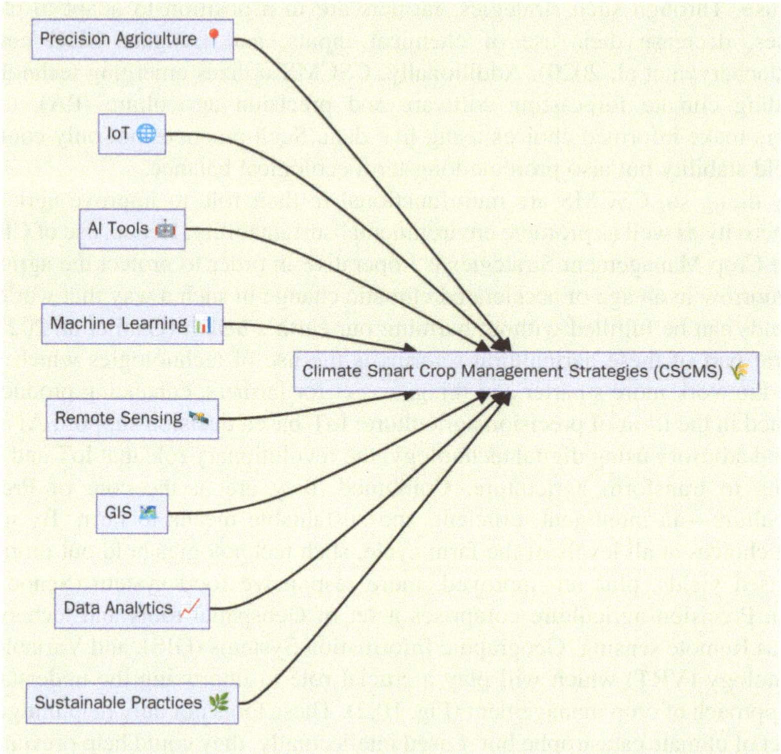

Fig. 10.2 Technological pillars supporting CSCMS for enhanced climate resilience and sustainable agriculture

apiculture, etc. These technologies help in real-time monitoring of livestock, fishes, and other animal species and thereby enhancing resources (Chang et al. 2024). An example of the aquaculture industry might make us realize the impact of these novel tools in agriculture. A Norwegian company AquaCloud established in 2017 works on generating meaningful data for fish farmers. They are working on several different projects like creating a centralized Data Platform for the fisheries network, Sensor data standards based on IoT and scalable sensors, Fish health standard reports for support and welfare of fishes, measuring environmental status in sea cages, and Sea Lice outbreak prediction in seas (Mustapha et al. 2021). Techniques such as hydroponics also deserve well-managed IoT sensors and AI-based solutions to control nutrients and early prediction of diseases. These tools monitor several parameters related to the growth and development of plants placed in a hydroponic system; since the hydroponic method mainly relies on water parameters, monitoring parameters such as pH, EC, and temperature of nutrient solution is mandatory for ensuring good yield (Shareef et al. 2024). Even though the possibility of diseases are low as compared to soil-based farming, there are other issues in hydroponics, such as waterborne disease and spread of disease in a well-connected hydroponics network, so in order to avoid the issue, we must continuously monitor the nutrient solution for contaminations. Other factors such as light, humidity, CO_2, and leaf conditions must also be monitored. Human monitoring is a difficult task, but sensors can execute this task with precision timing and accuracy. Further, these IoT sensors can be connected to a Machine learning model for training and evaluation, which will help predict the growth rate and yield of crops. Another advantage of ML is we can incorporate large datasets from the public domain such as FAO, World Data Bank, and Google Earth, analyze climate data and geospatial information with precision. IoT microcontrollers such as Raspberry Pi, Arduino, neural networks like DCNN, ResNet, Mobile Net, MRCNN, VGG16, Deep Lab V3, DenseNet 201, and Inception Module can be involved in hydroponic systems for smart automation (Rathor et al. 2024). Emerging trends in AI and other tools can make decision-making easier and scientific need to be implemented with a good strategy. Even though these technologies are robust and reliable, they cannot be implemented without proper field trials (Agarwal et al. 2024). For a sustainable future, making smart decisions in agriculture is the need of the hour. Otherwise, ensuring food security will become a dream, hence we need to analyze different aspects related to this issue, be it global warming, climate change impact, and also the disease resistance that is becoming prevalent in farm lands due to excessive use of chemical pesticides. Considering one or two parameters shall not make climate-smart crops successful, but a strategic implementation of Agri-Tech, along with a sustainable approach can be a good solution (Wahab et al. 2024).

10.2 Understanding Climate-Smart Crop Management

Farming practices need to be sustainable and at the same time, it should not lead to food insecurity, this is the challenge for scientists working on crop improvement. Demand for agricultural produce is increasing and the cultivable land area is decreasing. So, agriculture cannot stay in a historical way, we need to improve things in a sustainable way, so that it does not create further environmental issues (Saikanth et al. 2023). Even though climate change is a global issue, the impact of climate change varies for each region, each country differently. It is crucial that we make decisions according to the impact and the scenario of climate change in our region. This will help us in modeling approaches in an intellectual way. This modeling can be done considering various aspects such as size, agro-ecology, and sensitivity of the region toward climate change. For instance, crop improvement of maize grown in China will be different from sub Saharan Africa, depending on the climatic fluctuations and future predictions (Li et al. 2011). Climate-smart crop management must align itself with the core aspects of sustainability; otherwise it will not be smart in an actual sense. Our crops must be high yielding, with reduced carbon footprints and greenhouse gas emissions. This will not be possible until we do not ponder upon the root cause of environmental impact of agriculture, like soil degradation due to chemical fertilizers, excessive water consumption leading to decline in water table, eutrophication, and other major and minor impact of high-yield crops. The crop improvement technologies currently in action are mostly not sustainable and are a serious threat to both nature and human health. Impacts can be seen directly or indirectly on a large-scale analysis. Concern does not end in abiotic factors, but the negative impact also extends intensively in the biological aspect; for instance, the natural biological diversity is disrupted, herbicide overuse and its impact on weed ecology, gene pool problems, and the list extends long. The goal of sustainability cannot be compromised for short-term benefits (Pradhan et al. 2015). One of the major revolutions in the field of agriculture was The Green Revolution, even though it increased the productivity of crops, had several essential ecological and health issues. One of the main problems involved is the degradation of soils, specifically salinity and waterlogging caused by intensive irrigation systems. These issues impacted more than 45 million hectares of the world, degrading yields and often leaving the land unproductive. Monoculture as a continuous practice, as well as higher utilization of lands through HYVs, further contributed to the loss of essential nutrients from soils and the loss of fertility based on a long-term decline, notably so for wheat and rice crops. Another main issue involved higher chemical fertilizer and pesticide usage, with deep human health impacts (Cohen 2020). Millions of cases of pesticide poisoning occurred yearly, notably from developing nations free from good regulation and farmer education. Furthermore, nitrate contamination of groundwater through fertilizer run-off created serious threats, such as diseases such as stomach cancer as well as infant methemoglobinemia “blue baby syndrome” (Smith 2009). The Green Revolution further caused concern for genetic erosion, with traditional varieties being displaced by uniform HYVs, diminishing the genetic diversity essential for long-term crop resilience. While certain research contends that modern varieties still contain diverse germplasm, the perception of, and risk of, narrowed genetic bases still feature prominently. In addition, pest and disease epidemics became increasingly frequent as a consequence of monocultures’ susceptibility, encouraging greater reliance on pesticides. This further led to ecological imbalances as well as pesticide resistance. Another consequence of water resource stress was the consequence of excessive irrigation, including groundwater depletion, as well as raised water tables across many places. Long-term impacts of intensification further caused yield stagnation or decrease in some places, indicating short-term gains might have undermined sustainable productivity. Notably, much of the adverse consequences cannot be attributed exclusively to the technologies of the Green Revolution but stem from complementary socioeconomic, infrastructural, as well as policy, factors that fostered unregulated intensification (Assessment et al. 2001). In addition, enhancing the health of the soil is an essential component of sustainable agriculture. Researchers have identified and implemented a number of Soil and Crop Management Strategies (SCMS) or other mitigation strategies in order to raise awareness about the importance of maximizing the utilization of input resources while maintaining a particular emphasis on the preservation of the environment. The purpose of these techniques is to improve agricultural output while also halting the deterioration of land. This is accomplished by optimizing a number of soil’s physical, chemical, hydrological, and biological qualities through the utilization of optimal nutrient management to achieve the desired results. The synchronization of the quantity of input with crop need and the synchronization of application time with crop growth are the two fundamental criteria that should be adhered to in order to achieve a tight balance between the outputs and inputs of substances that include nutrients (Wu 2018). These have the dual benefits of increasing crop productivity and protecting the environment and soil resources. In order to conserve available plant nutrients and water, we can also use farm wastes, natural and mineral fertilizers, farmyard manure, crop residues, agroforestry, soil tillage, intercropping, crop rotations, fallows, irrigation, and drainage. According to Zhang et al. (2011), SCMS also incorporates cutting-edge techniques that can enhance nutrient uptake, like applying coated urea, urease inhibitors, and fertilizer at ground depth. Instead of concentrating just on the yield-contributed return, these interventions can encourage growers to prioritize long-term strategies and environmental sustainability (Esilaba et al. 2005). One of the most difficult tasks in the field is fertilizer usage reduction, which means using fertilizers as efficiently as possible to maximize crop yields and safeguard the environment. Because they are the main contaminants entering and leaving fields through fertilizers (both organic and inorganic), or other main sources of nutrition entering or leaving the field, such as dairy farm effluent management, nitrogen and phosphorus are primarily managed as plant nutrients (Vitousek et al. 2009). The excess nutrients, mostly N and P, that crops are unable to take can end up in the water table or other bodies of water, polluting the ecosystem. Nutrient management, according to Delgado et al. (2006), is the art and science of tillage, irrigation, and soil and water conservation in order to maximize crop fertilizer use, productivity, product quality, and net return while minimizing off-site nutrient movement and minimizing environmental impact. Another reported approach for conserving soil health is Site-Specific Nutrient Management (SSNM). It is a crop-oriented approach that provides growers with techniques, tools, guidelines, and principles to help them decide how much fertilizer to apply to a crop at a specific time and location during a specific production phase. Dobermann et al. (2002) define SSNM as the site-level, all-encompassing control of a crop's nutrition within a specific cropping period to balance nutrient supply and demand as characterized by differences in plant–soil cycle. These SSNMs aim to take advantage of (1) seasonal and regional variations in environmental production potential and crop nutrient demand, (2) spatial variability of fields with respect to intrinsic nutrient availability, (3) within-season dynamics of crop N demand at the farm level, and (4) cropping patterns and management practices at the site level (Peng et al. 2010). Fertilizer use efficiency relies totally on the soil, crop, and fertilizer interaction; if we consider practices like Integrated Nutrient Management (INM), it creates a fine balance between the essential factors for increase in crop production. INM integrates traditional and modern cropping system techniques that are environmentally friendly, economically viable, and make intelligent and efficient use of both natural and artificial manures for fertilizer application. INM manages the three main macronutrients N, P, and K as well as other macro and micronutrient inputs and outputs with the goal of nutrient cycling in close synchronization with the nutrient's requirements and soil use. INM saw a decrease in nitrogen losses from runoff, leaching, volatilization, and immobilization, which improved fertilizer usage efficiency (Wu and Ma 2015). In this era of climate change, water and soil management are foremost important for sustainable agriculture. The most crucial resource for the sustainable growth of agriculture is water, and the main challenge to agricultural water management is more effective water usage. Furthermore, climate change and increasing nonagricultural demands significantly exceed the amount of water available for agriculture. Water availability and crop demand must be balanced for greater quantity and higher quality at a fair price or without adverse environmental effects. This does not imply that the growing popularity and demand for micro-irrigation products, such as sprinklers or drips, is a result of their increased efficiency over time (Chartzoulakis and Bertaki 2015). Drip irrigation, for instance, delivers water straight to the root zone of the plant, reducing surface evaporation and potentially increasing agricultural output and WUE by at least 50%. Subsequent studies have shown that drip irrigation can improve the WUE and decrease salinization by reducing water evaporation. The salinity of our agricultural soil is a major factor limiting crop yields, thus scientists must focus on it, especially since it is linked to irrigation (Najafi and Tabatabaei 2007). All these strategies must aim for mitigation of carbon footprint, studies reveal a range of LCAs for organic farming in Europe. A cradle-to-grave system boundary approach to the carbon footprint of pumpkin production was in Germany, based on 1 ha of cultivated land and 1 kg of product as the functional units (Schäfer and Blanke 2012). Assessments in Italy included wheat farming and bread production, considering a functional unit of either 1 ha of wheat cultivation or 1 kg of bread, and a cradle-to-gate system boundary (Chiriacò et al. 2017). Belgium provided a study on environmental impacts from systems of wheat cultivation, using 1 ha and 1 kg of grain as systems units (Van Stappen et al. 2015). Environmental burdens from producing bread wheat, oilseed rape, and potatoes in England and Wales on a per-kilogram of product basis (Williams et al. 2010). Greece has also done studies based on lettuce and eggplant, based on land area (1 ha) and product weights (1 ton of lettuce) and 1000 m^2 cultivated area (Foteinis and Chatzisymeon 2016) and 1 ton of marketable eggplant, respectively (Foteinis et al. 2021). In Spain, greenhouse gas emissions were estimated for herbaceous cropping systems per 1 ha and per 1 kg of product (Aguilera et al. 2015). Slovenia assessed the carbon footprint of conventional and organic crop production (kg CO_2/kg crop) per hectare and Denmark compared different crop rotations with functional unit 1 ha and 1 kg crop, respectively (Knudsen et al. 2014).

10.3 The Complex Problems in Modern Agriculture

As discussed earlier in the introduction of this chapter, the complexity of concerns are wide and multidimensional in modern agriculture, so any kind of simple and narrow approach cannot be a better option. We need to analyze the multiple factors which impact the crops and design solutions according to the needs of the farmland. Otherwise, temporarily fixing issues is not our goal when it comes to CSCMS as discussed in previous sections; climate change impact is one of the kinds, but it is not the sole reason. Other factors such as soil degradation due to unregulated use of chemicals as a part of green revolution, increasing need for irrigation and depleting water resources, socio-economic barriers, and also the biotic stress factors like pest and disease outbreak are some of the other major concerns. The problems become more severe when they show these effects combining with one another and cause a combined negative impact (Kopecká et al. 2023). For instance, some of the abiotic stress like heat can trigger biotic stress by making the host plant susceptible to pathogen attack; this is one example where the effect can be different depending upon the conditions. Crops are subjected to multiple stress for example biotic and abiotic. Instead, a mix of biotic (such as infections) and abiotic (such as heat, salt, and drought) stressors may be encountered simultaneously. Information on plant responses to individual circumstances has been provided by traditional research, which have mostly focused on single-stress situations. With several stresses that might combine in unpredictable ways, field conditions are more dynamic. These interactions may have antagonistic, synergistic, or additive effects, with the end outcome being either the total of the impacts of the individual interactions, increased, or even diminished (Nadeem et al. 2023). The physiological effect of stress is also evident from various studies, for instance the optimal temperature of wheat will be 12 °C–22 °C for flowering and grain filling, but from head emergence to 10 days after anthesis, high temperatures (36 °C/31 °C) at 2-day intervals resulted in sterility of the grain and certain forms of morphological and cell damage that varied according to the grain's developmental stage upon transfer. Grain sterility was brought on by high temperatures 2–3 days prior to anthesis, and excessive humidity greatly accelerated this impact. High temperatures between anthesis and 3 days following anthesis resulted in shriveled, parthenocarpic, and abortion-prone grain. Between 6 and 10 days following anthesis, high temperatures resulted in notch-marked, split, and opaque grain (Tashiro and Wardlaw 1990). The second factor of concern is the effect of modern agricultural practices on soil quality, as discussed previously how proper scientific soil management can contribute to sustainable agriculture, it is clear that our existing practices are not beneficial for soil. Numerous causes contribute to soil degradation worldwide, such as excessive tillage, improper crop rotations, excessive grazing or removal of crop residue, deforestation, mining, building, and urban sprawl. In order to fulfill the demands of a growing world population, humanity must acknowledge and comprehend that the greatest way to mitigate and reverse present trends in soil degradation is to improve soil health via the adoption of sustainable agriculture and land management techniques. The degradation of "the thin layer covering the planet that stands between us and starvation" is consistently indicated by two common factors: soil erosion and the loss of soil organic matter (SOM). Although soil deterioration is not a recent issue, humanity can no longer afford to ignore the various contributing elements (Karlen and Rice 2015). After soil, it is obvious that the concern is for water shortage, crops generally grown are not much water efficient, so "more with less" in agriculture refers to using agricultural water more efficiently, which may be seen from a number of angles. "The cost return on the cost of investing in water supply" would be examined in a financial study of agricultural water consumption and application. The physiological basis for irrigation efficiency is the quantity of total plant growth, or of the harvested yield, per unit volume of water received by the plant from the soil. Efficiency is influenced by a wide range of elements, whether considered from a physiological or economical perspective. Financial efficiency of irrigation is closely reliant on the varying cost and pricing of crops and of water across locations. Physiological efficiency of irrigation is reliant on the plant, the type, and the range of agronomic practices used on the land (Assouline et al. 2015). From the standpoint of irrigation engineers, the main issue at the moment is irrigation efficiency. "The amount of water added to the root zone … as a fraction of the amount taken from some source" is the definition of what they refer to as "water application efficiency" or simply "irrigation efficiency." Since water is in a form that the plant can use when it is administered to its root zone, this type of efficiency is what we are concentrating on. In this regard, evaporation, deep penetration into the soil under the root zone, and runoff from excessive rapid application rates all reduce irrigation effectiveness (Ahmad and Khan 2017). Moving on to the brief account on biotic stress, the problem is evidently about the pest and disease outbreaks. Recent research indicates the need for early warning systems for diseases, early prediction can make the cultural, chemical, and biological control effective. Otherwise, treatment of disease involves so much economic and ecological losses in agriculture. Most of the effective treatments for diseases are generally chemical-based pesticides and insecticides that are a serious threat to the environment. Hence, it will be right to prevent diseases rather than to cure diseases; planting crops early in order to escape the infestation of pathogens, IPM, proper fertilization, and use of biopesticides can be the solution (Brown et al. 2022). Finally, overcoming these stress and climatic factors will only be possible when this issue becomes a concern in the society; many socioeconomic barriers can become a reason for unsolved problems in modern agriculture, the major barriers are gender, age, education, farm size, farming experience, and workforce available. Gender discrimination in agriculture can occur at various levels, ranging from wage disparities based on gender to lack of cooperation among genders, both of which can hinder the effective implementation of new ideas due to a reduced workforce. Additionally, age-related dynamics in the sector are influenced by the population's birth and death rates, with significantly lower participation of the younger generation in agriculture compared to other fields. Additionally, age-related dynamics in the sector are influenced by the population's birth and death rates, with significantly lower participation of the younger generation in agriculture compared to other fields (Apipoonyanon et al. 2021). Third is education, generally most farmers in developing countries are illiterate and this can cause barriers for the access to new technologies provided to them. Practicing agriculture is still considered a low-grade job in many parts of the world, besides it has a great socioeconomic benefit if done in an organized way. Other factors like farm size, experience can also lead to low output; engaging intellectual and healthy population in agriculture can be a solution for all the above issues (Takahashi et al. 2020). There is a need to increase potential farm trainers, who have access to current trends and information in agriculture, eager to learn and teach novel Agri-Tech to small farmers, and lead farmers toward profitable and sustainable farm practices. But

Fig. 10.3 Key sectors requiring Agri-Tech innovation

this can only be done through incentivizing farmer-trainers properly. This will motivate many agriculture graduates to participate and do this as a job (Dar et al. 2019). As discussed in the introductory part of this chapter, Agri-Tech offers numerous values to farmers, which includes increased productivity, better resource management, and improved decision-making, but the main practical issue in effective implementation of all the above discussed Agri-Tech such as Precision agriculture, remote sensing, GIS, and AI and ML tools relies on their adoption at all levels of agriculture. Agri-Tech's development and implementation in the agricultural production chain are hindered by a number of issues (Fig. 10.3). These obstacles can be divided into groups based on political, social, ethical, and economic considerations. Technological obstacles include things like reliability challenges, lack of infrastructure, energy management problems, component incompatibilities, and technical complexity. Economic barriers include things like the lack of easily accessible solutions for farmers and the high cost of skilled personnel, facility maintenance, and operational elements. From an ethical, social, and political perspective, some of the primary challenges are the lack of farm-oriented incentive policies, the limited applicability and accessibility of digital tools, concerns about the economic, ethical, and social implications of new technologies, and the need for comprehensive action plans for digital transformation. Additional obstacles include age group bias, political concerns, insufficient data infrastructure laws, a lack of support for research and business models, educational problems, and a shortage of skilled workers or digital capabilities (da Silveira et al. 2023).

10.4 Key Climate-Smart Crop Management Strategies

10.4.1 Soil Health Enhancement

10.4.1.1 Adoption of Conservation Agriculture to Maintain Soil Structure and Fertility

Conservation agriculture (CA) is a sustainable farming approach that minimizes soil disturbance, maintains permanent soil cover, and promotes crop diversification (Kassam et al. 2019). This system has gained global recognition for enhancing soil health, improving water retention, and increasing long-term agricultural productivity. The three core principles of CA—minimal tillage, cover cropping, and crop rotation—work synergistically to preserve soil structure, enhance organic matter, and reduce erosion (Thierfelder et al. 2017). Conventional tillage disrupts soil aggregates, accelerates organic matter decomposition, and increases erosion risks. In contrast, minimal tillage (or no-till farming) reduces mechanical disturbance, preserving soil porosity and microbial activity (Pittelkow et al. 2015). Studies indicate that no-till systems increase water infiltration rates by up to 30% compared to conventional tillage, reducing surface runoff and nutrient loss (Blanco-Canqui and Ruis 2018). Additionally, reduced tillage enhances carbon sequestration, as less soil disruption slows organic matter decomposition (West and Post 2018). Cover crops, such as legumes and grasses, play a crucial role in CA by protecting soil from erosion, suppressing weeds, and enhancing nutrient cycling (Abdalla et al. 2019). Leguminous cover crops, like clover and vetch, fix atmospheric nitrogen, reducing synthetic fertilizer dependency (Daryanto et al. 2018). Moreover, cover crop residues improve soil organic carbon (SOC), fostering microbial diversity and improving soil structure (McDaniel et al. 2014). Long-term studies demonstrate that cover cropping can increase SOC by 0.2%–0.5% annually, significantly enhancing soil fertility (Poeplau and Don 2015). Crop rotation breaks pest and disease cycles while improving nutrient use efficiency. Diverse rotations, including cereals, legumes, and oilseeds, enhance soil microbial diversity and reduce reliance on chemical inputs (Bowles et al. 2020). Research shows that diversified rotations increase yield stability and improve soil organic matter compared to monocropping (Gaudin et al. 2015). For instance, maize-soybean rotations exhibit higher nitrogen retention than continuous maize systems, reducing leaching losses (Syswerda et al. 2012). Despite its benefits, CA adoption faces barriers such as initial yield variability, weed management challenges, and the need for specialized equipment (Stevenson et al. 2014). However, long-term studies confirm that CA systems eventually match or surpass conventional yields while improving soil resilience (Pittelkow et al. 2015). Future research should focus on region-specific CA adaptations to maximize soil health benefits.

10.4.1.2 Use of Organic Amendments Like Compost and Biochar to Improve Soil Carbon Content

Organic amendments, including compost and biochar, are vital for enhancing soil carbon content, improving fertility, and mitigating climate change (Lehmann et al. 2015). These amendments restore degraded soils, increase water-holding capacity, and promote microbial activity, making them essential for sustainable agriculture. Compost, derived from decomposed organic waste, enriches soil with humus, nutrients, and beneficial microorganisms (Diacono and Montemurro 2015). Regular compost application increases SOC, enhancing soil aggregation and reducing compaction (Oldfield et al. 2018). Studies show that compost-amended soils exhibit 10%–20% higher carbon sequestration rates than unamended soils (Ryals and Silver 2018). Additionally, compost improves nutrient availability, reducing synthetic fertilizer requirements (Mondini et al. 2018). Biochar, a carbon-rich material produced through pyrolysis, is highly stable in soil, resisting decomposition for centuries (Wang et al. 2020). Its porous structure enhances water retention, nutrient exchange, and microbial habitats (Jeffery et al. 2017). Research indicates that biochar application can increase SOC by 20%–50% over a decade, significantly contributing to climate change mitigation (Woolf et al. 2016). Furthermore, biochar reduces greenhouse gas emissions by suppressing nitrous oxide (N_2O) release from soils (Cayuela et al. 2014). Combining compost and biochar maximizes soil health benefits. Biochar's adsorption capacity retains nutrients from compost, preventing leaching and enhancing long-term fertility (Schmidt et al. 2019). Field trials demonstrate that co-application increases crop yields by 15%–30% compared to single amendments (Agegnehu et al. 2016). Despite their advantages, large-scale adoption faces challenges such as high production costs, variability in feedstock quality, and lack of farmer awareness (Tan et al. 2017). Future research should optimize amendment formulations for different soil types and cropping systems.

10.4.2 Water Resource Management

10.4.2.1 Implementation of Efficient Irrigation Systems to Optimize Water Use

Water scarcity and inefficient irrigation practices pose significant challenges to global agriculture. The adoption of efficient irrigation systems, such as drip and sprinkler irrigation, has emerged as a sustainable solution to optimize water use, enhance crop productivity, and reduce wastage (Playán and Mateos 2019). These systems ensure precise water delivery, minimize evaporation losses, and improve nutrient use efficiency, making them essential for modern agriculture.

Drip Irrigation

Drip irrigation delivers water directly to the plant root zone through a network of pipes, emitters, and valves, significantly reducing water loss due to evaporation and runoff (Sivanappan 2018). Studies indicate that drip irrigation can improve water use efficiency (WUE) by 30%–60% compared to conventional flood irrigation (Pereira

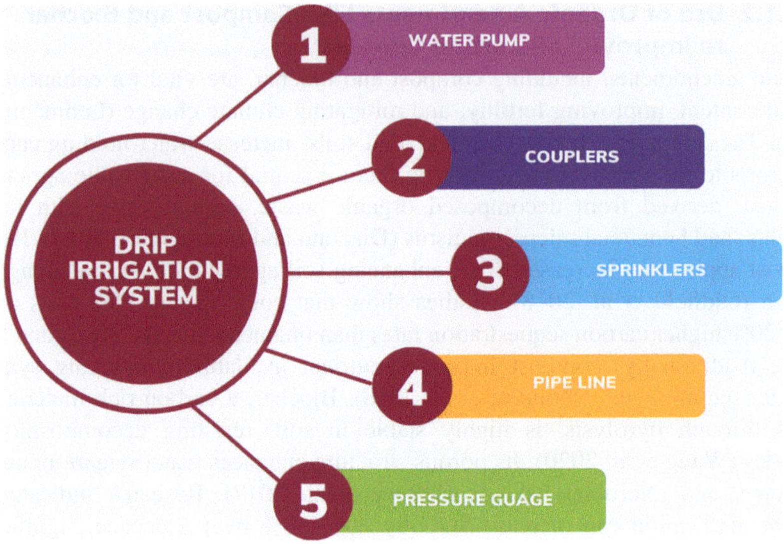

Fig. 10.4 Components of drip irrigation system

et al. 2020). This method is particularly beneficial for high-value crops, such as fruits and vegetables, where controlled moisture levels enhance yield and quality (Kulkarni et al. 2017). Additionally, drip irrigation facilitates fertigation—the simultaneous application of water and fertilizers—which improves nutrient uptake and reduces leaching losses (Singandhupe et al. 2016). Research in arid regions has demonstrated that drip-irrigated fields exhibit 20%–40% higher crop yields while using 50% less water (Darouich et al. 2017). However, challenges such as clogging of emitters, high initial costs, and the need for skilled maintenance limit widespread adoption in smallholder farming systems (Fig. 10.4; Yohannes and Ritsema 2018).

Sprinkler Irrigation

Sprinkler irrigation mimics natural rainfall by distributing water through pressurized nozzles, making it suitable for a wide range of crops and soil types (Zhao et al. 2019). This system is particularly effective in areas with uneven terrain where surface irrigation is impractical (Martínez-Romero et al. 2018). Studies show that sprinkler irrigation can reduce water consumption by 25%–50% compared to traditional flood methods while ensuring uniform moisture distribution (Smith et al. 2016). Modern advancements, such as low-pressure sprinklers and automated scheduling, further enhance efficiency by minimizing energy use and runoff (Dechmi et al. 2020). However, wind drift and evaporation losses remain significant drawbacks, particularly in hot and windy climates (Pardo et al. 2019). Despite these limitations, sprinkler systems are widely adopted in large-scale farming due to their scalability and adaptability to different cropping systems (Fig. 10.5). Both drip and sprinkler irrigation systems offer distinct advantages depending on crop type, soil conditions, and climatic factors. While drip irrigation excels in water-scarce regions with high-

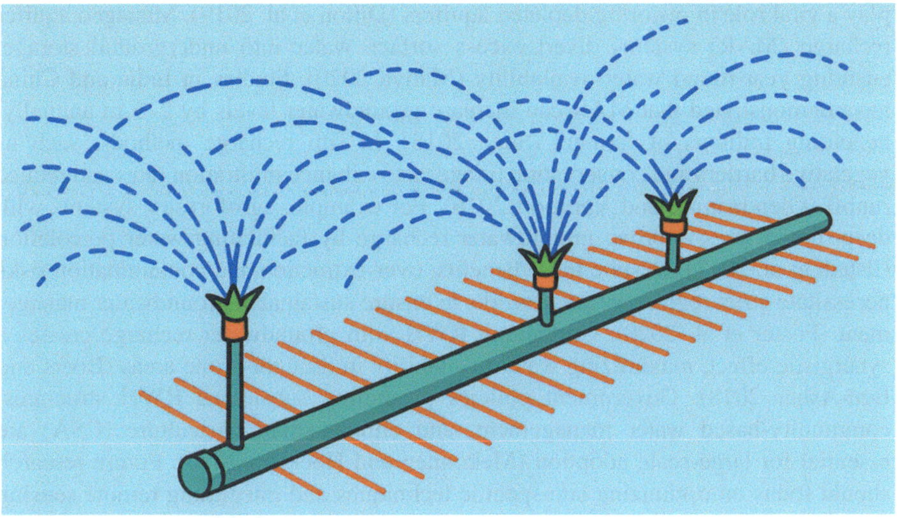

Fig. 10.5 Sprinkler irrigation system

value crops, sprinkler systems are more versatile for field crops like wheat and maize (Jat et al. 2016). Future research should focus on cost-effective innovations, such as solar-powered irrigation and sensor-based automation, to enhance accessibility for small-scale farmers (Narayanamoorthy 2020).

10.4.2.2 Rainwater Harvesting and Groundwater Recharge Techniques for Drought-Prone Areas

Rainwater harvesting (RWH) and groundwater recharge techniques are critical for mitigating water scarcity in drought-prone regions. These practices capture and store rainfall runoff, replenish aquifers, and enhance water availability for agricultural and domestic use (Rockström et al. 2018). With increasing climate variability, RWH systems have gained prominence as sustainable solutions to water stress. RWH techniques include rooftop collection, surface ponds, and check dams that store rainwater for irrigation and livestock (Kahinda et al. 2019). In arid and semiarid regions, small-scale RWH systems have been shown to increase crop yields by 30%–50% by extending growing seasons (Mwenge Kahinda et al. 2017). For instance, in sub-Saharan Africa, farm ponds and contour bunds have significantly improved maize and sorghum production during dry spells (Ngigi et al. 2016). Micro-catchment systems, such as pits and half-moon basins, are particularly effective in enhancing soil moisture retention in degraded lands (Reij et al. 2018). These low-cost interventions not only improve water availability but also reduce soil erosion and enhance groundwater recharge (Biazin et al. 2019). However, challenges such as siltation, high evaporation losses, and limited storage capacity hinder their long-term sustainability (Lasage et al. 2015). Artificial groundwater recharge techniques, including percolation tanks, recharge wells, and infiltration galleries, play a vital role in restoring depleted aquifers (Dillon et al. 2019). Managed aquifer recharge (MAR) systems divert excess surface water into underground storage, ensuring year-round water availability (Maliva 2020). Studies in India and China have demonstrated that MAR can increase groundwater levels by 2–5 m annually, benefiting millions of farmers (Shah 2018). In situ recharge methods, such as vegetative barriers and conservation tillage, also enhance infiltration rates and reduce runoff (Glendenning and Vervoort 2016). For example, agroforestry systems with deep-rooted trees improve groundwater recharge by facilitating water percolation (Ilstedt et al. 2017). Despite these benefits, over-extraction and contamination risks necessitate strict regulatory frameworks to ensure sustainable groundwater management (Foster et al. 2020). Combining RWH with groundwater recharge creates a synergistic effect, maximizing water availability in drought-prone areas (Boers and Ben-Asher 2018). Government policies promoting subsidized RWH structures, community-based water management, and climate-smart agriculture (CSA) are essential for large-scale adoption (Mekonnen and Hoekstra 2016). Future research should focus on optimizing site-specific techniques and integrating remote sensing for real-time monitoring of water resources (Karimi et al. 2018).

10.4.3 Crop Diversification and Climate-Resilient Varieties

10.4.3.1 Introduction of Drought-Tolerant, Heat-Resistant, and Pest-Resistant Crop Varieties

Climate change has intensified abiotic and biotic stresses on agricultural systems, threatening global food security. The development and adoption of climate-resilient crop varieties—including drought-tolerant, heat-resistant, and pest-resistant cultivars—have become crucial for sustainable agriculture (Lobell et al. 2015). These improved varieties help maintain productivity under extreme weather conditions while reducing dependency on chemical inputs (Ceccarelli et al. 2017). Drought-tolerant crops, developed through conventional breeding and biotechnology, exhibit enhanced water-use efficiency and deeper root systems (Vadez et al. 2018). Sorghum and millet varieties, for instance, have been successfully bred for arid regions, yielding 20%–40% more than traditional cultivars under water-limited conditions (Tardieu et al. 2018). Molecular markers have accelerated the identification of drought-responsive genes in staple crops like maize and wheat (Cattivelli et al. 2016). Field studies in sub-Saharan Africa demonstrate that drought-tolerant maize varieties can increase yields by 25% compared to local varieties during dry seasons (Fisher et al. 2015). However, farmer adoption remains limited by seed accessibility and affordability (Kassie et al. 2017). Public-private partnerships are essential to disseminate these varieties effectively (Langyintuo et al. 2018). With global temperatures projected to increase by 1.5 °C–2 °C, heat-resistant varieties are critical for maintaining yields (Asseng et al. 2015). Wheat cultivars with improved thermotolerance, developed through physiological breeding, show 15%–30% higher yields under heat stress (Reynolds et al. 2017). Similarly, heat-tolerant rice varieties like ‘Nagina 22’ have sustained productivity at temperatures above 35 °C (Jagadish et al. 2016). Challenges persist in breeding crops with simultaneous tolerance to multiple stresses (Mickelbart et al. 2015). Future research should integrate phenomics and genomic selection to accelerate cultivar development (Araus et al. 2018). Bt crops and host-plant resistance mechanisms have significantly reduced pesticide use while maintaining yields (Tabashnik et al. 2016). For example, Bt cotton has decreased insecticide applications by 40% in India (Kathage and Qaim 2016). However, pest resistance evolution necessitates continuous innovation in breeding strategies (Carrière et al. 2015).

10.4.3.2 Intercropping and Agroforestry Systems to Diversify Income Sources and Improve Ecosystem Resilience

Crop diversification through intercropping and agroforestry is a sustainable agricultural practice that enhances ecosystem resilience while supporting farmers' economic stability (Lin 2016). These systems promote biodiversity, optimize resource use, and improve long-term productivity compared to conventional monocropping (Table 10.1). Intercropping and agroforestry systems maximize the use of available resources such as light, water, and nutrients by combining crops with complementary growth patterns and root structures (Brooker et al. 2016). For example, deep-rooted plants can access nutrients from lower soil layers, while shallow-rooted species utilize surface resources, reducing competition. Nitrogen-fixing legumes intercropped with cereals enhance soil fertility, reducing the need for synthetic fertilizers. Diversified cropping systems enhance soil structure, organic matter content, and microbial activity (Brooker et al. 2016). The varied root exudates from different plant species support a more diverse soil microbiome, which improves nutrient cycling and suppresses soil-borne pathogens. Agroforestry systems, where

Table 10.1 Features of intercropping and agroforestry

Feature	Intercropping	Agroforestry
Definition	Growing two or more crops simultaneously on the same field	Integrating trees/shrubs with crops and/or livestock
Primary goal	Maximize land use efficiency, reduce pests, and improve yield stability	Enhance biodiversity, soil health, and long-term sustainability
Income diversification	Short-term income from multiple crops in a single season	Multiple revenue streams (timber, fruits, crops, livestock)
Ecosystem benefits	Improves soil fertility, reduces erosion, and disrupts pest cycles	Enhances carbon sequestration, biodiversity, and water retention
Timeframe for returns	Short to medium term (seasonal crop yields)	Long-term (tree products take years but provide sustained income)
Complexity and labor	Moderate (requires crop compatibility knowledge)	High (managing trees + crops/livestock needs planning)
Examples	Maize + beans, wheat + chickpeas	Silvopasture (trees + livestock), alley cropping
Risk mitigation	Spreads risk across crops but still vulnerable to market/weather	More resilient due to diversified and perennial components
Climate resilience	Moderate (depends on crop selection)	High (trees provide windbreaks, shade, and microclimate regulation)

trees are integrated with crops, contribute to reduced soil erosion and increased carbon sequestration, further enhancing long-term soil productivity. Intercropping and agroforestry modify the microclimate by providing shade, reducing wind speed, and maintaining higher humidity levels (Brooker et al. 2016). This buffering effect protects crops from extreme temperatures and water stress, improving yield stability under climate variability. For instance, shade from trees in agroforestry systems can reduce heat stress for understory crops, while intercropping tall and short plants can minimize soil moisture loss. By diversifying production, farmers reduce reliance on a single crop, spreading economic risks (Lin 2016). Intercropping and agroforestry can provide multiple harvests from the same land, ensuring more stable income streams. Additionally, reduced dependency on external inputs (e.g., fertilizers and pesticides) lowers production costs, improving profitability in the long term. Cereal-legume intercropping systems demonstrate improved nitrogen fixation and yield stability (Li et al. 2016). The maize-bean system in East Africa yields 30% higher land equivalent ratios than monocultures (Ojiem et al. 2017). Similarly, push-pull technology using desmodium and Napier grass reduces stem borer infestation in maize by 80% (Khan et al. 2016). Silvopastoral and alley cropping systems sequester 2–5 Mg C/ha/year while providing fodder and timber (Nair et al. 2017). In Central America, coffee agroforestry systems maintain 60% higher biodiversity than sun-grown plantations (Perfecto et al. 2016).

10.4.4 Integrated Pest and Nutrient Management

10.4.4.1 Use of Biological Pest Control Methods to Reduce Dependency on Chemical Pesticides

The increasing reliance on chemical pesticides in agriculture has led to numerous environmental and health concerns, including pesticide resistance, soil degradation, and water contamination (Pretty and Bharucha 2015). To mitigate these issues, biological pest control (BPC) has emerged as a sustainable alternative, leveraging natural predators, parasitoids, and microbial agents to manage pest populations. This review explores the efficacy of BPC methods, their advantages over synthetic pesticides, and their role in integrated pest management (IPM) systems. Biological control reduces chemical pesticide use, minimizing ecological disruption. Natural enemies such as *Trichogramma* wasps, lady beetles (*Coccinellidae*), and entomopathogenic fungi (*Beauveria bassiana*) effectively suppress pests like aphids, caterpillars, and whiteflies without harming nontarget species (Bale et al. 2008). Unlike chemical pesticides, which often lead to pest resistance, BPC maintains long-term efficacy by fostering ecological balance (Van Lenteren et al. 2018). Microbial biopesticides, including *Bacillus thuringiensis* (Bt), have been widely adopted due to their specificity and biodegradability (Glare et al. 2012). Studies indicate that Bt-based formulations reduce *Helicoverpa armigera* infestations in cotton by 60%–80%, with minimal impact on pollinators (Romeis et al. 2019). Similarly, nematodes (*Steinernema* spp.) control soil-dwelling pests like root borers, enhancing crop yields sustainably (Shapiro-Ilan et al. 2017). Despite its benefits, BPC faces challenges such as slower action compared to synthetic pesticides and susceptibility to environmental conditions (Eilenberg et al. 2018). Temperature, humidity, and UV exposure can reduce the viability of biocontrol agents, necessitating optimized application techniques (Lacey et al. 2015). Additionally, farmer awareness and training remain critical, as improper use can limit effectiveness (Wyckhuys et al. 2018). Combining BPC with cultural practices (crop rotation, intercropping) and pheromone traps enhances pest suppression (Pretty and Bharucha 2015). For instance, push-pull systems in East Africa, using *Desmodium* as an intercrop, reduce *Striga* weed and stem borer infestations while improving soil fertility (Khan et al. 2014). Such integrated approaches demonstrate the potential of BPC in sustainable agriculture.

10.4.4.2 Precision Agriculture Tools to Apply Fertilizers Efficiently, Reducing Environmental Runoff

Excessive fertilizer use in conventional farming has led to nutrient leaching, groundwater contamination, and eutrophication (Zhang et al. 2015). Precision agriculture (PA) technologies, including soil sensors, variable rate applicators (VRA), and remote sensing, optimize fertilizer use, minimizing environmental impact while enhancing crop yields. This review examines PA tools, their role in nutrient management, and their contribution to sustainable farming.

Soil and Crop Sensors

Real-time soil sensors measure nitrogen (N), phosphorus (P), and potassium (K) levels, enabling site-specific fertilizer application (Adamchuk et al. 2018). Optical sensors, such as greenseeker, assess crop nitrogen requirements by analyzing leaf reflectance, reducing N application by 20%–30% without compromising yield (Raun et al. 2016).

- *Precision Application*: Avoids over-fertilization by targeting deficient zones, reducing nutrient runoff and leaching.
- *Cost Savings*: Lowers fertilizer expenses by preventing unnecessary applications.
- *Improved Soil Health*: Prevents nutrient imbalances and long-term soil degradation.
- *Reduced Nitrogen Use*: Studies show that optical sensor-based recommendations can reduce N application by *20%–30%* without sacrificing yield (Raun et al. 2016).
- *Dynamic Adjustments*: Allows in-season N adjustments based on real-time crop health data.
- *Higher Efficiency*: Ensures optimal N utilization, minimizing losses due to volatilization or leaching.

Variable Rate Technology (VRT)

VRT adjusts fertilizer application based on spatial variability within fields. GPS-guided spreaders and drip irrigation systems deliver nutrients precisely, minimizing wastage (Gebbers and Adamchuk 2017). Studies in maize fields showed a 15% reduction in nitrogen runoff with VRT compared to uniform application (Miao et al. 2016)

- *Reduced Nutrient Wastage*: By applying fertilizers only where needed, VRT minimizes excess use, lowering costs and preventing pollution.
- *Improved Crop Performance*: Tailored nutrient application enhances yield potential in deficient areas while avoiding over-fertilization in high-fertility zones.
- *Environmental Protection*: Studies in *maize fields* demonstrated a *15% reduction in nitrogen runoff* with VRT compared to uniform application methods (Miao et al. 2016).

Remote Sensing and Drones

Satellite and drone-based multispectral imaging detect nutrient deficiencies early, allowing corrective measures before yield loss occurs (Zhang and Kovacs 2019). For example, NDVI (Normalized Difference Vegetation Index) maps help farmers apply fertilizers only where needed, reducing excess use (López-Granados et al. 2018)

- Farmers generate NDVI maps to identify low-vigor zones within fields.
- Fertilizer is applied only where needed, reducing excess use and minimizing environmental runoff.
- Studies show this approach can cut fertilizer costs by 10%–25% while maintaining or improving yields.

Remote sensing and drone technology enable early, precise detection of nutrient stress, facilitating smarter fertilizer management. By leveraging NDVI and multispectral imaging, farmers optimize inputs, cut costs, and enhance sustainability (Zhang and Kovacs 2019; López-Granados et al. 2018).

10.4.5 Digital and Technological Innovations

10.4.5.1 Leveraging Climate Information Services and Mobile-Based Advisory Systems to Support Timely Decision-Making

Climate variability poses significant risks to agricultural productivity, making timely decision-making crucial for farmers (Hansen et al. 2019). Climate information services (CIS) and mobile-based advisory systems have emerged as vital tools, providing real-time weather forecasts, pest alerts, and crop management recommendations (Table 10.2). Mobile platforms, such as SMS alerts and interactive voice response (IVR), deliver actionable climate and agronomic advice to smallholder farmers (Aker 2017). In India, the *mKisan* portal sends weather-based sowing advisories, helping farmers reduce water use by 20% while improving yields (Gakhar and Mehla 2020). Similarly, *Esoko* in Africa provides market prices and weather updates, reducing post-harvest losses by 15% (Fafchamps and Minten 2018).

Table 10.2 Overview of emerging Agri-Tech solutions, their benefits, challenges, and real-world applications

Technology	Benefits	Challenges	Example use cases
IoT-enabled sensors	Precision input application Reduced water/ fertilizer waste Improved yield predictability	High initial costs Sensor calibration/ maintenance Data security risks	Smart soil probes (e.g., Teralytic) Automated drip irrigation systems
AI-driven predictive analytics	Early risk detection Data-driven decision-making Reduced input costs	Requires large datasets Model interpretability issues Dependence on data quality	IBM Watson Decision Platform Climate FieldView AI insights
Blockchain for traceability	Enhanced transparency Reduced food fraud Faster recalls if contamination occurs	Scalability limitations High energy consumption (PoW) Adoption resistance in traditional systems	Walmart's leafy greens tracking IBM Food Trust blockchain network

10.4.5.2 Use of IoT-Enabled Sensors, AI-Driven Predictive Analytics, and Blockchain for Supply Chain Traceability

IoT-based soil moisture and weather sensors enable real-time monitoring, optimizing irrigation and reducing water wastage (Bwambale et al. 2022). For example, *Tensiometers* linked to mobile apps notify farmers when crops need watering, improving water-use efficiency by 30% (O'Shaughnessy et al. 2019). Drones equipped with thermal cameras detect drought stress early, allowing corrective measures before yield losses occur (Zhang and Kovacs 2019). Machine learning models analyze historical climate data to predict droughts, floods, and pest outbreaks (Liakos et al. 2018). *IBM's Watson Decision Platform* uses AI to recommend optimal planting dates, reducing climate-related risks in maize farming (Basso et al. 2016). Similarly, *PlantVillage Nuru*, an AI-powered app, diagnoses crop diseases via smartphone images, helping farmers take preventive actions (Ramcharan et al. 2020). Blockchain technology enhances traceability in agri-food supply chains, ensuring fair pricing and reducing fraud (Kamilaris et al. 2019). Projects like *IBM Food Trust* track produce from farm to fork, improving market access for smallholders (Tian 2020).

10.5 Implementation Framework for Climate-Smart Strategies

10.5.1 Community Involvement

Effective implementation of Climate-Smart Crop Management Strategies (CSCMS) requires active participation from farming communities to ensure solutions are tailored to local contexts (Chandra et al. 2017). Participatory approaches have proven critical in developing appropriate technologies that align with farmers' needs and indigenous knowledge systems. Farmer field schools (FFS) have emerged as a successful model for community engagement, particularly in Asia and Africa. In East Africa, the adoption of FFS approaches increased the uptake of drought-tolerant crop varieties by 35% compared to traditional extension methods (Nyasimi et al. 2017). These schools create platforms for experiential learning where farmers can test and adapt climate-smart practices under local conditions. The participatory varietal selection (PVS) approach in South Asia demonstrated that involving farmers in crop breeding programs led to 28% higher adoption rates of improved varieties (Tiwari et al. 2018). Community-based organizations (CBOs) play a pivotal role in scaling climate-smart agriculture. In Nicaragua, farmer cooperatives successfully implemented agroforestry systems that combined food crops with timber species, resulting in 40% higher incomes while enhancing carbon sequestration (Matocha et al. 2018). Similarly, watershed management committees in India's semiarid regions have enabled communities to collectively address water scarcity through rainwater harvesting structures (Barron and Noel 2018). However, challenges remain in ensuring equitable participation. Gender disparities often limit women's involvement in decision-making processes, despite their crucial role in agricultural production (Njuki et al. 2016). Successful programs like Kenya's Climate-Smart Villages have addressed this by establishing women-led farmer groups that specifically focus on climate-resilient kitchen gardens (Otzelberger 2018).

10.5.2 Capacity Building

Capacity development forms the backbone of successful CSCMS implementation, requiring comprehensive training programs that address both technical skills and knowledge gaps (Singh et al. 2018). Effective capacity building combines formal education, vocational training, and on-farm demonstrations to create a multiplier effect in rural communities. Extension services have evolved significantly to incorporate climate-smart messaging. The "Training of Trainers" (ToT) model in Ethiopia reached over 50,000 farmers with conservation agriculture techniques, resulting in 30% yield improvements for smallholders (Tesfaye et al. 2017). Digital platforms have augmented traditional extension methods; the Digital Green initiative in India used video-based learning to disseminate climate-smart practices to 1.2 million farmers (Gandhi et al. 2018). Vocational training programs have shown particular success with youth engagement. Rwanda's Farmer Field Schools for Young Entrepreneurs reported 65% of participants establishing climate-smart agribusinesses within 2 years (Rugege et al. 2019). Similarly, weather-indexed insurance training in Senegal improved farmers' risk management capacities, with insured farmers investing 25% more in climate-adaptive technologies (Carter et al. 2018). The integration of indigenous knowledge with scientific approaches has enhanced training effectiveness. In Bolivia, combining traditional forecasting methods with meteorological data improved farmers' climate prediction accuracy by 40% (Valdivia et al. 2017). However, persistent challenges include high trainerto-farmer ratios and the need for continuous skill upgrading as technologies evolve (Kristjanson et al. 2017).

10.5.3 Policy Support

Institutional frameworks and policy instruments are essential to create enabling environments for CSCMS adoption (Westermann et al. 2018). Effective policy integration requires alignment across agricultural, environmental, and economic sectors at multiple governance levels. Subsidy reforms have demonstrated significant impacts on sustainable practice adoption. India's policy shift from urea subsidies to soil health card-based nutrient management reduced fertilizer overuse by 18% while maintaining yields (Kumar et al. 2018). Similarly, Malawi's input subsidy program incorporated climate-smart packages, resulting in 35% higher maize productivity under drought conditions (Chibwana et al. 2017). National adaptation plans (NAPs) have emerged as comprehensive policy tools. Vietnam's NAP allocated 30% of agricultural R&D funding to climate-smart technologies, accelerating the development of saline-tolerant rice varieties (Trinh et al. 2018). Payment for ecosystem services (PES) schemes in Costa Rica have conserved 1.2 million hectares of farmland while increasing carbon stocks (Pagiola et al. 2016). However, policy implementation often faces challenges. In Brazil, conflicting agricultural and environmental policies created adoption barriers for integrated crop-livestock-forestry systems (Gil et al. 2017). Monitoring and evaluation systems remain weak in many countries, with only 35% of climate-smart policies having measurable indicators (Torquebiau et al. 2018).

10.5.4 Infrastructure Development

Climate-resilient infrastructure forms the physical foundation for effective CSCMS implementation, requiring investments in irrigation, storage, and transportation systems (Williams et al. 2018). Strategic infrastructure development can significantly reduce post-harvest losses and improve market access for climate-smart produce. Water management infrastructure has shown particularly high returns. Solar-powered drip irrigation in Niger increased vegetable production by 300% while reducing water use by 60% (Burney et al. 2017). In Bangladesh, community-managed flood shelters protected seed stocks during extreme weather events, enabling 90% faster post-disaster recovery (Rabbani et al. 2018). Cold chain development is critical for perishable commodities. India's Kisan Sampada scheme reduced fruit and vegetable post-harvest losses from 40% to 18% through integrated cold chain networks (Minten et al. 2018). Similarly, hermetic storage technologies in East Africa decreased grain storage losses from 30% to less than 2% (Tefera et al. 2017). Rural road networks have multiplier effects on CSA adoption. Ethiopia's road expansion program increased fertilizer use efficiency by 25% through improved market access (Stifel and Minten 2017). However, infrastructure gaps remain significant, with only 30% of sub-Saharan Africa's rural population having access to all-season roads (AUC 2018).

10.6 Benefits of Climate-Smart Crop Management

10.6.1 Economic Resilience

Climate-Smart Crop Management Strategies (CSCMS) have demonstrated significant potential to enhance economic stability for farming communities facing climate variability (Lipper et al. 2018). By integrating adaptive technologies with traditional farming practices, these systems provide a buffer against climate shocks while improving agricultural productivity. Field studies across multiple agro-ecological zones reveal that CSCMS adoption correlates strongly with increased farm incomes and reduced vulnerability to weather extremes. Improved yields constitute the most direct economic benefit of climate-smart practices. Conservation agriculture techniques, including minimum tillage and crop residue retention, have shown yield increases of 15%–25% for staple crops in semiarid regions (Thierfelder et al. 2017). In Zambia, the combination of drought-tolerant maize varieties with microdosing fertilization techniques boosted yields by 30%–50% during below-average rainfall seasons (Ngoma et al. 2018). Similarly, precision nutrient management systems in India's rice-wheat cropping systems reduced input costs by 20% while maintaining equivalent production levels (Jat et al. 2019). Risk mitigation represents another critical economic advantage. Weather-indexed insurance products linked to CSCMS have reduced income volatility for smallholders by 35%–40% in vulnerable regions (Greatrex et al. 2018). In Senegal, farmers adopting climate-smart millet production systems experienced 60% lower crop losses during drought years compared to conventional approaches (Sultan et al. 2019). These systems also demonstrate strong potential for value addition—in Kenya, climate-smart certification of French beans increased export prices by 22% while reducing rejection rates (Gambo et al. 2020). The economic resilience benefits extend beyond farm gates. CSCMS creates employment opportunities in rural areas, particularly through climate-smart value chains. A study in Vietnam documented 15% higher rural wages in areas with widespread adoption of climate-smart rice production systems (Trinh et al. 2018). However, realization of these economic benefits requires overcoming initial investment barriers through innovative financing mechanisms and targeted subsidies (Westermann et al. 2018).

10.6.2 Environmental Sustainability

The environmental benefits of CSCMS span multiple dimensions of ecosystem health, offering solutions to some of agriculture's most pressing sustainability challenges (Campbell et al. 2018). These systems demonstrate particular efficacy in addressing soil degradation, biodiversity loss, and greenhouse gas emissions—

three critical indicators of agricultural environmental performance. Soil health improvement emerges as a consistent outcome across diverse CSCMS implementations. Integrated soil fertility management practices in sub-Saharan Africa have increased soil organic carbon by 0.3%–0.5% annually while reducing erosion rates by 40%–60% (Vanlauwe et al. 2017). Agroforestry systems in Central America show even more dramatic results, with 1.2%–1.8% annual increases in soil carbon stocks (Somarriba et al. 2018). These improvements translate directly into enhanced water retention capacity—studies in India's semiarid regions document 25%–30% improvements in soil moisture retention under climate-smart practices (Jat et al. 2018). Biodiversity conservation represents another significant environmental benefit. Diversified cropping systems incorporating legumes and cover crops support 30%–50% higher pollinator populations compared to monocultures (Garibaldi et al. 2019). In Brazil, integrated crop-livestock-forestry systems have doubled avian biodiversity indices while maintaining agricultural productivity (Gil et al. 2018). These biodiversity enhancements create positive feedback loops, with improved pollination and natural pest control further boosting system resilience. Greenhouse gas mitigation potential varies by practice but can be substantial. Alternate wetting and drying in rice systems reduces methane emissions by 30%–50% without yield penalties (Sander et al. 2017). At landscape scales, climate-smart grazing management has demonstrated potential to sequester 0.5–1.5 t CO_2-equivalent per hectare annually (Ayantunde et al. 2018). However, realizing these environmental benefits at scale requires addressing knowledge gaps among farmers and aligning incentives with conservation objectives (Rosenstock et al. 2019).

10.6.3 Social Equity

The social dimensions of CSCMS implementation have gained increasing recognition, particularly regarding their potential to empower marginalized groups and reduce rural inequalities (Kristjanson et al. 2017). These systems demonstrate unique capacity to address gender disparities, youth unemployment, and indigenous community marginalization through inclusive design and targeted interventions. Gender equity improvements emerge across multiple impact pathways. Climate-smart technologies that reduce labor burdens disproportionately benefit women farmers—mechanized planting equipment in Ethiopia decreased women's drudgery by 35% while increasing planting efficiency (Teka et al. 2019). In Nepal, women's groups managing community seed banks improved access to climate-resilient varieties, resulting in 25% higher adoption rates among female-headed households (Paudel et al. 2018). However, persistent cultural barriers require addressing—only 15% of extension services in developing countries specifically target women farmers (Ragasa et al. 2019). Youth engagement represents another critical social benefit. Climate-smart agribusiness incubators in Rwanda have enabled 40% of participating youth to establish viable enterprises, reversing rural-urban migration trends (Rugege et al. 2019). Digital platforms particularly resonate with younger farmers—in Nigeria, the use of climate information apps increased by 300% among farmers aged 18–35 (Adekunle et al. 2018). These interventions simultaneously address unemployment and knowledge transfer challenges in aging farming populations. Indigenous knowledge integration strengthens both social equity and climate adaptation outcomes. Participatory research with Quechua farmers in Peru yielded potato varieties with 20% better drought tolerance than commercial alternatives (Scurrah et al. 2019). Similar collaborations with Maori communities in New Zealand have improved the climate resilience of traditional food systems (Lambert et al. 2018). These approaches validate traditional knowledge while creating economic opportunities for indigenous populations. The equity impacts extend to broader community resilience. In Bangladesh, climate-smart villages reported 40% faster recovery from extreme weather events compared to conventional villages (Islam et al. 2020). However, maximizing these social benefits requires intentional program design that addresses structural inequalities and power dynamics within rural communities (Taylor 2018).

10.7 Challenges in Adoption and How to Overcome Them

While Climate-Smart Crop Management Strategies (CSCMS) offer transformative potential for sustainable agriculture, their adoption on a wide scale remains limited. This disparity is particularly visible in developing countries, where smallholder farmers face multiple socioeconomic and infrastructural barriers. Bridging this adoption gap is not merely about disseminating new technology, but it also requires a systemic understanding of ground realities, inclusive policy making, and grassroots engagement. This section of the chapter delves into the main challenges hindering the adoption of CSCMS and explores actionable solutions to address these challenges (Fig. 10.6).

10.7.1 Financial Constraints

A significant barrier to adopting CSCMS is the high initial investment required. Technologies such as precision irrigation systems, soil health sensors, and resilient seed varieties often entail costs that are unaffordable for small and marginal farmers. According to the Food and Agriculture Organization (FAO), over 70% of farmers in Sub-Saharan Africa and South Asia operate on less than two hectares of land and have limited access to formal credit institutions (FAO 2021). Adopting climate-smart agriculture (CSA) technologies such as laser land levelers and happy seeders is widely recognized for its potential to enhance sustainability, improve resilience, and mitigate emissions in agriculture. However, one of the most significant barriers to their uptake is the high upfront cost, particularly for smallholder farmers. Villalba et al. (2024) found that despite promising results, such as a 20% increase in profitability and a 78% reduction in emissions from using the Happy Seeder, the actual ownership rates remain low (only 7% for laser land levelers and 21% for Happy Seeders) among farmers in Haryana. Instead, many rely on Custom Hiring

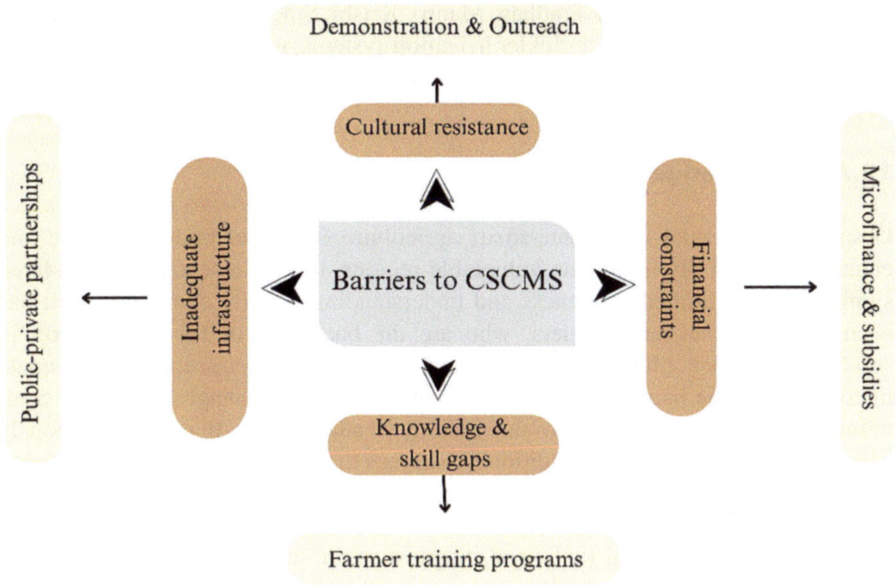

Fig. 10.6 Barriers and solutions in Climate-Smart Crop Management Strategies (CSCMS)

Centers (CHCs) due to the prohibitive costs of purchasing these machines. The study revealed that while 77% of farmers used laser land levelers and 52% used happy seeders, the vast majority accessed these through rental services, not ownership. The preference for hiring over purchase stems from the capital-intensive nature of these technologies, which makes upfront investment unattainable for many (Villalba et al. 2024). Similarly, the cost of implementing some of these (CSA) strategies remains prohibitively high, particularly for small-scale farmers who lack the financial capacity or access to affordable credit mechanisms (Kumar et al. 2024). Even when subsidies or incentives are available, the lack of timely disbursement and bureaucratic hurdles in financial systems further deters farmers from investing in CSA. This economic constraint is further exacerbated by the fact that many benefits of CSA technologies, such as improved soil health, water efficiency, or yield stabilization, are often realized in the medium to long term, whereas costs are immediate and substantial. This mismatch in cost-benefit timing presents a significant psychological and practical hurdle to adoption. In response to these challenges, indirect finance pathways, such as loans from family, savings, or informal moneylenders, are often preferred over formal credit institutions, which are perceived as slow and inaccessible due to bureaucratic procedures (Villalba et al. 2024). Yet, such informal avenues are often insufficient to cover the cost of expensive machinery, making widespread adoption of CSA technologies limited and inequitable. These financial constraints are exacerbated by the perceived risks of investing in unfamiliar technologies. The fear of incurring losses further discourages adoption. Government subsidies, low-interest agricultural loans, and microfinance initiatives can be instrumental here. For instance, India's Pradhan Mantri Krishi Sinchayee Yojana (PMKSY) offers subsidies on drip and sprinkler irrigation systems, making them more accessible to resource-poor farmers (Government of India 2020).

10.7.2 Knowledge Gaps

Despite the potential of climate-smart agriculture (CSA) to build resilience and enhance productivity in climate-vulnerable regions, its implementation is hindered significantly by limited awareness and understanding in rural communities. It has been observed that smallholders, who are the backbone of food production in sub-Saharan Africa, often rely on traditional knowledge systems that are inadequate in coping with the increased extremes brought by climate change. Collective community knowledge is limited to anticipating weather events that have previously been experienced, but not the additional extremes that climate change is predicted to bring (Jellason et al. 2020). Even when technologies are financially accessible, a lack of knowledge and technical skills can impede their implementation. Many farmers are unfamiliar with digital tools, soil diagnostics, or sustainable nutrient management practices. A study by the International Food Policy Research Institute (IFPRI) found that less than 30% of surveyed farmers in rural India had access to reliable agricultural extension services (IFPRI 2020). The study, conducted in Zango and Kofa communities in North-Western Nigeria, highlights how knowledge gaps around CSA practices remain a critical barrier. While participatory learning and action (PLA) models increased farmer confidence in specific CSA strategies like water and fertilizer use, the initial lack of awareness and technical understanding limited their ability to engage meaningfully in CSA before such interventions. Moreover, extension services, which are the key channels for delivering climate-smart technologies, are themselves limited in reach. Publicly provided agricultural extension is likely to support the uptake of technologies such as community-supported agriculture (CSA), yet the ratio of extension agents to farmers remains a challenge (Jellason et al. 2020). Without adequate exposure to training or practical demonstrations, many rural farmers are unaware of how to integrate CSA into their cropping systems or how these practices align with their socioeconomic contexts. The importance of gender-responsive CSA strategies in enhancing awareness and participation is also noteworthy. Given that women produce up to 70% of food in SSA, targeting CSA education through inclusive and culturally sensitive approaches could significantly improve adoption rates. So, for climate-smart crop management to be successful in rural communities, there must be sustained investment in awareness-building, locally tailored training, and inclusive extension services. Without this foundational knowledge, the adoption and scaling of CSA practices will continue to lag behind the urgency of climate adaptation needs. Bridging this knowledge gap requires a participatory approach. Farmer Field Schools (FFS), peer learning models, and collaborations with nongovernmental organizations (NGOs) can empower farmers with practical skills. Digital platforms offering localized advice in regional languages can also significantly enhance outreach. For example, apps like Kisan Suvidha in India provide real-time weather forecasts, market prices, and crop advisories in several Indian languages (Ministry of Agriculture 2021).

10.7.3 Cultural Resistance

Agriculture is deeply interwoven with cultural traditions. Consequently, farmers often exhibit a strong attachment to conventional practices passed down through generations. Despite increasing exposure to climate change, smallholder farmers in Northern Ghana often exhibit reluctance to adopt climate-smart agriculture (CSA) practices, preferring instead to rely on traditional or indigenous farming methods. This reluctance is informed by a combination of cultural familiarity, economic constraints, and a perception of higher reliability based on generational knowledge.

Trust in Indigenous Knowledge and Practices Many farmers trust indigenous knowledge systems, which have been refined over generations to suit the local agro-ecological conditions. These practices include traditional weather forecasting, the use of local seed varieties, application of organic inputs like manure and ash, and intercropping. Such practices are not only deeply rooted in the local cultural fabric but are also viewed as cost-effective and accessible in the absence of formal support systems (File and Nhamo 2023).

Economic and Institutional Constraints Another significant reason for farmers' reluctance to shift from traditional methods is the limited access to financial resources, infrastructure, and institutional support. CSA often requires initial investment in tools, improved seed varieties, fertilizers, or irrigation facilities, resources that smallholder farmers in the region struggle to access. Consequently, they opt for low-cost indigenous strategies that do not depend on external inputs (File and Nhamo 2023).

Perceived Risks and Uncertainty with CSA Farmers also associate CSA with uncertainty and risk, especially when unfamiliar technologies or methods are involved. This perception is heightened by inconsistent access to agricultural extension services, poor dissemination of CSA information, and a lack of training, which contribute to a fear of potential failure if traditional methods are replaced (File and Nhamo 2023).

Socio-cultural Alignment of Indigenous Practices Indigenous practices are also preferred because they align with the farmers' worldviews, religious beliefs, and social norms. For instance, traditional rainmaking ceremonies and land preparation rituals are part of the farming calendar. CSA practices introduced externally may not align with these norms, creating resistance (File and Nhamo 2023).

Overcoming this resistance requires sensitivity and respect for traditional knowledge systems. Demonstration plots, success stories, and farmer-to-farmer training can foster trust and illustrate the tangible benefits of CSCMS. Incorporating indigenous practices into modern frameworks, such as combining traditional composting with precision nutrient application, can also ease the transition (Altieri 2009).

10.7.4 Infrastructure Limitations

In many rural areas, the impact of Climate-Smart Crop Management Strategies (CSCMS) is limited by poor infrastructure, such as inadequate roads, cold storage facilities, and market access. Even when advanced techniques are used, perishable crops often suffer post-harvest losses due to the lack of cold chains or processing units (World Bank 2019). These issues are especially evident in smallholder systems across sub-Saharan Africa.

Poor Infrastructure and Market Access CSA benefits are greatly hindered by weak infrastructure and limited market connectivity. Many smallholder farmers, particularly in remote areas, lack access to reliable roads, storage, and market linkages, which prevents them from profiting from surplus production or obtaining necessary inputs. The absence of proper storage leads to post-harvest losses and discourages the adoption of yield-boosting CSA practices. This creates a cycle where infrastructure gaps reduce both productivity and the incentive to invest in CSA technologies (Makate et al. 2020).

Wealth-Related Inequalities in CSA Access CSA uptake is often shaped by preexisting socioeconomic inequalities. Evidence from Malawi and Ethiopia reveals that wealthier farmers are more likely to adopt advanced CSA practices such as minimum tillage, planting pits, and the use of improved maize varieties. They also have better access to fertilizers, irrigation, storage infrastructure, and extension services, which are all crucial for CSA effectiveness (Makate et al. 2020). These disparities are explained by heterogeneity in asset ownership, landholding size, education levels, and political connectivity. Such inequalities create an uneven playing field, where poor farmers, despite being the most vulnerable to climate risks, are also the least equipped to adopt adaptive strategies.

Institutional and Policy Implications The findings point to an urgent need for targeted policy interventions that prioritize disadvantaged groups. This includes investing in rural infrastructure, improving access to extension services, and subsidizing CSA inputs for resource-poor farmers. Equally important is the decentralization of institutional support and strengthening local institutions to ensure inclusive governance in CSA implementation (Makate et al. 2020). Investments in rural infrastructure are thus essential to support climate-smart agriculture. Public-private partnerships can play a pivotal role. In Kenya, the “Last Mile” infrastructure initiative has improved market access for remote farmers, enabling them to fetch better prices for their produce and invest back into sustainable practices (USAID 2020).

10.8 Agri-Tech and Digital Solutions for Climate-Smart Agriculture

The convergence of agriculture and digital technology has given rise to a new frontier Agri-Tech. By leveraging tools such as artificial intelligence (AI), remote sensing, blockchain, and mobile platforms, Agri-Tech is revolutionizing how farmers make decisions, manage resources, and cope with climate variability (Fig. 10.7). These digital innovations are key enablers of climate-smart agriculture (CSA), enhancing productivity, ensuring sustainability, and building resilience. This chapter explores how various Agri-Tech solutions are supporting climate-smart practices around the world.

10.8.1 Precision Farming

The integration of Artificial Intelligence (AI), Internet of Things (IoT) sensors, and remote sensing technologies has revolutionized modern agricultural practices, particularly in the domain of climate-smart precision farming. These tools are at the heart of India's transition toward digital agriculture, enhancing decision-making, improving resource efficiency, and fostering environmental sustainability. According to the comprehensive review by Pathak (2023), real-time crop monitoring has gained momentum with the adoption of AI-powered analytics, which enables farmers to process vast amounts of heterogeneous agricultural data. These insights are drawn from multiple sources, including satellite imagery, sensor networks, and historical crop performance, allowing for data-driven predictions regarding yield estimation, pest outbreaks, and optimal irrigation windows. IoT-based sensor networks, which are deployed across farms, are instrumental in capturing microclimatic data such as soil moisture, temperature, humidity, and nutrient content in real-

Fig. 10.7 Agri-Tech applications and their benefits in climate-smart agriculture (CSA)

time. These sensors facilitate site-specific crop management, a critical component of precision agriculture. The collected data is then transmitted to centralized platforms where AI algorithms analyze patterns and generate actionable insights for farmers and agronomists. Moreover, remote sensing technologies such as satellite and drone-based imaging systems provide macro-level views of farmland. These systems help monitor crop health, identify water stress zones, and assess biomass and chlorophyll levels. Integration with Geographic Information Systems (GIS) allows for the visualization of spatial patterns, thereby improving decision-making related to crop rotation, fertilization, and disease management. One significant innovation discussed in the report is the role of Decision Support Systems (DSS), powered by AI and sensor data, which offer real-time alerts and customized advisories. These systems are especially valuable in climate-resilient agriculture, enabling adaptive strategies in response to unpredictable weather events, drought, and other stressors. Thus, the confluence of AI, IoT, and remote sensing not only optimizes farm productivity and input efficiency but also contributes to broader climate adaptation goals. By embracing these digital tools, Indian agriculture is becoming more responsive, resilient, and resource-conscious, aligning with the vision of climate-smart agriculture. Precision agriculture utilizes data analytics, geospatial technology, and IoT devices to optimize field-level management. It allows farmers to apply water, fertilizers, and pesticides in precise amounts only where needed, thus reducing waste and environmental impact. For instance, the use of drones and satellite imagery helps monitor crop health, detect pest infestations early, and assess soil moisture variability. AI algorithms can predict yield potential and provide real-time insights. According to a report by McKinsey & Company, farms employing precision agriculture saw input savings of up to 20% and yield increases of up to 15% (McKinsey 2022). In Brazil, large-scale soybean farms use platforms like Climate FieldView to make informed decisions, while in India, smallholders are beginning to adopt similar tech via mobile-integrated low-cost sensors and analytics (Bayer Crop Science 2021).

10.8.2 Smart Irrigation Systems

Water is a critical resource under climate stress, and smart irrigation systems offer a sustainable solution. These systems use soil moisture sensors, evapotranspiration models, and weather forecasts to automate irrigation schedules. As a result, they minimize water use while maximizing crop yield. The convergence of sensor-based technologies and intelligent data systems has significantly transformed modern agriculture. One of the standout developments in this domain is smart irrigation, which refers to a technique that utilizes automated and sensor-based systems to optimize water use in real-time, reducing waste and improving productivity. These systems operate through the integration of Internet of Things (IoT) devices, such as soil moisture sensors, weather stations, and remote monitoring tools, which collect high-resolution spatiotemporal data to support precision farming efforts. Smart irrigation systems rely on continuous monitoring of variables like soil moisture, temperature, and rainfall forecasts, feeding this data into AI-driven platforms or Decision Support Systems (DSS). These platforms then analyze the data to determine the precise timing and volume of irrigation required, ensuring that water is applied only when and where it is needed. Such real-time, data-driven irrigation strategies not only conserve water but also enhance yield quality and farm profitability, making them indispensable tools for future-ready agriculture (Gebresenbet et al. 2023). Israel, a pioneer in drip irrigation, has developed advanced solutions that are now being adopted globally. Indian sugarcane farmers using Netafim's drip systems report up to 40% water savings and a 25% increase in yield (Netafim 2020). Integrating these systems with solar-powered pumps and mobile alerts further enhances their efficiency in off-grid areas.

10.8.3 Blockchain for Supply Chain Transparency

Blockchain technology ensures traceability, transparency, and trust across the agricultural supply chain. It allows all stakeholders to access immutable records of crop production, handling, and transportation. This is especially valuable for organic and fair-trade certification, reducing food fraud and empowering farmers to receive fair compensation. Climate-smart agriculture (CSA) is grounded in improving agricultural productivity and resilience while reducing greenhouse gas emissions. Integrating Information and Communication Technologies (ICTs) into CSA efforts can significantly enhance these goals by addressing core inefficiencies in the agricultural value chain, particularly in traceability, food waste reduction, and farmer-market connectivity.

Ensuring Traceability Traceability is vital to improving food system transparency, safety, and trust. ICTs like blockchain, digital tagging, and cloud-based platforms can track products across the entire supply chain, from farm to table, thereby ensuring authenticity and enabling more informed consumer choices. According to the article, such digital tools allow the seamless collection and sharing of production and transaction data, which ensures the traceability of agricultural products, helping to improve transparency and accountability in the agri-food system (Arif et al. 2024). These systems not only empower consumers with knowledge of product origins but also help farmers gain recognition for sustainable practices, which can lead to premium market opportunities.

Reducing Food Waste Food waste is a major contributor to greenhouse gas emissions and resource inefficiencies. ICTs are instrumental in tackling this challenge by improving supply chain coordination and forecasting. Real-time data on demand, logistics, and storage conditions, which are enabled by sensors, mobile apps, and artificial intelligence that can minimize spoilage and overproduction. ICTs help minimize post-harvest losses through real-time monitoring and efficient logistics (Arif et al. 2024). These systems can alert stakeholders to storage issues (such as temperature fluctuations) or market gluts, allowing timely action that prevents wastage. Furthermore, ICT-enabled supply chains allow stakeholders to dynamically adjust their operations based on accurate, location-specific data. This capability reduces the ecological footprint of agricultural production by minimizing inputs used for excess or spoiled produce and curbing emissions related to food disposal.

Improving Farmer-Market Linkages A recurring limitation in traditional agricultural systems is the weak linkage between farmers and markets, which often results in exploitation, poor price realization, and lack of access to demand information. Digital agriculture platforms and mobile-based marketplaces are transforming this landscape. ICTs enhance access to market information, enabling farmers to make more informed production and sales decisions (Arif et al. 2024). These platforms allow farmers to access real-time data on prices, buyer availability, and demand trends, giving them greater bargaining power and reducing reliance on intermediaries. Moreover, such tools create opportunities for smallholders to participate in broader value chains by matching them with buyers, processors, or cooperatives beyond their immediate geography. This connectivity improves income stability and encourages adoption of sustainable practices by providing market incentives for quality and traceability.

Platforms like IBM Food Trust and Agri 10× in India are already piloting blockchain for coffee, rice, and horticulture. Farmers using blockchain can directly connect with buyers, eliminating intermediaries and improving their income (IBM 2021).

10.8.4 Mobile-Based Advisory Services

In areas with limited internet or infrastructure, mobile technology can serve as a powerful tool for disseminating climate-smart information. SMS-based services, interactive voice response systems, and smartphone apps offer weather forecasts, pest alerts, crop calendars, and market prices. In the face of climate variability and agricultural uncertainty, mobile-based advisory services have emerged as a transformative tool for supporting climate-smart agriculture (CSA), especially among smallholder farmers. These services, typically delivered via SMS messages or mobile applications, offer crucial real-time information, including weather forecasts, pest and disease alerts, and market trends, which are essential for informed decision-making on farms. Digital tools, including mobile-based platforms, enable smallholder farmers to adapt to climatic stresses by delivering localized and timely insights (Mujeyi and Mujeyi 2023). This includes information on soil moisture, weather predictions, and market prices, which help reduce input costs and minimize yield losses by facilitating better crop management decisions. Mobile phones, particularly, serve as the most accessible gateway for these services, reaching even those in remote areas where advanced infrastructure may be lacking. As the study notes, predictive technologies powered by artificial intelligence and cloud computing can alert farmers to potential pest and disease outbreaks, helping them prepare or act in advance. A complementary study by Mittal and Mehar (2016) provides further empirical backing. Their work underscores that SMS-based and voice-enabled advisories not only improve awareness but also directly influence farming practices. Farmers who received weather forecasts, agronomic tips, or price trends were able to optimize planting dates, choose resilient crop varieties, and time their access to markets more effectively. This enhances both resilience and productivity, key pillars of CSA. The study also notes that two-way communication platforms allowed farmers to interact with experts, further strengthening the advisory ecosystem and tailoring the recommendations to local contexts. Despite their promise, these services are not without challenges. It is noteworthy that low digital literacy, cost of smartphones, and patchy internet coverage can limit the scalability of these tools. Addressing these barriers through targeted subsidies, infrastructure upgrades, and digital skill-building programs could substantially boost adoption and impact. Thus, mobile-based advisory services are proving vital for scaling CSA practices by turning phones into powerful tools for risk management, resource optimization, and knowledge dissemination. They bridge the information gap for smallholder farmers, enabling them to farm smarter, not harder, in the face of climate uncertainty. One notable example is WeFarm, an SMS platform in East Africa that enables peer-to-peer agricultural advice without needing an internet connection. Similarly, India's Kisan Suvidha app offers localized, actionable insights in multiple languages, helping farmers make informed decisions (WeFarm 2021; Ministry of Agriculture 2021).

10.9 Case Studies/Real-World Applications

Global food security is seriously threatened by the growing frequency and severity of droughts, which are made worse by climate change. As a result, efficient management techniques that are adapted to the unique circumstances of various agro-ecological zones must be developed (Aguilera et al. 2020; Mulugeta 2023; Nyaga 2021; Paudel et al. 2020). Thus, effective drought control techniques are essential. These tactics need to be comprehensive, integrating long-term resilience-building initiatives with short-term relief activities. Improving water consumption efficiency is one of the best short-term solutions. Conservation of limited resources can be achieved by drastically lowering the quantity of water needed for agriculture through the use of drought-resistant crop types and drip irrigation. For example, drip irrigation reduces evaporation and runoff by delivering water straight to the plant's root zone. This approach has been effective in nations where water shortage is a recurring problem, such as Israel (Kassa and Andualem 2020; Mirlas et al. 2020). Farmers can sustain agricultural production even during times of less rainfall by managing water management. Drought control also requires the collection and storage of water. When there is an abundance of rainfall, methods like rooftop rainwater harvesting, building tiny dams, and using subterranean storage tanks can collect and store water, making it available during dry seasons. In order to supplement water resources for agriculture, rooftop rainwater collection has been heavily encouraged in India (Hussain et al. 2021; Mohanty et al. 2020). This strategy lessens the strain on groundwater supplies, which are frequently overused in many regions of the world, in addition to offering a dependable supply of water during dry spells. Another essential component of drought mitigation is soil management and conservation. The use of organic mulches, cover crops, and minimal tillage are among techniques that can improve soil health, decrease erosion, and increase soil moisture retention (Ngosong et al. 2019; Vincent et al. 2019). In areas with deteriorated soils and poor water penetration rates, these techniques are especially crucial. Conservation agriculture, for instance, has been widely used in Brazil's Cerrado area to improve drought resistance and fight soil erosion (da Silva et al. 2021). Even in arid environments, these techniques promote good crop development by preserving a permanent soil cover and reducing soil disturbance. In order to prepare for and respond to drought, early warning systems are essential. These systems forecast drought conditions and give farmers, policymakers, and other stakeholders timely information by utilizing weather data, remote sensing technology, and prediction models (Sharafi et al. 2021; Van Ginkel and Biradar 2021; Tripathi et al. 2025). Flash floods or persistent flooding have a detrimental effect on the rice ecologies in Asia and Africa, which range from lowland to deep-water environments. These floods affect around 20 million hectares in Asia and sizable portions of Africa (Bailey-Serres et al. 2010; Dar et al. 2017; Panda and Barik 2021). Rice production is facing major obstacles due to climate change, particularly in lowland coastal areas where it might result in huge losses (Ismail et al. 2012; Kato et al. 2019). A major scientific advance was made when the submergence tolerance gene SUB1A on chromosome 9 was discovered in FR13A, an Indian flood-tolerant rice variety descended from the traditional landrace Dhalputtia (Xu and Mackill 1996; Xu et al. 2000). SUB1A was identified as the primary gene responsible for submergence tolerance among them. Through this innovation, SUB1A was effectively incorporated into a number of important rice cultivars, giving them a 2-week tolerance for flash floods (Septiningsih et al. 2009; Bailey-Serres et al. 2010; Ismail et al. 2013a; Singh et al. 2016; Emerick and Ronald 2019; Anumalla et al. 2025). As a result of soil deterioration and climate change, conservation agriculture (CA) has emerged as a potentially more sustainable option than traditional tillage-based agricultural systems (Hermans et al. 2021). Crop rotation, crop variety, and minimal mechanical soil disturbance are all components of conservation agriculture (FAO 2022; Powlson et al. 2016). According to Šarauskis et al. (2018), little tillage enables natural water filtering and root growth by reducing soil layer compaction. Research conducted in Zambia has demonstrated that CA can improve soil quality and increase agricultural production. The soil organic carbon (0.74% vs. 0.57%), soil pH (6.30 vs. 4.95), and cation exchange capacity (3.80 vs. 1.60 cmolc kg^{-1}) inside planting basins were significantly higher after 7 years of CA in an acrisol at Mkushi, Zambia, than outside of planting basins (Martinsen et al. 2019). Mazvimavi (2017) found that in Zambia's Chipata, Sinazongwe, and Monze districts, maize grain yields in CA were noticeably greater than in traditional fields (Phiri et al. 2025). Agroforestry systems mitigate the effects of climate change by incorporating trees into agricultural landscapes and providing a number of other environmental advantages. According to Mbow et al. (2014), agroforestry systems store significant amounts of carbon through soil and vegetation sequestration, which lowers atmospheric CO2 concentrations. Ethiopia is particularly susceptible to the effects of climate change since it is situated in the Horn of Africa. Temperatures are already rising, rainfall is unpredictable, and droughts are occurring more often across the nation, especially in the lowland and semiarid areas. The effects of climate change are making poverty, food insecurity, and land degradation worse. Nonetheless, agroforestry has been acknowledged by Ethiopia as a crucial instrument for climate change adaptation and mitigation. According to Garrity et al. (2010), Ethiopia has a wide variety of agroforestry systems, from more organized ones like parklands and alley cropping to more traditional home gardens and boundary plantings. Trees are frequently incorporated with agriculture and cattle in Ethiopia, especially in highland regions. For smallholder farmers, home gardens that include fruit trees like avocado, banana, and papaya in addition to vegetables and legumes help to diversify their sources of income and provide food security. In regions where droughts are occurring more often, these agroforestry systems have been demonstrated to increase soil fertility and improve water retention (Sileshi et al. 2014; Bogale and Daba 2025). Agroforestry is one of the best ways to mitigate a variety of abiotic stressors that affect plant yields, such as drought, metal, and salinity brought on by climate change and excessive use of chemical fertilizers and pesticides (Kumari et al. 2022). According to Sanches (2019), 87% of the agricultural area in the southeast of the Brazilian state of Bahia was used for cocoa cultivation in 2015. Combining forests with agricultural practices, integrated cacao agroforestry is a sustainable way to lessen soil erosion and runoff, store carbon to slow down climate change, provide commodities for people, and support biodiversity and conservation. The CO_2 emissions in cacao-AFS (agroforestry systems) vary according to the kind of plant cover and soil properties, in contrast to the protected forest (Costa et al. 2018: François et al. 2023). Farmers' agricultural decision-making can be revolutionized by the application of information and communication technologies (ICTs) in Climate Information Services (CIS). Various new ICT platforms, such as websites, Facebook bulletins, YouTube videos on weather information, WhatsApp weather forecasts in the form of flyers, short message service (SMS), audio WhatsApp weather forecasts, and weather apps, are important tools for providing farmers with weather and climate forecasts. The results highlight the significance of ICTs in improving access to critical climatic information for better farming practices, with significant ramifications for corporations, governments, and agricultural extension services. In Ghana's agricultural industry, the younger population opens up potential for the application of modern ICTs. According to Nyamekye et al. (2021), the Ghanaian government's digitalization ambition has also made it easier to draw in venture capital and investors (Sarku et al. 2025). Smartphones are being actively adopted by farmers in developing nations, making it simple to design and make available mobile applications that cater to their requirements. There have been several applications for individual farmers in recent years. Giving farmers useful information to aid them with planting, growing, and harvesting is the aim of these mobile applications. Weather forecasts, disease outbreak statistics, and best practices for agriculture are a few examples of this data. These apps may also teach novice farmers and let seasoned farmers try new techniques. Additionally, these mobile applications may offer details on agricultural goods like seeds, pesticides, fertilizers, etc. In order to evaluate the pricing of various crops in the marketplace and gain insight into the true supply and demand situation of a certain crop, farmers also require some market knowledge (Armoogum et al. 2023). The benefits of SMS-based agricultural information, including weather and market data, for Indian farmers were covered by Fafchamps and Minten (2012). According to research, small farmers need to be informed about the many technologies that may significantly increase their output because there is a low level of awareness regarding the use of e-agriculture (Thankachan et al. 2014). In the same way, Zakar and Zakar (2009), stress the value of using ICT to inform planters about market prices, weather, how to choose the right fertilizers, seed quality, pesticides, crop intermixture, water management, land preparation, harvesting, and the general adoption of current ICT to enhance agricultural practices. Glood et al. (2018) discusses the importance that mobile phones play in agriculture development among Iraqi farmers. The smartphone application "Mokaro," which is the first of its type, was introduced in Mauritius in June 2019 with the goal of assisting planters in their agricultural endeavors. Farmers and planters will use the "Mokaro" software as a tool to better organize their farming operations, manage resources, and reduce losses. They will receive guidance on future planting, irrigation, and other field operations from the "Mokaro" application after crop analysis and supply and demand assessments. Additionally, farmers will receive details on Mauritius' climate, up-to-date agricultural news, and alerts that might be helpful for managing and producing crops (Table 10.3; Fig. 10.8) (Armoogum et al. 2023).

10.10 Future Directions and Innovations

10.10.1 Emerging Technologies in Agriculture

10.10.1.1 Artificial Intelligence (AI) and Machine Learning (ML)

Crop yields are negatively impacted by extraordinary weather variations, and climate change has a significant impact on agricultural production. Temperature changes, frosts, precipitation events, and drought are examples of weather variations that can drastically reduce agricultural output and quality. These climatic extremes' increasing frequency and unpredictability is concerning for future food security (Harkness et al. 2020). Pests and diseases cause significant agricultural post-harvest losses all over the world. For example, wheat, a key food crop, was reported to have yield losses of about 21.5% globally in 2019 (Savary et al. 2019). Thanks to technical advancements made over the past century, it is possible to increase agricultural productivity while overcoming a variety of obstacles. Agriculture is undergoing a change because of emerging technologies like AI, ML, and IoT, which improve production efficiency and solve conventional agricultural problems. Advanced sensor technology, smaller hardware, and increased precision have all helped the Internet of Things (IoT) expand over the past century, boosting supply chains and

Table 10.3 Real-world applications and regional case studies of climate-smart agriculture

Region/ country	Adaptation strategy/case	Impact/benefits	References
India (Maharashtra)	Drip irrigation and crop diversification in semiarid zones	Improved water efficiency and resilience to prolonged droughts	Bhagat et al. (2024)
Bangladesh	Use of flood-tolerant rice varieties (e.g., Swarna-Sub1)	Reduced crop losses during flash floods; ensured food security	Ismail et al. (2013)
Nepal	Climate-smart agriculture (CSA) led by women cooperatives in hill regions	Empowered women farmers; improved livelihoods through diversified practices	Chhetri et al. (2023)
Kenya	Mobile-based weather advisories via Shamba Shape Up	Enhanced farmer preparedness and decision-making	Caine et al. (2015)
Ethiopia	Agroforestry with tree buffer zones and integrated watershed management	Increased soil fertility, carbon sequestration, and water retention	Bogale and Daba (2025)
Zambia	Conservation agriculture (e.g., minimum tillage, crop rotation)	Improved soil health and maize productivity	Mhlanga et al. (2021)
Brazil	Cacao agroforestry systems in Bahia	Enhanced carbon storage, biodiversity, and sustainable livelihoods	Salas-Macías et al. (2025)
Honduras	Contour farming and stone barriers	Controlled erosion and improved water retention	Rivera et al. (2011)
Peru	Indigenous climate forecasting with traditional potato farming	Sustained crop diversity and resilience through ancestral knowledge	Giamberso and Lisa Rose (2014)
Mauritius	"Mokaro" mobile app for crop planning and resource management	Facilitated informed decision-making and reduced farming losses	Armoogum et al. (2023)

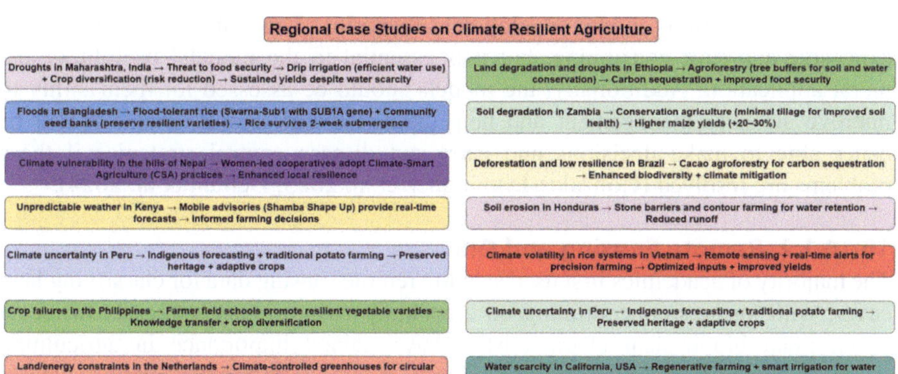

Fig. 10.8 Regional case studies on climate-resilient agriculture

guaranteeing safe, high-quality agricultural goods. IoT-based devices and machine learning algorithms are being used to monitor soil moisture and nutrient levels in order to optimize crop fertilizer requirements (Sharma et al. 2020). Accurate diagnosis and recognition of agricultural production illnesses and pests is made possible by advances in computers using machine learning (ML), a component of artificial intelligence. Accurate disease identification in rice plants is made possible using deep learning (DL), a subclass of machine learning (ML), in combination with image processing (Haridasan et al. 2023). AI-enabled drones help with disease detection, pesticide application, and field monitoring (Sharma et al. 2020). In order to increase the precision and effectiveness of contemporary farming systems, the use of AI and ML may be expanded to manage other facets of agricultural production, such as crop yield prediction, intelligent harvesting, and enhanced irrigation systems. For example, Sharma et al. gave several instances of AI applications using data from diverse locations to make dynamic decisions for agricultural production prediction in wheat, barley, corn, rice, silage maize, and corn (Sharma et al. 2020). One study that was discussed was conducted by Kamir et al. (2020), who identified yield gap hotspots in wheat production using machine learning algorithms. They created extremely high-resolution yield maps by combining information from many sources, including temperature and rainfall data, time-series data from the Normalized Difference Vegetation Index (NDVI), and observed grain yield maps gathered by intelligent harvesting equipment. Growing knowledge of plant disease dynamics and improvements in data collection and analysis methods are opening the door to more effective, affordable, and useful forecasting systems, which will help prevent and lessen the effects of crop and plant diseases on food production. For those crops that provide millions of people living in the least developed nations with staple meals, this is especially important. The necessity of integrating many data sources, creating models that can capture and forecast the variability of illness in the field, and obtaining high-quality, labeled data were also mentioned (Fenu and Malloci 2021; Kaur et al. 2022). Even though this discipline has advanced significantly, challenges like correct data validation for the models still need to be resolved for trustworthy predictions in various model development scenarios. This validation procedure is expensive and time-consuming, but it is essential to the models' usefulness and moral application. Additionally, the models should be assessed for their ability to improve sickness management and help users. Farmers and agronomists will be better able to make decisions and carry out disease control strategies if these problems are resolved (Gonzalez-Dominguez et al. 2023; Delfani et al. 2024).

10.10.1.2 Remote Sensing and GIS

The majority of academics first used satellite remote sensing data for classifying land cover, with farmers concentrating on crop kinds as a primary use case. In recent years, plant biophysical characteristics have gained importance in agricultural remote sensing. Agriculture has traditionally made use of remote sensing. One of the largest advancements in this subject is the categorization of crop canopies using image processing. The benefit of using remote sensing in precision agriculture is that it may deliver relevant information repeatedly without causing crop damage through destructive sampling. Remote sensing offers a cost-effective substitute for conventional data collecting techniques in wide geographic areas (De Beurs et al. 2008). When remote sensing and GIS are used together, spatial variables of interest can be produced that can be used in a number of different fields, such as mapping flood plains, urban development, land use changes, surface energy flux, crop growth monitoring, stress detection, and hydrological modeling (Kingra et al. 2016). In order to follow changes in crop and health area time series across geographical units, remote sensing technology can gather data and extract information with spatial precision. By examining the amount of chlorophyll or structural alterations in crops, it may therefore document variations in crop phenology throughout the course of the growing season. It can map and categorize crops and evaluate their health using satellite image analysis; it does this by analyzing the electromagnetic spectrum (EM) and the varying interaction of various healthy crop coverings (Caballero et al. 2020; Hakam et al. 2023). Remote sensing is incredibly helpful in identifying and controlling a range of agricultural issues, including biotic and abiotic stressors, even on small farms. To effectively use agricultural data for economic improvement, state or district level information systems based on accessible remote sensing and GIS crop data are needed. Governments can utilize data from remote sensing to solve national agricultural challenges or enact policies. Using a novel and unconventional remote sensing method, nanochips made of plant and seed tissue may be implanted in near real time to monitor crops. It is obvious that remote sensing will become increasingly significant in evaluating agricultural research in the future (Adhikary et al. 2022).

10.10.2 Sustainable Value Chains

The Sustainable Development Goals (SDGs) provide a thorough framework for sustainable adaptation to the effects of climate change on nature-based value chains. The rangeland ecology and biodiversity, as well as socioeconomic and related sectors including water resources, cattle, crops, and food security, will all be significantly impacted by climate change (Kalele et al. 2021). It is predicted that changes in rainfall patterns may result in flooding and/or severe water shortages (Nyika 2022). Due to changes in pest and disease vector distribution and agricultural growing seasons brought on by rising temperatures, more people will be at risk (Muema et al. 2018). Increases in temperature have the potential to drastically hasten the extinction of several favored plant species and their environments (Ndiritu et al. 2021). This might upset the majority of value chains, increasing the vulnerability of pastoralists (Kirui et al. 2022). According to Fanzo et al. (2017) and Farmery et al. (2021), a value chain is the complete collection of operations necessary to take a product or service from its inception to its ultimate disposal following usage, encompassing physical transformation and the input of numerous producer services. For rangeland value chains, it encompasses all rangeland actors involved from production to the end user who transact a specific product as it moves through the value chain, including input, farmers, traders, processors, transporters, wholesalers, retailers, and final consumers. As a result, assessing climate risk for value chains is a critical stage in any climate change adaptation process. Understanding current and future climate risks, considering the consequences for value chain activities, and examining potential solutions for mitigating negative impacts are all part of the process for value chain actors. They can use this information to engage in forward-thinking and flexible planning processes that maximize their business's potential while accounting for climate concerns. Therefore, prompt and effective service delivery is a key factor in facilitating the expansion of Rangelands' value chain. Among the most important services that facilitate the development of sustainable value chains in rangelands are market information, financial services, and climate information services. Climate information services are essential to all decision-makers, including banks, government agricultural investment programs, and agro-pastoralists. Climate risk assessment is supported by history and present climate trends, future estimates, and seasonal weather forecasts, even if early warnings for climate extremes may increase preparedness for shocks (Ngigi and Muange 2022). Many value chains rely heavily on commercial banks and microfinance institutions that offer financial services including loans, savings, and insurance (Isako and Kimindu 2019). Accessible finance facilitates planning, diversified revenue streams and activities, and serves as a safety net in times of shock. Furthermore, it encourages innovation and experimentation, two crucial elements of adaptive management. In order to give value chain participants fast and reliable data so they can make educated decisions, it is imperative to invest in climate information services. Research and extension services are also essential for sharing best practices and expertise, which helps value chain participants become more adaptable. Value chains that can endure the effects of climate change can be strengthened by such concentrated efforts. The larger objectives of reducing poverty and ensuring food security in places that are at risk can then benefit from this. In arid and semiarid regions, creating climate-resilient value chains requires a comprehensive strategy that includes improved access to climate information, more financial assistance, supportive legislation, strong research, and easier collaboration (Lutta et al. 2024).

10.10.3 Youth Engagement

One of the most significant economic sectors in India is agriculture, which employs over 60% of the labor force and accounts for roughly 17% of the GDP. Low productivity, climate change, market instability, and a lack of infrastructure are just a few of the numerous issues facing the industry. Furthermore, the new generation, who frequently move to cities in quest of better possibilities, finds the industry unappealing. The proportion of young people (15–29 years old) in rural areas fell from 35.5% in 2001 to 30.5% in 2011, according to the 2011 census. Forecasts indicate that 73 million young people will be unemployed globally in 2022, with a notable decline in the number of people in agriculture between the ages of 25 and 29 (ILO 2022). Encouraging young people to participate in agriculture is essential to overcoming the challenges. This will contribute to the employment and compensation of youth and transform the agriculture sector into one that is more innovative, profitable, and dynamic. Involving young people in agriculture and entrepreneurship may be summed up as their starting and managing their own agribusinesses and doing a range of duties related to agricultural production, processing, marketing, and value addition (Vasanthi et al. 2024). Food security and agricultural production are seriously threatened by climate change and environmental deterioration. Natural catastrophes, extended droughts, and irregular weather patterns can all have a negative impact on agricultural production and livelihoods, especially for young people and smallholder farmers. Sustainable techniques and creative solutions are needed to adapt to climate change and increase agricultural resilience. Precision farming, smartphone apps, and online platforms are examples of technological innovations that give young people access to information, financial services, and market connections, increasing agricultural output and efficiency. Young farmers may monitor crop health, maximize resource usage, and make data-driven decisions by adopting technology (Nazneen et al. 2024; Sulochna et al. 2023). The importance of youth-led projects and entrepreneurship in promoting innovation and sustainable growth in the agricultural sector is being increasingly acknowledged. Social companies, entrepreneurs, and youth-led groups are using creativity and technology to solve problems in agriculture and generate jobs. Development partners and governments may enable young people to be change agents in agriculture by encouraging an entrepreneurial culture (Azra et al. 2024). Youth play a vital, varied, and dynamic role in the food system that is essential to reaching sustainable development objectives, eradicating hunger, advancing social justice, and creating resilient food systems that feed both people and the environment. Youth can be fully utilized as change agents in agriculture, food security, and rural transformation if societies invest in their education, skill development, and entrepreneurship, harness their energy, creativity, and talent, and create environments that allow them to participate meaningfully (Kote et al. 2024). The desire to pursue chances outside of the farm but along the agricultural value chain is seen in Kenyan youth's attitude to food systems participation, which is closely tied to the allure of agripreneurship. Although agripreneurship also involves turning the focus of on-farm activities into a business, this unique approach shows how eager Kenyan youth are to work in other value chain roles, such as agro dealers and extension agents, Agri-Tech and digital services, processing, packaging, and storage, transportation and logistics, trade and market facilitators, food and produce retailers, financial service providers, food bloggers, and agri-media (Fig. 10.9) (Mugo et al. 2023).

10.11 Conclusion

Climate-Smart Crop Management Strategies (CSCMS) provide an innovative and integrated solution to the complex problems of contemporary agriculture, such as climate-related stresses like unpredictable rainfall, extended drought, and extreme temperature, in addition to soil degradation, water limitations, pest and disease outbreaks, and socioeconomic constraints that imperil global food security and

Fig. 10.9 Future directions and innovations

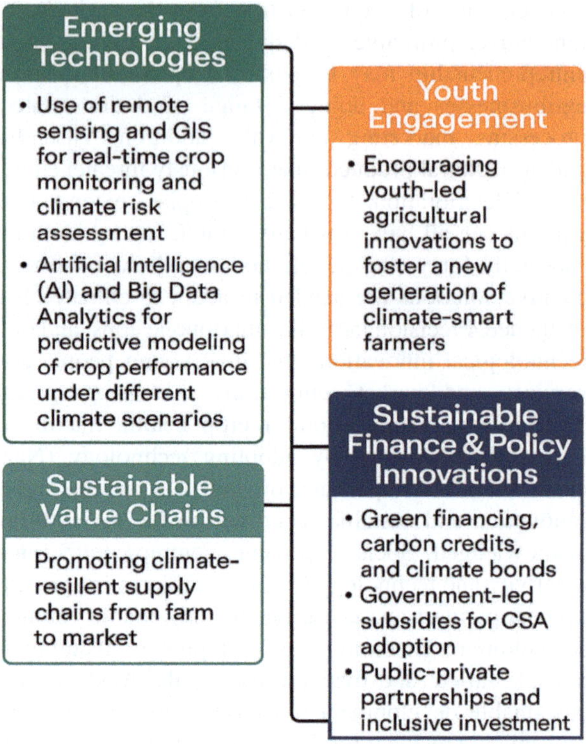

livelihoods of farmers. By their commitment to the basic principles of sustainably enhancing productivity, climate shock resilience, and greenhouse gas emission reduction, CSCMS adopt innovative practices such as conservation agriculture with minimal tillage, cover crops, and organic manure to restore soil fertility; efficient irrigation technologies such as drip and sprinkler and rainwater harvesting to save water; and drought-tolerant, heat-resistant crop varieties, intercropping, and agroforestry to diversify income and enhance ecosystem resilience. The integration of Agri-Tech technologies—like IoT-enabled sensors for real-time monitoring of crops, AI-powered predictive analytics for accurate decision-making, blockchain for traceability of supply chains, and mobile-based advisory services providing weather and pest information alerts—equips farmers to respond to climate variability with less environmental damage. Farm-level achievements, including drought-tolerant farming in the semiarid regions, agroforestry with enhanced soil health and income, and climate information enabling timely adjustments in planting schedule, highlight the pragmatic benefits of CSCMS: increased yields, reduced crop loss, economic resilience, enhanced biodiversity, and reduced carbon emissions, together with social equity via empowerment of the smallholder farmers, particularly women and the vulnerable. In spite of such challenges as high initial costs, low awareness, cultural resistance to giving up conventional practices, and poor rural infrastructure, these can be overcome through participatory community involvement, extensive training and field demonstrations, supportive policies with subsidies, and investments in climate-resilient supply chains and market access. In the future, technologies like remote sensing, geographic information systems, and big data analytics, and sustainable value chains and youth's agricultural innovation, will continue to support CSCMS in their scalability and sustainability. By creating partnership among farmers, policymakers, researchers, and technologists and linking with global sustainability goals, CSCMS can make food systems secure, enhance livelihoods, and build resilient agricultural landscapes, which form the cornerstone of a sustainable, equitable, and prosperous world agriculture future provided that urgent and collective action is taken.

References

Abdalla M et al (2019) Cover crops and ecosystem services: insights from studies in temperate soils. Agric Ecosyst Environ 258:1–12

Adamchuk VI, Hummel JW, Morgan MT, Upadhyaya SK (2018) On-the-go soil sensors for precision agriculture. Comput Electron Agric 44(1):71–91

Adekunle AA et al (2018) Youth engagement with digital agriculture. Inf Technol Int Dev 14:33–47

Agarwal R, Bhardwaj I, Sharma AK, Sanghi A, Agarwal G (2024) Innovations in Agri-tech: a review of artificial intelligence applications and challenges in modern agriculture. In: Second international conference on advanced computing & communication technologies (ICACCTech), pp 599–604. https://doi.org/10.1109/ICACCTech65084.2024.00101

Agegnehu G et al (2016) Biochar and compost effects on soil properties. Plant Soil 409(1–2):1–20

Aguilera E, Guzmán G, Alonso A (2015) Greenhouse gas emissions from conventional and organic cropping systems in Spain. II. Fruit tree orchards. Agron Sustain Dev 35(2):725–737. https://doi.org/10.1007/s13593-014-0265-y

Ahmad A, Khan S (2017) Water and energy scarcity for agriculture: is irrigation modernization the answer? Irrig Drain 66(1):34–44. https://doi.org/10.1002/ird.2021

Altieri MA (2009) Agroecology, small farms, and food sovereignty. Mon Rev 61(3):102–113

Apipoonyanon C, Szabo S, Tsusaka TW, Leeson K, Gunawan E, Kuwornu JK (2021) Socio-economic and environmental barriers to increased agricultural production: new evidence from Central Thailand. Outlook on Agric 50(2):178–187. https://doi.org/10.1177/0030727020967361

Araus JL et al (2018) Translating high-throughput phenotyping into genetic gain. Trends Plant Sci 23(5):451–466

Arif J, Samadhiya A, Naz F, Kumar A (2024) Exploring the application of ICTs in decarbonizing the agriculture supply chain: a literature review and research agenda. Heliyon 10(8):e29564

Assouline S, Russo D, Silber A, Or D (2015) Balancing water scarcity and quality for sustainable irrigated agriculture. Water Resour Res 51(5):3419–3436. https://doi.org/10.1002/2015WR017071

Ayantunde AA et al (2018) Climate-smart grazing in West Africa. Agric Syst 161:61–72

Azenzem R, Eddomairi H (2025) Internet of things (IoT) and artificial intelligence (AI) in agriculture: applications for sustainable crop protection. In: Dutta PK, Hamad A, Haghi AK, Prabhakar PK (eds) Food and industry 5.0: transforming the food system for a sustainable future. Springer, pp 73–86. https://doi.org/10.1007/978-3-031-76758-6_6

Bale JS, van Lenteren JC, Bigler F (2008) Biological control and sustainable food production. Philos Trans R Soc B 363(1492):761–776

Barron J, Noel S (2018) Assessing the effectiveness of participatory watershed management in India. Agric Syst 158:76–86

Basso B, Dumont B, Cammarano D et al (2016) Environmental and economic benefits of variable rate nitrogen fertilization in a nitrate vulnerable zone. Sci Total Environ 545–546:227–235

Bayer Crop Science (2021) Digital farming tools

Bhatnagar S, Chaudhary R, Sharma S, Janjhua Y, Thakur P, Sharma P, Keprate A (2024) Exploring the dynamics of climate-smart agricultural practices for sustainable resilience in a changing climate. Environ Sustainability Indic 100535:100535

Bhattacharyya P, Pathak H, Pal S (2020) Crop Management for Climate-Smart Agriculture. In: Bhattacharyya P, Pathak H, Pal S (eds) Climate smart agriculture: concepts, challenges, and opportunities. Springer, pp 85–111. https://doi.org/10.1007/978-981-15-9132-7_7

Biazin B et al (2019) Rainwater harvesting for dryland agriculture in Ethiopia. Agric Water Manag 212:5–15

Blanco-Canqui H, Ruis SJ (2018) No-tillage and soil physical environment. Geoderma 326:164–200

Bland, R., Ganesan, V., Hong, E., & Kalanik, J. (2023). Trends driving automation on the farm. McKinsey Global Publishing

Brooker RW et al (2016) Improving intercropping. Adv Agron 137:45–91

Brown ME, Mugo S, Petersen S, Klauser D (2022) Designing a pest and disease outbreak warning system for farmers, agronomists and agricultural input distributors in East Africa. Insects 13(3): Article 3. https://doi.org/10.3390/insects13030232

Burney J et al (2017) Solar-powered irrigation in Niger. Environ Res Lett 12(7):074010

Carter M et al (2018) Index insurance and climate risk management. World Dev 101:58–71

Cattivelli L et al (2016) Drought tolerance improvement in crop plants. Field Crop Res 105(1–2): 1–14

Cayuela ML et al (2014) Biochar reduces N_2O emissions from soils. Glob Chang Biol 20(8): 2753–2764

CGIAR Assessment Standing Panel on Impact Assessment, Maredia MK, Pingali PL (2001). Environmental impacts of productivity-enhancing crop research: a critical review. https://hdl.handle.net/10947/491

Chandra A et al (2017) Community-based adaptation to climate change. Clim Dev 9(5):392–402

Chang K-R, Lee C-C, Hsieh Y-L, Hsieh P-C, Kuo Y-F (2024) Strategic short note: application of smart machine vision in aquaculture and animal husbandry. In: Ahamed T (ed) IoT and AI in agriculture: smart automation systems for increasing agricultural productivity to achieve SDGs and society 5.0. Springer, pp 447–451. https://doi.org/10.1007/978-981-97-1263-2_27

Chartzoulakis K, Bertaki M (2015) Sustainable water Management in Agriculture under climate change. Agric Agric Sci Procedia 4:88–98. https://doi.org/10.1016/j.aaspro.2015.03.011

Chibwana C et al (2017) Malawi's climate-smart input subsidies. Food Policy 69:84–96

Chiriacò MV, Grossi G, Castaldi S, Valentini R (2017) The contribution to climate change of the organic versus conventional wheat farming: a case study on the carbon footprint of wholemeal bread production in Italy. J Clean Prod 153:309–319. https://doi.org/10.1016/j.jclepro.2017.03.111

Cohen M (2020) Environmental toxins and health: the health impact of pesticides. Aust Fam Physician 36(12). https://doi.org/10.3316/informit.355467860496836

da Silveira F, da Silva SLC, Machado FM, Barbedo JGA, Amaral FG (2023) Farmers' perception of the barriers that hinder the implementation of agriculture 4.0. Agric Syst 208:103656. https://doi.org/10.1016/j.agsy.2023.103656

Dar, M. H., de Janvry, A., Emerick, K., Kelley, E. M., & Sadoulet, E. (2019). Endogenous information sharing and the gains from using network information to maximize technology adoption. https://escholarship.org/uc/item/8qx7m4zq

Darouich H et al (2017) Drip irrigation efficiency in Mediterranean agroecosystems. Agric Water Manag 183:158–168

Daryanto S et al (2018) Quantifying the benefits of reduced tillage on soil erosion in maize-based systems. Soil Tillage Res 182:1–10

Dechmi F et al (2020) Improving sprinkler irrigation performance in semi-arid regions. Irrig Sci 38(2):145–160

Delgado JA, Shaffer M, Hu C, Lavado RS, Wong JC, Joosse P, Li X, Rimski-Korsakov H, Follett R, Colon W, Sotomayor D (2006) A decade of change in nutrient management: a new nitrogen index. J Soil Water Conserv 61. https://www.tandfonline.com/doi/abs/10.1080/00224561.2006.12435859

Diacono M, Montemurro F (2015) Long-term effects of compost on soil fertility. Agron Sustain Dev 35(2):401–422

Dillon P et al (2019) Global review of managed aquifer recharge applications. Hydrogeol J 27(1): 1–18

Dobermann A, Witt C, Dawe D, Abdulrachman S, Gines HC, Nagarajan R, Satawathananont S, Son TT, Tan PS, Wang GH, Chien NV, Thoa VTK, Phung CV, Stalin P, Muthukrishnan P, Ravi V, Babu M, Chatuporn S, Sookthongsa J et al (2002) Site-specific nutrient management for intensive rice cropping systems in Asia. Field Crop Res 74(1):37–66. https://doi.org/10.1016/S0378-4290(01)00197-6

Eilenberg J, Hajek A, Lomer C (2018) Suggestions for unifying the terminology in biological control. BioControl 46(4):387–400

Esilaba AO, Byalebeka JB, Delve RJ, Okalebo JR, Ssenyange D, Mbalule M, Ssali H (2005) On farm testing of integrated nutrient management strategies in eastern Uganda. Agric Syst 86(2): 144–165. https://doi.org/10.1016/j.agsy.2004.09.005

FAO (2021) The state of food and agriculture 2021. Food and Agriculture Organization, Rome

File DJMB, Nhamo G (2023) Farmers' choice for indigenous practices and implications for climate-smart agriculture in northern Ghana. Heliyon 9(11):e22162

Foteinis S, Chatzisymeon E (2016) Life cycle assessment of organic versus conventional agriculture. A case study of lettuce cultivation in Greece. J Clean Prod 112:2462–2471. https://doi.org/10.1016/j.jclepro.2015.09.075

Foteinis S, Hatzisymeon M, Borthwick AGL, Chatzisymeon E (2021) Environmental impacts of conventional versus organic eggplant cultivation systems: influence of electricity mix, yield, over-fertilization, and transportation. Environments 8(3):Article 3. https://doi.org/10.3390/environments8030023

Gandhi R et al (2018) Digital green's video-based extension. Inf Technol Int Dev 14:1–15

Gani RA, Shofiati R, Pramudia A (2022) A review of the use of web-based climate information systems for agriculture purposes. Jurnal Sumberdaya Lahan 16(2):153–148

Garibaldi LA et al (2019) Farming approaches for pollination services. Sci Adv 5(7):1–8

Gebresenbet G, Bosona T, Patterson D, Persson H, Fischer B, Mandaluniz N et al (2023) A concept for application of integrated digital technologies to enhance future smart agricultural systems. Smart Agric Technol 5:100255

Gil J et al (2017) Policy conflicts in Brazil. Land Use Policy 68:441–451

Gil J et al (2018) Biodiversity in integrated systems. Agric Ecosyst Environ 266:106–115

Glare T, Caradus J, Gelernter W et al (2012) Have biopesticides come of age? Trends Biotechnol 30(5):250–258

Government of India (2020) Pradhan Mantri Krishi Sinchayee Yojana

Greatrex H et al (2018) Scaling up index insurance for smallholder farmers. Glob Environ Chang 28:238–249

Hussain S, Hussain S, Guo R, Sarwar M, Ren X, Krstic D, Aslam Z, Zulifqar U, Rauf A, Hano C, El-Esawi MA (2021) Carbon sequestration to avoid soil degradation: a review on the role of conservation tillage. Plants (Basel) 10(10): https://doi.org/10.3390/plants10102001. PMID: 34685810; PMCID: PMC8539297

IBM (2021) IBM Food Trust

IFPRI (2020) Agricultural extension services in rural India

Islam MM et al (2020) Community resilience in Bangladesh. Clim Dev 12(3):261–274

Jagadish SVK et al (2016) Heat stress during flowering in cereals. J Exp Bot 66(5):1145–1156

Jat ML et al (2016) Precision water management in cereal systems. Adv Agron 137:159–210

Jat ML et al (2019) Precision nutrient management in South Asia. Field Crop Res 230:83–94
Jellason NP, Conway JS, Baines RN (2020) Understanding impacts and barriers to adoption of climate-smart agriculture (CSA) practices in North-Western Nigerian drylands. J Agric Educ Ext 27(1):55–72
Kahinda JM et al (2019) Rainwater harvesting for resilience in sub-Saharan Africa. Water Int 44(3): 275–290
Karlen DL, Rice CW (2015) Soil degradation: will humankind ever learn? Sustainability 7(9): Article 9. https://doi.org/10.3390/su70912490
Kassam A et al (2019) Global spread of conservation agriculture. Int J Environ Stud 76(1):29–51
Khan ZR, Midega CA, Pittchar JO et al (2014) Achieving food security for one million sub-Saharan African poor through push-pull innovation. PLoS One 9(8):e105108
Khan ZR et al (2016) Push-pull technology for pest management. Annu Rev Entomol 61:405–425
Knudsen MT, Meyer-Aurich A, Olesen JE, Chirinda N, Hermansen JE (2014) Carbon footprints of crops from organic and conventional arable crop rotations – using a life cycle assessment approach. J Clean Prod 64:609–618. https://doi.org/10.1016/j.jclepro.2013.07.009
Kopecká R, Kameniarová M, Černý M, Brzobohatý B, Novák J (2023) Abiotic stress in crop production. Int J Mol Sci 24(7):Article 7. https://doi.org/10.3390/ijms24076603
Krishnababu ME, Devi BR, Soni A, Panigrahi CK, Sudeepthi B, Rathi A, Shukla A (2024) A review on precision agriculture navigating the future of farming with AI and IoT. Asian J Soil Sci Plant Nutr 10(2):336–349. https://doi.org/10.9734/ajsspn/2024/v10i2291
Kristjanson P et al (2017) Agricultural extension challenges. Global Food Secur 12:63–72
Kumar V, Sharma KV, Kedam N, Patel A, Kate TR, Rathnayake U (2024) A comprehensive review on smart and sustainable agriculture using IoT technologies. Smart Agric Technol 100487
Lehmann J et al (2015) Biochar effects on soil biota. Soil Biol Biochem 43(9):1812–1836
Li Y, Conway D, Xiong W, Gao Q, Wu Y, Wan Y, Li Y, Zhang S (2011) Effects of climate variability and change on Chinese agriculture: A review. Clim Res 50(1):83–102. https://doi.org/10.3354/cr01038
Lobell DB et al (2015) The shifting geography of agricultural productivity. Science 349(6249): 238–240
Makate C, Mutenje M, Ketema M (2020) Inequalities in access to climate-smart agriculture technologies, infrastructure, and institutional Services in Africa: evidence from Malawi and Ethiopia. In: Handbook of climate change management: research, leadership, transformation, pp 1–29
Matocha J et al (2018) Agroforestry adoption in Nicaragua. Agrofor Syst 92(4):1057–1067
McKinsey & Company (2022) Agriculture and technology: shaping the future
Miao Y, Mulla DJ, Robert PC (2016) Identifying important factors influencing corn yield and grain quality variability using artificial neural networks. Precis Agric 17(6):711–725
Ministry of Agriculture (2021) Kisan Suvidha App
Minten B et al (2018) Cold chains in India. Food Policy 77:50–61
Mittal S, Mehar M (2016) Socio-economic factors affecting adoption of modern information and communication technology by farmers in India: analysis using multivariate probit model. J Agric Educ Ext 22(2):199–212
Mujeyi A, Mujeyi K (2023) Digitalization options for scaling climate smart agriculture in smallholder farming systems: lessons and opportunities. FARA Res Report 7(4):10–21
Mustapha UF, Alhassan A-W, Jiang D-N, Li G-L (2021) Sustainable aquaculture development: A review on the roles of cloud computing, internet of things and artificial intelligence (CIA). Rev Aquac 13(4):2076–2091. https://doi.org/10.1111/raq.12559
Nadeem H, Khan A, Gupta R, Hashem M, Alamri S, Siddiqui MA, Ahmad F (2023) Stress combination: when two negatives may become antagonistic, synergistic or additive for plants? Pedosphere 33(2):287–300. https://doi.org/10.1016/j.pedsph.2022.06.031
Nair PKR et al (2017) Agroforestry systems and carbon sequestration. Curr Opin Environ Sustain 6: 8–14

Najafi P, Tabatabaei SH (2007) Effect of using subsurface drip irrigation and ET-HS model to increase WUE in irrigation of some crops. Irrig Drain 56(4):477–486. https://doi.org/10.1002/ird.322
Netafim (2020) Smart irrigation solutions
Ngoma H et al (2018) Drought-tolerant maize in Zambia. Agric Syst 160:57–66
Nyasimi M et al (2017) Farmer field schools in East Africa. Clim Risk Manag 16:208–221
Pagiola S et al (2016) Costa Rica's PES program. Ecol Econ 126:62–73
Pathak H (2023) Impact, adaptation, and mitigation of climate change in Indian agriculture. Environ Monit Assess 195(1):52
Paudel GP et al (2018) Gender and seed systems in Nepal. World Dev 108:57–68
Peng S, Buresh RJ, Huang J, Zhong X, Zou Y, Yang J, Wang G, Liu Y, Hu R, Tang Q, Cui K, Zhang F, Dobermann A (2010) Improving nitrogen fertilization in rice by sitespecific N management. A review. Agron Sustain Dev 30(3):649–656. https://doi.org/10.1051/agro/2010002
Pereira LS et al (2020) Improved indicators for water use efficiency in agriculture. Agric Syst 178:1–12
Perfecto I et al (2016) Shade coffee and biodiversity conservation. Bioscience 66(5):354–363
Pittelkow CM et al (2015) Productivity limits of conservation agriculture. Nat Sustain 1(12):772–779
Playán E, Mateos L (2019) Modernization of irrigation systems: a global review. Water Resour Res 55(4):2345–2360
Pradhan, P., Fischer, G., Velthuizen, H. van, Reusser, D. E., & Kropp, J. P. (2015). Closing yield gaps: how sustainable can we be? PLoS One, 10(6), e0129487. https://doi.org/10.1371/journal.pone.0129487
Pretty J, Bharucha ZP (2015) Integrated pest management for sustainable intensification of agriculture in Asia and Africa. Insects 6(1):152–182
Rabbani G et al (2018) Flood shelters in Bangladesh. Disasters 42(2):345–366
Rathor AS, Choudhury S, Sharma A, Nautiyal P, Shah G (2024) Empowering vertical farming through IoT and AI-driven technologies: A comprehensive review. Heliyon 10(15):e34998. https://doi.org/10.1016/j.heliyon.2024.e34998
Richard B, Qi A, Fitt BDL (2022) Control of crop diseases through integrated crop management to deliver climate-smart farming systems for low- and high-input crop production. Plant Pathol 71(1):187–206. https://doi.org/10.1111/ppa.13493
Rockström J et al (2018) Sustainable intensification of rainwater management. Nat Sustain 1(9):458–465
Saikanth DRK, Gupta K, Srivastava P, Saryam M, Rani K, Jena P, Rout S (2023) Environmental sustainability and food security of traditional agricultural practices in India: a review. Int J Environ Clim Change 13:1847. https://doi.org/10.9734/ijecc/2023/v13i82138
Sander BO et al (2017) Methane reduction in rice systems. Glob Chang Biol 23(10):3998–4006
Schäfer F, Blanke M (2012) Farming and marketing system affects carbon and water footprint – a case study using Hokaido pumpkin. J Clean Prod 28:113–119. https://doi.org/10.1016/j.jclepro.2011.08.019
Schröder JJ, Smit AL, Cordell D, Rosemarin A (2017) Improved phosphorus use efficiency in agriculture: a key requirement for its sustainable use. Chemosphere 84(6):822–831
Scurrah M et al (2019) Indigenous potato varieties in Peru. Food Secur 11(4):885–897
Senoo EEK, Anggraini L, Kumi JA, Karolina LB, Akansah E, Sulyman HA, Mendonça I, Aritsugi M (2024) IoT solutions with artificial intelligence Technologies for Precision Agriculture: definitions, applications, challenges, and opportunities. Electronics 13(10):1894. https://doi.org/10.3390/electronics13101894
Shah T (2018) Groundwater governance and recharge strategies in South Asia. Water Policy 20(2):197–214
Shareef U, Rehman AU, Ahmad R (2024) A systematic literature review on parameters optimization for smart hydroponic systems. AI 5(3):Article 3. https://doi.org/10.3390/ai5030073

Singh NP et al (2018) Capacity building for climate-smart agriculture. Clim Dev 10(3):275–288

Smith RP (2009) Macroscope: the blue baby syndromes. Am Sci 97(2):94–96

Stifel D, Minten B (2017) Market access and agricultural productivity. World Dev 96:411–424

Takahashi K, Muraoka R, Otsuka K (2020) Technology adoption, impact, and extension in developing countries' agriculture: A review of the recent literature. Agric Econ 51(1):31–45. https://doi.org/10.1111/agec.12539

Tashiro T, Wardlaw IF (1990) The response to high temperature shock and humidity changes prior to and during the early stages of grain development in wheat. Funct Plant Biol 17(5):551–561. https://doi.org/10.1071/pp9900551

Teka AM et al (2019) Gender and agricultural mechanization. Agric Syst 173:263–275

Thierfelder C et al (2017) Conservation agriculture in southern Africa. Agric Ecosyst Environ 187: 87–105

Tiwari TP et al (2018) Participatory varietal selection in South Asia. Field Crop Res 213:114–125

Trinh TQ et al (2018) Economic impacts of climate-smart rice in Vietnam. Food Policy 77:19–32

USAID (2020) Kenya last mile infrastructure project

Valdivia C et al (2017) Indigenous and scientific forecasting in Bolivia. Agric Syst 155:126–138

Van Stappen F, Loriers A, Mathot M, Planchon V, Stilmant D, Debode F (2015) Organic *versus* conventional farming: the case of wheat production in Wallonia (Belgium). Agric Agric Sci Procedia 7:272–279. https://doi.org/10.1016/j.aaspro.2015.12.047

Vanlauwe B et al (2017) Soil fertility management in Africa. Geoderma 285:1–14

Villalba R, Joshi G, Daum T, Venus TE (2024) Financing climate-smart agriculture: a case study from the Indo-Gangetic plains. Mitig Adapt Strateg Glob Chang 29(5):33

Vitousek PM, Naylor R, Crews T, David MB, Drinkwater LE, Holland E, Johnes PJ, Katzenberger J, Martinelli LA, Matson PA, Nziguheba G, Ojima D, Palm CA, Robertson GP, Sanchez PA, Townsend AR, Zhang FS (2009) Nutrient Imbalances in agricultural development. Science 324(5934):1519–1520. https://doi.org/10.1126/science.1170261

Wahab SA, Nor NM, Rashid AA (2024) Adopting innovations in Agri-tech sector for a sustainable future. Environ Behav Proc J 9(SI19):Article SI19. https://doi.org/10.21834/e-bpj.v9iSI19.5773

WeFarm (2021) Connecting farmers through mobile

Westermann O et al (2018) Policy coherence for CSA. Glob Environ Chang 49:73–83

Williams AG, Audsley E, Sandars DL (2010) Environmental burdens of producing bread wheat, oilseed rape and potatoes in England and Wales using simulation and system modelling. Int J Life Cycle Assess 15(8):855–868. https://doi.org/10.1007/s11367-010-0212-3

Williams TO et al (2018) Infrastructure for CSA. Agric Syst 164:206–215

Woolf D et al (2016) Sustainable biochar for climate change mitigation. Nat Commun 7:1–9

World Bank (2019) Enabling the business of agriculture

Wu W (2018) Sustainable crop rotation for improving crop productivity and environmental safety: A book review. J Clean Prod 176:555–556. https://doi.org/10.1016/j.jclepro.2017.12.146

Wu W, Ma B (2015) Integrated nutrient management (INM) for sustaining crop productivity and reducing environmental impact: a review. Sci Total Environ:415–512–513, 427. https://doi.org/10.1016/j.scitotenv.2014.12.101

Zhang C, Kovacs JM (2019) The application of small unmanned aerial systems for precision agriculture: a review. Precis Agric 13(6):693–712

Zhang F, Cui Z, Fan M, Zhang W, Chen X, Jiang R (2011) Integrated soil–crop system management: reducing environmental risk while increasing crop productivity and improving nutrient use efficiency in China. J Environ Qual 40(4):1051–1057. https://doi.org/10.2134/jeq2010.0292

Next-Generation Natural Farming Practices for Crop Improvement

11

Suganthy Mariappan, Sowmiya Arumugam, Janaki Ponnusamy, Ravisankar Natesan, and Sharma Shanti Kumar

Abstract

The Green Revolution revolutionized Indian agriculture, shifting the nation from subsistence to surplus production by introducing high-yielding crop varieties, synthetic inputs, and improved irrigation systems. However, its unintended consequences, including soil degradation, groundwater depletion, and escalating cultivation costs, have raised concerns about its sustainability. As a response, natural farming has emerged as a viable alternative, emphasizing ecological harmony, resource conservation, and sustainable livelihoods. This chapter examines the concepts and methodologies of natural farming, offering a detailed analysis of its evolution and growth globally as well as in India. Key features such as *jeevamrit, beejamrit, acchadana,* and *whapasa*, foundational to its success, are discussed alongside policy initiatives, institutional support, and its current status in India. While challenges such as initial yield reduction, labour-intensive processes, and market limitations persist, innovative next-generation strategies hold promise for scaling their adoption. These include advancements in precision agriculture, regenerative soil management, climate-resilient practices, and digital innovations like blockchain for traceability. By integrating traditional knowledge with modern technologies, natural farming presents an opportunity to address

S. Mariappan (✉) · S. Arumugam · J. Ponnusamy
Tamil Nadu Agricultural University, Coimbatore, Tamil Nadu, India
e-mail: suganthy@tnau.ac.in

R. Natesan
Indian Institute of Farming System Research, Modipuram, Meerut, Uttar Pradesh, India

S. S. Kumar
Krishi Anusandhan Bhavan, ICAR, New Delhi, India

Y. Ahlawat et al. (eds.), *Next-Generation Strategies for Crop Improvement*,
https://doi.org/10.1007/978-981-95-0309-4_11

critical agricultural challenges sustainably. This chapter provides an insightful overview of its practices, challenges, and future directions, highlighting its potential as a cornerstone of sustainable agricultural development.

Keywords

Natural farming · Sustainability · Four wheels · Challenges · Next-generation strategies

11.1 Introduction

The Green Revolution, propelled by the adoption of high-yielding varieties, widespread use of synthetic and chemical inputs, and improved irrigation infrastructure, brought a significant transformation to Indian agriculture (Eliazer Nelson et al. 2019; Miglani et al. 2025). It propelled the country from subsistence farming to a surplus-generating agricultural economy. However, this transformation came at a cost. The widespread adoption of these technologies has raised concerns about their long-term sustainability, owing to their impact on natural resources, soil health, and the environment (Gamage et al. 2023).

Issues such as groundwater depletion, declining soil fertility, heavy dependence on nonrenewable inputs, escalating costs of cultivation, and diminishing economic returns are now raising questions about the viability of these methods (Hans and Prabhu 2024). As a result, attention has turned toward alternative approaches that are not only productive but also resource-conserving, environmentally friendly, and capable of providing safe, nutritious food sustainably. Among these alternatives, nonchemical farming systems, particularly organic farming, have emerged as a promising solution. Since the early 2000s, organic farming has been recognized as an umbrella term encompassing various nonchemical farming practices (Ortiz and Peris 2022).

Natural farming embodies a sustainable agricultural approach that emphasizes synergy with ecological systems, conservation of resources, and environmental well-being (Devarinti 2016). With its cost-effective, chemical-free approach, natural farming offers the potential to boost agricultural productivity while tackling critical issues like climate change, soil degradation, and pollution. By fostering soil biodiversity, recycling biomass, and utilizing indigenous inputs, natural farming serves as a viable alternative to conventional methods, paving the way for long-term agricultural sustainability (Fukuoka 2009). Natural farming is an agricultural approach that minimizes human intervention and synthetic inputs while maximizing ecological harmony. It emphasizes soil health, biodiversity, and the utilization of natural resources to sustain crop production. This approach minimizes environmental deterioration while enhancing sustainable livelihoods for farmers (Verma et al. 2023).

Natural farming has garnered global attention as a solution to the adverse effects of traditional farming, such as soil erosion, loss of biodiversity, and groundwater depletion. This chapter examines the core principles of natural farming, traditional practices, and innovative strategies for the future, backed by relevant literature and research findings.

11.2 Understanding the Concept of Natural Farming

The term natural farming refers to farming practices that harmonize with natural processes. It is grounded in agroecological principles, emphasizing improved soil quality and physical structure through organic matter management and biological activity. Key elements include diversified cropping systems, enhanced biomass recycling, and fostering beneficial biological interactions (Duddigan et al. 2022).

Natural farming replicates natural ecosystems by integrating crops and livestock, promoting sustainable agroecosystems. It excludes the use of synthetic pesticides and fertilizers, relying instead on crop rotations, residues, animal manures, legumes, green manures, organic wastes, rock powders, and indigenous pest control methods (Ravisankar et al. 2017). These components support soil fertility, nutrient supply, and pest management, making natural farming a sustainable approach aligned with ecological balance.

11.2.1 Carbon Sequestration in Soil

Regenerative farming practices are crucial in sequestering carbon in the soil, providing a natural solution to help mitigate climate change. By focusing on techniques like cover cropping, crop rotation, minimal soil disturbance, and the incorporation of organic matter, these practices boost soil organic carbon (SOC) levels and enhance overall soil health (Nazir et al. 2024). Cover crops and diverse rotations increase biomass production, which, upon decomposition, enriches the soil with organic carbon. Minimal tillage reduces soil disruption, thereby preventing carbon release into the atmosphere. Moreover, integrating agroforestry and grazing systems further promotes carbon sequestration by enhancing root biomass and microbial activity (Haughey et al. 2023). These methods contribute to carbon storage while enhancing soil fertility and moisture retention, fostering sustainable agriculture. Such approaches are increasingly recognized as essential tools in the global effort to combat climate change (Lal 2004b).

Carbon sequestration offers dual benefits of mitigating climate change and enhancing soil health, making it a cornerstone of sustainable agricultural practices. By capturing atmospheric carbon dioxide and storing it in soil nutrients and plant biomass, carbon storage contributes to reducing greenhouse gas levels, helping to address climate change (Kumara et al. 2023). Moreover, the added carbon content in soils improves soil composition, fertility, and water retention, encouraging plant growth and enhancing resistance to environmental challenges. Enhanced microbial activity in carbon-rich soils further promotes nutrient cycling, reducing dependency on synthetic fertilizers (Oldfield et al. 2022). Additionally, carbon sequestration contributes to erosion control and biodiversity conservation, creating a synergistic effect that bolsters both agricultural productivity and ecological sustainability. These combined advantages highlight the importance of carbon storage as a crucial approach in tackling global environmental issues (Lal 2004a).

11.2.2 Regenerative Grazing and Crop Systems

Integrated livestock and crop systems foster soil regeneration and biodiversity by mimicking natural ecosystems and creating a mutually beneficial relationship between plants, animals, and microorganisms (Rempelos et al. 2023). These systems enhance nutrient cycling as animal manure enriches the soil with organic material, boosting its structure, water retention, and fertility. Grazing animals help stimulate plant growth and diversify the root systems, enhancing soil carbon storage and microbial activity (Schreefel et al. 2022; Musto et al. 2023). Additionally, rotational grazing and diverse crop rotations help prevent soil erosion, disrupt pest and pathogen cycles, and reduce reliance on chemical inputs. This approach also enhances biodiversity by creating environments for diverse soil organisms, pollinators, and wildlife, ultimately fostering a more resilient agroecosystem. Such integrated systems exemplify a sustainable model that boosts productivity while regenerating degraded soils (Herrero et al. 2023).

Rotational grazing, a livestock management practice where animals are moved between pastures to allow for recovery periods, has notable benefits for soil carbon sequestration and plant health. By alternating grazing areas, this practice prevents overgrazing, allowing vegetation to regrow and deepen root systems, which enhances carbon storage within the soil matrix. The enhanced plant cover and root systems minimize soil erosion, enhance water absorption, and boost soil organic content, promoting healthier microbial communities (Qamar et al. 2024). Moreover, manure distribution in rotational systems enriches soil nutrients, promoting plant growth and resilience to stress. This method also diversifies plant species in grazed areas, contributing to ecosystem stability and productivity. Rotational grazing thus proves to be a sustainable practice for enhancing soil health, promoting plant vitality, and addressing climate change through carbon storage in the soil (Teague et al. 2011).

11.2.3 Agroecological Systems for Sustainable Agriculture

Combining ecological principles with farming practices is a transformative approach to regenerating ecosystems while maintaining agricultural productivity. This method leverages natural processes such as biodiversity enhancement, nutrient recycling, and pest management to create resilient and sustainable farming systems. Practices like agroforestry, intercropping, and conservation tillage align with ecological principles by mimicking natural ecosystems, promoting soil regeneration, and increasing carbon sequestration. Integrating cover crops and organic amendments enhances soil fertility and moisture retention, reducing the reliance on chemical inputs (Deguine et al. 2023). Additionally, fostering biodiversity through habitat restoration and pollinator-friendly practices strengthens ecosystem functions, such as pollination and pest regulation. This synergy between ecological understanding and agricultural techniques not only restores degraded landscapes but also enhances farm productivity and resilience to climate change (Thamarai et al. 2024).

Permaculture and regenerative agroforestry systems are crucial for fostering long-term sustainability by integrating farming methods with natural ecosystems. Permaculture designs integrate principles of ecology, resource efficiency, and sustainable living to create self-sustaining agricultural landscapes (Hirschfeld and Van Acker 2021). Similarly, regenerative agroforestry incorporates trees and shrubs into crop and livestock systems, enhancing biodiversity and ecosystem services. These systems enhance soil health by boosting organic matter and carbon storage, minimize soil erosion, and enhance water infiltration and retention (Diwan et al. 2021). Additionally, they foster diverse habitats that support pollinators, wildlife, and beneficial microorganisms (Azhar et al. 2024). By reducing reliance on synthetic inputs and promoting a closed-loop system of nutrient recycling, these approaches contribute to resilient food production and climate change mitigation (Vidadala 2024). Together, permaculture and regenerative agroforestry provide a blueprint for sustainable land use that balances ecological integrity with human needs (Ferguson and Lovell 2014).

11.3 Farming System Diversification

India, often regarded as the birthplace of nonchemical farming, has a rich history of diverse farming systems, each with unique principles, management practices, and inputs. Despite their differences, these systems share common features such as the integration of livestock, biodiversity enhancement, on-farm recycling of biomass, promotion of natural nutrient cycles, and a strong focus on building soil health (Mortensen and Smith 2020).

Some systems rely on external inputs, while others focus exclusively on managing on-farm resources using plant and cattle-based formulations. Over time, many of these systems have been standardized based on the experiences of practitioners, gaining increasing acceptance among farmers. One such system is natural farming, which is positioned as a step beyond organic farming. It aims to free farmers from external inputs, lower cultivation costs, and achieve comparable productivity over time (Sánchez et al. 2022; Puech and Stark 2023).

11.4 Key Features of Natural Farming

Key natural farming practices include microbial enhancers, beejamrit (for seeds), jeevamrit, and ghanjeevamrit to boost microbial activity. Multicropping, acchadana (mulching), and whapasa (humus development through alternate furrow irrigation) aid in improving soil health despite adoption challenges. The key components of

Table 11.1 Components of natural farming

Components/practice	Details	Remarks
Solid and liquid concoctions (prepared using cow dung, cow urine, pulse flour, undisturbed soil, jaggery, water, and lime)	Beejamrit (seed treatment) Jeevamrit (1500 liters/ha) Ghanjeevamrit (1000 kg/ha)	Scientifically proven for enhancing soil health
Multicropping/intercropping	Mandatory 365 days of soil cover, live mulch, and bund cropping	Scientifically proven for enhancing soil health and pest management
Acchadana	Mulching (cover crops or crop residues, or through azolla/blue-green algae)	Scientifically proven for weed management and soil moisture conservation
Whapasa	Improves soil aeration through alternate furrow irrigation and alternate wetting and drying in the case of rice	Scientifically proven for enhancing the soil health and soil microbes
Pest management	Foliar sprays of agniastra, dashparni ark, brahmastra, neemastra, dry ginger—milk extract, sour butter milk	Scientifically proven for pest and disease management

natural farming given by Suganthy et al. (2023a, b) have been explained in Table 11.1.

11.4.1 Soil Mulching

Soil mulching in natural farming entails layering the soil with organic materials that decompose into humus, fostering microbial activity, and enhancing soil structure. Mulching aids in moisture retention, temperature regulation, and weed suppression by blocking sunlight (Ngosong et al. 2019). It also protects against erosion and nutrient runoff while fostering beneficial microorganisms essential for nutrient cycling (Iqbal et al. 2020). Studies indicate that organic mulches minimize soil evaporation and enhance water use efficiency, contributing to sustainable farming practices. Mulching thus supports soil health and enhances crop productivity without synthetic inputs.

11.4.2 Minimal Soil Disturbance

Minimal soil disturbance in natural farming avoids plowing or tilling, preserving soil structure and organic matter. This practice enhances water retention, supports microbial biodiversity, and reduces erosion, replicating conditions in undisturbed ecosystems. Additionally, it promotes carbon sequestration, reduces soil compaction, and minimizes the disturbance of beneficial organisms, like earthworms and mycorrhizal fungi, which are essential for soil fertility and nutrient cycling (Borrelli et al. 2017; Somasundaram et al. 2020). It also helps in lowering greenhouse gas emissions by reducing fuel use and maintaining stable soil organic carbon levels.

11.4.3 Biocultures

Biocultures, such as farm-made bioinputs like jeevamrit and beejamrit, are essential in natural farming. These formulations, prepared using small quantities of cow dung and urine from indigenous breeds, are packed with beneficial microbes that foster microbial diversity and enhance soil fertility (Suganthy et al. 2023a, b). Studies show that jeevamrit significantly boosts nutrient cycling, promoting plant growth and improving soil health. These biocultures also stimulate enzymatic activity in the soil, reducing dependency on synthetic fertilizers. Their application supports sustainable agriculture by fostering a healthy, self-sufficient soil ecosystem (Kalarani et al. 2023).

11.4.4 Reduced Cultivation Cost

Reduced cultivation costs in natural farming arise from eliminating synthetic fertilizers, pesticides, and heavy machinery, thereby lowering expenses. Utilizing low-cost inputs like jeevamrit and beejamrit enhances soil fertility and crop yields sustainably. Studies indicate that farmers practicing natural farming save up to 50% on input costs while maintaining or improving productivity (Korav et al. 2020; Kumar et al. 2023a). Additionally, by fostering soil health and reducing environmental degradation, this method ensures long-term agricultural sustainability, improving farmers' profitability and resilience.

11.4.5 Desi Cow Dependency

The utilization of cow dung and urine from native breeds of India, such as the Gir, Sahiwal, and Red Sindhi, is integral to natural farming. These organic inputs are believed to contain a superior microbial content that enhances soil health and nutrient availability. Research has shown that dung and urine from desi cows have higher concentrations of beneficial microbes, which improve fertility and the structure of the soil (Sharma et al. 2022; Sagar et al. 2024). The traditional belief is that these breeds produce more potent bioinputs due to their adaptation to local environmental conditions, fostering a more resilient and sustainable agricultural framework (Behera et al. 2024).

11.4.6 No External Fertilizers

In natural farming, no synthetic or even conventional organic fertilizers are added to the soil. Instead, nutrients are supplied directly on the surface of the soil through the breakdown of organic matter, including mulches, crop residues, and farmyard manure. This process replicates natural nutrient cycling, where microorganisms decompose organic matter into vital nutrients for plants (Prasad et al. 2021). Research suggests that this approach enhances soil microbial diversity and improves soil structure, fostering long-term fertility and minimizing the reliance on external inputs (Dotaniya et al. 2020). This method promotes sustainable agriculture by depending on natural processes to sustain soil health and productivity.

11.4.7 Eco-Friendly Pest Management

Eco-friendly pest management in natural farming depends on utilizing natural pesticides like neem-based formulations, such as *Astra*, which are efficient in managing pests and diseases. Neem, rich in compounds like azadirachtin, disrupts the hormonal systems of pests, preventing feeding, growth, and reproduction (Patel et al. 2020; Wahjono et al. 2024). Studies have demonstrated that neem-based pesticides offer a safe alternative to synthetic chemicals, reducing environmental contamination and promoting biodiversity (Gupta and Dikshit 2010). Additionally, these biopesticides support the health of beneficial organisms in the soil and on plants, thus contributing to the sustainable management of insect pests in natural farming systems.

11.4.8 Weed Management

In natural farming, weeds are not viewed as a threat but as a valuable resource. They are utilized as living or dead mulch to enhance soil quality, strengthen soil structure, and prevent erosion. By leaving weeds in place or using them as organic mulch, their decomposition adds nutrients and organic material to the soil, enhancing its fertility (Kushal et al. 2024; Rosenstock et al. 2025). This practice also reduces the necessity for synthetic herbicides, promoting a healthier, more sustainable farming system. Research suggests that this approach encourages biodiversity and supports soil health while managing weed growth naturally (Kumar et al. 2023a).

11.4.9 Multicropping System

The multicropping system in natural farming emphasizes cultivating diverse crops together, as opposed to monoculture practices. This system promotes biodiversity, improves soil fertility, and enhances pest and disease resistance. By growing multiple crops, nutrient cycling is more efficient, and the likelihood of pest outbreaks is minimized due to diverse plant species that attract different beneficial organisms (Yin and Song 2024). Studies also show that multicropping increases overall productivity and improves soil health by preventing nutrient depletion (Waha et al. 2020). This approach supports ecological balance and contributes to sustainable farming practices.

11.5 Other Key Principles of Natural Farming

11.5.1 Intercropping

Intercropping in natural farming allows farmers to offset cultivation costs by growing multiple crops simultaneously, thus generating diverse income streams (Mazzafera et al. 2021). This practice maximizes land use and can improve overall farm productivity, making it closer to a zero-cost activity. By diversifying crops, farmers can lower the risk of complete crop failure caused by pest infestations and diseases, or climate extremes, thus enhancing resilience (Huss et al. 2022). Intercropping also improves soil health through the complementary relationships between crops, such as nitrogen fixation by legumes, which benefits neighboring plants (Maitra and Gitari 2020).

11.5.2 Contours and Bunds

Contours and bunds are essential techniques for rainwater conservation in natural farming. Shaping the land into contours helps slow down water runoff, allowing rainwater to be absorbed by the soil, which enhances moisture retention and reduces erosion (Walia and Kaur 2023). Bunds, which are embankments of soil, further aid in holding water, preventing it from flowing away. These techniques not only retain water but also boost soil fertility by reducing nutrient loss, ultimately boosting crop productivity (Homeshwari-Devi et al. 2024). By maintaining optimal moisture levels in the soil, contours and bunds contribute to sustainable and resilient agriculture (Kurdekar et al. 2023).

11.5.3 Local Earthworms

Promoting native soil-dwelling earthworms in natural farming is essential for boosting soil fertility. These earthworms decompose organic matter, like crop residues, into nutrient-rich compost, enhancing soil structure and stimulating microbial activity (Singh et al. 2020). Their burrowing also aerates the soil, enhancing water infiltration and root penetration. Earthworms help cycle nutrients efficiently, making them easier for the plants to absorb. By fostering a healthy earthworm population, farmers can naturally boost soil health and fertility, reducing the requirement of synthetic/chemical fertilizers and contributing to agricultural sustainability (Bertrand et al. 2015).

11.5.4 Cow Dung

Cow dung from indigenous *Bos indicus* cattle, such as Gir or Sahiwal, is highly valued in Zero Budget Natural Farming due to its superior microbial content compared to European breeds like Holstein. These native cows are believed to produce dung that is richer in beneficial microorganisms, which enhance the fertility of the soil, thus supporting microbial diversity (Sagar et al. 2024). The use of cow dung from indigenous breeds is fundamental for Zero Budget Natural Farming, as it not only improves soil health but also minimizes external input costs (Sagar et al. 2023). Historically, the Indian cow has been pivotal in rural agricultural practices, symbolizing sustainability and self-reliance in farming systems.

11.6 Natural Farming: Global and Indian Perspective

11.6.1 Global Perspective

Natural farming is an ecosystem-based farming practice that emphasizes the integration of natural biodiversity into agricultural systems. It fosters soil biological activity by nurturing the intricate network of living organisms, including plants and animals (Cassman and Grassini 2020). The goal is to create a thriving ecosystem that harmoniously supports food production while sustaining natural ecological processes. This approach seeks to replicate natural ecosystems, enhancing resilience, reducing dependency on synthetic inputs, and promoting long-term sustainability.

11.6.2 Indian Context

In India, natural farming is characterized by practices that prioritize plant diversity, soil health, and sustainable resource use. Key features include promoting soil aeration, recycling biomass through mulching, and rejuvenating soil with microbial formulations derived from the dung and urine of indigenous cows. Pest and disease management relies on farm-made botanical preparations, eliminating the need for synthetic chemicals. Importantly, natural farming in the Indian context strictly prohibits the use of external inputs, focusing instead on self-sufficiency by utilizing resources available on the farm (Shyam et al. 2020).

11.7 Four Wheels of Natural Farming

The four wheels of natural farming are beejamrit (seed treatment), jeevamrit (soil enrichment), acchadana (mulching), and whapasa (soil moisture management) and are illustrated in Fig. 11.1. These principles enhance soil fertility, microbial activity, and sustainable crop production.

11.7.1 Jeevamrit

Jeevamrit is a microbial solution in fermented form that serves as a nutrient supplier and microbial catalyst, enhancing the activity of earthworms and soil microbes. Throughout its fermentation period for around 48 h, both the aerobic and anaerobic bacteria in cow dung and urine thrive by metabolizing organic materials like pulse flour. Undisturbed soil at a very small amount is added to introduce native microbial species, enriching the solution. Jeevamrit also helps in preventing fungal and bacterial diseases of plants. Jeevamrit is applied bimonthly or as needed, either mixed with water used for irrigation or as a 10% foliar spray (Figs. 11.2 and 11.3) (Palekar 2005a, 2005b).

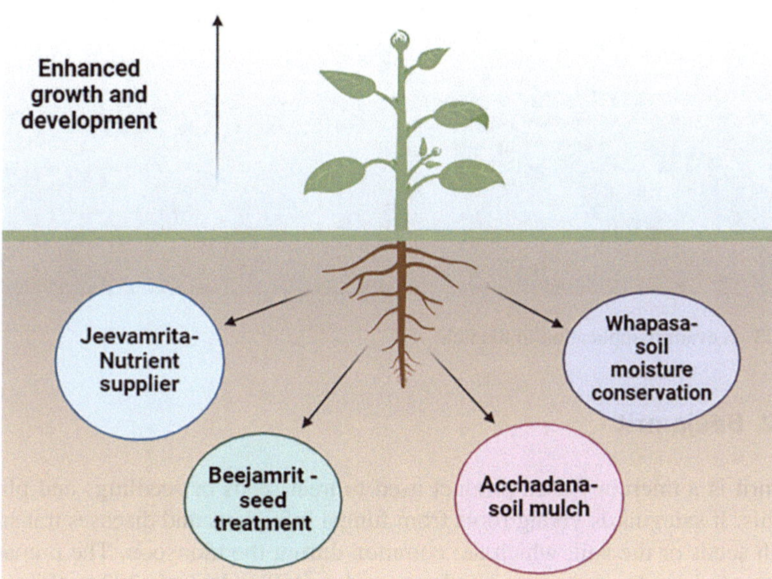

Fig. 11.1 Four wheels of natural farming

Fig. 11.2 Preparation of jeevamrit

Fig. 11.3 Jeevamrit application in the field

11.7.2 Beejamrit

Beejamrit is a microbe-based product used to treat seeds or seedlings and planting materials. It safeguards young roots from fungal infections and diseases transmitted through seeds or the soil, which are common during the monsoon. The preparation consists of cow dung, cow urine, lime, and soil (Patel et al. 2024) (Fig. 11.4). Beejamrit is applied to seeds by hand and left to dry before sowing. For legumes, a quick dip followed by drying is sufficient.

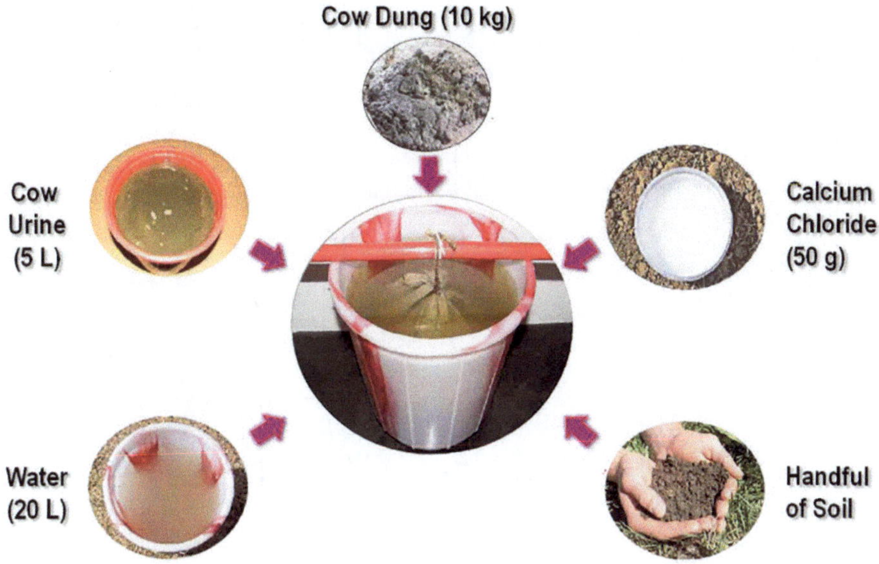

Fig. 11.4 Preparation of beejamrit

11.7.3 Mulching (Acchadana)

Mulching involves soil covering to conserve and retain moisture, improve fertility, and promote biodiversity. Soil mulch maintains soil health by preventing topsoil erosion and enhancing aeration and water retention. Deep ploughing is avoided. Straw mulch utilizes crop residues or any decomposing organic material, enriching the soil via microbial processes (Fig. 11.5). Live mulch combines symbiotic intercrops of monocots and dicots to supply nutrients like nitrogen, potash, and sulfur (Ranjan et al. 2017; Ranjan and Sow 2021).

11.7.4 Whapasa

Whapasa is a soil condition where both air and water molecules coexist, reducing excessive irrigation needs. It prevents over-saturation, allowing plant roots to access both moisture and oxygen efficiently. This principle enhances microbial activity, improving soil health and nutrient availability (Kumar et al. 2023a). Irrigation is performed selectively, such as at midday and in alternate furrows, significantly lowering water requirements in natural farming (Khan et al. 2022). Additionally, whapasa helps mitigate soil erosion, conserves groundwater, and reduces dependency on external irrigation sources, making it a key practice for sustainable agriculture in water-scarce regions (Mandla and Sharma 2023).

Fig. 11.5 Paddy straw mulching in the natural farming cotton field

11.8 Development of Natural Farming in India

Natural farming gained recognition in the 1980s, pioneered by Subhash Palekar, an Indian farmer and agricultural scientist. After extensively researching Vedic knowledge, organic farming practices, and conventional agricultural science, he introduced Zero Budget Natural Farming (ZBNF), following experiments on his own farm (Khadse et al. 2018). Shri Acharya Devvrat, the former Governor of Himachal Pradesh and current Governor of Gujarat, played a pivotal role in promoting the natural farming movement. Under his leadership, natural farming practices reached all villages and panchayats in Himachal Pradesh within 3 years (Laishram et al. 2022).

11.8.1 Policy and State-Level Initiatives

Natural farming has gained significant policy support at both the state and national levels, reflecting its potential for sustainable agriculture. In Andhra Pradesh, Zero Budget Natural Farming (ZBNF) emerged as a grassroots movement and was later institutionalized as a major state policy, promoting chemical-free farming across millions of hectares (Veluguri et al. 2021; Duddigan et al. 2022). Himachal Pradesh launched the Prakritik Kheti Khushhal Kisan Yojana in 2018 to scale up Subhash Palekar Natural Farming (SPNF), encouraging farmers to adopt cost-effective, ecological practices (Chandel et al. 2021). At the national level, the Government of India officially recognized natural farming under the Bhartiya Prakritik Krishi Paddhati (BPKP), drawing inspiration from traditional knowledge systems like *Vrikshayurveda*, which emphasizes farming in harmony with nature (Sharma et al. 2023). The importance of natural farming was further underscored in the Union Budget (2019–2020), where the Finance Minister of India highlighted its role in sustainable agriculture. Prime Minister Narendra Modi reinforced this vision in a national conference on December 16, 2021, urging farmers to transition toward natural farming. Various states, including Gujarat, Madhya Pradesh, and Uttar Pradesh, have also introduced pilot programs and incentives to encourage their adoption. These policy initiatives aim to mainstream natural farming by integrating it with existing agricultural schemes, ensuring long-term sustainability and farmer welfare (Darjee 2023).

11.8.2 Budgetary and Institutional Support

In the 2022 budget, the finance minister proposed the promotion of chemical-free farming, starting with 5 km-wide corridors along the Ganga River. To bolster the adoption of natural farming practices, the *Prakrutik Krishi Vikas Board* was established in Gujarat in 2021–2022 (Singh et al. 2024b). From 2023–2024, the government introduced the *National Mission on Natural Farming (NMNF)* as a separate initiative to expand the BPKP (Panda et al. 2022). This initiative aims to encourage behavioral change among farmers, replacing chemical-based inputs with cow-based, locally produced alternatives through training and capacity building.

11.8.3 Current Status of Natural Farming in India

11.8.3.1 Expansion Across States

Natural farming is expanding across India through grassroots movements, civil society efforts, and government programs. Farmers and civil society have advanced natural farming in Karnataka, Tamil Nadu, Maharashtra, and other states. The movement often overlaps with organic farming, and terms like "organic" and "natural" farming are used interchangeably by communities. In Karnataka, Zero Budget Natural Farming (ZBNF), promoted by Subhash Palekar, is widely adopted, reducing chemical use (Khadse et al. 2018; Naik 2021). Tamil Nadu has also embraced natural farming, focusing on restoring native forests and improving biodiversity (Paramasivam et al. 2022). In Maharashtra, farmers in drought-prone areas like Vidarbha and Marathwada have adopted ZBNF to conserve water and cut costs (Kumar et al. 2024b). Gujarat, Rajasthan, Himachal Pradesh, and Andhra Pradesh have launched large-scale natural farming programs under government schemes like Bharatiya Prakritik Krishi Paddhati (BPKP) (Ravisankar 2023) and Paramparagat Krishi Vikas Yojana (PKVY) (Yadav et al. 2022). Madhya Pradesh farmers in Chhindwara have shifted from conventional to organic cotton farming with support from organizations like WWF and SRIJAN, improving soil health and income (Keller et al. 2024). These efforts show India's commitment to sustainable farming and soil conservation.

11.8.3.2 Area Under Organic and Natural Farming

As of 2024, India has made significant progress in organic and natural farming, with a total of 73.0 lakh hectares under organic certification. This includes 17.64 lakh hectares under certified organic farming, 36.27 lakh hectares under conversion to organic farming, and 19.09 lakh hectares under wild harvest areas. The distribution of organic farming across various government schemes comprises 26.57 lakh hectares under the National Program for Organic Production (NPOP), 6.19 lakh hectares under the Paramparagat Krishi Vikas Yojana (PKVY), 4.09 lakh hectares under the Bharatiya Prakritik Krishi Paddhati (BPKP) for natural farming, and 1.23 lakh hectares under the Namami Gange Programme, with an additional 34.92 lakh hectares falling under other organic initiatives.

State-wise, Madhya Pradesh leads with over 11.4 lakh hectares under organic cultivation, followed by Maharashtra with over 12.0 lakh hectares, while Rajasthan, Gujarat, and Karnataka also contribute significantly to the expansion of organic farming. To further promote sustainable agriculture, the Government of India has set a target to bring an additional 7.5 lakh hectares under natural farming by implementing the initiative in 15,000 clusters across willing Gram Panchayats, aiming to reach 1 crore farmers. This initiative prioritizes the adoption of biofertilizers, organic inputs, and certification programs to enhance soil health, improve biodiversity, and reduce dependency on synthetic chemicals, thereby reinforcing India's commitment to eco-friendly and sustainable agricultural practices.

11.8.3.3 Area Under Regenerative Agriculture Practices

As of 2025, regenerative agriculture in India has expanded significantly, with around 14.99 lakh hectares under organic farming and 12 lakh hectares under the Bharatiya Prakritik Krishi Paddhati (BPKP). BPKP promotes traditional practices like biomass recycling and cow dung-urine formulations while excluding synthetic inputs, with ₹12,200 per hectare financial support for 3 years. Andhra Pradesh, Karnataka, Himachal Pradesh, Gujarat, Uttar Pradesh, and Kerala have widely adopted BPKP, enhancing sustainability and rural livelihoods. The National Mission on Natural Farming aims to form 15,000 clusters, covering 7.5 lakh hectares and 1 crore farmers, further strengthening India's shift toward sustainable agriculture.

11.9 Challenges and Opportunities in Natural Farming

A recent report by NITI Aayog (2022) highlights that despite increasing awareness, less than 1% of India's cropped land currently utilizes biofertilizers, and only about 25% relies on organic fertilizers. This underscores the significant potential for scaling up the use of bio- and organic fertilizers to reduce dependence on subsidized inorganic inputs. Historical data shows that dung manure's contribution to the total value of fertilizers declined from 43% in 1970–1971 to 13% by the 1990s, and recent estimates suggest a continued decline due to reduced livestock numbers and changing farming practices. Addressing these challenges through improved supply chains, research, and financial incentives can enhance soil health and accelerate the adoption of natural farming.

11.9.1 Initial Yield Reduction

Transitioning from conventional to natural farming often leads to temporary yield declines (Brewer et al. 2024; Mponella et al. 2024). Natural farming, emphasizing sustainable and chemical-free agricultural practices, offers environmental and health benefits. However, transitioning from conventional farming often faces a significant hurdle of initial yield reduction. This challenge, deeply rooted in soil, ecological, and socioeconomic factors, can deter farmers from adopting these methods despite their long-term potential (Smith et al. 2020; Kurniawati et al. 2023).

11.9.1.1 Soil Transition Period

The shift from chemical fertilizers to organic or bio-based inputs disrupts existing soil chemistry and microbial activity. Soils previously reliant on synthetic inputs often experience nutrient imbalances and a decline in immediate nutrient availability, leading to a temporary drop in crop yields (Bharadwaj et al. 2025). The decomposition of accumulated synthetic residues and the slow adaptation of microbial communities can further delay soil fertility restoration. Additionally, soil structure changes during this transition, potentially affecting water infiltration and root development. The unpredictability of microbial succession and nutrient mineralization rates poses a significant challenge, making the transition period a critical bottleneck in adopting natural farming (Santos et al. 2012; Sacco et al. 2015).

11.9.1.2 Input Availability and Efficiency

Natural farming relies on bioinputs such as compost, biopesticides, and microbial inoculants. However, limited access to high-quality inputs and the lack of standardized production protocols can compromise their effectiveness, leading to yield reductions in the initial stages (Lohr and Park 2007; Shahrajabian et al. 2021). Variability in input composition, storage conditions, and application methods further affects nutrient availability and pest control efficacy. Dependence on locally sourced materials may result in inconsistent input quality, creating challenges in large-scale adoption (Röös et al. 2018). Moreover, high initial costs and logistical constraints in input distribution can hinder widespread implementation, delaying the transition to natural farming (Arthurson and Jäderlund 2011).

11.9.1.3 Knowledge and Labour Requirements

Natural farming requires specialized knowledge and intensive manual labour, posing challenges for widespread adoption. Farmers must learn techniques such as composting, biofertilizer preparation, and soil health management, which involve a steep learning curve. The absence of structured training programs and extension services can lead to suboptimal implementation, affecting productivity (Durham and Mizik 2021; Sapbamrer and Thammachai 2021). Additionally, the high labour demand for activities like manual weeding and mulching increases workload, making it less appealing for smallholders and aging farming populations (Singh and Thakur 2022). Without adequate knowledge transfer and skill development, the efficiency and scalability of natural farming remain limited.

11.9.1.4 Economic Pressures

The initial yield reduction in natural farming can create economic stress for farmers, particularly those dependent on high-yield, market-driven crops like paddy and wheat. Lower productivity in the transition phase reduces income, making it difficult to cover input costs, repay loans, or sustain household expenses (Durham and Mizik 2021). Additionally, the lack of premium pricing or guaranteed procurement for naturally grown produce further discourages adoption. Farmers relying on government incentives and subsidies for conventional agriculture may find it financially challenging to shift without adequate policy support, financial assistance, or alternative income sources during the transition (Milestad and Darnhofer 2003; Kledal et al. 2006).

11.9.2 Strategies to Overcome Initial Yield Reductions

11.9.2.1 Soil Health Management

Soil health management is crucial in natural farming to restore fertility and maintain long-term productivity. Practices like mulching, cover cropping, and incorporating nitrogen-fixing plants improve soil structure, enhance microbial activity, and increase organic matter content (Shahane and Shivay 2022). In arid regions like Andhra Pradesh, mulching has proven effective in reducing soil erosion, conserving moisture, and stabilizing yields under Zero Budget Natural Farming (ZBNF) (Kumar et al. 2023b). Additionally, crop rotations and compost applications help replenish essential nutrients, ensuring sustainable soil health without reliance on synthetic inputs (Tully and McAskill 2020).

11.9.2.2 Supportive Policies and Incentives

Governments can offset initial yield losses through subsidies, grants, and premium pricing for organic produce. Programs like Godhan Nyay Yojana facilitate access to affordable organic inputs, reducing dependency on costly synthetic fertilizers (Łuczka et al. 2021). Additionally, certification support, tax benefits, and direct procurement by government agencies can enhance market stability, encouraging more farmers to adopt sustainable practices (Yang et al. 2024).

11.9.2.3 Education and Capacity Building

State-led initiatives, such as those in Andhra Pradesh, have demonstrated the effectiveness of community-based learning and peer-to-peer knowledge sharing in overcoming adoption barriers (Zuin et al. 2021). Providing farmers with technical training and advisory services ensures the effective implementation of natural farming methods (Alotaibi et al. 2021). Digital platforms and mobile advisory services can further enhance access to real-time guidance, ensuring continuous support for farmers transitioning to sustainable practices (Kassem et al. 2021; Klerkx 2021). Additionally, workshops, field demonstrations, and farmer-led innovation hubs play a crucial role in strengthening practical knowledge, making natural farming more accessible and scalable (Nelson et al. 2023; Pircher et al. 2024).

11.9.2.4 Research and Evidence-Based Practices

Expanding research into the long-term impacts of natural farming on soil health, yield, and economic viability can help address skepticism and inform policy frameworks. Current studies suggest that yield penalties can be minimized through effective resource management and adaptation to local conditions (Möhring et al. 2024). Strengthening collaborations between research institutions, agricultural extension agencies, and farmers can facilitate evidence-based recommendations, ensuring more reliable and scalable natural farming practices.

While initial yield reduction presents a significant challenge to natural farming, it is not insurmountable. A combination of policy support, resource availability, education, and innovation can help farmers transition smoothly. Emphasizing the enduring benefits notably, reduced input costs, improved soil health, and environmental sustainability, can inspire confidence in adopting natural farming practices, ultimately leading to a more resilient agricultural system (Tran-Nam and Tiet 2022). By addressing these challenges holistically, natural farming can transition from a niche practice to a sustainable solution for global agricultural challenges (Duddigan et al. 2022).

11.9.3 Labour-Intensive Practices

Natural farming, focusing on sustainability and ecological balance, often necessitates labour-intensive practices. These methods aim to work in synergy with nature, eschewing synthetic inputs for more organic alternatives like mulching, intercropping, and manual weeding. While this approach benefits long-term soil health and biodiversity, it poses significant challenges due to its high labour demands (Daum et al. 2023; Tshotsho et al. 2023).

11.9.3.1 Dependence on Manual Labour

Natural farming depends heavily on practices such as manual weeding, mulching, and pest control. These activities, while environmentally friendly, are time-consuming and physically demanding compared to the mechanized methods used in conventional farming (Raj et al. 2022). For example, farmers engaged in natural farming spend a substantial portion of their time managing weeds without synthetic herbicides, increasing their workload significantly (Karunathilake et al. 2023).

11.9.3.2 Shortage of Skilled Labour

The rural-urban migration trend has exacerbated labour shortages, particularly in farming sectors. Younger generations often avoid agriculture due to its perceived lack of financial stability and demanding nature. This trend has left many natural farmers struggling to find workers willing to engage in labour-intensive tasks, further complicating operations (Prause 2021).

11.9.3.3 High Labour Costs

Rising labour costs also strain natural farming operations. For instance, wages for farm labour in the United States increased by 6% in 2023, adding financial pressure on farmers who rely on hired labour for time-intensive activities. Programs like the H-2A visa, designed to address labour shortages, involve additional costs for housing and transportation, creating challenges for small-scale farmers to implement natural methods (Orsini et al. 2018; Gamage et al. 2023).

11.9.3.4 Complex Transition Processes

Transitioning from conventional/traditional to natural farming requires a steep learning curve for farmers and their workforce. Adopting eco-friendly methods demands a mindset shift, specialized training, and adaptation to techniques such as crop rotation and composting, which are more labour-intensive than conventional approaches (Lamine et al. 2014; Verburg et al. 2022).

11.9.4 Strategies to Overcome Labour-Intensive Practices

11.9.4.1 Mechanization and Innovation

Mechanization and innovation are easing the labour challenges of natural farming. Precision tools like automated weeders, mulchers, and drones improve efficiency while reducing manual effort. Small-scale mechanization, including power tillers and seeders, supports marginal farmers (Sims and Kienzle 2017). AI-driven monitoring and IoT-enabled soil sensors optimize resource use, making natural farming more scalable and sustainable. These advancements bridge traditional practices with modern efficiency, reducing labour dependency (Rathor et al. 2024).

11.9.4.2 Community Support Models

Community support models, such as cooperative farming and labour-sharing networks, help smallholders reduce costs and workload (Durham and Mizik 2021). Farmer Producer Organizations (FPOs) and Self-Help Groups (SHGs) enable bulk procurement of organic inputs and market access (Shree and Vaishnavi 2022; Ghosh and Ghosh 2024). Community-supported agriculture (CSA) strengthens direct farmer-consumer linkages, ensuring stable income (Sulistyowati et al. 2023; Wu et al. 2024). Digital platforms facilitate knowledge exchange and collaboration, making natural farming more viable (Borrero and Mariscal 2022). These collective approaches enhance sustainability, efficiency, and economic resilience for small farmers.

11.9.4.3 Policy Support

Policy support plays a crucial role in promoting natural farming by providing financial incentives, training, and infrastructure. Governments can offer subsidies for labour costs, organic inputs, and certification processes to ease the transition for farmers. Skill development programs and educational initiatives can encourage youth participation, ensuring long-term adoption (Stolze and Lampkin 2009). Strengthening Farmer Producer Organizations (FPOs) and cooperative models can enhance market access and price stability. Additionally, research and extension services can help integrate scientific advancements with traditional knowledge, making natural farming more efficient and scalable (Vijayakumar 2021; Gurung et al. 2024).

11.9.4.4 Integration of Technology

Digital platforms can also facilitate real-time monitoring of soil health, weather patterns, and crop conditions using AI and IoT-based solutions (AlZubi and Galyna 2023). Additionally, blockchain technology can enhance supply chain transparency, ensuring better market linkages and fair pricing for farmers practicing natural farming (Tegeltija et al. 2022; Manteghi et al. 2023).

Natural farming has great potential for sustainable agriculture, but overcoming its labour-intensive aspects calls for joint efforts from policymakers, researchers, and farming communities. By investing in supportive infrastructure and education, the lasting advantages of natural farming can be fully realized.

11.9.5 Knowledge Gaps

Natural farming is an eco-friendly, sustainable, and resource-efficient alternative to conventional farming, which has gained traction worldwide. However, significant knowledge gaps hinder its widespread adoption, particularly in understanding best practices, ecological dynamics, and resource optimization. These gaps present a formidable challenge to farmers, researchers, and policymakers (Sahu et al. 2010).

11.9.5.1 Lack of Access to Scientific Research

Natural farming practices often rely on traditional knowledge and ecological principles, but the scientific foundation behind these methods is still underdeveloped (Papzan and Shiri 2012). Comprehensive research is required to understand soil biology, nutrient cycles, and pest management strategies specific to natural farming systems. Bridging this gap demands more investments in agricultural research and experimental studies, as highlighted by recent analyses of natural farming initiatives (Ponzio et al. 2013; Sapbamrer and Thammachai 2021).

11.9.5.2 Inadequate Knowledge Dissemination

Farmers face difficulties accessing information about natural farming techniques due to limited extension services and training programs (Gikunda et al. 2021). While traditional methods are valuable, a blend of traditional and modern approaches through farmer-to-farmer knowledge transfer and extension services could address this gap effectively (Onunwa et al. 2022). Governments and agricultural institutions must prioritize education campaigns and workshops to ensure farmers are equipped with evidence-based knowledge (Lone and Rashid 2024).

11.9.5.3 Challenges in Ecosystem Understanding

The interactions within natural farming ecosystems, such as the role of biodiversity in pest control and nutrient recycling, remain underexplored (Zerssa et al. 2021; McPhearson et al. 2022; Boix-Fayos and de Vente 2023). A detailed understanding of these interactions can enable farmers to optimize their farming systems. Existing studies suggest that improving knowledge of such ecological functions could significantly enhance productivity and sustainability.

11.9.5.4 Policy and Institutional Support

Farmers require strong institutional support to adopt natural farming effectively. Policies that promote research, technical training, and reliable data collection are essential for informed decision-making. Collaborative platforms connecting farmers, researchers, and policymakers can help reduce uncertainties and align efforts toward sustainable agricultural goals (Krishnan et al. 2021). Government incentives, such as subsidies for natural inputs, certification support, and market linkages, can accelerate the adoption of natural farming (Lv et al. 2023). Strengthening extension services and farmer cooperatives will enhance knowledge-sharing and grassroots capacity-building (Bhatt and John 2023). Additionally, financial assistance, crop insurance, and risk-mitigation frameworks can provide farmers with economic security during the transition from conventional to natural farming. Developing supply chain infrastructure and dedicated organic markets will further boost profitability and ensure the long-term sustainability of natural farming practices (Zhang and Georgescu 2022).

11.9.5.5 Need for Context-Specific Solutions

Natural farming practices must be adapted to local climatic, soil, and ecological conditions to ensure their effectiveness. Region-specific research is essential to identify suitable crop varieties, soil amendments, and water management techniques that align with local needs (Zollet and Maharjan 2021). Traditional knowledge from indigenous farming communities can provide valuable insights into resilient, location-specific practices (Diogo et al. 2022). Case studies such as those in Madhya Pradesh demonstrate the importance of localized strategies, including integrated nutrient management and pest control methods tailored to regional biodiversity (Tscharntke et al. 2021). Additionally, collaborative research between agricultural institutions and local farmers can refine natural farming models, ensuring higher productivity and sustainability (Carton et al. 2022). Encouraging adaptive policy frameworks that allow flexibility in implementation will further enhance the success of natural farming across diverse agroecosystems (Fruscella et al. 2021).

11.9.6 Strategies to Overcome Knowledge Gaps in Natural Farming

To overcome the knowledge shortages in natural farming, a multipronged approach is essential. This includes advancing scientific research, fostering farmer-to-farmer networks, expanding educational outreach, and implementing robust policy frameworks. By confronting these difficulties, natural farming can realize its potential to contribute to sustainable farming, rural development, and environmental conservation (Adefila et al. 2024).

11.9.7 Market Limitations

Natural farming offers significant ecological and health benefits, but market limitations present formidable challenges to its widespread adoption. The economic sustainability of natural farming often depends on a supportive market structure, including demand for organic produce, competitive pricing, and robust supply chains (Singh et al. 2024a). Addressing these market challenges is essential to enable farmers to transition successfully to natural farming.

11.9.7.1 Limited Consumer Awareness and Demand

Despite growing awareness of organic products, the market for natural farm produce remains niche in many regions. Consumers often confuse natural farming with organic farming, leading to inconsistent demand. Additionally, the higher prices of natural products due to labour-intensive practices discourage price-sensitive consumers, especially during periods of economic hardship, such as inflation (Jensen et al. 2011; Rasool et al. 2021).

11.9.7.2 Inadequate Market Access

Farmers practicing natural farming frequently face difficulty in accessing mainstream markets. Limited certification processes and poor infrastructure for the storage and transport of perishable goods exacerbate this problem (Nandi et al. 2021). For instance, natural farmers in developing regions often lack cold storage facilities or efficient logistics networks, which lead to increased wastage and reduced profitability (Azam et al. 2019).

11.9.7.3 Certification and Trust Issues

The absence of universally recognized certification standards for natural farming complicates market penetration. Without reliable certification, consumers are reluctant to spend more on natural and organic products, and farmers struggle to differentiate their produce from conventional goods. This undermines farmer confidence and reduces incentives for adopting sustainable practices (Watanabe et al. 2021; Murphy et al. 2022).

11.9.7.4 Price Volatility and Competition

Farmers transitioning to natural farming often compete with large-scale conventional producers who benefit from economies of scale. Moreover, price volatility in organic markets can make natural farming financially unstable. In some cases, global competition further complicates market conditions, as imported organic produce may undercut locally produced goods (Su and Cook 2015; Brzezina et al. 2016).

11.9.8 Strategies to Overcome Market Limitations

11.9.8.1 Policy Interventions

Governments and agricultural organizations must establish reliable certification systems and support infrastructure development for processing and storage (Akkaya et al. 2021). Policies should also incentivize local procurement of natural produce by retailers and food companies (Dabbert et al. 2004). Price stabilization mechanisms, such as minimum support prices (MSPs) or direct benefit transfers, can protect farmers from market fluctuations (Musselli 2017; Akkaya et al. 2021). Strengthening farmer cooperatives and producer groups will improve collective bargaining power and facilitate better market access (Haldar and Damodaran 2022). Additionally, digital marketplaces and e-commerce platforms can bridge the gap between farmers and consumers, ensuring fair pricing and wider reach for natural produce (Rani et al. 2022; Dalali et al. 2023; Matos et al. 2024).

11.9.8.2 Building Consumer Awareness

Campaigns that emphasize the health benefits and environmental advantages of natural farming could help expand the customer base. Educational initiatives, transparent labeling, and certification standards can build consumer trust and differentiate natural products in the market (Mkhize and Ellis 2020). Public awareness programs, social media outreach, and collaborations with health professionals can further highlight the nutritional value of chemical-free produce (Rehber 2002). Additionally, farm-to-table initiatives, organic food festivals, and partnerships with restaurants and retailers can create direct linkages between farmers and consumers, fostering demand for naturally grown products (Soni et al. 2022).

11.9.8.3 Strengthening Supply Chains

Partnerships between farmers, cooperatives, and distributors can streamline logistics and improve market access. Localized markets, such as farmers' markets and direct-to-consumer models, can reduce reliance on intermediaries and ensure better prices for farmers (Mendes et al. 2024). Investing in cold storage, processing units, and efficient transportation networks can minimize postharvest losses and maintain product quality (Sharma et al. 2024). Additionally, e-commerce platforms and digital marketplaces can expand reach, allowing farmers to connect directly with consumers and institutional buyers, further enhancing profitability and sustainability (Onwude et al. 2020).

11.9.8.4 Encouraging Innovation

Integrating technology into the supply chain, such as blockchain for transparency or AI for demand forecasting, can enhance the efficiency and reliability of natural farming markets. Digital traceability systems can build consumer trust by ensuring authenticity and fair pricing (Bai et al. 2023). Mobile apps and online platforms can facilitate direct transactions between farmers and buyers, reducing dependency on intermediaries. Additionally, smart logistics solutions, automated warehousing, and IoT-enabled monitoring can optimize storage conditions and minimize wastage, making natural farming more competitive and scalable (Arun Kumar and Karthikeyan 2022; Rajabzadeh and Fatorachian 2023).

By addressing these market challenges, natural farming can achieve its potential to contribute to sustainable agriculture and rural development. Stakeholders must collaborate to create an enabling ecosystem that rewards the environmental and societal gains of natural farming practices.

11.10 Natural Farming Practices

11.10.1 Summer Cropping or Fallow Management

Natural farming promotes premonsoon dry sowing and mixed cropping to maintain soil cover, enhance microbial activity, and suppress weeds. A diverse seed mix, broadcasted at 25% of sole crop rates, improves nutrient cycling and pathogen suppression through allelopathy and competition (Thallam 2023). It has been reported that integrating green gram, black gram, cowpea, and millets in summer cropping under Zero Budget Natural Farming (ZBNF) enhanced soil fertility and moisture retention. Continuous soil cover reduces erosion, minimizes evaporation, and improves soil structure (Garnier et al. 2024). Studies on intercropping legumes with pearl millet have increased drought resilience by optimizing soil moisture. Deep-rooted crops like pigeon pea and sorghum improve aeration and water infiltration, mitigating compaction (Kumar et al. 2019). By suppressing weeds and regulating pest populations, summer cropping fosters a stable and self-sustaining agroecosystem (Martinez et al. 2020).

11.10.2 Seed Treatment

Seed treatment in natural farming enhances germination, protects against soil-borne pathogens, and strengthens early root development using bioinoculants. Beejamrit, a microbial-rich formulation made from desi cow dung, cow urine, lime, virgin soil, and water, is widely used in organic paddy cultivation to suppress fungal pathogens like *Rhizoctonia solani* and *Fusarium* spp. (Sharma et al. 2023). Neem leaf extract, traditionally applied to pulses such as chickpea and pigeon pea, acts as a natural pesticide, reducing seed-borne fungal infections and deterring pests like *Callosobruchus chinensis* (Reddy and Chowdary 2021). Fermented formulations such as Panchagavya, used in tomato and chili farming, enhance seed vigour and early seedling establishment by supplying essential micronutrients and growth-promoting hormones (Vijayakumary et al. 2023). Mycorrhizal fungi inoculation, particularly in crops like maize and groundnut, improves phosphorus uptake, enhancing drought tolerance and root proliferation. To ensure better microbial adhesion, jaggery or pulse flour is often added, a common practice in cotton and sorghum farming (George and Ray 2023; Soti et al. 2023). These holistic treatments minimize chemical dependency, leading to healthier seedlings and improved crop resilience in diverse agroecosystems.

11.10.3 Nutrient Management

Natural farming relies on microbial inputs like jeevamrit, beejamrit, and ghanjeevamrit to enhance soil fertility and nutrient availability (Thallam 2023; Rosenstock et al. 2025). These formulations, made from cow dung, cow urine, pulse flour, jaggery, soil, and water, improve soil microbial diversity, accelerate nutrient cycling, and support plant health. In organic paddy cultivation, farmers apply jeevamrit bimonthly at 500 L/ha via irrigation, significantly improving soil aeration and nutrient absorption (Saharan et al. 2023). Beejamrit is widely used in seed treatment for crops like wheat and chickpea, reducing fungal infections and enhancing early seedling vigour (Patel et al. 2024). Ghanjeevamrit, a dry alternative, is particularly effective in rainfed millet farming in Rajasthan, where it is applied at 250 kg/ha to sustain soil fertility and moisture retention (Das et al. 2023). These practices have been observed to reduce synthetic fertilizer dependency, improve crop resilience to drought and pests, and support long-term soil health (Darjee et al. 2024).

11.10.4 Weed Management Through Mulching (Acchadana)

Natural farming utilizes mulching as an effective weed management technique, minimizing competition for nutrients and moisture while enhancing soil health (Merfield 2023). Soil mulching through shallow tillage (10–15 cm) disrupts early weed growth while maintaining soil integrity (Khan et al. 2022). Straw mulching, widely used in organic paddy and vegetable cultivation, involves spreading crop residues or straw to suppress weeds, conserve moisture, and promote microbial activity (Krishnan 2023). Live mulching, practiced in both rainfed and irrigated systems, involves growing cover crops like cowpea, sunn hemp, horse gram and mustard alongside the main crop to suppress weed growth, improve soil structure, and enhance nutrient availability (Sportelli et al. 2022). In agroforestry-based mulching, deep-rooted nitrogen-fixing trees such as *Gliricidia sepium* provide shade, outcompete weeds, and recycle nutrients, making it a common approach in orchards and perennial cropping systems (Japheth et al. 2023). These integrated mulching strategies reduce reliance on manual weeding and chemical herbicides, making weed control more sustainable, labour-efficient, and cost-effective (Sarma and Dutta 2024).

11.10.5 Water Management (Whapasa)

This practice aims to maintain an optimal balance of water and air in the soil, reducing water dependency while ensuring efficient moisture use. Key methods include reduced tillage, crop residue mulching, and avoiding irrigation during peak evaporation hours, such as noon (Rosenstock et al. 2025). In Sikkim's organic cardamom plantations, farmers use leaf litter mulching to conserve soil moisture and suppress weeds, reducing irrigation needs (Wang et al. 2021; El-Beltagi et al. 2022). Drip irrigation combined with mulching is widely practiced in organic vegetable farms (Tiwari et al. 2023), where moisture sensors help regulate water supply, minimizing wastage (Ray and Majumder 2024). Rainwater harvesting techniques, such as farm ponds and percolation pits, are effectively used in organic wheat and millet cultivation to improve groundwater recharge (Huang et al. 2021; Srivastav et al. 2021). Additionally, intercropping with deep-rooted crops like pigeon pea in rainfed systems enhances soil moisture retention, benefiting neighboring shallow-rooted crops (Homulle et al. 2022). These approaches collectively reduce water stress, enhance soil microbial activity, and improve overall crop resilience in organic farming systems.

11.10.6 Pest Management

In natural farming, pest and disease management relies on promoting predators and parasitoids, withdrawing chemical pesticides, and using practices like trap crops, border crops, sticky traps (yellow, white, blue), and pheromone traps. The

Table 11.2 Effect of the intercropping system on pest levels

Crop		Pest reduced	Reference
Main crop	Intercrop		
Sorghum	Red gram	Earhead bug	Raheja (1974)
Sorghum	Cowpea	Stem borer	Omolo and Ogango (2001)
Pigeon pea	Sorghum	Leafhopper	Sekhar et al. (1997)
Green gram	Sorghum	Leafhopper	Sekhar et al. (1997)
Ground nut	Sorghum	Leafhopper	Sekhar et al. (1997)
Pigeon pea	Sorghum	Pod borer	Mohammed and Rao (1998)
Chickpea	Wheat/mustard/safflower	Pod borer	Visalakshmi et al. (2005)
Sugarcane	Green gram/black gram	Early shoot borer	Rajendran (2002)

trap crops for major crops controlling important pests are described in Table 11.2. Creating habitats for beneficial insects and applying biological formulations to seeds further support pest management. Botanical concoctions like neem oil, neem seed kernel extract, neemastra, brahmastra, agniastra, dashparni ark, dry ginger + milk extract are key inputs in pest and disease management (Mochiah et al. 2011; Reddy and Chowdary 2021). Fermented cow urine-based formulations like neemastra and agniastra have shown efficacy in controlling sucking pests like aphids and whiteflies in cotton and okra fields (Suganthy and Ganesan 2023). Companion planting, such as marigold with tomato, repels nematodes and aphids, while coriander in vegetable fields attracts parasitoid wasps that control caterpillar pests (Sarkar et al. 2018). In rice ecosystems, duck farming is practiced to naturally manage brown planthopper and stem borer by feeding on insect larvae (Teng et al. 2016; Qian et al. 2022). Agroforestry systems incorporating *Gliricidia sepium* provide shelter for predatory birds that help control rodent and insect populations (Rosati et al. 2021). Additionally, pheromone traps for fruit fly control in mango orchards (Ganie et al. 2022; Mani 2022) and yellow sticky traps for whitefly suppression in organic vegetable farms reduce pest damage without chemical inputs (Abubakar et al. 2022). These integrated methods strengthen plant defense mechanisms, reduce pesticide residues, and support biodiversity in natural farming systems, and the field representing IPM components has been given in Fig. 11.6.

Fig. 11.6 IPM components for pest management under natural farming system

11.10.7 Standards and Certification

Standards and certification for natural farming are still in the developmental stage, with no formal regulations currently in place. However, efforts are underway to establish a structured certification system to ensure authenticity and compliance. The Indian government has initiated pilot programs under the Bharatiya Prakritik Krishi Paddhati (BPKP) to promote and standardize natural farming practices (Courville and Iseal Alliance 2006). Additionally, participatory guarantee systems (PGS) have been adopted in some regions, enabling farmer groups to self-certify their produce through peer reviews and adherence to prescribed guidelines (Vogl et al. 2005). Some private organizations and state governments are also working on developing region-specific certification frameworks to facilitate market recognition. Once implemented, these standards will help consumers identify genuine natural farm produce, promote transparency, and support farmers in accessing premium markets (Jahanban and Davari 2012).

11.11 Next-Generation Strategies in Natural Farming

To overcome the challenges and scale up natural farming practices, innovative strategies are being integrated into traditional frameworks, and the diagrammatic representation of next-generation natural farming practices has been illustrated (Fig. 11.7).

11.11.1 Technological Advancements

11.11.1.1 Precision Agriculture

IoT (Internet of Things) devices have revolutionized precision agriculture by enabling real-time monitoring and control of various farming parameters (Reddy and Masila 2025). Devices like CropX provide real-time temperature and soil moisture, and electrical conductivity data allows farmers to maximize irrigation and fertilization efficiency, cutting down on water and chemical consumption while maximizing yield (Kumar et al. 2024a). Weather stations such as those from allMETEO or Pycno track temperature, humidity, and rainfall, enabling farmers to adjust planting and harvesting schedules (Shaikh et al. 2022). Systems like Growlink use IoT to monitor and automatically adjust temperature, light, and humidity in greenhouses, ensuring optimal growth conditions (Chai et al. 2023).

11.11.1.2 AI and Data Analytics

Artificial intelligence and data analytics enhance decision-making in agriculture. AI platforms like IBM's Watson Decision Platform for Agriculture use historical weather patterns and pest outbreak data to forecast conditions and provide actionable insights for planting schedules (Redhu et al. 2022; AlZubi and Galyna 2023). AI tools analyze drone or satellite imagery to identify early risks of nutrient deficiencies

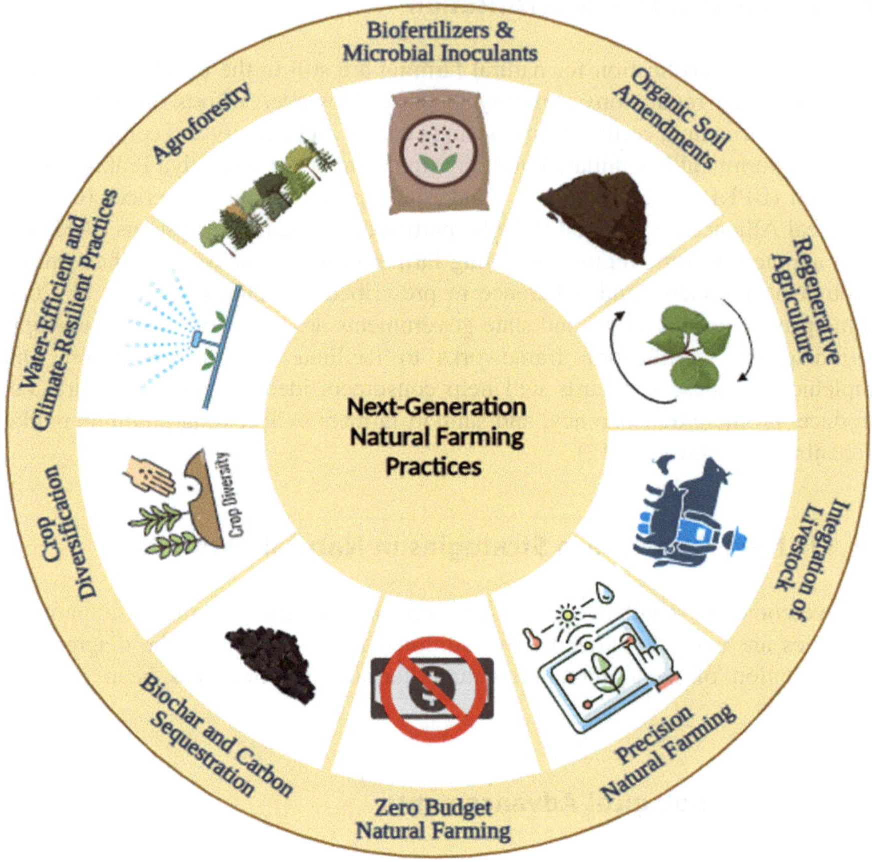

Fig. 11.7 Next-generation natural farming practices

and plant diseases (Goswami and Nayak 2022; Catala-Roman et al. 2024). For instance, Microsoft AI has been deployed to detect pests like the fall armyworm in African countries (Kariyanna and Sowjanya 2024). Big data systems analyze historical and real-time data to predict the best crop rotation and planting practices, improving soil productivity and soil health (Kamilaris et al. 2019; Osinga et al. 2022).

11.11.1.3 Automated Tools

Automation addresses labour shortages and increases efficiency. Robots like those from FarmWise can identify and remove weeds without damaging crops, reducing reliance on herbicides (Remus et al. 2024). Machines such as John Deere's self-driving tractors use GPS and RTK (Real-Time Kinematic) systems for precise seeding, minimizing waste and labour (Kadam et al. 2023). Robotic harvesters, like those used in strawberry farms, ensure high-speed, high-accuracy harvesting with minimal human input (Kondoyanni et al. 2022; Karkee and Silwal 2023). These technologies collectively help optimize resource use, reduce environmental impact, and increase agricultural productivity, showcasing the transformative potential of smart farming practices (Paul et al. 2022).

11.11.2 Enhanced Soil Health Management

Both microbial biofertilizers and regenerative agriculture are pivotal in advancing sustainable farming practices, addressing environmental challenges while boosting productivity. These innovations ensure long-term agricultural viability.

11.11.2.1 Microbial Biofertilizers

Microbial biofertilizers utilize beneficial microbes, namely, algae, bacteria, and fungi, to enrich soil fertility and enhance crop durability (Vessey 2003). Plant growth-promoting Rhizobacteria improve nutrient availability by fixing nitrogen, releasing phosphorus, and synthesizing growth hormones. For instance, strains of *Azospirillum* and *Bacillus* are widely used for nitrogen fixation and root growth promotion (Chandran et al. 2021). Research highlights the potential of marine algae like *Chlorella* and *Scenedesmus* for nitrogen fixation and stress tolerance. These are effective in crops like wheat and sugar beet (Kapoore et al. 2021; Ammar et al. 2022). Advances in biofertilizer formulations include nano-biofertilizers and biofilm-based systems (Kumar et al. 2024a). These approaches address specific soil and crop needs, improving the uptake of nutrients and plant health (Ibáñez et al. 2023; Prisa et al. 2023). The formulations and combinations of novel biofertilizers given by Suganthy et al. (2023a, b) were depicted in Table 11.3.

Table 11.3 Proposed new biofertilizer formulations and related combinations

Biofertilizers	Organism	Role
Sulfur-oxidizing bacteria	*Thiobacillus* sp. *Dyella thiooxydans*	Increases available soil sulfate and stabilizes soil pH
Magnesium-oxidizing bacteria	*Serratia* sp.	Recycles magnesium in the soil
Calcium-solubilizing bacteria	*Pseudomonas* sp.	Solubilizes complex calcium and converts it into an assimilatory form
Silicate-solubilizing bacteria	*Bacillus* sp. and *Pseudomonas* sp.	Solubilizes silica in soil
Methylotrophic bacteria	*Methylobacterium radiotolerans*	Phyllospheric plant growth promoter
Bio NPK Zn	Bacterial consortia	Fixes N and solubilizes and mobilizes P, K, and Zn

11.11.2.2 Regenerative Agriculture

This holistic approach restores degraded lands through practices that enhance soil health and biodiversity. Combining trees with crops improves carbon sequestration and microclimatic conditions. For instance, integrating leguminous trees in tropical farms boosts nitrogen levels and reduces erosion (Rosati et al. 2021). Cover crops like rye and clover enhance soil organic matter, prevent erosion, and enhance water retention (Quintarelli et al. 2022). They are instrumental in mitigating extreme weather impacts (Jordon et al. 2022). Rotational grazing and manure application recycle nutrients, improve soil structure, and reduce reliance on synthetic fertilizers (Diwan et al. 2021). Practices like minimal tillage and diverse crop rotations elevate soil organic carbon content, contributing to climate change reduction and enriched soil fertility (Lal 2020).

11.11.3 Climate-Resilient Practices

11.11.3.1 Drought-Resistant Crops

Modern approaches to breeding drought-resistant crops combine traditional breeding techniques with advanced biotechnology (Nguyễn et al. 2024). For example, scientists have developed new durum wheat and barley varieties in Morocco, specifically bred for arid regions (Bozalmat et al. 2024). These varieties exhibit enhanced resilience to drought, improved yields, and increased nutritional value, crucial for food security in dry lands (Achli et al. 2025). Additionally, molecular studies on crops like quinoa and rice have identified specific genes and pathways responsible for drought tolerance, which are being used to design more robust plants using genetic engineering and marker-assisted selection (Samuel et al. 2025).

11.11.3.2 Water Harvesting Systems

Rainwater harvesting techniques, such as constructing check dams, ponds, and contour bunding, have been revitalized with modern materials and engineering. These methods are increasingly employed in semi-arid regions to store rainwater and recharge aquifers (Yazar and Ali 2017). For example, innovative projects in India have combined traditional check dams with modern geotextiles to enhance water retention and prevent erosion. Such systems improve water availability for crops during drought (Rautaray 2021).

11.11.3.3 Agroforestry

Integrating trees with crops and livestock in agroforestry systems creates microclimates that mitigate the effects of extreme weather while enhancing biodiversity. Recent advancements include planting nitrogen-fixing trees like *Gliricidia sepium* alongside crops to improve soil fertility and moisture retention (Rosati et al. 2021). This approach has been widely adopted in parts of South Asia and Sub-Saharan Africa, where it helps buffer crops from drought and extreme temperatures, increasing resilience to climate variability (Nair et al. 2021). These strategies are crucial for adapting agriculture to a changing climate, enhancing both sustainability and productivity in vulnerable regions.

11.11.3.4 Next-Generation Natural Farming (NGNF) in Different Agroclimatic Regions

The implementation of Next-Generation Natural Farming (NGNF) practices varies across agroclimatic regions, offering benefits like improved soil health, reduced chemical dependency, and increased climate resilience while facing challenges such as nutrient leaching, pest outbreaks, and resource limitations. In tropical regions, high microbial activity and natural pest control methods enhance productivity, but excessive rainfall leads to nutrient loss and fungal diseases (Paramesh et al. 2022; Nallasamy et al. 2025). Arid and semi-arid regions benefit from water conservation techniques like mulching and drip irrigation, yet slow organic matter decomposition and limited inputs hinder large-scale adoption (Panda et al. 2025). In temperate regions, NGNF fosters soil carbon accumulation and biodiversity while reducing pest pressure, but short growing seasons and slow microbial activity delay nutrient availability (Wangchu et al. 2024). Coastal and deltaic regions see improved soil salinity management through organic amendments and salt-tolerant crops, though high salinity and frequent cyclones challenge sustainability (Wu et al. 2024). While NGNF practices show promise, adapting them to local conditions through research and innovation is key to maximizing their impact on sustainable agriculture.

11.11.4 Digital and Social Innovations

11.11.4.1 Blockchain for Traceability

Blockchain technology is transforming agriculture by enhancing supply chain transparency. It allows all stakeholders, including farmers, distributors, and consumers, to ensure the secure tracing of agricultural products' origin and movement (Lin et al. 2021). For instance, blockchain-based systems have been successfully implemented to ensure food safety and traceability in high-value supply chains, like organic and fair trade coffee (Tegeltija et al. 2022). These systems reduce fraud and ensure that products meet the standards claimed on their labels. Additionally, blockchain's decentralized and immutable nature helps build consumer trust and compliance with international trade requirements (Kamilaris et al. 2019).

11.11.4.2 Farmer Cooperatives

Farmer cooperatives empower smallholder farmers to combine resources, secure better prices, and reach larger markets. For example, dairy cooperatives in India have provided rural farmers with improved access to technology, shared resources, and fairer market terms (Zhong et al. 2018). These cooperatives can be pivotal in introducing innovations like shared agricultural equipment or cooperative processing facilities, significantly reducing individual costs and risks (Manimozhi and Krishnamoorthy 2025).

11.11.4.3 Participatory Research

Participatory research involves farmers in designing and testing agricultural innovations, ensuring solutions are tailored to local contexts. This approach has been effective in projects like the Participatory Plant Breeding initiative, where farmers and researchers collaborate to develop crop varieties suited to specific climatic and soil conditions (Soto et al. 2021). Such collaborations increase adoption rates and ensure the relevance of technological advancements (Ceccarelli and Grando 2007).

These innovations combine technology and social cooperation to address complex agricultural challenges, ensuring sustainable and inclusive growth in the sector.

11.12 Case Studies

11.12.1 Zero Budget Natural Farming (ZBNF) in Andhra Pradesh, India

Zero Budget Natural Farming (ZBNF) is a chemical-free farming approach that reduces input costs while maintaining soil health and productivity. Initially promoted by Subhash Palekar, this method has been widely adopted in Andhra Pradesh through government-supported initiatives. ZBNF principles include using jeevamrit (a microbial inoculant made from cow dung and urine), beejamrit (a natural seed treatment), acchadana (mulching), and whapasa (soil aeration) to enhance soil structure and biological activity (Rosenstock et al. 2025). Studies indicate that ZBNF has led to improved yields in rice, pulses, and millets, alongside significant cost savings due to reduced dependence on synthetic fertilizers and pesticides (Khadse et al. 2018). According to Walker et al. (2025), farmers practicing ZBNF in Andhra Pradesh have reported increased soil microbial diversity and resilience to drought, making it a sustainable alternative to conventional chemical-based farming.

11.12.2 Subhash Palekar Natural Farming (SPNF) in Maharashtra, India

Subhash Palekar Natural Farming (SPNF) is a refined version of natural farming methods that emphasize self-sufficiency in farm inputs. This system integrates microbial cultures like Jeevamrutha and promotes minimal soil disturbance to enhance aeration and microbial activity. In Maharashtra, farmers adopting SPNF techniques have reported lower input costs, improved organic matter in soil, and better crop resilience to pests and diseases (Sharma et al. 2023). A long-term study by Hiremath et al. (2025) found that SPNF increased the net profitability of smallholder farmers by reducing reliance on chemical inputs and enhancing soil fertility through natural composting and mulching methods.

11.12.3 Natural Farming in Sikkim, India

Sikkim became the first state in India to achieve 100% organic agriculture, banning synthetic fertilizers and pesticides in 2016. The government promoted organic alternatives such as biofertilizers, composting, and integrated pest management to enhance soil health and crop productivity. According to Gurung and Choubey (2021), this transition led to increased biodiversity, better soil fertility, and improved market access for organic produce. Farmers in Sikkim experienced increased profitability due to premium prices for organic products in both domestic and international markets. However, challenges such as declining yields in the initial transition phase and the need for enhanced farmer training were also observed (Das and Bhattacharyya 2018).

11.12.4 Korean Natural Farming (KNF) in Hawaii, USA

Korean Natural Farming (KNF), developed by Master Cho Han-Kyu, is a system that utilizes indigenous microorganisms (IMOs) to enhance soil fertility and plant growth. KNF techniques focus on fermenting plant and animal-based nutrients, using naturally occurring microbes to improve soil structure and reduce synthetic input dependence. In Hawaii, several farmers have successfully implemented KNF, particularly in fruit and vegetable farming. Research by Keliikuli et al. (2019) demonstrated that KNF improved soil microbial diversity and water retention capacity, leading to higher crop yields compared to conventional farming methods. Farmers practicing KNF also reported significant reductions in pest infestations due to enhanced natural pest control mechanisms (Wang et al. 2013).

11.13 Future Scope

Natural farming represents a transformative shift in agriculture, promoting practices that harmonize with nature while addressing the limitations of conventional farming. By integrating traditional wisdom with modern innovations, natural farming enhances soil health, conserves water, and fosters biodiversity. Key principles such as *jeevamrit, beejamrit, acchadana,* and *whapasa* offer a low-cost, eco-friendly alternative to input-intensive farming. Future strategies, including precision agriculture, climate-resilient cropping, and regenerative soil management, can further strengthen the relevance and scalability of natural farming. Digital innovations like blockchain for supply chain transparency and participatory research initiatives can create new opportunities for farmers while maintaining ecological integrity.

11.14 Conclusion

Despite its potential, natural farming faces challenges such as initial yield reductions, labour-intensive practices, and gaps in knowledge and market infrastructure. Overcoming these barriers requires collaborative efforts from farmers, researchers, and policymakers. Investments in education, capacity-building programs, and market support systems are essential to facilitate wider adoption. Ultimately, natural farming is more than an agricultural method; it serves as a pathway to environmental resilience, economic viability, and nutritional security. As global and national efforts to transition toward sustainable farming intensify, natural farming provides a sustainable model that aligns agricultural productivity with ecological stewardship, ensuring food security for future generations.

References

Abubakar M, Koul B, Chandrashekar K, Raut A, Yadav D (2022) Whitefly (*Bemisia tabaci*) management (WFM) strategies for sustainable agriculture: a review. Agriculture 12(9):1317

Achli S, Ongoma V, Epule TE, Dhiba D, Salih W, Ousayd L et al (2025) Exploring the key drivers of crop yields in Morocco–a systematic review. Front Agron 7:1515938

Adefila AO, Ajayi OO, Toromade AS, Sam-Bulya NJ (2024) Integrating traditional knowledge with modern agricultural practices: a sociocultural framework for sustainable development. J Sustain Agric Dev

Akkaya D, Bimpikis K, Lee H (2021) Government interventions to promote agricultural innovation. Manuf Serv Oper Manag 23(2):437–452

Alotaibi BA, Yoder E, Brennan MA, Kassem HS (2021) Perception of organic farmers towards organic agriculture and role of extension. Saudi J Biol Sci 28(5):2980–2986

AlZubi AA, Galyna K (2023) Artificial intelligence and internet of things for sustainable farming and smart agriculture. IEEE Access 11:78686–78692

Ammar EE, Aioub AA, Elesawy AE, Karkour AM, Mouhamed MS, Amer AA et al (2022) Algae as bio-fertilizers: between current situation and future prospective. S J Biol Sci 29(5):3083–3096

Arthurson V, Jäderlund L (2011) Utilization of natural farm resources for promoting high energy efficiency in low-input organic farming. Energies 4(5):804–817

Arun Kumar K, Karthikeyan J (2022) Review on implementation of IoT for environmental condition monitoring in the agriculture sector. J Amb Intel Human Comput 13:1–18

Azam MS, Shaheen M, Narbariya S (2019) Marketing challenges and organic farming in India—does farm size matter? Int J Nonprofit Volunt Sect Mark 24(4):e1654

Azhar MF, Ali E, Aziz A (2024) Regenerative agroforestry for soil restoration, biodiversity protection, and climate change mitigation. Regenerative agriculture for sustainable food systems. Springer, pp 423–451

Bai L, Liu M, Sun Y (2023) Overview of food preservation and traceability technology in the smart cold chain system. Foods 12(15):2881

Behera SR, Pandey R, Golui K, Sahoo S, Jakhwal R, Pal R (2024) Application of Panchagavya, a cow-based liquid formulation, as a lever for sustainable and enhanced vegetable crop production: a review. Int J Environ Clim Change 14(5):214–232

Bertrand M, Barot S, Blouin M, Whalen J, de Oliveira T, Roger-Estrade J (2015) Earthworm services for cropping systems. A review. Agron Sustain Dev 35:553–567

Bharadwaj M, Jain M, Sharma M (2025) Organic farming: a socio-ecological approach to long-term survival. RNT J Agric Allied Sci:101

Bhatt A, John J (2023) Including farmers' welfare in a government-led sector transition: the case of Sikkim's shift to organic agriculture. J Clean Prod 411:137207

Boix-Fayos C, de Vente J (2023) Challenges and potential pathways towards sustainable agriculture within the European Green Deal. Agric Syst 207:103634

Borrelli P, Robinson DA, Fleischer LR, Lugato E, Ballabio C, Alewell C et al (2017) An assessment of the global impact of 21st century land use change on soil erosion. Nat Commun 8(1):2013

Borrero JD, Mariscal J (2022) A case study of a digital data platform for the agricultural sector: a valuable decision support system for small farmers. Agriculture 12(6):767

Bozalmat W, Alaoui SB, Sidikou AAH, Abouabdillah A (2024) The management of fungal diseases in organic production systems through a mixture of durum wheat varieties. Sustainability 16(21):9304

Brewer J, Larsen A, Noack F (2024) The land use consequences of rural to urban migration. Am J Agric Econ 106(1):177–205

Brzezina N, Kopainsky B, Mathijs E (2016) Can organic farming reduce vulnerabilities and enhance the resilience of the European food system? A critical assessment using system dynamics structural thinking tools. Sustainability 8(10):971

Carton N, Swiergiel W, Tidåker P, Röös E, Carlsson G (2022) On-farm experiments on cultivation of grain legumes for food–outcomes from a farmer–researcher collaboration. Ren Agric Food Syst 37(5):457–467

Cassman KG, Grassini P (2020) A global perspective on sustainable intensification research. Nat Sustain 3(4):262–268

Catala-Roman P, Segura-Garcia J, Dura E, Navarro-Camba EA, Alcaraz-Calero JM, Garcia-Pineda M (2024) AI-based autonomous UAV swarm system for weed detection and treatment: enhancing organic orange orchard efficiency with agriculture 5.0. Internet Things 28:101418

Ceccarelli S, Grando S (2007) Decentralized-participatory plant breeding: an example of demand driven research. Euphytica 155:349–360

Chai K, Azizi MMF, Romeli S, Lau HY, Lau HY (2023) Deployment of IoT sensors in the revolution of agriculture. RA J Appl Res 9

Chandel R, Gupta M, Sharma S, Sharma P, Verma S, Chandel A (2021) Impact of Palekar's natural farming on farmers' economy in Himachal Pradesh. Indian J Ecol 48(3):872–877

Chandran H, Meena M, Swapnil P (2021) Plant growth-promoting rhizobacteria as a green alternative for sustainable agriculture. Sustainability 13(19):10986

Chari MS, Ramanjaneyulu GV, Zakir Hussain G, Sekhar C, Rajashekar G (2015) Managing pests without using pesticides. Knowledge:94

Courville S, Iseal Alliance U (2006) Organic standards and certification. Organic agriculture. BioOne, p 201

Dabbert, S., Häring, A.M. and Zanoli, R. (2004) Organic farming: Policies and prospects, London, Zed Books Ltd., London. 192p. ISBN: 9781848137462

Dalali S, Adarsh C, Abhishek B, Akshay K (eds) (2023) Organic mart: E commerce web site for agriculture. In: ICDSMLA 2021: Proceedings of the 3rd international conference on data science, machine learning and applications. Springer

Darjee DK (2023) A review on policy initiatives, institutional mechanism and support for organic farming in India with special reference to north-east states: an exploratory study. Int J Res Analyt Rev:2348–1269

Darjee S, Singh R, Dhar S, Pandey R, Dwivedi N, Sahu PK et al (2024) Empirical observation of natural farming inputs on nitrogen uptake, soil health, and crop yield of rice-wheat cropping system in the organically managed Inceptisol of Trans Gangetic plain. Front Sustain Food Syst 8:1324798

Das J, Bhattacharyya D (2018) An enquiry into the challenges of organic farming in Sikkim. Bus Stud 39(1)

Das SK, Mishra V, Laha R, Devi E, Bhutia T (2023) Integrated organic nutrient management for soil and plant health management under organic and natural farming. Indian Farm 73(5):28–31

Daum T, Baudron F, Birner R, Qaim M, Grass I (2023) Addressing agricultural labour issues is key to biodiversity-smart farming. Biol Conserv 284:110165

Deguine J-P, Aubertot J-N, Bellon S, Côte F, Lauri P-E, Lescourret F et al (2023) Agroecological crop protection for sustainable agriculture. Adv Agron 178:1–59

Devarinti S (2016) Natural farming: eco-friendly and sustainable. Agrotechnology 5(2):1000147

Diogo V, Helfenstein J, Mohr F, Varghese V, Debonne N, Levers C et al (2022) Developing context-specific frameworks for integrated sustainability assessment of agricultural intensity change: an application for Europe. Environ Sci Pol 137:128–142

Diwan A, Harke S, Pande B, Panche A (2021) Regenerative agriculture farming. Indian Farm 71:12

Dotaniya C, Yashona D, Aher S, Rajput P, Doutaniya R, Lata M et al (2020) Crop performance and soil properties under organic nutrient management. Int J Curr Microbiol App Sci 9(4): 1055–1065

Duddigan S, Collins CD, Hussain Z, Osbahr H, Shaw LJ, Sinclair F et al (2022) Impact of zero budget natural farming on crop yields in Andhra Pradesh, SE India. Sustainability 14(3):1689

Durham TC, Mizik T (2021) Comparative economics of conventional, organic, and alternative agricultural production systems. Economies 9(2):64

El-Beltagi HS, Basit A, Mohamed HI, Ali I, Ullah S, Kamel EA et al (2022) Mulching as a sustainable water and soil saving practice in agriculture: a review. Agronomy 12(8):1881

Eliazer Nelson ARL, Ravichandran K, Antony U (2019) The impact of the green revolution on indigenous crops of India. J Ethnic Food 6(1):1–10

Ferguson RS, Lovell ST (2014) Permaculture for agroecology: design, movement, practice, and worldview. A review. Agron Sustain Develop 34:251–274

Fruscella L, Kotzen B, Milliken S (2021) Organic aquaponics in the European Union: towards sustainable farming practices in the framework of the new EU regulation. Rev Aquac 13(3): 1661–1682

Fukuoka M (2009) The one-straw revolution: an introduction to natural farming: New York review of books

Gamage A, Gangahagedara R, Gamage J, Jayasinghe N, Kodikara N, Suraweera P et al (2023) Role of organic farming for achieving sustainability in agriculture. Farm Syst 1(1):100005

Ganie SA, Rehman SA, Nisar T, Paray M, Bano P, Khurshid R (2022) Fruit fly management and control strategies: a review. Biopesticides. International 18(2)

Garnier J, Casquin A, Mercier B, Martinez A, Gréhan E, Azougui A et al (2024) Six years of nitrous oxide emissions from temperate cropping systems under real-farm rotational management. Agric For Meteorol 354:110085

George NP, Ray JG (2023) The inevitability of arbuscular mycorrhiza for sustainability in organic agriculture—a critical review. Front Sustain Food Syst 7:1124688

Ghosh A, Ghosh A (2024) Empowering rural women in India: self-help groups as catalysts for development. Futuristic Trend Soc Sci 3:1–16

Gikunda RM, Lawver DE, Baker M, Boren-Alpizar AE, Guo W (2021) Extension education needs for improved adoption of sustainable organic agriculture in Central Kenya. American Journal of Geographic Information System, 10(2): 61-71, doi:10.5923/j.ajgis.20211002.01

Goswami B, Nayak P (2022) Optimization in agricultural growth using AI and satellite imagery. Data science in societal applications: concepts and implications. Springer, pp 107–125

Gupta S, Dikshit A (2010) Biopesticides: an ecofriendly approach for pest control. J Biopest 3 (Special Issue):186

Gurung R, Choubey M (2021) Integrating farmer producer organisations in Sikkim organic mission: opportunities, challenges and policy measures. South Asian J Soc Stud Econ 9(1): 39–49

Gurung R, Choubey M, Rai R (2024) Economic impact of farmer producer organisation (FPO) membership: empirical evidence from India. Int J Soc Econ 51(8):1015–1028

Haldar T, Damodaran A (2022) Can cooperatives influence farmer's decision to adopt organic farming? Agri-decision making under price volatility. Environ Dev Sustain 24(4):5718–5742

Hans V, Prabhu S (2024) Revisiting the green revolution in India: assessing achievements, challenges, and future prospects. Seema, revisiting the green revolution in India: assessing achievements, challenges, and future prospects (May 2, 2024)

Haughey E, Neogi S, Portugal-Pereira J, Van Diemen R, Slade R (2023) Sustainable intensification and carbon sequestration research in agricultural systems: a systematic review. Environ Sci Pol 143:14–23

Herrero M, Mason-D'Croz D, Thornton PK, Fanzo J, Rushton J, Godde C et al (2023) Livestock and sustainable food systems: status, trends, and priority actions. Sci Innov Food Syst Transform:375

Hiremath R, Mahajan Y, Sanjeev M, Sharma V, Kapse M (2025) Diffusing responsible innovation: seeding sustainability through Subhash Palekar Krushi (SPK). SAGE

Hirschfeld S, Van Acker R (2021) Ecosystem services in permaculture systems. Agroecol Sustain Food Syst 45(6):794–816

Homeshwari-Devi M, Haokip IC, Kalidas-Singh S. Impact of traditional land management practices on soil fertility and crop productivity. Sustainable land management in India: Springer; 2024. p. 179–200

Homulle Z, George TS, Karley AJ (2022) Root traits with team benefits: understanding below-ground interactions in intercropping systems. Plant Soil 471:1–26

Huang Z, Nya EL, Rahman MA, Mwamila TB, Cao V, Gwenzi W et al (2021) Integrated water resource management: rethinking the contribution of rainwater harvesting. Sustainability 13(15):8338

Huss C, Holmes K, Blubaugh C (2022) Benefits and risks of intercropping for crop resilience and pest management. J Econ Entomol 115(5):1350–1362

Ibáñez A, Garrido-Chamorro S, Vasco-Cárdenas MF, Barreiro C (2023) From lab to field: biofertilizers in the 21st century. Horticulture 9(12):1306

Iqbal R, Raza MAS, Valipour M, Saleem MF, Zaheer MS, Ahmad S et al (2020) Potential agricultural and environmental benefits of mulches—a review. Bull Natl Res Cent 44:1–16

Jahanban L, Davari M (2012) Standards and certification procedures in organic agriculture: an overview focusing on organic crop production. Int Res J Appl Basic Sci 3(9):1825–1836

Japheth DH, Agera SIN, Dachung G, Amonum IJ (2023) Perspective chapter: mitigating modern agriculture's problems through agroforestry system and organic farming. Updates on organic farming. IntechOpen

Jensen KOD, Denver S, Zanoli R (2011) Actual and potential development of consumer demand on the organic food market in Europe. NJAS Wageningen J Life Sci 58(3–4):79–84

Jordon MW, Smith P, Long PR, Bürkner P-C, Petrokofsky G, Willis KJ (2022) Can regenerative agriculture increase national soil carbon stocks? Simulated country-scale adoption of reduced tillage, cover cropping, and ley-arable integration using RothC. Sci Total Environ 825:153955

Kadam S, Shinde AS, Bari AM, Gujar J (2023) The role of artificial intelligence and the internet of things in smart agriculture towards green engineering. Applied computer vision and soft computing with interpretable AI. Chapman and Hall/CRC, pp 135–152

Kalarani M, Karthikeyan R, Sivakumar R, Umapathi M (2023) Natural farming: scope and importance. Nat Farm Sustain Agric 1

Kamilaris A, Fonts A, Prenafeta-Boldύ FX (2019) The rise of blockchain technology in agriculture and food supply chains. Trends Food Sci Technol 91:640–652

Kapoore RV, Wood EE, Llewellyn CA (2021) Algae biostimulants: a critical look at microalgal biostimulants for sustainable agricultural practices. Biotechnol Adv 49:107754

Kariyanna B, Sowjanya M (2024) Unravelling the use of artificial intelligence in management of insect pests. Smart Agric Technol 8:100517

Karkee M, Silwal A (2023) Robotic fruit harvesting. Encyclopedia of digital agricultural technologies. Springer, pp 1145–1160

Karunathilake E, Le AT, Heo S, Chung YS, Mansoor S (2023) The path to smart farming: innovations and opportunities in precision agriculture. Agriculture 13(8):1593

Kassem HS, Alotaibi BA, Ghoneim YA, Diab AM (2021) Mobile-based advisory services for sustainable agriculture: assessing farmers' information behavior. Inf Dev 37(3):483–495
Keliikuli A, Smith K, Li Y, Lee C (2019) Natural farming: the development of indigenous microorganisms using korean natural farming methods. Sustain Agric SA-19:1–19
Keller C, Joshi S, Joshi T, Goldmann E, Riar A (2024) Challenges for crop diversification in cotton-based farming systems in India: a comprehensive gap analysis between practices and policies. Front Agron 6:1370878
Khadse A, Rosset PM, Morales H, Ferguson BG (2018) Taking agroecology to scale: the zero budget natural farming peasant movement in Karnataka, India. J Peasant Stud 45(1):192–219
Khan MR, Raja W, Bhat TA, Mir MS, Naikoo NB, Amin Z et al (2022) Zero budget natural farming: a way forward towards sustainable agriculture. Curr J Appl Sci Technol 41(13):31–43
Kledal PR, Kjeldsen C, Refsgaard K, Söderbaum P (2006) Ecological economics and organic farming. Global development of organic agriculture: challenges and prospects. CABI Publishing, Wallingford, pp 113–133
Klerkx L (2021) Digital and virtual spaces as sites of extension and advisory services research: social media, gaming, and digitally integrated and augmented advice, vol 27. Taylor & Francis, pp 277–286
Kondoyanni M, Loukatos D, Maraveas C, Drosos C, Arvanitis KG (2022) Bio-inspired robots and structures toward fostering the modernization of agriculture. Biomimetics 7(2):69
Korav S, Dhaka A, Chaudhary A, YS M. (2020) Zero budget natural farming a key to sustainable agriculture: challenges, opportunities and policy intervention. Ind J Pure App Biosci 8(3): 285–295
Krishnan R (2023) Crop residue and weed management under natural farming. Nat Farm Sustain Agric 28
Krishnan R, Yen P, Agarwal R, Arshinder K, Bajada C (2021) Collaborative innovation and sustainability in the food supply chain-evidence from farmer producer organisations. Resour Conserv Recycl 168:105253
Kumar R, Kumar S, Yashavanth B, Meena P (2019) Natural farming practices in India: its adoption and impact on crop yield and farmers' income. Indian J Agric Econ 74(3):420–432
Kumar R, Kumar S, Yashavanth B, Venu N, Meena P, Dhandapani A et al (2023a) Natural farming practices for chemical-free agriculture: implications for crop yield and profitability. Agriculture 13(3):647
Kumar R, Kumari S, Chauhan S, Chaudhary PL (2023b) Zero-budget natural farming. Just. Agriculture 3:309–314
Kumar MS, Nair A, Kalaivanan D, Hebbar S, Prabhakar M, Sridevi V (2024a) Organic farming practices in vegetable crops. Advances in organic farming: apple. Academic Press, pp 381–403
Kumar D, Kumar R, Sakshi, Deepshikha, Kumar A (2024b) Commercialization and market perspectives of biofertilizers through advanced approaches. In: Metabolomics, proteomics and gene editing approaches in biofertilizer industry, vol II. Springer, pp 339–357
Kumara K, Pal S, Chand P, Kandpal A (2023) Carbon sequestration potential of sustainable agricultural practices to mitigate climate change in Indian agriculture: a meta-analysis. Sustain Prod Consump 35:697–708
Kurdekar AK, Ashoka P, Dambale AS, Zaman MI (2023) Modern techniques in soil and water conservation. Adv Farm Technol 1
Kurniawati A, Toth G, Ylivainio K, Toth Z (2023) Opportunities and challenges of bio-based fertilizers utilization for improving soil health. Org Agric 13(3):335–350
Kushal GER, Boraiah B, Sanjay M (2024) A review on challenges and emerging opportunities for weed management in organic agriculture. Organization 70
Laishram C, Vashishat RK, Sharma S, Rajkumari B, Mishra N, Barwal P et al (2022) Impact of natural farming cropping system on rural households—evidence from Solan District of Himachal Pradesh, India. Front Sustain Food Syst 6:878015
Lal R (2004a) Soil carbon sequestration to mitigate climate change. Geoderma 123(1–2):1–22

Lal R (2004b) Soil carbon sequestration impacts on global climate change and food security. Science 304(5677):1623–1627

Lal R (2020) Regenerative agriculture for food and climate. Soil Water Conservation, Ankeny, pp 123A-124A

Lamine C, Navarrete M, Cardona A (2014) Transitions towards organic farming at the farm and at the local scales: the role of innovative production and organisational modes and networks. Organic farming, prototype for sustainable agricultures: prototype for sustainable agricultures. Springer, pp 423–438

Lin X, Chang S-C, Chou T-H, Chen S-C, Ruangkanjanases A (2021) Consumers' intention to adopt blockchain food traceability technology towards organic food products. Int J Environ Res Public Health 18(3):912

Lohr L, Park TA (2007) Efficiency analysis for organic agricultural producers: the role of soil-improving inputs. J Environ Manag 83(1):25–33

Lone AH, Rashid I (2024) Harvesting insights: a qualitative exploration of facilitators, inhibitors and information seeking in family-based sustainable organic farm businesses. Kybernetes

Łuczka W, Kalinowski S, Shmygol N (2021) Organic farming support policy in a sustainable development context: a polish case study. Energies 14(14):4208

Lv N, Liu F, Zhu H, Wang G (2023) Effect of government intervention and market incentives on farmer organic fertilizer application behavior and agricultural emission reduction. Nat Hazar Rev 24(1):04022035

Maitra S, Gitari H (2020) Scope for adoption of intercropping system in organic agriculture. Indian J Nat Sci 11(63):28624–28631

Mandla I, Sharma S (2023) Compendium of success stories of farmers practising natural farming in Himachal Pradesh. Indian Farm 73(8):11–15

Mani M (2022) Organic pest management in horticultural crops. Trends Hortic Entomol:211–241

Manimozhi R, Krishnamoorthy G (2025) Innovative techniques in agriculture: transitioning from traditional farming to precision and hydroponic agriculture. Environ Qual Manag 34(3):e70047

Manteghi Y, Arkat J, Mahmoodi A (2023) The competition between conventional and organic food production in the presence of the blockchain technology. Trends Food Sci Technol 136:282–294

Martinez L, Soti P, Kaur J, Racelis A, Kariyat RR (2020) Impact of cover crops on insect community dynamics in organic farming. Agriculture 10(6):209

Matos F, Catapan E, Prado G, Zanuzzi C, Selig P (2024) Exploring the key success factors: a case study of a digital marketplace platform for Brazilian small farmers. Proc Comp Sci 232:159–168

Mazzafera P, Favarin JL, Andrade SAL (2021) Intercropping systems in sustainable agriculture. Front Media SA:634361

McPhearson T, Cook EM, Berbés-Blázquez M, Cheng C, Grimm NB, Andersson E et al (2022) A social-ecological-technological systems framework for urban ecosystem services. One Earth 5(5):505–518

Mendes CMI, Demattê Filho LC, Gameiro AH (2024) Incorporating sustainability in the food supply chain: the development of a private standard of nature farming poultry production in Brazil. Res Globaliz 8:100185

Merfield CN (2023) Integrated weed management in organic farming. Organic farming. Elsevier, pp 31–109

Miglani R, Rawat G, Joshi S, Singh V, Parveen N, Siddiqui U et al (2025) The green revolution reimagined: agricultural biodiversity and sustainable food security. CRC Press, Biotechnological Innovations for Sustainable Biodiversity and Development, pp 331–342

Milestad R, Darnhofer I (2003) Building farm resilience: the prospects and challenges of organic farming. J Sustain Agric 22(3):81–97

Mkhize S, Ellis D (2020) Creativity in marketing communication to overcome barriers to organic produce purchases: the case of a developing nation. J Clean Prod 242:118415

Mochiah M, Banful B, Fening K, Amoabeng B, Ekyem S, Braimah H et al (2011) Botanicals for the management of insect pests in organic vegetable production. J Entomol Nematol 3(6):85–97

Mohammed G, Rao AS (1998) Influence of intercrops on the incidence of *Helicoverpa armigera* (Hübner) in postrainy season pigeonpea. pp 46–48
Mohan S, G Balasubramanian, M Gopalan and S Jayaraj (1987) Solar heat treatment-a novel method to check rice weevil and red flour beetle infestation in sorghum during storage. Stewart Postharv Rev: 235–236
Möhring N, Muller A, Schaub S (2024) Farmers' adoption of organic agriculture—a systematic global literature review. Eur Rev Agric Econ 51(4):1012–1044
Mortensen DA, Smith RG (2020) Confronting barriers to cropping system diversification. Front Sustain Food Syst 4:564197
Mponella PKTAP, Timothy J, Lopez KNSS. Outmigration and the rural-urban transition: challenges and opportunities to crop-livestock interactions in Nepal. 2024
Murphy B, Martini M, Fedi A, Loera BL, Elliott CT, Dean M (2022) Consumer trust in organic food and organic certifications in four European countries. Food Control 133:108484
Musselli I (2017) Agriculture, price stabilisation and trade rules: a principled approach. Brill
Musto G, Swanepoel P, Strauss J (2023) Regenerative agriculture v. conservation agriculture: potential effects on soil quality, crop productivity and whole-farm economics in Mediterranean-climate regions. J Agric Sci 161(3):328–338
Naik A (2021) Ashokkumar B. Pharm Innov J, Status of zero budget natural farming: an overview, pp 343–347
Nair PR, Kumar BM, Nair VD, Nair PR, Kumar BM, Nair VD (2021) Classification of agroforestry systems. An introduction to agroforestry: four decades of scientific developments. pp 29–44
Nallasamy T, Chandana P, KR L, SP S. (2025) Integrated organic fish+ poultry+ horticulture farming systems for sustaining the income for small and marginal farmers in tropical and sub-tropical India. Biol Agric Hortic 41:1–16
Nandi R, Nedumaran S, Ravula P (2021) The interplay between food market access and farm household dietary diversity in low and middle income countries: a systematic review of literature. Glob Food Sec 28:100484
Nazir MJ, Li G, Nazir MM, Zulfiqar F, Siddique KH, Iqbal B et al (2024) Harnessing soil carbon sequestration to address climate change challenges in agriculture. Soil Tillage Res 237:105959
Nelson E, Hargreaves S, Muldoon D. Farmer knowledge as formal knowledge: a case study of farmer-led research in Ontario, Canada. J Agric Food Syst Commun Dev 2023;12(4):15–38–15–38
Ngosong C, Okolle JN, Tening AS (2019) Mulching: a sustainable option to improve soil health. In: Soil fertility management for sustainable development, pp 231–249
Nguyễn CT, Gammatantrawet N, Susawaengsup C, Tandee K, Ramli ANM, Tongkoom K et al (2024) Drought-ready plant resilience: harnessing nano-biotechnology techniques for swift screening and selection of organic crop varieties. S Afr J Bot 169:553–566
NITI Aayog (2022) Compendium of success stories of natural farming. Government of India
Oldfield EE, Eagle AJ, Rubin RL, Rudek J, Sanderman J, Gordon DR (2022) Crediting agricultural soil carbon sequestration. Science 375(6586):1222–1225
Omolo, E and Ogango, K.O. (2001) Use of intra-cropping systems for the control of stem borer (Chilo partellus) in sorghum and cowpea aphid (Aphis craccivora) in cowpea. East African Agricultural and Forestry Journal. 67: 77–81. https://doi.org/10.1080/00128325.2001.11663339
Onunwa A, Uko I, Ndukwe O, Okeke C, Meludu N (2022) Organic agriculture research output: dissemination, awareness and implementation in selected South-Eastern and South-South Nigerian States. International Journal of Agriculture, Food and Biodiversity 1 (1): 70–77.
Onwude DI, Chen G, Eke-Emezie N, Kabutey A, Khaled AY, Sturm B (2020) Recent advances in reducing food losses in the supply chain of fresh agricultural produce. Processes 8(11):1431
Orsini S, Padel S, Lampkin N (2018) Labour use on organic farms: a review of research since 2000. Organic Farm 4(1):7–15

Ortiz R, Peris J (2022) The role of farmers' umbrella organizations in building transformative capacity around grassroots innovations in rural Agri-food systems in Guatemala. Sustainability 14(5):2695

Osinga SA, Paudel D, Mouzakitis SA, Athanasiadis IN (2022) Big data in agriculture: between opportunity and solution. Agric Syst 195:103298

Palekar S (2005a) The philosophy of spiritual farming Amravati: zero budget natural farming research. Dev Exten Mov Amravati

Palekar S (2005b) The philosophy of spiritual farming, 2nd edn. Zero Budget Natural Farming Research. Development & Extension Movement, Amravati

Panda A, Jakhar P, Sahoo L, Prusty M, Pattanaik S (2022) Natural farming: win-win for nature and farmer. Vigyan Varta 3(2):47–52

Panda S, Tripathi VK, Yadav VS (2025) A synergetic approach for multidimensional drought assessment in the Indian agro-climatic zone using coherency, propagation and AHP techniques. Environ Sci Pollut Res 32(7):4354–4371

Papzan A, Shiri N (2012) Study of barriers toward development of organic farming. Space Econ Rural Dev 1(1):113–126

Paramasivam S, Henry P, Seethapathy P, Rajamohan T (2022) A strategic model for empowering farmers by improving livelihood security through organic farming practices in Tamil Nadu, India. J Agric Sci Sri Lanka 17(3)

Paramesh V, Kumar P, Shamim M, Ravisankar N, Arunachalam V, Nath AJ et al (2022) Integrated farming systems as an adaptation strategy to climate change: case studies from diverse agro-climatic zones of India. Sustainability 14(18):11629

Patel SK, Sharma A, Singh GS (2020) Traditional agricultural practices in India: an approach for environmental sustainability and food security. Energy Ecol Environ 5(4):253–271

Patel M, Islam S, Glick BR, Choudhary N, Yadav VK, Bagatharia S et al (2024) Zero budget natural farming components Jeevamrit and Beejamrit augment Spinacia oleracea L.(spinach) growth by ameliorating the negative impacts of the salt and drought stress. Front Microbiol 15: 1326390

Paul K, Chatterjee SS, Pai P, Varshney A, Juikar S, Prasad V et al (2022) Viable smart sensors and their application in data driven agriculture. Comput Electron Agric 198:107096

Pircher T, Nertinger M, Goss L, Hilger T, Karungi-Tumutegyereize J, Waswa L et al (2024) Farmer-centered and structural perspectives on innovation and scaling: a study on sustainable agriculture and nutrition in East Africa. J Agric Educ Ext 30(1):137–158

Ponzio C, Gangatharan R, Neri D (2013) The potential and limitations of farmer participatory research in organic agriculture: a review. Afr J Agric Res 8(32):4285–4292

Prasad S, Malav LC, Choudhary J, Kannojiya S, Kundu M, Kumar S et al (2021) Soil microbiomes for healthy nutrient recycling. In: Current trends in microbial biotechnology for sustainable agriculture, pp 1–21

Prause L (2021) Digital agriculture and labor: a few challenges for social sustainability. Sustainability 13(11):5980

Prisa D, Fresco R, Spagnuolo D (2023) Microbial biofertilisers in plant production and resistance: a review. Agriculture 13(9):1666

Puech T, Stark F (2023) Diversification of an integrated crop-livestock system: Agroecological and food production assessment at farm scale. Agric Ecosyst Environ 344:108300

Qamar R, Ashraf S, Javeed HMR, Yaseen M, Khan BA, Abbas T et al (2024) Regenerative organic farming for encouraging innovation and improvement of environmental, social, and economic sustainability. Regenerative agriculture for sustainable food systems. Springer, pp 175–216

Qian P, Quais MK, Zhou W, Ye J, Wang G, Wu M et al (2022) Effects of integrated rice-duck farming on weed pressure, herbivore-predator interactions and economic benefits. Biocontrol Sci Tech 32(6):715–730

Quintarelli V, Radicetti E, Allevato E, Stazi SR, Haider G, Abideen Z et al (2022) Cover crops for sustainable cropping systems: a review. Agriculture 12(12):2076

Raheja AI (1974) Report on the insect pests of grain legumes in northern Nigeria. In: 1st IITA grain legume improvement workshop 1973. International Institute of Tropical Agriculture, Ibadan, pp 295–299
Raj EFI, Appadurai M, Athiappan K (2022) Precision farming in modern agriculture. Smart agriculture automation using advanced technologies: data analytics and machine learning, cloud architecture, automation and IoT. Springer, pp 61–87
Rajabzadeh M, Fatorachian H (2023) Modelling factors influencing IoT adoption: with a focus on agricultural logistics operations. Smart Cities 6(6):3266–3296
Rajendran B (2002) Cultural controls. Encyclopedia of Pest management. Dekker, New York, pp 174–178
Rani S, Arya V, Kataria A (eds) (2022) Dynamic pricing-based e-commerce model for the produce of organic farming in India: a research roadmap with main advertence to vegetables. In: Proceedings of data analytics and management: ICDAM 2021. Springer
Ranjan S, Sow S (2021) A way towards sustainable agriculture through zero budget natural farming. Food Sci Rep 2(3):30–32
Ranjan P, Patle G, Prem M, Solanke K (2017) Organic mulching-a water saving technique to increase the production of fruits and vegetables. Curr Agric Res J 5(3):371
Rasool S, Cerchione R, Salo J, Ferraris A, Abbate S (2021) Measurement of consumer awareness of food waste: construct development with a confirmatory factor analysis. Br Food J 123(13): 337–361
Rathor AS, Choudhury S, Sharma A, Nautiyal P, Shah G (2024) Empowering vertical farming through IoT and AI-Driven technologies: a comprehensive review. Heliyon 10:e34998
Rautaray SK (2021) Field design for enhancing water productivity in waterlogged areas with efficient water harvesting and farming system. Agric Res 10(2):255–261
Ravisankar N (2023) Organic farming: an integrated approach for sustainable livelihood. Futuristic strategies and options for profitable organic farming, vol 9
Ravisankar N, Panwar A, Prasad K, Kumar V, Bhaskar S (2017) Organic farming crop production guide, vol 586. Network Project on Organic Farming, ICAR-Indian Institute of Farming Systems Research, Modipuram, p 586
Ray S, Majumder S (2024) Water management in agriculture: innovations for efficient irrigation. Modern Agron:169–185
Reddy DS, Chowdary NM (2021) Botanical biopesticide combination concept—a viable option for pest management in organic farming. Egypt J Biol Pest Control 31(1):1–10
Reddy L, Masila G (2025) Analysis of IoT applications in highly precise agriculture farming. Recent Patent Eng 19(2):E180923221207
Redhu NS, Thakur Z, Yashveer S, Mor P (2022) Artificial intelligence: a way forward for agricultural sciences. In: Bioinformatics in agriculture. Elsevier, pp 641–668
Rehber E (2002) Tackling the market obstacles for organic products. Cahiers Options Méditerranéennes 61:97–118
Rempelos L, Kabourakis E, Leifert C (2023) Innovative organic and regenerative agricultural production. MDPI 13:1344
Remus OM, Florin N, Iulian V, Dan C, Cătălin P, Maria-Diana B et al (2024) Emerging agricultural equipment using artificial intelligence and spectroscopy for the agricultural systems of the future. Acta Technica Corviniensis Bull Eng 17(4):87–94
Röös E, Mie A, Wivstad M, Salomon E, Johansson B, Gunnarsson S et al (2018) Risks and opportunities of increasing yields in organic farming. A review. Agron Sustain Dev 38:1–21
Rosati A, Borek R, Canali S (2021) Agroforestry and organic agriculture. Agrofor Syst 95:805–821
Rosenstock TS, Mayzelle M, Namoi N, Fantke P (2025) Climate impacts of natural farming: a cradle to gate comparison between conventional practice and Andhra Pradesh community natural farming. Res Rev J Agric Allied Sci
Sacco D, Moretti B, Monaco S, Grignani C (2015) Six-year transition from conventional to organic farming: effects on crop production and soil quality. Eur J Agron 69:10–20

Sagar S, Singh A, Bala J, Chauhan R, Kumar R, Walia A. Comparative analysis of cow dung microflora of different cow breeds for plant growth-promoting traits and antifungal properties for prospective use in organic and natural farming system. 2023

Sagar S, Singh A, Bala J, Chauhan R, Kumar R, Bhatia RK et al (2024) Insights into cow dung-based bioformulations for sustainable plant health and disease management in organic and natural farming system: a review. J Soil Sci Plant Nutr 24(1):30–53

Saharan BS, Tyagi S, Kumar R, Vijay, Om H, Mandal BS et al (2023) Application of Jeevamrit improves soil properties in zero budget natural farming fields. Agriculture 13(1):196

Sahu R, Singh RJ, Singh K (2010) Knowledge gap about organic farming practices of farmers of Bageshwar district of Uttarakhand. Indian J Ext Educ 46(3 and 4):135–136

Samuel J, Rao CR, Raju B, Lokesh S, Singh V (2025) Managing drought risk in rainfed agriculture. Indian Farm 75(01):98–101

Sánchez AC, Kamau HN, Grazioli F, Jones SK (2022) Financial profitability of diversified farming systems: a global meta-analysis. Ecol Econ 201:107595

Santos VB, Araújo AS, Leite LF, Nunes LA, Melo WJ (2012) Soil microbial biomass and organic matter fractions during transition from conventional to organic farming systems. Geoderma 170: 227–231

Sapbamrer R, Thammachai A (2021) A systematic review of factors influencing farmers' adoption of organic farming. Sustainability 13(7):3842

Sarkar SC, Wang E, Wu S, Lei Z (2018) Application of trap cropping as companion plants for the management of agricultural pests: a review. Insects 9(4):128

Sarma HH, Dutta SK (2024) Method of preparation of various inputs used in natural farming. Vigyan Varta 5(7):88–94

Schreefel L, De Boer I, Timler C, Groot J, Zwetsloot M, Creamer R et al (2022) How to make regenerative practices work on the farm: a modelling framework. Agric Syst 198:103371

Sekhar, J.C., Singh, K.M., Singh, R.N., Singh,Y. and Malik, K.S.1997. Impact of intercropping on the incidence of greenjassid, Empoasca kerri on pigeonpea. Indian Journal of Entomology, 59 (1):119–123

Shahane AA, Shivay YS (2022) Soil health management in organic production system–a review. Int J Bio-Res Stress Manag 13(11):1186–1200

Shahrajabian MH, Chaski C, Polyzos N, Petropoulos SA (2021) Biostimulants application: a low input cropping management tool for sustainable farming of vegetables. Biomolecules 11(5):698

Shaikh FK, Karim S, Zeadally S, Nebhen J (2022) Recent trends in internet-of-things-enabled sensor technologies for smart agriculture. IEEE Internet Things J 9(23):23583–23598

Sharma PB, Kumar P, Tiwari S, Kumari S, Singh M (2022) The potential of cow's dung in sustainable agriculture and environmental health. Animal manure: agricultural and biotechnological applications. Springer, pp 107–122

Sharma SK, Ravisankar N, Jain N, Sarangi S (2023) Natural farming: current status, research and case studies. Indian J Agron 68(22):1–15

Sharma K, Kumar R, Kumar A, Balabantaray S, Arora M (2024) A digital ecosystem for sustainable fruit supply chain in Uttarakhand: a comprehensive review. Environ Dev Sustain 26(5): 13217–13252

Shree D, Vaishnavi P (2022) Challenges faced by farmer producer organisations (FPOs)-a review. J Agric Exten Manag 23(1):131–138

Shyam DM, Dixit S, Nune R, Gajanan S, Chander G (2020) Zero budget natural farming-an empirical analysis. Green Farm 10(6):661–667

Sims B, Kienzle J (2017) Sustainable agricultural mechanization for smallholders: what is it and how can we implement it? Agriculture 7(6):50

Singh A, Thakur RK (2022) Understanding the perception, constraints and reasons for the adoption of organic farming. Indian Res J Exten Educ 22(5):110–117

Singh A, Karmegam N, Singh GS, Bhadauria T, Chang SW, Awasthi MK et al (2020) Earthworms and vermicompost: an eco-friendly approach for repaying nature's debt. Environ Geochem Health 42:1617–1642

Singh V, Singh PD, Chauhan A, eds (2024a) The future of pest control: a smart phone app that can identify pests in real time. AIP Conference Proceedings. AIP Publishing
Singh S, Krishna H, Sharma S, Singh R, Tripathi A, Behera T (2024b) Organic farming in vegetable crops: challenges and opportunities. Vegetable Sci 51:1–10
Smith J, Yeluripati J, Smith P, Nayak DR (2020) Potential yield challenges to scale-up of zero budget natural farming. Nat Sustain 3(3):247–252
Somasundaram J, Sinha N, Dalal RC, Lal R, Mohanty M, Naorem A et al (2020) No-till farming and conservation agriculture in South Asia–issues, challenges, prospects and benefits. Crit Rev Plant Sci 39(3):236–279
Soni R, Gupta R, Agarwal P, Mishra R (2022) Organic farming: a sustainable agricultural practice. Vantage J Thematic Analy 3(1):21–44
Soti P, Kariyat R, Racelis A (2023) Effective farm management promotes native AMF and benefit organic farming systems. Agric Ecosyst Environ 342:108240
Soto RL, de Vente J, Padilla MC (2021) Learning from farmers' experiences with participatory monitoring and evaluation of regenerative agriculture based on visual soil assessment. J Rural Stud 88:192–204
Sportelli M, Frasconi C, Fontanelli M, Pirchio M, Gagliardi L, Raffaelli M et al (2022) Innovative living mulch management strategies for organic conservation field vegetables: evaluation of continuous mowing, flaming, and tillage performances. Agronomy 12(3):622
Srivastav AL, Dhyani R, Ranjan M, Madhav S, Sillanpää M (2021) Climate-resilient strategies for sustainable management of water resources and agriculture. Environ Sci Pollut Res 28(31): 41576–41595
Stolze M, Lampkin N (2009) Policy for organic farming: rationale and concepts. Food Policy 34(3): 237–244
Su Y, Cook ML (2015) Price stability and economic sustainability–achievable goals? A case study of Organic Valley®. Am J Agric Econ 97(2):635–651
Suganthy M, Ganesan K (2023) Bio-inputs for plant protection in natural farming. Nat Farm Sustain Agric:101
Suganthy M, Renuka Rani B, Sharma N (eds) (2023a) Natural farming for sustainable agriculture. Nammazhvar Organic Farming Research Centre, Tamil Nadu Agricultural University, Coimbatore & National Institute of Agricultural Extension Management (MANAGE), Hyderabad, pp 122–138
Suganthy M, P Janaki, E Parameswari, R Krishnan, P S Kavitha, M Ramasubramanian and M K Kalarani (2023b) Natural farming: present status and future prospects.
Sulistyowati CA, Afiff SA, Baiquni M, Siscawati M. Challenges and potential solutions in developing community-supported agriculture: a literature review. Agroecol Sustain Food Syst 2023;47(6):834–856
Teague W, Dowhower S, Baker S, Haile N, DeLaune P, Conover D (2011) Grazing management impacts on vegetation, soil biota and soil chemical, physical and hydrological properties in tall grass prairie. Agric Ecosyst Environ 141(3–4):310–322
Tegeltija S, Dejanović S, Feng H, Stankovski S, Ostojić G, Kučević D et al (2022) Blockchain framework for certification of organic agriculture production. Sustainability 14(19):11823
Teng Q, Hu X-F, Cheng C, Luo Z, Luo F, Xue Y et al (2016) Ecological effects of rice-duck integrated farming on soil fertility and weed and pest control. J Soils Sediments 16:2395–2407
Thallam VK (2023) Community-managed natural farming in Andhra Pradesh, India. Biological approaches to regenerative soil systems. CRC Press, pp 431–448
Thamarai P, Deivayanai V, Saravanan A, Vickram A, Yaashikaa P (2024) Carbon mitigation in agriculture: pioneering technologies for a sustainable food system. Trends Food Sci Technol 147:104477
Tiwari AK, Mishra H, Nishad DC, Pandey A (2023) Sustainable water management in agriculture: irrigation techniques and water conservation. Dr Ajay B Jadhao. p 53
Tran-Nam Q, Tiet T (2022) The role of peer influence and norms in organic farming adoption: accounting for farmers' heterogeneity. J Environ Manag 320:115909

Tscharntke T, Grass I, Wanger TC, Westphal C, Batáry P (2021) Beyond organic farming–harnessing biodiversity-friendly landscapes. Trends Ecol Evol 36(10):919–930
Tshotsho, Lippert C, Feuerbacher A (2023) Organic agriculture, labour exchange, and social networks: a case study of smallholder farming in Bhutan. Organic. Agriculture 13(1):83–98
Tully KL, McAskill C (2020) Promoting soil health in organically managed systems: a review. Org Agric 10(3):339–358
Veluguri D, Bump JB, Venkateshmurthy NS, Mohan S, Pulugurtha KT, Jaacks LM (2021) Political analysis of the adoption of the zero-budget natural farming program in Andhra Pradesh, Indai. Agroecol Sustain Food Syst 45(6):907–930
Verburg RW, Verberne E, Negro SO (2022) Accelerating the transition towards sustainable agriculture: the case of organic dairy farming in The Netherlands. Agric Syst 198:103368
Verma NS, Kuldeep DK, Chouhan M, Prajapati R, Singh SK (2023) A review on eco-friendly pesticides and their rising importance in sustainable plant protection practices. Int J Plant Soil Sci 35(22):200–214
Vessey JK (2003) Plant growth-promoting rhizobacteria as biofertilizers. Plant Soil 255:571–586
Vidadala R (2024) Reimagining agroforestry: climate-resilient landscapes for regenerative agriculture. Agroforestry solutions for climate change and environmental restoration. Springer, pp 171–201
Vijayakumar A (2021) Farmer producer organization: an emerging business model for empowering small and marginal farmers in India-a case study. Int J Indian Cult Bus Manag 24(4):481–502
Vijayakumary P, Sriramajayam S, Ramesh D (2023) Alternate source of energy and utilization of biomass under natural farming system. Futur Strateg Opt Profit Org Farm:173
Visalakshmi V, Ranga Rao GV, Arjuna P, Rao. (2005) Integrated pest management strategy against *Helicoverpa armigera* Hubner in chickpea. Indian J Plant Protect 33:17–22
Vogl CR, Kilcher L, Schmidt H (2005) Are standards and regulations of organic farming moving away from small farmers' knowledge? J Sustain Agric 26(1):5–26
Waha K, Dietrich JP, Portmann FT, Siebert S, Thornton PK, Bondeau A et al (2020) Multiple cropping systems of the world and the potential for increasing cropping intensity. Glob Environ Chang 64:102131
Wahjono TE, Suhatman A, Wihermanto W, Hadiyanto H (2024) Neem (*Azadirachta indica* A. Juss; Meliaceae) as the source for plant-based pesticides as an effective and sustainable biocontrol alternative. J Bioresour Environ Sci 3(3):128–141
Walia SS, Kaur T (2023) Concept of farming system in relation to conservation of natural resources. Basics of integrated farming systems. Springer, pp 165–171
Walker G, Osbahr H, Duddigan S, George J, Ponnolu S, Anisetti H et al (2025) "It feels like we're doing something good." Mapping farmer perceptions of zero budget natural farming onto crop yields in Andhra Pradesh. World Dev Perspect 37:100665
Wang K-H, DuPonte M, Chang K (2013). Use of Korean natural farming for vegetable crop production in Hawai 'i Hanai'Ai/The Food Provider
Wang B, Niu J, Berndtsson R, Zhang L, Chen X, Li X et al (2021) Efficient organic mulch thickness for soil and water conservation in urban areas. Sci Rep 11(1):6259
Wangchu L, Angami T, Jini D, Bam J, Singh R, Tasung A, et al (2024) Natural farming: scope and prospective in Arunachal Pradesh. ICAR (Research Complex) for NEH Region, Umiam
Watanabe EAM, Alfinito S, Barbirato LL (2021) Certification label and fresh organic produce category in an emerging country: an experimental study on consumer trust and purchase intention. Br Food J 123(6):2258–2271
Wu S, Guo Z, Askar A, Li X, Hu Y, Li H et al (2024) Dynamic land cover and ecosystem service changes in global coastal deltas under future climate scenarios. Ocean Coastal Manag 258: 107384
Yadav K, Yadav J, Yadav S (2022) Knowledge level of farmers about agriculture development activities carried out under Paramparagat Krishi Vikas Yojana (PKVY). Indian J Ext Educ 58(1):116–121

Yang X, Dai X, Zhang Y (2024) The government subsidy policies for organic agriculture based on evolutionary game theory. Sustainability 16(6):2246
Yazar A, Ali A (2017) Water harvesting in dry environments. Innovations in dryland agriculture. Springer, pp 49–98
Yin X, Song Z (2024) Multiple cropping systems for improving crop yield and soil quality. MDPI 14:2966
Zerssa G, Feyssa D, Kim D-G, Eichler-Löbermann B (2021) Challenges of smallholder farming in Ethiopia and opportunities by adopting climate-smart agriculture. Agriculture 11(3):192
Zhang H, Georgescu P (2022) Sustainable organic farming, food safety and pest management: an evolutionary game analysis. Mathematics 10(13):2269
Zhong Z, Zhang C, Jia F, Bijman J (2018) Vertical coordination and cooperative member benefits: case studies of four dairy farmers' cooperatives in China. J Clean Prod 172:2266–2277
Zollet S, Maharjan KL (2021) Overcoming the barriers to entry of newcomer sustainable farmers: insights from the emergence of organic clusters in Japan. Sustainability 13(2):866
Zuin VG, Eilks I, Elschami M, Kümmerer K (2021) Education in green chemistry and in sustainable chemistry: perspectives towards sustainability. Green Chem 23(4):1594–1608

Speed Breeding Technology for Accelerating Crop Improvement

12

Mutian Gao, Ying Xu, and Cheng Huang

Abstract

Speed breeding represents a transformative advancement in crop improvement, effectively addressing the limitations of traditional breeding methods while responding to the growing demands for food security and climate adaptation. By utilizing controlled environmental systems, this technology significantly accelerates crop generation cycles, enabling the production of multiple cycles per year and substantially reducing the time required to develop improved varieties. It integrates cutting-edge methodologies such as genomic selection, genome editing, and high-throughput phenotyping, collectively enhancing the precision and efficiency of trait selection. The versatility of speed breeding has been demonstrated across a wide range of crops, including staple cereals like wheat and rice, as well as various horticultural species. These advancements have driven critical improvements in areas such as yield optimization, improved nutritional quality, enhanced disease resistance, and increased tolerance to abiotic stresses. Furthermore, its capacity to address pressing challenges, including those posed by climate change, underscores its potential to contribute to the development of climate-resilient crops. Despite its significant contributions, the adoption of speed breeding is not without challenges. This technology requires substantial investment in infrastructure, faces scalability issues, and necessitates rigorous

M. Gao
Key Laboratory of the Ministry of Education for Crop Physiology and Molecular Biology, College of Agronomy, Hunan Agricultural University, Changsha, China

Y. Xu · C. Huang (✉)
Key Laboratory of the Ministry of Education for Crop Physiology and Molecular Biology, College of Agronomy, Hunan Agricultural University, Changsha, China

Yuelushan Laboratory, Changsha, China
e-mail: hc66@hunau.edu.cn

Y. Ahlawat et al. (eds.), *Next-Generation Strategies for Crop Improvement*,
https://doi.org/10.1007/978-981-95-0309-4_12

field testing to ensure the applicability of results under real-world conditions. However, integrating speed breeding with emerging innovations, such as machine learning, synthetic biology, and advanced imaging technologies, presents promising opportunities for further progress. As efforts to enhance its accessibility continue to evolve, speed breeding is poised to play a pivotal role in transforming crop improvement practices and advancing sustainable agriculture.

Keywords

Speed breeding · Crop improvement · Controlled environmental system · Generation cycle · Climate-resilient crop

12.1 Introduction

12.1.1 Limitations of Traditional Breeding Methods

Plant breeding has been a cornerstone of agricultural innovation, playing a pivotal role in feeding an ever-growing global population. Traditional methods such as crossbreeding and selection have achieved significant success in enhancing crop yield, quality, and resistance to various biotic and abiotic stresses (Acquaah 2015). However, these traditional methods face inherent limitations that hinder their ability to meet the escalating demands for food security and to address the multifaceted challenges posed by climate change.

One of the primary limitations of traditional breeding lies in its time-intensive nature. Developing a new crop variety through these methods typically requires 8–10 years, owing to the prolonged generation cycles of crops and the necessity of multiple generations to stabilize desirable traits (Ahmar et al. 2020). The dependence on natural growing seasons further aggravates these delays, making it difficult to respond swiftly to emerging threats such as novel pests, diseases, and climatic changes. Another critical limitation is the low precision of traditional breeding techniques. The traits of interest are often governed by complex interactions among numerous genes, and during the recombination process, beneficial traits can be unintentionally lost, while undesirable ones may be introduced (Holland 2007). Addressing this imprecision demands extensive field trials and repeated backcrossing, which further extend the breeding timeline. Moreover, traditional methods struggle to meet the growing need for resilient and diverse crop varieties. As climate change exacerbates environmental challenges, there is an urgent demand for crops capable of withstanding extreme conditions such as drought, heat, and salinity (Yang et al. 2024). However, traditional methods are often too slow to introduce adaptive traits, particularly in crops with narrow genetic diversity or those lacking natural resistance to specific stresses. Additionally, reliance on phenotypic selection makes traditional breeding labor-intensive and resource-demanding (Wang et al. 2019), requiring substantial investments in land, workforce, and time. This dependency poses significant challenges in resource-constrained regions, particularly in developing countries, where limited access to these resources hampers breeding efforts. These limitations underscore the pressing need for advanced breeding technologies that can more effectively accelerate crop improvement and address the complex demands of modern agriculture.

12.1.2 Speed Breeding Technology

Speed breeding represents a revolutionary technique in plant breeding, offering a novel method to accelerate crop generation turnover. This method employs controlled environmental systems to promote rapid plant growth and development, significantly increasing the pace of breeding cycles. By fine-tuning variables such as light quality, temperature, day length, humidity, and CO_2 concentration (see Fig. 12.1), researchers have achieved up to six generations per year for certain crops, a remarkable improvement compared to the one or two generations possible through traditional methods (Watson et al. 2018; Ghosh et al. 2018; Hickey et al. 2019).

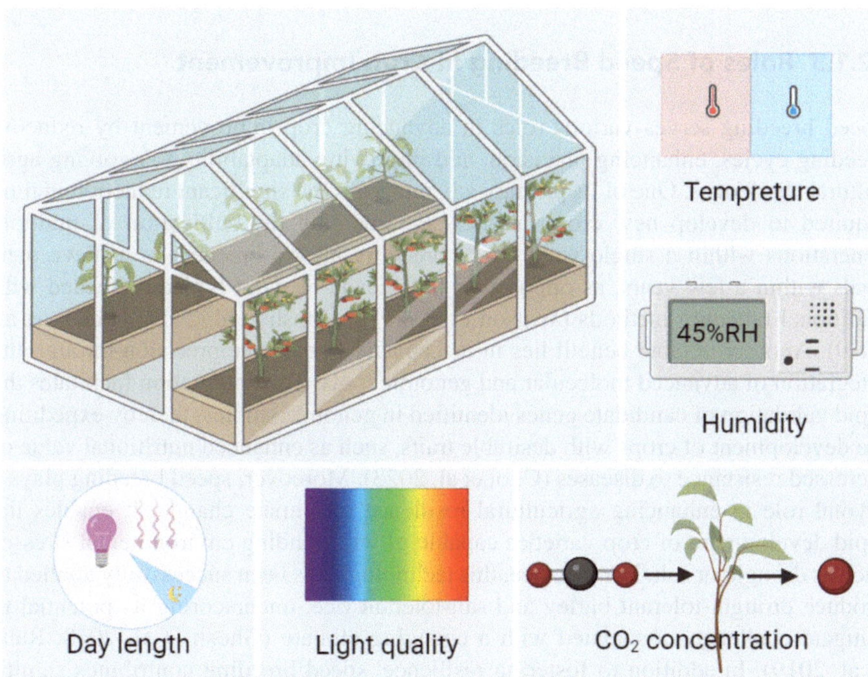

Fig. 12.1 Speed breeding accelerating crop improvement through controlled environmental systems. Key environmental variables include light quality, temperature, day length, humidity, and CO_2 concentration. These environmental variables are continuously monitored and precisely adjusted to facilitate speed breeding, enabling accelerated generation turnover, shortened growth cycles, and improved outcomes in crop breeding programs

The concept of speed breeding originated from NASA's experiments in the 1980s, which utilized artificial lighting to cultivate wheat in space. Building on these pioneering efforts, modern speed breeding integrates advancements in controlled-environment agriculture and plant science. Notable innovations include the use of high-intensity LED lighting, which provides specific wavelengths optimized for photosynthesis, and climate-controlled chambers that maintain ideal temperature and humidity conditions (Watson et al. 2018; Ghosh et al. 2018; Hickey et al. 2019). In addition, speed breeding leverages cutting-edge techniques to maximize efficiency. Techniques such as high-throughput phenotyping, genomic selection, and genome editing enable the precise identification and propagation of plants with desirable traits (Araus et al. 2018; Krishnappa et al. 2021; Hussain et al. 2023). A key advantage of speed breeding lies in its adaptability and scalability. This technology can be applied to a diverse range of crops, including cereals, legumes, and horticultural species (Xu et al. 2023), and can be customized to achieve specific breeding goals, such as enhancing yield, optimizing nutritional quality, improving disease resistance, or increasing tolerance to abiotic stresses (Al-Ashkar et al. 2023). This remarkable versatility establishes speed breeding as a critical technology in addressing the multifaceted challenges of modern agriculture.

12.1.3 Roles of Speed Breeding in Crop Improvement

Speed breeding serves various roles in advancing crop improvement by reducing breeding cycles, enhancing precision, and improving adaptability to evolving agricultural challenges. One of its primary advantages is the significant reduction in time required to develop new crop varieties. By enabling the cultivation of multiple generations within a single year, speed breeding allows breeders to achieve their goals within a few years, as opposed to the decade or more typically needed with traditional breeding methods (Watson et al. 2018; Ghosh et al. 2018; Hickey et al. 2019). Another notable benefit lies in its capacity to enhance precision through the integration of advanced molecular and genomic tools. This integration facilitates the rapid validation of candidate genes identified in genomic studies, thereby expediting the development of crops with desirable traits, such as enhanced nutritional value or increased resistance to diseases (Choi et al. 2023). Moreover, speed breeding plays a pivotal role in enhancing agricultural resilience to climate change. It enables the rapid development of crop varieties capable of withstanding environmental stresses such as drought or salt. For instance, this technology has been successfully applied to produce drought-tolerant barley and salt-tolerant rice, underscoring its potential to mitigate challenges associated with a changing climate (Ghosh et al. 2018; Rana et al. 2019). In addition to fostering resilience, speed breeding contributes significantly to food security and sustainability. By accelerating the introduction of high-yielding and resilient crop varieties, it boosts agricultural productivity while alleviating pressure on natural resources (Collard et al. 2017). The reliance on controlled environments in speed breeding further minimizes the dependence on arable land and water, providing a more sustainable alternative to traditional breeding methods.

12.2 Applications of Speed Breeding in Accelerating Crop Improvement

Speed breeding technology has rapidly emerged as a transformative innovation in crop improvement, enabling the accelerated development of plant varieties with desirable traits. By leveraging controlled environmental conditions and advanced breeding techniques, this technology substantially reduces the duration of traditional breeding cycles, thereby creating new opportunities to enhance crop performance. In the following section, we will summarize the applications of speed breeding in advancing crop improvement, with a particular focus on increasing yields, improving quality traits, improving disease resistance, enhancing tolerance to abiotic stresses, and supporting the development of climate-resilient crops.

12.2.1 Increasing Yields

Achieving higher crop yields remains a fundamental objective in modern crop breeding. Speed breeding technology has emerged as a transformative innovation, dramatically reducing the breeding cycle and enabling the rapid identification of high-yielding genotypes. By employing controlled environments, such as light-regulated growth chambers or greenhouses, plants can mature significantly faster than through traditional field-based methods. This breakthrough allows multiple generations to be cultivated within a single year, thereby accelerating the selection process for yield improvement.

In wheat, researchers have successfully implemented speed breeding protocols to achieve up to six generations annually, greatly surpassing the typical two generations achieved under conventional field conditions (Watson et al. 2018). This rapid generational turnover enables rigorous selection for yield-related traits, including grain size, grain number, and harvest index, ultimately enhancing yield potential (Guo et al. 2023). Comparable advancements have been observed in other crops such as barley and rice, where accelerated breeding has facilitated the rapid identification and validation of high-yielding varieties within significantly shorter timeframes (Hickey et al. 2017; Kabade et al. 2024). Furthermore, speed breeding accelerates the introgression of yield-enhancing traits from wild relatives or other genetic resources into elite cultivars (Cha et al. 2022; Li et al. 2024a). Traits that traditionally required years to incorporate can now be integrated with much greater efficiency. This streamlined process allows breeders to introduce beneficial alleles that boost productivity, enhance resilience to environmental challenges, and ultimately increase agricultural output on a broader scale.

12.2.2 Improving Quality Traits

Crop quality, encompassing traits such as nutritional value, flavor, texture, and post-harvest durability, remains a pivotal focus of crop improvement efforts. Speed breeding technology has transformed the development of high-quality cultivars by significantly accelerating the enhancement of nutritional and processing traits (Potts et al. 2023). This innovation is particularly critical for staple crops like rice, wheat, and maize, where both yield and quality are essential for sustaining market competitiveness (Ray et al. 2013; Marijan et al. 2019; Zhao et al. 2022).

In wheat, speed breeding has been employed to develop varieties with superior protein content, enhanced gluten strength, and improved baking performance (Rai and Han 2023). By shortening breeding cycles, researchers can perform additional rounds of selection for these traits, expediting the production of high-quality cultivars. Similarly, in rice, speed breeding has been applied to enhance grain quality attributes such as cooking properties, texture, and nutrient density, aligning with evolving consumer preferences and nutritional guidelines (Sah et al. 2021). This technology also plays a pivotal role in improving the nutritional composition of crops by increasing the levels of essential vitamins, minerals, and antioxidants. For example, speed breeding facilitates the rapid development of golden rice, which is engineered to contain higher levels of provitamin A (Majumder 2024). These advancements are especially significant in combating micronutrient deficiencies, which disproportionately affect populations in developing countries. By mitigating food losses caused by poor quality, this technology contributes to reducing food waste, a pressing challenge within global food systems.

12.2.3 Improving Disease Resistance

Pathogen-induced diseases caused by fungi, bacteria, and viruses continue to pose a significant threat to global food security. Speed breeding has emerged as a transformative technology to accelerate the development of cultivars with enhanced resistance to these diseases. By enabling rapid screening of extensive plant populations in controlled environments, this technology facilitates the efficient identification of genetic resistance to various pathogens.

In wheat, speed breeding has notably advanced the detection of resistance to rust diseases, including stripe rust (*Puccinia striiformis*) and leaf rust (*Puccinia triticina*) (Hickey et al. 2012; Riaz et al. 2016). These programs enable the swift identification of resistance alleles and their efficient integration into high-yielding breeding lines. Similarly, this technology has proven effective in identifying rice blast disease, a severe threat caused by the fungus *Magnaportheoryzae*. By accelerating the selection of resistant varieties, the time required to produce cultivars with durable resistance has been significantly shortened (Wang et al. 2022). Furthermore, when combined with genomic selection tools, speed breeding facilitates the identification and incorporation of novel resistance genes into elite cultivars (Watson et al. 2018). This integrated approach enhances the efficiency of developing crops with robust and durable resistance to a wide range of pathogens.

12.2.4 Enhancing Tolerance to Abiotic Stresses

Abiotic stresses, including drought, salinity, heat, and frost, increasingly threaten global food production, particularly in the context of a changing climate. Developing crops with enhanced tolerance to these stresses through traditional breeding methods has been a time-consuming process. This delay is primarily due to the intricate genetic architecture of stress tolerance and the requirement for multiple breeding cycles (Fujita et al. 2006; Zhang et al. 2022; Waadt et al. 2022). Fortunately, speed breeding provides an innovative solution by significantly accelerating the development of stress-tolerant crop varieties.

In wheat, speed breeding has been utilized to rapidly evaluate and select for drought tolerance. Large plant populations are screened under controlled water-deficient conditions, enabling the swift identification of tolerant lines (Alahmad et al. 2018). Advanced phenotyping tools, including drone-based imaging and soil moisture sensors (Watt et al. 2020), further enhance this process by identifying drought-resistant plants within just a few generations. These shortened breeding cycles not only expedite the development of drought-tolerant cultivars but also facilitate the testing of various combinations of drought-resistance genes, thereby increasing the likelihood of success. Similarly, speed breeding demonstrates substantial potential in developing salt-tolerant rice varieties. Rapid phenotyping under saline conditions allows for the efficient identification of genotypes capable of thriving in high-salinity soils, where conventional crops typically fail (Ranaet al. 2019; Basit et al. 2024). This approach is particularly beneficial in regions with extensive saline soils, as it accelerates the breeding process and increases the probability of identifying successful candidates. By minimizing the time required for screening and selection, speed breeding enables the rapid development of crops equipped to withstand abiotic stresses. This advancement plays a pivotal role in enhancing food security in regions frequently impacted by environmental challenges.

12.2.5 Supporting the Development of Climate-Resilient Crops

The increasing frequency and intensity of climate-related challenges, including heatwaves, droughts, and extreme rainfall, present significant risks to global agriculture (Robinson et al. 2021). Developing climate-resilient crops has become a pressing priority, with speed breeding emerging as a critical tool in addressing this need. By expediting the breeding process, this technology supports the creation of crops better adapted to changing environmental conditions, enabling farmers to cope with climate variability.

Speed breeding has demonstrated remarkable effectiveness in developing heat-tolerant crops. Heat stress during flowering is a critical factor that limits yields in crops and can be mitigated through the rapid selection of heat-resilient genotypes. This technology significantly shortens the time required to produce varieties with improved heat tolerance (Gaur et al. 2013). Moreover, integrating speed breeding with genomic tools allows researchers to identify genetic loci associated with heat resistance, thereby accelerating the development of climate-resilient cultivars (Ayenan et al. 2019). Similarly, speed breeding has advanced the development of crops with enhanced water-use efficiency, a crucial trait for regions experiencing water scarcity. By enabling plants to sustain productivity with lower water inputs, improved water-use efficiency supports agricultural sustainability in arid and semi-arid areas. For instance, in rice, speed breeding has been employed to select traits that enhance water-use efficiency, facilitating the creation of drought-resistant varieties (Sharma et al. 2024).

12.3 Integrating Speed Breeding with Other Modern Technologies

The adoption of speed breeding has revolutionized crop improvement by enabling the rapid generation of plant populations. However, its full potential is realized when combined with other cutting-edge agricultural technologies. This section explores the synergistic integration of speed breeding with high-throughput phenotyping, genomic selection, and genome editing, highlighting their collective impact on expediting the development of superior crop varieties (see Fig. 12.2).

12.3.1 Integrating Speed Breeding with High-Throughput Phenotyping

High-throughput phenotyping technologies have transformed plant trait assessment by facilitating rapid, precise, and non-invasive data collection (Mir et al. 2019). When integrated with speed breeding, high-throughput phenotyping substantially enhances the efficiency of identifying superior genotypes, thereby accelerating the process of trait evaluation.

(1) *Automating Trait Selection for Large Populations*

Speed breeding significantly accelerates plant generation cycles, generating large populations that demand efficient and reliable evaluation methods. Traditional phenotyping approaches, being labor-intensive and susceptible to human error, are insufficient to manage the extensive datasets produced by speed breeding programs (Li et al. 2021). High-throughput phenotyping, utilizing advanced sensors, imaging systems, and robotic platforms, offers a solution by automating the measurement of key traits such as plant height, leaf area, biomass, water-use efficiency, and

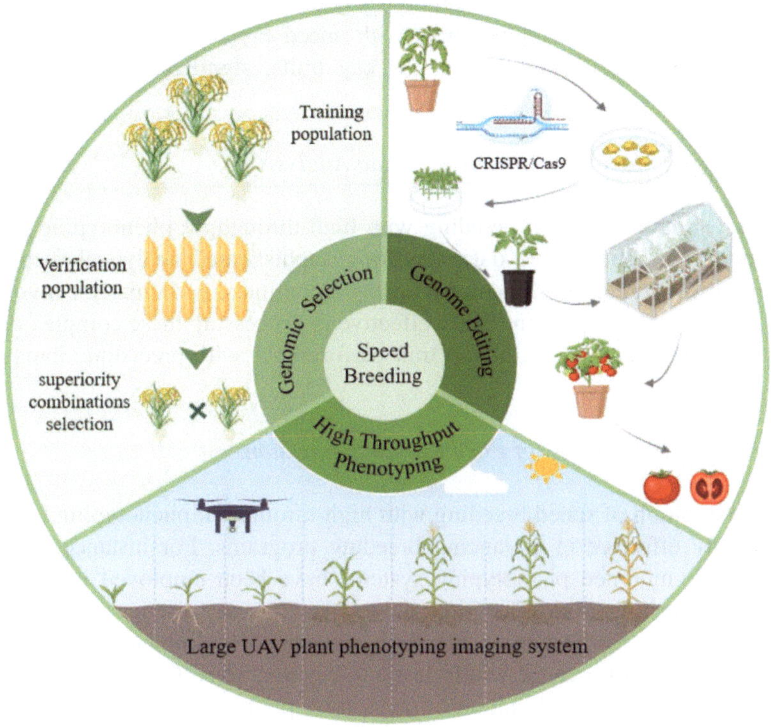

Fig. 12.2 Integrating speed breeding with other modern technologies for accelerating crop improvement. The integration of speed breeding with other modern technologies provides a synergistic framework for accelerating crop improvement. Speed breeding is driven by three interconnected components: genomic selection, genome editing, and high-throughput phenotyping. Together, these elements significantly enhance the precision and efficiency of modern breeding processes. By integrating these advanced technologies, the breeding cycle is dramatically accelerated, enabling the swift development of superior crop varieties

photosynthetic capacity (Shakoor et al. 2017). This integration of automated phenotypic data collection with the accelerated breeding cycles ensures precise and timely trait evaluation, effectively supporting the demands of modern crop improvement.

(2) *Precision Phenotyping in Controlled Environments*

Growth chambers and greenhouses employed in speed breeding provide controlled environments that are highly conducive to the implementation of high-throughput phenotyping systems. For example, hyperspectral imaging facilitates the quantification of physiological and biochemical parameters, such as chlorophyll content and stress responses, thereby delivering valuable insights into plant health (Zhang et al. 2024). Similarly, thermal imaging enables the detection of canopy temperature variations under drought conditions, which aids in the rapid identification of drought-tolerant genotypes (Ludovisi et al. 2017). The integration of these advanced phenotyping tools with speed breeding allows researchers to efficiently screen large populations for key traits, significantly accelerating the process of trait selection.

(3) *Leveraging Machine Learning for Data Analysis*

The integration of speed breeding with high-throughput phenotyping produces vast amounts of phenotypic data, requiring sophisticated analytical methods for meaningful interpretation. Machine learning techniques, particularly convolutional neural networks, have proven highly effective in processing these complex datasets, identifying patterns, and predicting trait performance with precision (Jiang and Li 2020).

(4) *Practical Applications of Phenotyping Integration*

The integration of speed breeding with high-throughput phenotyping has proven to be highly effective in advancing breeding programs. For instance, unmanned aerial vehicle-mounted phenotyping systems have been employed in conjunction with speed breeding to develop drought-resistant wheat varieties. This innovative approach enabled researchers to identify genotypes with enhanced water-use efficiency and drought tolerance within just a few generations (Condorelli et al. 2018). Similarly, in rice, the combination of phenomics platforms and speed breeding has significantly expedited the evaluation of grain quality traits, leading to the accelerated release of superior cultivars (Kumar and Walia 2024).

12.3.2 Integrating Speed Breeding with Genomic Selection

Genomic selection represents an advanced predictive breeding methodology that leverages genome-wide molecular markers to estimate the genetic potential of individuals (Heslot et al. 2015). When integrated with speed breeding, this synergistic approach enables the rapid identification and selection of superior genotypes, thereby expediting the breeding process and achieving notable genetic advancements.

(1) *Leveraging Genomic Data for Accelerated Selection*

Speed breeding accelerates the generation of breeding populations, while genomic selection enhances the precision of the selection process by predicting an individual's genetic performance based on its genomic profile (Anilkumar et al. 2022). This synergy reduces the dependence on labor-intensive phenotypic evaluations during the initial stages of breeding. Instead, selection efforts are concentrated on individuals with the highest genomic-estimated breeding values, enabling a more efficient allocation of resources (Crossa et al. 2017). As a result, the integration of genomic selection with speed breeding significantly shortens the time and lowers the costs associated with developing improved crop varieties.

(2) *Enhancing Genetic Gains Through Integration*

The integration of genomic selection with speed breeding significantly accelerates the accumulation of favorable alleles across generations. By applying genomic selection in each breeding cycle, genetic gains per unit of time are maximized, outperforming the efficiency of traditional breeding methods (Pandey et al. 2022). For example, maize breeding programs that combine genomic selection with speed breeding have dramatically reduced the time required to develop drought-tolerant hybrids (Nepolean et al. 2018). This approach has facilitated the rapid deployment of climate-resilient varieties, effectively addressing challenges associated with water scarcity.

(3) *Practical Applications in Breeding Programs*

The integration of genomic selection with speed breeding has yielded significant advancements in practical applications. For instance, in wheat, this combination has been pivotal in expediting the development of rust-resistant varieties by significantly shortening the time required to integrate resistance genes into elite lines (Jamil et al. 2020). Likewise, barley breeding programs have successfully employed genomic selection to expedite the development of cultivars with improved malting quality, catering to the specialized requirements of the brewing industry (Schmidt et al. 2016).

12.3.3 Integrating Speed Breeding with Genome Editing

Genome editing technologies, especially the CRISPR-Cas9 system, have transformed plant breeding by enabling precise alterations to genetic material (Bortesi and Fischer 2015). When combined with speed breeding, this approach significantly accelerates the evaluation and incorporation of beneficial traits, leading to faster development of improved crop varieties.

(1) *Rapid Validation of Gene Functions*

Speed breeding provides an efficient platform for producing multiple generations of genome-edited plants in a short period of time. By integrating CRISPR-mediated genome editing with accelerated growth cycles, this approach enables the rapid functional validation of target genes. Researchers can swiftly evaluate the phenotypic effects of genetic modifications, facilitating the identification of traits such as disease resistance or tolerance to abiotic stressors like drought and salinity (Varshney et al. 2021). These modifications and assessments are conducted in controlled environments, ensuring precision and reliability.

(2) *Generation of Novel Alleles*

One of the primary advantages of genome editing lies in its capacity to generate genetic variations that are improbable to arise naturally through conventional breeding methods (Schleif et al. 2021). When combined with speed breeding, these novel alleles can be efficiently incorporated into elite breeding lines, significantly reducing the time required to develop commercially viable cultivars. For example, in tomato breeding programs, CRISPR-Cas9 has been utilized to modify genes associated with yield traits. Integrated with speed breeding, these genetic modifications have facilitated the rapid production of high-yielding tomato varieties, exemplifying the potential of this synergistic approach (Sahu et al. 2024).

12.4 Challenges and Limitations of Speed Breeding

While speed breeding has emerged as a transformative innovation in crop improvement, its widespread adoption is impeded by several significant challenges. These include technical complexities, resource limitations, and biological constraints. Overcoming these barriers is crucial to fully realizing the potential of this advanced technology.

12.4.1 Technical Challenges

(1) *Customizing Speed Breeding Protocols for Diverse Crop Species*

The successful application of speed breeding protocols relies heavily on the biological and environmental requirements of individual crop species, including their photoperiod sensitivity, growth rates, and optimal climatic conditions (Chaudhary and Sandhu 2024). While early advancements in speed breeding predominantly focused on well-studied staple crops such as wheat, rice, and barley, extending this technology to a wider variety of crops poses considerable challenges.

For crops with prolonged growth cycles, such as perennial fruit trees, implementing speed breeding is particularly complex (Flachowsky et al. 2011). These species often require several years to reach reproductive maturity, rendering accelerated growth cycles impractical without fundamentally altering traditional breeding methodologies. Similarly, orphan crops, which are cultivated in limited regions and have received minimal research attention, often lack the foundational data required to develop effective speed breeding protocols. Establishing such protocols demands precise adjustments to variables including light intensity, photoperiod length, and temperature. This process is both time-consuming and resource-intensive, which hinders the widespread adoption of speed breeding across diverse agricultural systems. Furthermore, certain crop species exhibit unique developmental traits that complicate their adaptation to speed breeding methods. For instance, legumes such as soybeans and cowpeas are highly sensitive to photoperiod manipulation. Abrupt changes in day length can disrupt critical processes, including flowering and seed formation, thereby significantly reducing breeding efficiency (Weller and Ortega 2015). Addressing these complexities often requires extensive experimentation to identify optimal conditions for each crop, necessitating substantial investments in time, expertise, and resources.

(2) *Challenges in Controlled Environments for Targeted Trait Development*

Speed breeding heavily relies on controlled environments, such as growth chambers and greenhouses, to regulate variables like temperature, humidity, and light cycles. While this approach enables rapid plant development, it often falls short in replicating the dynamic conditions of natural field environments. This limitation becomes particularly pronounced when breeding for traits influenced by environmental interactions, such as drought tolerance, nutrient efficiency, or resistance to biotic stressors. For example, breeding for drought resistance typically involves assessing plant performance under well-defined water stress conditions within controlled environments. However, such settings fail to account for real-world complexities, including fluctuating weather patterns, variations in soil composition, and simultaneous biotic pressures like pest infestations. As a result, crop varieties selected in these environments may perform suboptimally when exposed to actual field conditions (Langstroff et al. 2022). To address this shortcoming, field-based trials must complement speed breeding efforts to ensure that selected genotypes are resilient across diverse environmental conditions. Nonetheless, incorporating field testing adds both time and cost to the breeding process, partially offsetting the efficiency gains provided by speed breeding. Another significant challenge lies in the limited scalability of controlled environments. Growth chambers and greenhouses are typically constrained in size, restricting the number of plants that can be evaluated simultaneously. This limitation is particularly critical for crops like hybrid maize, which require large populations for effective selection (Ćeran et al. 2024). Expanding these facilities to accommodate greater plant numbers would necessitate substantial infrastructure investments, further complicating the implementation of large-scale breeding programs.

12.4.2 Resource Constraints

(1) *High Initial Investment in Infrastructure and Equipment*

Establishing speed breeding systems requires substantial financial investment in infrastructure, including cutting-edge growth chambers, advanced lighting systems, and sophisticated temperature regulation equipment (Watson et al. 2018). These facilities are meticulously designed to create the precise environmental conditions necessary for accelerating plant development. For example, they often incorporate specialized technologies such as LED lighting systems with customizable spectra, automated irrigation setups, and advanced climate control mechanisms. However, the initial costs of setting up such systems can be prohibitively high, posing significant financial challenges, particularly for institutions in developing countries with constrained budgets for agricultural research. Additionally, speed breeding programs heavily depend on advanced phenotyping tools to generate detailed data on plant traits. These tools encompass imaging technologies, hyperspectral sensors, and unmanned aerial vehicles for efficient data collection (Song et al. 2021). While these technologies greatly improve the accuracy and efficiency of trait evaluation, they also contribute to the already substantial costs of implementing speed breeding on a large scale. Consequently, the significant financial resources required to establish and maintain these systems limit their accessibility, particularly for small-scale breeders and underfunded research institutions.

(2) *Dependence on Electricity and Advanced Tools in Resource-Limited Regions*

The reliance on consistent electricity supplies presents a significant challenge for the implementation of speed breeding programs, as controlled environments require uninterrupted power to sustain optimal conditions. In regions with unreliable electricity, power outages can disrupt these controlled settings, undermining accelerated plant growth (Potts et al. 2023). Moreover, many resource-limited regions face additional barriers due to a lack of access to advanced tools and technical expertise essential for effective speed breeding. The absence of sophisticated equipment, such as automated growth chambers and digital phenotyping systems, exacerbates the technological divide between developed and developing regions. This disparity significantly hampers the capacity of underfunded research institutions to adopt speed breeding, despite its potential to enhance food security in areas most vulnerable to climate change and resource scarcity.

Addressing these resource constraints is crucial for ensuring equitable access to speed breeding technology. Developing affordable growth chambers, utilizing simplified lighting systems, and establishing training programs for researchers in resource-limited settings represent practical strategies to democratize this technology. By enhancing accessibility, such initiatives can bridge the existing technological gap, enabling broader implementation of speed breeding in regions where it is most urgently needed.

12.4.3 Biological Constraints

(1) *Potential Trade-Offs Between Accelerated Growth and Natural Development*

While speed breeding significantly reduces the time required to complete breeding cycles, it can inadvertently affect plant physiology and growth patterns. Accelerated growth under artificial conditions often leads to altered traits, such as smaller seed size, weaker root systems, or delayed flowering in subsequent generations (Hamidon and Ahamed 2023). These changes are primarily attributed to the unnatural photoperiods and heightened light intensities utilized during speed breeding, which can disrupt the hormonal and metabolic equilibrium critical for normal plant development.

(2) *Challenges in Gene–Environment Interactions Under Artificial Conditions*

The interaction between genetic traits and environmental factors is crucial in shaping plant performance, particularly for complex traits such as stress tolerance, yield stability, and pest resistance. However, the artificial conditions employed in speed breeding, which are characterized by controlled temperature, light, and water availability, often fail to mimic the intricate environmental challenges encountered in natural field conditions. Consequently, crops bred in these controlled environments may exhibit performance discrepancies when introduced to actual agricultural ecosystems. For instance, drought-tolerant varieties selected under strictly regulated water stress conditions may underperform in the field if exposed to concurrent stresses such as heat or salinity (Li et al. 2024b). Similarly, pathogen resistance evaluated in pathogen-free environments may not adequately reflect a plant's resilience against the multiple pest pressures typically encountered in natural environments (Bradshaw 2021). Such discrepancies can result in either an overestimation or underestimation of a genotype's potential, complicating the identification of robust, high-performing varieties.

12.5 Future Prospects

12.5.1 Expanding the Application of Speed Breeding to Diverse Crops

The implementation of speed breeding has primarily focused on staple crops like wheat, barley, and rice due to their significance in global food systems. However, expanding its scope to include a wider variety of crops, such as legumes, horticultural crops, and underutilized species, can significantly broaden the impact of this technology. Incorporating these diverse crops into speed breeding programs will contribute to food security, agricultural sustainability, and economic growth in many regions.

(1) *Orphan Crops*

Orphan crops, often referred to as underutilized crops, such as millet and sorghum, are vital for the livelihoods of communities in many developing regions. These crops are inherently resilient to challenging environmental conditions, such as drought, low soil fertility, and temperature extremes. Despite their importance, they have historically received minimal attention in research and development initiatives

(Tadele 2019). Tailoring speed breeding protocols for these species can unlock their potential by improving yield, adaptability, and nutritional value. For instance, millet and sorghum are staples in arid and semiarid regions where water scarcity is a persistent challenge (Chaudhary et al. 2024). Speed breeding can be adapted to rapidly introduce drought tolerance traits into these crops, enhancing their productivity under water-limited conditions. Expanding research efforts to include these underutilized species could play a pivotal role in strengthening regional food security and reducing dependency on major staple crops.

(2) *Perennial and Long-Life Cycle Crops*

Perennial crops, such as fruit trees, pose distinct challenges for speed breeding due to their prolonged growth cycles and delayed reproductive maturity. Traditional breeding methods for these species often require decades to produce tangible results, rendering them less compatible with conventional speed breeding approaches. Nonetheless, the integration of speed breeding with complementary technologies presents promising opportunities to overcome these limitations. For instance, combining tissue culture techniques with juvenile-phase reduction strategies can significantly accelerate the breeding process in perennials (Lebedev 2024). Methods such as early flowering induction, grafting, and the use of controlled environments for rapid propagation can drastically reduce the time required to complete multiple breeding cycles (Sharma et al. 2023). Furthermore, the application of genomic selection tools to identify desirable traits at early growth stages offers additional efficiency gains. This integrated approach is particularly valuable for crops like apples, almonds, and citrus, where breeding efforts are often focused on enhancing pest resistance and improving fruit quality.

12.5.2 Integration with Advanced Computational Tool

The extensive datasets produced by speed breeding programs require the integration of advanced computational technologies, such as machine learning. This tool holds immense potential to enhance breeding efficiency by refining trait selection processes and accurately forecasting plant performance across varying environmental conditions.

Machine learning has become an indispensable tool for analyzing complex phenotypic and genomic datasets, facilitating the identification of patterns and the prediction of genotype performance across diverse environments. Advanced algorithms, such as convolutional neural networks, can efficiently process large datasets and extract critical phenotypic traits from high-throughput phenotyping platforms (Kumar et al. 2024). The integration of machine learning with speed breeding streamlines the selection of superior genotypes, significantly reducing the reliance on labor-intensive and time-consuming field trials (Rai 2022). This synergy not only accelerates decision-making in breeding programs but also expedites the identification of high-performing cultivars, thereby enhancing the overall efficiency of crop improvement initiatives.

12.5.3 Adapting Speed Breeding for Climate-Resilient Agriculture

The escalating frequency of extreme weather events, coupled with rapidly shifting climatic conditions, poses substantial challenges to global agriculture. To mitigate these impacts, the adaptation and optimization of speed breeding techniques are imperative for developing crops capable of thriving in these dynamic and unpredictable environments.

(1) *Breeding for Abiotic Stress Tolerance*

Speed breeding represents a transformative approach for expediting the development of crops with enhanced tolerance to abiotic stresses, including drought, heat, salinity, and frost. By integrating speed breeding with cutting-edge technologies such as genomic selection and genome editing, researchers can efficiently identify and incorporate stress-resilient traits into crop varieties. For example, the CRISPR-Cas9 system can be employed to precisely target and modify genes associated with drought tolerance (Sami et al. 2021). Traits identified and edited through this technology can subsequently be validated and rapidly deployed using accelerated breeding cycles, thereby substantially reducing the time required to develop stress-tolerant crops (Raza et al. 2023). Moreover, genomic selection facilitates the prediction of stress-resilient genotypes' performance based on genetic markers, enabling faster and more accurate selection of desired traits within speed breeding programs (Krishnappa et al. 2021). The integration of these advanced tools into speed breeding programs not only expedites the development of climate-resilient crops but also enhances their genetic diversity and adaptability to a wide range of environmental stressors.

(2) *Climate-Smart Breeding Platforms*

Future applications of speed breeding stand to gain significantly from integrating advanced environmental simulations that closely mimic real-world climatic scenarios. Controlled environments that replicate dynamic factors such as fluctuating temperatures, variable humidity, and inconsistent water availability provide a more accurate representation of the agricultural challenges expected in the future. Environmental chambers or growth facilities equipped with programmable systems can simulate daily and seasonal temperature fluctuations as well as changes in precipitation patterns. These simulations enable researchers to assess the performance of speed-bred crops under stress combinations that mirror anticipated climatic realities. By subjecting crops to these dynamic conditions, breeders can identify genotypes with stable performance across diverse environments, ensuring that newly developed varieties are both high-yielding and resilient to climate change.

12.5.4 Increasing Accessibility and Global Adoption

To fully unlock the potential of speed breeding, it is crucial to address key challenges associated with infrastructure, funding, and technical expertise, particularly in resource-constrained regions. Enhancing cost efficiency, building local capacity, and enacting supportive policies are essential strategies to facilitate the global adoption and integration of this transformative technology.

(1) *Development of Low-Cost Speed Breeding Systems*

Making speed breeding technology more affordable is a pivotal step toward enhancing its accessibility. Developing cost-effective solutions, such as modular growth chambers and energy-efficient LED lighting systems, can empower researchers in developing regions to adopt speed breeding without incurring substantial financial burdens. For example, modular chambers that are simple to assemble and maintain can significantly reduce barriers to entry for smaller research facilities. To address infrastructure constraints, innovative approaches such as solar-powered growth chambers can help minimize dependency on traditional electricity sources. These systems are particularly advantageous in regions with unstable power supplies, where frequent outages disrupt the maintenance of controlled environmental conditions. By integrating solar or other renewable energy technologies, researchers in remote or resource-limited areas can achieve greater reliability in conducting speed breeding, thereby fostering agricultural innovation in underserved communities.

(2) *Building Local Capacity and Promoting Knowledge Sharing*

Establishing a solid foundation for speed breeding in resource-limited countries needs significant investments in capacity building. Global partnerships and collaborative initiatives are critical to developing local expertise and broadening the adoption of this transformative technology. Targeted strategies, such as structured training programs, immersive workshops, and the development of open-access knowledge-sharing platforms, can empower researchers and practitioners to implement speed breeding practices effectively.

For instance, collaborations between international agricultural research organizations and local institutions can facilitate the transfer of technical knowledge, resources, and best practices. These partnerships ensure that researchers in developing nations gain access to the most recent advancements in speed breeding. Additionally, digital tools such as online training modules and webinars provide scalable and flexible solutions for disseminating education and technical guidance to diverse audiences. Furthermore, fostering regional networks of breeders and scientists can promote a sense of community and facilitate the exchange of region-specific innovations and adaptations. Such networks serve as a catalyst for sustained capacity building, enabling local researchers to independently maintain and expand speed breeding initiatives over the long term.

(3) *Supporting Adoption Through Policy and Funding*

Government policies and financial support are critical for incorporating speed breeding into mainstream agricultural systems. Strategic investments in infrastructure and training programs can enable countries to overcome initial adoption challenges. National governments can further promote the use of innovative breeding technologies by offering subsidies for specialized equipment or providing grants to research institutions.

International funding organizations, such as the Food and Agriculture Organization and the International Fund for Agricultural Development, can also play a pivotal role by allocating resources to support collaborative speed breeding initiatives. These efforts can prioritize the development of resilient crop varieties tailored to the specific needs of diverse regions, addressing both food security and climate adaptation objectives. Moreover, policymakers can facilitate the integration of advanced breeding technologies by establishing supportive regulatory frameworks. For example, incorporating speed breeding into national agricultural development strategies can solidify its role as a cornerstone of comprehensive food security and sustainability efforts.

12.6 Conclusion

Speed breeding has emerged as a transformative approach for accelerating crop development, addressing the urgent need for resilient, high-yielding, and sustainable crop varieties in the face of challenges such as climate change and global food insecurity. By significantly shortening breeding cycles through controlled environmental conditions and integrating advanced technologies such as genomic selection, genome editing, and high-throughput phenotyping, speed breeding has revolutionized traditional crop improvement methodologies. This innovation has achieved notable success in enhancing crop yields, improving resistance to diseases, and promoting tolerance to abiotic stresses. Its role in developing climate-resilient crops highlights its potential to mitigate the adverse effects of environmental variability. Furthermore, when combined with emerging technologies like artificial intelligence, digital simulations, and synthetic biology, speed breeding offers unparalleled precision and efficiency, heralding a new era in crop enhancement. Despite its promise, the implementation of speed breeding faces several challenges. Technical and biological hurdles, such as adapting protocols to a wide range of crop species and ensuring that traits developed in controlled environments translate effectively to field conditions, remain critical. In addition, resource limitations, particularly in developing regions, emphasize the need for cost-efficient solutions and equitable access. Addressing these obstacles will require collaborative efforts, significant investments in infrastructure, and the establishment of supportive policy frameworks. Looking ahead, the integration of speed breeding with advanced computational tools and sustainable agricultural practices offers immense potential to revolutionize global food systems. This holistic approach could help achieve a critical balance between meeting the nutritional demands of a growing global population and preserving environmental integrity. Continued research and strategic investments will be vital in refining and optimizing this transformative technology to address the complex challenges of modern agriculture.

Acknowledgments This work was supported by the Natural Science Foundation for Distinguished Young Scholars of Changsha City (kq2209016), the Science and Technology Innovation Program of Hunan Province (2023RC3153), the Natural Science Foundation of Hunan Province (2025JJ50165), and the Training Plan for Young Backbone Teachers in Colleges and Universities of Hunan Province (202210537001gg). Figures were created using BioRender (https://biorender.com/).

References

Acquaah G (2015) Conventional plant breeding principles and techniques. In: M. A-KJ, Jain SM, Johnson DV (eds) Advances in plant breeding strategies: breeding, biotechnology and molecular tools. Springer, pp 115–158

Ahmar S, Gill RA, Jung K-H et al (2020) Conventional and molecular techniques from simple breeding to speed breeding in crop plants: recent advances and future outlook. Int J Mol Sci 21(7):2590

Alahmad S, Dinglasan E, Leung KM et al (2018) Speed breeding for multiple quantitative traits in durum wheat. Plant Methods 14:36

Al-Ashkar I, Al-Doss A, Ullah N (2023) Accelerating crop improvement through speed breeding. In: Hasanuzzaman M (ed) Climate-resilient agriculture, volume I: crop responses and agroecological perspectives. Springer, pp 821–847

Anilkumar C, Sunitha N, Harikrishna et al (2022) Advances in integrated genomic selection for rapid genetic gain in crop improvement: a review. Planta 256(5):87

Araus JL, Kefauver SC, Zaman-Allah M et al (2018) Translating high–throughput phenotyping into genetic gain. Trends Plant Sci 23(5):451–466

Ayenan MAT, Danquah A, Hanson P et al (2019) Accelerating breeding for heat tolerance in tomato (*Solanum lycopersicum* L.): an integrated approach. Agronomy 9(11):720

Basit F, Feng X, Bhat JA (2024) Integration of genomics-assisted and speed breeding for enhancement of plant salt tolerance. In: Ganie SA, Wani SH (eds) Genetics of salt tolerance in plants: a central dogma perspective and strategies for enhancement. CABI, pp 119–135

Bortesi L, Fischer R (2015) The CRISPR/Cas 9 system for plant genome editing and beyond. Biotechnol Adv 33(1):41–52

Bradshaw JE (2021) Improving resistance to diseases and pests: a dynamic situation. In: Bradshaw JE (ed) Potato breeding: theory and practice. Springer, pp 247–337

Ćeran M, Miladinović D, Đorđević V et al (2024) Genomics-assisted speed breeding for crop improvement: present and future. Front Sustain Food Syst 8:1383302

Cha J-K, O'Connor K, Alahmad S et al (2022) Speed vernalization to accelerate generation advance in winter cereal crops. Mol Plant 15(8):1300–1309

Chaudhary N, Sandhu R (2024) A comprehensive review on speed breeding methods and applications. Euphytica 220:42

Chaudhary S, Chadha M, Tyagi B et al (2024) Contribution of sorghum and finger millets for sustainable food and nutritional security. In: Thakur M (ed) Sustainable food systems (volume I); sfs: framework, sustainable diets, traditional food culture & food production. Springer, pp 297–319

Choi H, Back S, Kim GW et al (2023) Development of a speed breeding protocol with flowering gene investigation in pepper (*Capsicum annuum*). Front Plant Sci 14:1151765

Collard BC, Beredo JC, Lenaerts B et al (2017) Revisiting rice breeding methods–evaluating the use of rapid generation advance (RGA) for routine rice breeding. Plant Prod Sci 20(4):337–352

Condorelli GE, Maccaferri M, Newcomb M et al (2018) Comparative aerial and ground based high throughput phenotyping for the genetic dissection of NDVI as a proxy for drought adaptive traits in durum wheat. Front Plant Sci 9:349736

Crossa J, Pérez-Rodríguez P, Cuevas J et al (2017) Genomic selection in plant breeding: methods, models, and perspectives. Trends Plant Sci 22(11):961–975

Flachowsky H, Roux PML, Peil A et al (2011) Application of a high-speed breeding technology to apple (Malus × domestica) based on transgenic early flowering plants and marker-assisted selection. New Phytol 192(2):364–377

Fujita M, Fujita Y, Noutoshi Y et al (2006) Crosstalk between abiotic and biotic stress responses: a current view from the points of convergence in the stress signaling networks. Curr Opin Plant Biol 9(4):436–442

Gaur PM, Jukanti AK, Samineni S et al (2013) Climate change and heat stress tolerance in chickpea. In: Tuteja N, Gill SS (eds) Climate change plant abiotic stress tolerance. Wiley, pp 837–856

Ghosh S, Watson A, Gonzalez-Navarro OE et al (2018) Speed breeding in growth chambers and glasshouses for crop breeding and model plant research. Nat Protoc 13(12):2944–2963

Guo X, Xue X, Chen L et al (2023) Effects of LEDs light spectra on the growth, yield, and quality of winter wheat (*Triticum aestivum* L.) cultured in plant factory. J Plant Growth Regul 42:2530–2544

Hamidon MH, Ahamed T (2023) Artificial lighting systems for plant growth and development in indoor farming. In: Ahamed T (ed) IoT and AI in agriculture: self-sufficiency in food production to achieve society 5.0 and SDG's globally. Springer, pp 25–46

Heslot N, Jannink JL, Sorrells ME (2015) Perspectives for genomic selection applications and research in plants. Crop Sci 55(1):1–12

Hickey LT, Wilkinson PM, Knight CR et al (2012) Rapid phenotyping for adult-plant resistance to stripe rust in wheat. Plant Breed 131(1):54–61

Hickey LT, Germán SE, Pereyra SA et al (2017) Speed breeding for multiple disease resistance in barley. Euphytica 213:64

Hickey LT, Hafeez AN, Robinson H et al (2019) Breeding crops to feed 10 billion. Nat Biotechnol 37(7):744–754

Holland JB (2007) Genetic architecture of complex traits in plants. Curr Opin Plant Biol 10(2): 156–161

Hussain K, Mahrukh M, Nisa RT et al (2023) The utilization of speed breeding and genome editing to achieve zero hunger. In: Prakash CS, Fiaz S, Nadeem MA, Baloch FS, Qayyum A (eds) Sustainable agriculture in the era of the OMICs revolution. Springer, pp 1–15

Jamil S, Shahzad R, Ahmad S et al (2020) Role of genetics, genomics, and breeding approaches to combat stripe rust of wheat. Front Nutr 7:580715

Jiang Y, Li C (2020) Convolutional neural networks for image-based high-throughput plant phenotyping: a review. Plant Phenomics 2020:4152816

Kabade PG, Dixit S, Singh UM et al (2024) SpeedFlower: a comprehensive speed breeding protocol for indica and japonica rice. Plant Biotechnol J 22(5):1051–1066

Krishnappa G, Savadi S, Tyagi BS et al (2021) Integrated genomic selection for rapid improvement of crops. Genomics 113(3):1070–1086

Kumar KP, Walia P (2024) Speed breeding in cereal crops: accelerating genetic improvement for rapid agricultural advancement. J Exp Agric Int 46(6):465–477

Kumar P, Senthilselvi A, Manju I et al (2024) HUMRC-PS: revolutionizing plant phenotyping through regional convolutional neural networks and pelican search optimization. Evol Syst 15: 2211–2230

Langstroff A, Heuermann MC, Stahl A et al (2022) Opportunities and limits of controlled–environment plant phenotyping for climate response traits. Theor Appl Genet 135(1):1–16

Lebedev V (2024) Speed breeding in fruit trees. In: Chen JT (ed) Plant speed breeding and high-throughput technologies. CRC Press

Li D, Quan C, Song Z et al (2021) High-throughput plant phenotyping platform (HT3P) as a novel tool for estimating agronomic traits from the lab to the field. Front Bioeng Biotechnol 8:623705

Li H, Zhu L, Fan R et al (2024a) A platform for whole-genome speed introgression from *Aegilops tauschii* to wheat for breeding future crops. Nat Protoc 19(2):281–312

Li L, Li H, Liu N et al (2024b) Water use characteristics and drought tolerant ability of different winter wheat cultivars assessed under whole growth circle and at seedling stage. Agric Water Manag 300:108921

Ludovisi R, Tauro F, Salvati R et al (2017) UAV-based thermal imaging for high-throughput field phenotyping of black poplar response to drought. Front Plant Sci 8:1681

Majumder P (2024) Golden opportunities: harnessing bioinformatics to revolutionize plant research and unleash the power of golden rice in crop breeding. In: Chen JT (ed) Bioinformatics for plant research and crop breeding. Wiley, pp 505–537

Marijan J, Samobor V, Mui M et al (2019) Improving global competitiveness of China in wheat export by increasing wheat quality. Agric Stud 3:135–145

Mir RR, Reynolds M, Pinto F et al (2019) High-throughput phenotyping for crop improvement in the genomics era. Plant Sci 282:60–72

Nepolean T, Kaul J, Mukri G et al (2018) Genomics-enabled next-generation breeding approaches for developing system-specific drought tolerant hybrids in maize. Front Plant Sci 9:361

Pandey S, Singh A, Parida SK et al (2022) Combining speed breeding with traditional and genomics-assisted breeding for crop improvement. Plant Breed 141(3):301–313

Potts J, Jangra S, Michael VN et al (2023) Speed breeding for crop improvement and food security. Crops 3(4):276–291

Rai KK (2022) Integrating speed breeding with artificial intelligence for developing climate-smart crops. Mol Biol Rep 49:11385–11402

Rai A, Han SS (2023) Recent developments on the contribution of glutenin and puroindoline proteins to improve wheat grain quality. Cereal Chem 100:56–71

Rana MM, Takamatsu T, Baslam M et al (2019) Salt tolerance improvement in rice through efficient SNP marker–assisted selection coupled with speed-breeding. Int J Mol Sci 20(10):2585

Ray DK, Mueller ND, West PC et al (2013) Yield trends are insufficient to double global crop production by 2050. PLoS One 8(6):e66428

Raza A, Mubarik MS, Sharif R et al (2023) Developing drought-smart, ready-to-grow future crops. Plant Genome 16(1):e20279

Riaz A, Periyannan S, Aitken E et al (2016) A rapid phenotyping method for adult plant resistance to leaf rust in wheat. Plant Methods 12:1–10

Robinson A, Lehmann J, Barriopedro D et al (2021) Increasing heat and rainfall extremes now far outside the historical climate. npj Clim Atmos Sci 4:45

Sah R, Kumar A, Dash S et al (2021) Speed breeding-a unique tool for higher genetic gain in rice. In: Pradhan SK, Das SR, Patra BC (eds) Advances in rice breeding: stress tolerance, climate resilience, quality & high yield. ICAR-NRRI, pp 316–327

Sahu S, Barik M, Mohanty JN et al (2024) Genome editing systems for speed breeding of tomato (*Solanum lycopersicon* L.). In: Chen JT (ed) Plant speed breeding and high-throughput technologies. CRC Press

Sami A, Xue Z, Tazein S et al (2021) CRISPR-Cas9-based genetic engineering for crop improvement under drought stress. Bioengineered 12(1):5814–5829

Schleif N, Kaeppler SM, Kaeppler HF (2021) Generating novel plant genetic variation via genome editing to escape the breeding lottery. In Vitro Cell Dev Biol Plant 57:627–644

Schmidt M, Kollers S, Maasberg-Prelle A et al (2016) Prediction of malting quality traits in barley based on genome-wide marker data to assess the potential of genomic selection. Theor Appl Genet 129(2):203–213

Shakoor N, Lee S, Mockler TC (2017) High throughput phenotyping to accelerate crop breeding and monitoring of diseases in the field. Curr Opin Plant Biol 38:184–192

Sharma S, Kumar A, Dhakte P et al (2023) Speed breeding opportunities and challenges for crop improvement. J Plant Growth Regul 42:46–59

Sharma A, Hazarika M, Heisnam P et al (2024) Controlled environment ecosystem: a cutting-edge technology in speed breeding. ACS Omega 9(27):29114–29138

Song P, Wang J, Guo X et al (2021) High-throughput phenotyping: breaking through the bottleneck in future crop breeding. Crop J 9(3):633–645

Tadele Z (2019) Orphan crops: their importance and the urgency of improvement. Planta 250(3): 677–694

Varshney RK, Bohra A, Yu J et al (2021) Designing future crops: genomics-assisted breeding comes of age. Trends Plant Sci 26(6):631–649

Waadt R, Seller CA, Hsu P-K et al (2022) Plant hormone regulation of abiotic stress responses. Nat Rev Mol Cell Biol 23(10):680–694

Wang C, Tang S, Zhan Q et al (2019) Dissecting a heterotic gene through GradedPool-Seq mapping informs a rice–improvement strategy. Nat Commun 10(1):2982

Wang F, Liu Y, Zhang A et al (2022) Breeding an early maturing, blast resistance water-saving and drought-resistance rice (WDR) cultivar using marker-assisted selection coupled with rapid generation advance. Mol Breed 42(8):46

Watson A, Ghosh S, Williams MJ et al (2018) Speed breeding is a powerful tool to accelerate crop research and breeding. Nat Plants 4(1):23–29

Watt M, Fiorani F, Usadel B et al (2020) Phenotyping: new windows into the plant for breeders. Annu Rev Plant Biol 71:689–712

Weller JL, Ortega R (2015) Genetic control of flowering time in legumes. Front Plant Sci 6:207

Xu Y, Luo H, Zhang H et al (2023) Feeding the world using speed breeding technology. Trends Plant Sci 28(3):372–373

Yang Y, Tilman D, Jin Z et al (2024) Climate change exacerbates the environmental impacts of agriculture. Science 385(6713):eadn3747

Zhang H, Zhu J, Gong Z et al (2022) Abiotic stress responses in plants. Nat Rev Genet 23(2): 104–119

Zhang K, Yan F, Liu P (2024) The application of hyperspectral imaging for wheat biotic and abiotic stress analysis: a review. Comput Electron Agric 221:109008

Zhao M, Feng Y, Shi Y et al (2022) Yield and quality properties of silage maize and their influencing factors in China. Sci China Life Sci 65(8):1655–1666

The Nanoparticle Revolution in Agronomy: A New Frontier for Crop Enhancement

13

Jaya Gangwar and Kadanthottu Sebastian Joseph

Abstract

Rapid population expansion and environmental concerns in agriculture necessitate innovative and long-term solutions to fulfill the world's increasing food demand. Recent nanotechnological breakthroughs have demonstrated extensive use in agricultural protection and postharvest. Engineered nanoparticles are critical for plant control, seed germination, and genetic modification. Nanoparticles might be used to distribute fertilizers, insecticides, herbicides, and plant growth regulators in a more focused and controlled manner. In the food industry, this improves crop growth and production capacity by reducing chemical-based pesticides and their water solubility, protecting agrochemicals from deterioration, enhancing soil health, and naturally controlling crop pathogens, weeds, and insects. Despite the potential benefits of nanomaterials, their full adoption in agriculture has yet to be realized, owing primarily to concerns about nanoparticle absorption, translocation, bioavailability, and ecotoxicity. Understanding the underlying mechanisms and reactions caused by nanoparticles using molecular techniques is crucial for evaluating nanomaterials' biological potential. This chapter discusses nanoparticles' role in enhancing plant development and soil fertility, managing plant diseases, and addressing plant and soil toxicity. It discusses nanotechnology's potential to combat abiotic stress in crops, improve growth, absorption, and translocation, and highlights critical applications for sustainable agriculture and food systems, including stress management. The importance and significance of nanotechnology in addressing agricultural and food sector difficulties and concerns, as well as the limitations and future potential of nanoparticles, are critically explored.

J. Gangwar · K. S. Joseph (✉)
Department of Life Sciences, Christ University, Bangalore, Karnataka, India
e-mail: joseph.ks@christuniversity.in

Y. Ahlawat et al. (eds.), *Next-Generation Strategies for Crop Improvement*,
https://doi.org/10.1007/978-981-95-0309-4_13

Keywords

Agricultural sustainability · Crop production · Nanoparticles · Plant growth · Stress resilience

13.1 Introduction

Crop production in agriculture has been hampered by various concerns, including climate change, slower development, viable raw resource consumption, diseases that affect plants, and environmental issues such as pesticide misuse and fertilizer building. These issues are exacerbated by the requirement to feed 6–9 billion people by 2050. New opportunities are emerging, such as the generation of power from agricultural waste, but they must be backed by viable financial strategies and favorable regulations. Agriculture is a country's primary source of economic power; therefore, emerging nations must deal with a fast-expanding and increasingly complicated agricultural sector. A shortage of cropland, the loss of current agricultural land due to competing economic expansion activities, dependency on goods, shortages, and malnutrition are just a few of the significant problems that developing countries must deal with; these problems will take considerable effort and time to fix. The agronomic industry has suffered substantial organizational alterations as a result of the last few decades' rapid technological innovations. In the twenty-first century, every aspect of the global economy, industry, and public's lives will be significantly impacted by the wide and fascinating scientific tool that includes nanotechnology and engineering. In the nanoscale sciences, characteristics, methods, and matter phenomena at the nanometer (1–100 nm) scale are revealed (Aamir Iqbal 2020).

Nanotechnology is thought to be the fastest-growing technology of the past ten years, and it has potential applications in agriculture. Precision farming has replaced conventional farming practices with nanotools like nanopesticides and nanofertilizers. The oxides of copper, iron, manganese, molybdenum, silicon, silver, and zinc, and traditional agricultural inputs like validamycin, azadirachtin, phosphorus, sulfur, tebuconazole, urea, and sulfur have been converted into nanopesticides and nanofertilizers via nanoformulation (Konappa et al. 2021). In terms of seed, plant development, and production, the nanoforms have demonstrated encouraging outcomes at the ideal concentrations. Likewise, the control of plant diseases was positively impacted by nanopesticides (Hazarika et al. 2022).

Utilizing biological processes to produce nanomaterials, such as those involving bacteria, algae, yeast, fungi, and plants, has a significant chance of producing environmentally benign nanomaterials (Sharma et al. 2019). The soil is considered fertile when all the necessary plant nutrients are in sufficient quantities for crop growth. There should not be a toxic or inadequate amount of these vital plant nutrients. Natural disasters like floods and the ensuing erosion must not exist in the soil for it to be productive. Fertile soil is the secret to thriving agriculture in the long run (Ghosh and Bera 2021). Nanotechnology has increased human control over soil conditions through a variety of applications, including nanosensors for measuring soil conditions, chemical penetration, ecological contamination, and safety assurance, as well as contributing to the sustainability of agriculture and the environment. Potential benefits of nanoscience include improved food security, grading, product traceability, nutrition delivery, packaging appearance, agriculture and food processing, and more sustainable farming (Liu et al. 2021).

Metal nanoparticles (NPs) directly impact plant transcript processing and transcriptional regulation via chromatin arrangement modifications or binding to particular promoter regions; they can interact with DNA molecules to change the expression of genes (Kang et al. 2022). Furthermore, metal nanoparticles can affect RNA-binding proteins and splicing factors to modify transcript processing, which can alter gene expression patterns and eventually affect plant growth and improvement (Francis et al. 2024). Metal nanoparticles are vital for the biosynthesis and modification of proteins during translation and after. The effectiveness of protein synthesis and the quantity of particular proteins can be influenced by their ability to modulate ribosomal activity. Metal nanoparticles have also been demonstrated to cause posttranslational changes in proteins, including ubiquitination, acetylation, and phosphorylation, which can change the protein's stability, activity, and localization. Certain alterations play a crucial role in controlling multiple biological functions, such as hormone transmission, reaction to stress, and control of plant growth (Al-Khayri et al. 2023).

Increased plant productivity has been studied extensively through genetic modification of stress signaling pathways. In order to boost productivity and yield while boosting resistance to a variety of biotic and abiotic stresses, plants have been genetically modified (Parmar et al. 2017). The tissue culture method, which is laborious and not always practical for plants like cotton, is the main method used in traditional genetic engineering. Even though the various physical methods for transforming pollen are more efficient in comparison, issues like a decreased success rate and pollen viability loss are still common (Wijerathna-Yapa et al. 2022). Concerns about food security are another issue with genetically modified crops, and it will take more time for them to be tested and approved for marketable use, especially in India (Shukla et al. 2018). Today, a common practice for increasing crop yield and protecting crops from pathogen attacks is the application of chemical fertilizers. However, the overuse and aimless application of fertilizers disrupt the balance between soil and nutrients, ultimately raising global temperatures even more (Tudi et al. 2021). The majority of macronutrients in soil are insoluble, so plants only use about half of the fertilizer. Consequently, these agrochemicals over-accumulate in the soil and leach into water in large quantities (Aamir Iqbal 2020).

This book chapter discusses the role of nanoparticles in controlling plant diseases, improving soil fertility, and reducing toxicity to plants and soil. It talks about how nanotechnology can help crops grow, absorb, and relocate better, and it emphasizes important uses for stress management in sustainable agriculture and food systems. It also addresses how nanotechnology can help crops fight abiotic stress. The necessity of nanotechnology, its role in addressing the issues facing the food industry and agriculture, as well as the limitations and potential applications of nanoparticles, are all critically examined.

13.2 Conventional Approaches for Crop Improvement

Crop productivity is a consequence of an increase in human population. Due to environmental factors and human activity, farms are losing soil fertility. Conventional crop production techniques increase the resource and cost of production. The plants grow abnormally because they do not have enough micronutrients (Fu et al. 2020). Eutrophication is brought on by the leftover nutrients that seep into the aquatic system. Commercial fertilizers are necessary to boost crop yields, but improper fertilizer management can negatively affect the environment and finances. Nitrate leakage into marine ecosystems and N-oxide releases into the atmosphere are just two examples of the serious natural consequences that arise from the nitrogen fertilizer supplied to farms being lost to abiotic factors and other processes (Khan and Mohammad 2014). The low efficiency of phosphorus utilization poses a significant issue as phosphorus is a finite resource, and overspills worsen eutrophication in marine surroundings. It is impossible to overlook the enormous economic impact of ineffective fertilization (Withers et al. 2015). Traditional pesticide management techniques have hurt the environment and the financial interests of agronomists because 90% of the pesticides used to combat diseases and pests are wasted. Unrestrained use of pesticides decreases soil biodiversity, encourages pesticide resistance, and hinders N_2 fixation (Gangwar et al. 2023).

13.3 Source, Synthesis, and Delivery of Nanoparticles

The unique features of nanoparticles have led to a wide variety of applications in several industries. On the other hand, nanoparticles are not new to nature; they are found in the atmosphere, soils, seas, and biota. Nanoparticles are found naturally in combustion byproducts from wildfires, ocean spray, ash from volcanoes, and cosmic dust (Altammar 2023). The key influences on customized nanoparticles, which are synthesized using both bottom-up and bottom-down methods, are their regulated size-shape, bioactivity, and bioavailability. Fundamentally, the production of metal and metal oxide nanoparticles requires the reduction of metal salts utilizing reducing agents available from a variety of sources (Kumari et al. 2023). In the last few years, for instance, it has been demonstrated that microorganisms such as fungi and bacteria can synthesize metal nanoparticles intracellularly and extracellularly; the latter method has drawn greater interest due to its ability to do so without requiring further processing. A variety of techniques for biological nanoparticle production involving microbial biomass, supernatant, and derived components have been described in recent literature (Singh et al. 2021). Microorganisms with diverse reductase enzymes can aggregate and detoxify heavy metals by reducing metal salts to metallic nanoparticles with low polydispersity and a limited size distribution (Campaña et al. 2023). Nanoparticles selectively adsorb biomolecules in complicated biological fluids, creating a layer capable of interacting with biological systems more efficiently than naked phyto-NPs (Abbasi et al. 2023). The physical characteristics of nanoparticles are strongly influenced by concentration, pH, temperature, and time. Changes in pH can affect the form and production of

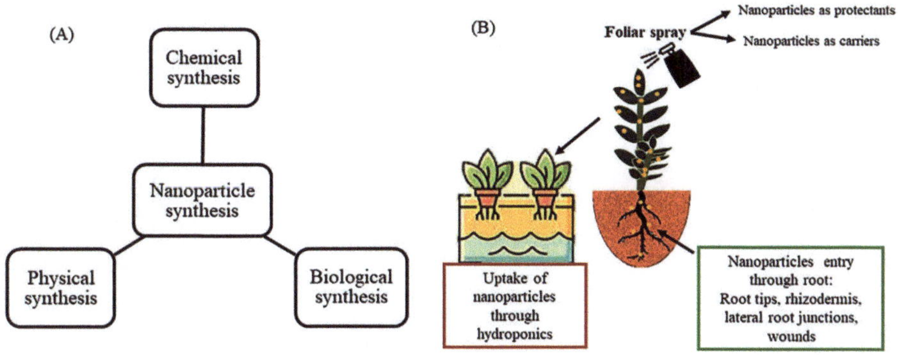

Fig. 13.1 (**a**) Different methods for nanoparticle synthesis. (**b**) Different modes for uptake of nanoparticles in plants

nanoparticles by affecting the charge of plant secondary metabolites. This can affect their reduction and binding capabilities during nanoparticle synthesis. Elevated temperatures may increase the activity of enzymes involved in nanoparticle formation (Rahman et al. 2021).

NPs enter the conductive system (xylem, phloem) depending on the complex structural properties of the cuticle, epidermis, and endodermis (Mohapatra et al. 2023). The process of NP entry into leaves happens through stomatal openings, which allow for penetration through the leaf based on the aperture. However, the degree to which leaves respond to light, moisture, temperature, CO_2 concentrations, and plant-specific stomata positions determines whether the leaves remain open or closed (Melotto et al. 2008). The cuticle in leaves acts as a significant barrier, limiting nanoparticles to less than 5 nm in size. NPs larger than 10 nm move through stomata through symplastic and apoplastic channels (Mohapatra et al. 2023). The roots provide another entrance route for nanoparticle absorption into the plant system. Root endodermis, particularly passage cells, are critical to NP penetration and transport throughout the plant. The secondary root emerging sites and damaged roots might be potential entrance routes for nanoparticles. The transport velocity of nanoparticles in the vascular arrangement is determined by sap flow rates in xylem, phloem, and cell wall interstices, as well as hydrostatic pressure gradients in restricted flow channels (Wu and Li 2022). Figure 13.1 briefly describes different methods for nanoparticle synthesis and different modes for the uptake of nanoparticles in plants.

13.4 Mechanism of Nanoparticle Uptake in Plants

The nanoparticles can enter a plant system through various channels and mechanisms. These pathways may be influenced by the plant species, the surrounding setting, the exposure settings, and the properties of the nanoparticles. The following are some typical points of entry for metal nanoparticles into plant systems (Francis et al. 2024) (Fig. 13.2).

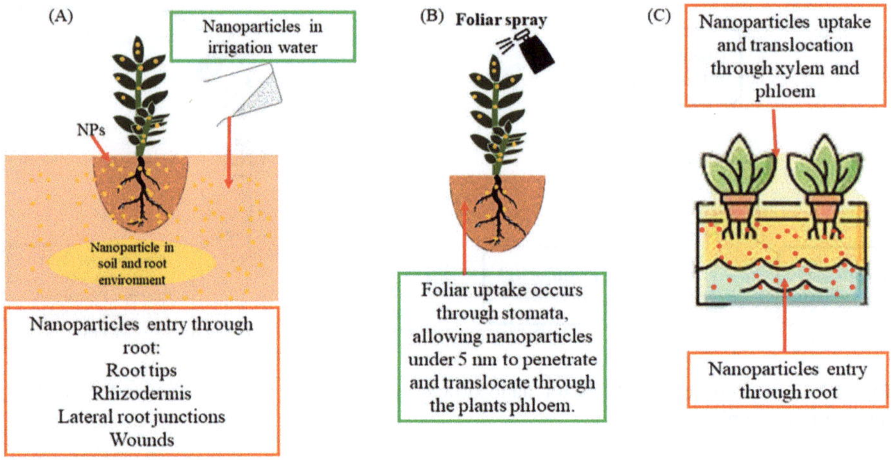

Fig. 13.2 Mechanism of nanoparticle uptake in plants:.(**a**) Uptake by roots, (**b**) foliar spray, and (**c**) hydroponics

13.4.1 Root Uptake

The complex initial interaction, adsorption, and ultimate internalization of nanoparticles in plant roots are controlled by a complex web of pathways and mechanisms (Wang et al. 2023). The way nanoparticles attach themselves to plant roots begins with adsorption on the root surface. Most remarkably, the root surface is negatively charged due to the presence of root hairs, which release chemicals like mucus and organic acids. The ability of positively charged nanoparticles to accumulate and then be absorbed by the root surface is enhanced by this electrostatic phenomenon (Singh et al. 2024). A further way nanoparticles can interact with one another and enter the root architecture is by developing lateral roots. The movement of nanoparticles in plant roots occurs in a dynamic symphony that involves two essential pathways: the apoplastic and symplastic pathways (Francis et al. 2024). The apoplastic route meanders across the extracellular spaces, taking advantage of the intricate maze of cell walls and intercellular gaps to facilitate nanoparticle transport. To get to the xylem tissue of the root, nanoparticles must pass through the porous cell walls. If the exodermis layer is absent from the root, the range of possible transport for the nanoparticles is expanded if they enter the xylem or central column (Wang et al. 2023).

On the other hand, the plasmodesmata channels allow nanoparticles to pass through the living components of plant cells in the symplastic pathway. Starting at the root epidermis, these microscopic channels enable the nanoparticles to pass through the cytoplasm of the root cells. Nanoparticles with symplastic movement can investigate different cellular compartments, which could impact their absorption and effects on plant physiology (Sembada and Lenggoro 2024).

Nanoparticle size, chemical makeup, and surface charge are three important factors that affect how well plant roots absorb nanoparticles. Of these, the size of the nanoparticles is the most important factor. *Vicia faba* L. can absorb nanoparticles smaller than 10 nm, including gold nanoparticles (3–5 nm) and CeO_2 nanoparticles ($<$25 nm), plus corn roots. On the other hand, bigger particles show less absorption, and particles bigger than 100 nm have trouble being absorbed by roots (Rasheed et al. 2022). Remarkably, nanoparticles containing silicon and natural polymers that are larger than 100 nm have demonstrated the capacity to be absorbed. According to confocal and transmission electron microscopy, for example, after 6 weeks of treatment, *Arabidopsis* roots absorbed silicon oxide NPs (200 nm) and sugarcane roots absorbed zein nanoparticles (135 $\pm$ 3 nm) (Prasad et al. 2018).

The mechanism underpinning nanoparticle–microbe interactions has significant consequences for plant absorption. These contacts begin when nanoparticles come into contact with microbial surfaces in the rhizosphere. They are based on intricate physicochemical processes. The nanoparticles can gather with microbial products or stick to microbial cells. Ag nanoparticles bind to microbial surfaces in a complex way, providing insight into the various adsorption mechanisms (Mgadi et al. 2024). These nanoparticle–microbe relationships may have an effect on the general mobility and bioavailability of nanoparticles in the soil matrix, which may affect their accessibility to plant roots. Additionally, these relationships might change the chemical cues that attract nanoparticles by affecting root exudation patterns. Such changes to the rhizosphere environment may facilitate or hinder plant roots' ability to absorb nanoparticles (Rasheed et al. 2022). The key background for understanding larger ecosystem dynamics is by giving insight on how nanoparticle–microbe interactions influence nanoparticle bioavailability in plants. Recognizing the complicated interplay between nanoparticles and bacteria in the rhizosphere broadens our understanding of the processes driving nanoparticle absorption by plants. It offers up possibilities for using these interactions to boost (Francis et al. 2024).

13.4.2 Foliar Uptake

Plant leaves can accumulate nanoparticles, particularly those suspended in water or air. From there, they may enter the leaf cuticle or be taken up by stomata, which are tiny pores on the leaf surface. This mechanism is especially pertinent for nanoparticles that are small enough to be carried by air or water droplets (Avellan et al. 2021). (a) Stomatal uptake: Stomata, which are tiny pores that control gas exchange, are the entry points for nanoparticles that are deposited on leaf surfaces. Smaller nanoparticles can pass through open stomata. (b) Cuticular penetration: A small number of nanoparticles may pass through the leaf surface's waxy cuticle. For nanoparticles with surface modifications that improve penetration, this is more typical (Sembada and Lenggoro 2024). When nanoparticles enter mesophyll cells through the stomata and cuticle, they begin a complex series of transport pathways, allowing them to travel large distances within plants. These pathways, which are involved in the transport of nanoparticles, comprise the plastid and extracellular pathways (Wu and Li 2022).

Plants transport nanoparticles in a complex process that is influenced by several critical features, including the properties of the nanoparticles themselves, the characteristics of the plant species, and the surrounding environment (Avellan et al. 2021). Notably, the absorption of nanoparticles is highly dependent on their size. Numerous investigations have demonstrated that metal-based nanoparticles can penetrate plant leaves via the stomatal pathway when their diameter is less than 50 nm (Wang et al. 2023). The absorption capacity of leaves decreases with increasing particle size. For instance, research has shown that the stomatal pathway is how ZnO nanoparticles (30 nm) enter wheat leaves, and watermelon leaves absorbed larger MgO nanoparticles via the cuticular pathway (Francis et al. 2024). In addition to their metal-based counterparts, nonmetal equivalents such as SiO_2.etc. was also studied. In addition, it was discovered that polystyrene nanoplastics (93–106 nm) were absorbed. Remarkably, silicon nanoparticles (166 nm) based on chitosan were able to penetrate rice leaves (Yee et al. 2021). Nanoparticle entry is further influenced by shape and surface charge. Both positively and negatively charged nanoparticles were absorbed by leaves, and rod-shaped nanoparticles showed higher absorption rates than spherical ones. Negatively and positively charged nanoparticles are attracted to each other electrostatically (Avellan et al. 2021).

13.4.3 Uptake of Nanoparticles in Hydroponics

Plants grown in nutrient solutions without soil or hydroponic systems offer a controlled environment to study the dynamics of the interactions between nanoparticles and plants. This methodology allows for targeted investigation of the interactions between NPs and plants while avoiding the confounding effects of soil properties (Francis et al. 2024). Metal nanoparticle-containing nutrient solutions are applied to plants in hydroponic systems. During this crucial phase of exposure, nanoparticles and root surfaces can directly interact, starting a sequence of events similar to those described in root uptake mechanisms (Ahmad et al. 2022). When nanoparticles come into contact with root structures, they may trigger the previously mentioned set of uptake pathways, which include binding with cell membrane proteins, endocytosis, and adsorption. These processes control the possibility of metal nanoparticles entering the plant's vascular system and then dispersing throughout its different sections (Francis et al. 2024).

13.5 Nanoparticles for Improved Crop Improvement

Plant development and growth have been shown to be improved by nanoparticles. They are crucial for photosynthesis, crop yield, disease control, and seed germination. They also help crops develop resistance to diseases. Modern agricultural methods have focused on nanostructures as nanofertilizers for physiology, metabolite formation, and general growth dynamics—all of which are critical to the important activities of plants (Mittal et al. 2020). Chemical applicators exploit nanofertilizers' enhanced features, such as smaller particle size, higher surface area, and more particles per volume, to promote release and penetration into the plant system. This increases utilization, bioavailability, performance, and effectiveness (Al-Juthery et al. 2021). Furthermore, slow-release nanofertilizers are employed to counterbalance the loss of inorganic fertilizers. Plants would be able to absorb the majority of their nutritional needs as a consequence without losing any vital nutrients. In order to improve the absorption of nutrients, amino acids and sugars like mannose are frequently added to nanofertilizers. Moreover, the amount of nanocomposites grows (Ahmed et al. 2021; Akhtar et al. 2022).

13.5.1 Role of Nanoparticles in Plant Growth

Delivering agrochemicals based on nanoparticles to particular crops is crucial, and this can be done by employing highly stable nanocapsules with degradable structures. By improving targeted delivery, welfare, infection, and pesticide adherence, as well as lowering plant protection costs, the growing use of nanoparticles in agriculture boosts the efficacy of chemical insecticides (Kutawa et al. 2021). More environmentally friendly techniques have been developed using ionizing radiation chemistry in aqueous solutions and plant extracts. Because of their superior catalytic, electrical, and light absorption properties, oxidized nanoparticles—such as titanium dioxide nanomaterials and magnesium, calcium, and zinc oxides—have also been frequently promoted. One technique to increase the biocompatibility of nanoparticles is the use of polymeric shells. A porous channel with a larger surface area that is highly functionalizable has been used to create nanostructures such as silica nanoparticles (Shang et al. 2019) (Table 13.1).

Nanoparticles enter the plant through the shoots and roots, where the roots and soil subsequently take them up. Different plant structures impose different size exclusion limits (SEL) on the movement of nanoparticles (NPs). The SEL of 5–20 nm is imposed by the cell wall, 1 nm by the casparian strip, and 3–50 nm by the plasmodesmata. The NPs' modified surface has an impact on their mobility as well. NPs can affect plants in various ways, mainly depending on their physicochemical properties (Fig. 13.3).

13.5.2 Role of Nanoparticles in Soil Fertility

There are several agroeconomic strategies for increasing soil fertility using microorganisms. Eco-friendly nanofertilizer promotes soil growth, moisture retention, and carbon retention. It is safe to use on agricultural crops like vegetables, horticulture, and grains. Mesoporous aluminosilicate-based nanoparticles offer more possibilities than mineral salt for the efficient and controlled release of both macro

Table 13.1 Application of different nanoparticles in plant growth

Nanoparticle type	Application	References
Carbon nanotubes	Assisting in pollutant filtration from the soil, promoting root elongation, and stimulating plant and seedling growth	Fritea et al. (2021)
Graphene oxide	Enhancing seed germination processes	Shang et al. (2019)
Copper oxide	Encouraging overall plant growth and seed germination in *Arabidopsis thaliana*	Wang et al. (2016)
Zinc oxide nanorods	Boosting broccoli plant growth	Skiba et al. (2021)
Silver	Stimulating plant growth and inhibiting harmful bacterial growth	Abasi et al. (2022)
Multiwalled carbon nanotubes	Improving kernel germination	Fritea et al. (2021)
Aluminum oxide	Increasing root length in various plant species	Sembada and Lenggoro (2024)
Zinc and cerium oxides	Supporting general plant growth improvement	Skiba et al. (2021)
Silver and iron oxides	Contributing to enhanced plant growth	Rehman et al. (2024)
Silicon	Increasing plant resistance to environmental stresses like drought and heat	Rastogi et al. (2019)
Titanium oxide	Promoting better soil health and boosting plant productivity	Francis et al. (2024)
Iron oxide	Assisting in the elimination of heavy metals from contaminated soils	Atrak et al. (2020)

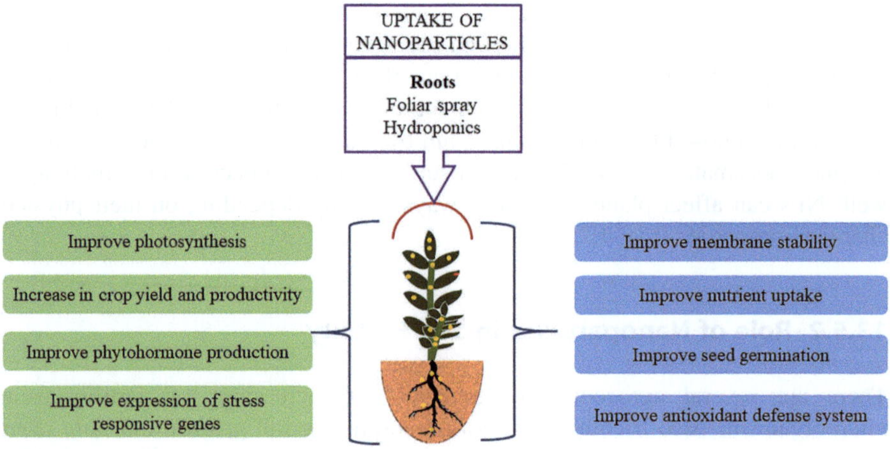

Fig. 13.3 Role of nanoparticles in soil fertility and plant growth

and micronutrients in the soil, improving plant development and production (Zulfiqar et al. 2019). Under aerobic and submerged conditions, using a nano-sized $MnCO_3$ hollow core-shell system resulted in a gradual and consistent release of zinc in rice compared to zinc sulfate. Nitrate reductase activity was increased in soybeans through the application of silicon oxide–titanium nanomaterials, improving nutrient absorption. In a similar vein, applying Fe_2O_3 nanoparticles instead of chelated iron enhanced nutrient uptake, which in turn enhanced peanut growth and productivity (Aamir Iqbal 2020). Because they are bioavailable to plants, metallic nanoparticles may greatly affect food crops. The zinc oxide nanoparticles caused significant amounts of metal to be translocated into the leaves and beans, and the soybean plants bioaccumulated metals from the soil (Fig. 13.3). Plants cultivated with high zinc oxide nanoparticle concentration showed no change in aboveground biomass yield (Table 13.3).

13.5.3 Nanoparticles in Controlling Plant Diseases

Plant growth is promoted through the implantation of beneficial bacteria onto seeds. Nano-agrochemicals are chemical pesticide and fungicide reformulations found to preserve plants over time while using fewer pesticides (Alghuthaymi et al. 2021). Pathogen-induced plant infections lead to decreased yield, lower product value, and shorter shelf life, which cause financial loss. Using nanomaterials raises additional concerns about the potential for trophic-level transfer and nanoparticle bioaccumulation in crops (Cao et al. 2021). The most popular nanoparticles used to manage plant diseases are carbon, silver, silica, and aluminosilicate nanoparticles. Growth in plants is stimulated by silver. The potent bactericidal and broad-spectrum antibacterial properties of nanosilver have been demonstrated. There is evidence that plants absorb silicon to reduce their susceptibility to disease and stress. Illness is unaffected by silica because it has no direct effect on dangerous bacteria. It is also being developed to spray aluminum silicate nanotubes on plant surfaces, which are readily absorbed by insect hairs. Grooming and eating pesticide-filled nanotubes is an aggressive behavior (Sagar et al. 2022). It has been observed that exposed soybean roots store cerium oxide nanoparticles before they are transferred to edible tissue. Agriculture can benefit from the use of titanium oxide nanoparticles for disease prevention.

The use of traditional agricultural production techniques to manage infections and pests has raised pathogen and pest resistance, bio-accumulated chemicals, reduced soil biodiversity, and decreased the incidence of insect pests and their products in agricultural areas. Innovative delivery systems and nanocarriers, designed for enhanced efficiency, creative and targeted distribution, and only function, can address these environmental concerns when necessary. This shows that nanoparticles play an important role in the control of host–pathogen interactions and insect pests. Several studies have found that nanopesticides are more effective against a broader spectrum of plant diseases. Nano-derived agrochemicals have been shown to be effective against bacterial and fungal infections (Droby et al. 2022).

Nanoencapsulation, which resembles microencapsulation, is used to enhance the targeted chemical's superiority over the biological process. Lately, chemical companies have started openly advertising the sale of nanoscale insecticides under the name "microencapsulated insecticides." Products from Syngenta (Switzerland), like BASF's microencapsulated insecticides, Penncap-M, Ospray's Chyella, and Karate ZEON, may be prepared to battle at the nanoscale (Mishra et al. 2018). It has been discovered that spinach plants are adversely affected by copper-based nanopesticides, which change their metabolic profiles and reduce antioxidant molecules such as ferulic acid, ascorbic acid, alpha-tocopherol, threonic acid, and 4-hydroxybutyric acid.

Nano and synthetic pesticides have proven to be valuable weapons in the fight against plant disease vectors. Nevertheless, the ecosystem and nontargeted organisms have suffered from these pesticides. Furthermore, it could result in bioaccumulation, which would be extremely dangerous for wildlife and plants. Various nanoparticles, including alumina, Ag, silica, ZnO, CuO, and Mn_2O_3 are employed to address this problem and have demonstrated insecticidal effects (Konappa et al. 2021). Pesticide delivery systems (PDS) use nanoparticles to deliver an active compound to a target at a specific concentration and time for maximum biological efficacy. This minimizes the detrimental impacts on nontarget insects while focusing on the targeted creatures. Insecticide compounds can be linked to nanoparticles, which act as nanocarriers. PEG (polyethylene glycol) polymer nanoformulation-based pesticides, which employ polymer nanoparticles as nanocarriers, have exhibited a significant gradual release of active ingredients compared to chemical pesticides, as well as improved performance against worms and beetles (Chaud et al. 2021). Effective insecticide delivery has been achieved through the use of inorganic nanoparticles, such as silica nanoparticles. Pesticide formulations based on silica nanoparticles demonstrated higher efficacy and a delayed release of chlorfenapyr (Al-Juthery et al. 2021).

Numerous types of nanoparticles have been reported to have been used as nanofertilizers, including those based on nitrogen, phosphate, iron, zinc, and titanium. Even polymers acting as dendrimers and serving as nanocarriers have been reported to have been used as nanofertilizers, as well as gold nanorods, core-shell quantum dots (QDs), and gold nanorods (Mahapatra et al. 2022). Owing to their many advantageous qualities, particularly their targeted and timely release, nanofertilizers have been shown to be highly significant in contemporary agriculture. These nanoparticles are typically applied in vivo (soil and foliar application) or in vitro (hydroponics and aeroponics) (Rajput et al. 2021). The size, chemical makeup, and physiology of the plant determine how well the nanoparticles are absorbed once they are delivered. Herbicide-resistant weeds have emerged as a result of ongoing herbicide application. Even with resistant management techniques, such as integrated weed control, overuse of herbicides can harm the environment and make these techniques outdated. Herbicides with nanostructures may offer a more effective and efficient option for lowering herbicide consumption while also improving solubility, lowering toxicity rates, and raising productivity rates (Ramanand et al. 1993).

Biosensors can also be developed using many of the physical–chemical properties of materials at the nanoscale. The novel signal transduction methods and the use of nanomaterials can enhance the sensitivity and functionality of biosensors, due to the significant need for quick, accurate, and affordable nanobiosensor systems in key anthropogenic activity areas (Hotessa Halake and Muda Haro 2022). Due to the presence of a transducer that acts together with a target molecule to generate a signal and a baroreceptor, which is a naturally occurring element, biosensor systems exhibit higher specificity and sensitivity when compared to conventional techniques. Recently, a wide range of applications have used bioreceptors, including thin films, dendrimers, enzymes, and others. Biosensors are analytical tools that convert an electrical signal from a biological reaction (Bhalla et al. 2016). It is centered on biological elements such as constituents, proteins, enzymes, antibodies, and nucleic acids. Numerous nanomaterials, including carbon, silver, gold, and quantum dots, among others have been employed to create nanosensors. Because biosensors can sense residual chemicals like pesticides and fertilizers, monitor pathogens and their toxins, and detect heavy metals in agronomic fields, they can be used to increase sustainable agronomic efficiency (Fritea et al. 2021) (Table 13.2).

13.5.4 Role of Nanoparticles in Alleviating Abiotic Stress

NPs not only affect the growth and development but also have the ability to lessen the harmful effects of a number of important abiotic stresses that plants face. These ever-growing abiotic factors include heat stress, waterlogging, salinity, drought, and toxicity from heavy metals. Significant restrictions on plant productivity and yield are caused by exposure to these abiotic factors (Al-Khayri et al. 2023). The NPs assist in reducing such stressors by activating certain genes, aggregating osmolytes, boosting phytonutrients, and creating photopigments. One of the inescapable indications of these abiotic stressors is the generation of ROS due to the plant cell's restricted oxygen supply (Hasanuzzaman et al. 2021). Although plants have the enzyme mechanism to combat oxidative stress from the surroundings, the stress often overwhelms the plant defense system, leading to detrimental effects at the molecular and physiological levels. However, because of their minuscule size, NPs differ from their bulk counterparts in a few special ways that open up new possibilities across a varied range of applications. The application of nanoparticles against these stresses has been shown in earlier research to increase further the plant's intrinsic pool of antioxidant enzymes (Mishra et al. 2023) (Table 13.3).

Numerous significant impacts on plant physiology and stress-related adaptations have been investigated and identified in recent years. One such abiotic stressor linked to osmotic stress is that it can potentially upset cellular homeostasis and ion distribution (Waadt et al. 2022). Heat stress can have adverse effects that collectively impede the photosynthetic process, such as the decline in chlorophyll content, disturbance in photosystem II, disturbance in electron flow, and reduction in carbon

Table 13.2 Applications of nanoparticles in plant disease management

Nanoparticle type	Disease/pathogen targeted	References
Copper	Effective against insect pests and pathogenic microbes	Alghuthaymi et al. (2021)
Copper oxide	Antifungal agent against *Poria hypolateritia*	Siwach and Verma (2021)
Copper oxide, manganese oxide, titanium oxide, zinc oxide	Protect watermelon from *Fusarium* wilt infection	Elmer et al. (2018)
Encapsulated nanosilver	Removes unwanted microbes from soil	Akhtar et al. (2022)
Magnesium oxide	Antifungal action on different species of fungus	Koka et al. (2019)
Manganese and titanium oxide	Enhance development of tomatoes and eggplants in pathogen-infested soil	Elmer and White (2016)
Myconanoparticles (gold, silver, selenium, palladium)	Control the spreading of plant diseases	Alghuthaymi et al. (2015)
Nanocarbon, silica, alumina silicates	Control crop diseases	Alghuthaymi et al. (2021)
Copper	Controls bacterial spots in tomatoes	
Nickel chitosan nanoconjugate	Antifungal agent for fighting *Fusarium* rot of wheat	Chouhan et al. (2022)

Table 13.3 Role of nanoparticles in enhancing soil fertility and alleviating abiotic stress

Nanoparticle type	Application	References
Zinc oxide	Enhanced plant growth and improved soil fertility	Mishra et al. (2018)
Titanium oxide	Increased photosynthetic activity and decreased oxidative stress in plants under drought environments	Akhtar et al. (2022)
Silicon oxide	Increased plant salinity tolerance by regulating osmotic balance and improving antioxidant enzyme activity	Rasheed et al. (2022)
Iron oxide	Increased soil fertility by enhancing nitrogen fixation and encouraging root growth	Liu et al. (2021)
Copper	Alleviating fungal stress by aiding as an antifungal agent and reducing stress from pathogens	Akhtar et al. (2022)
Silver	Mitigating stress and improving soil health	Abasi et al. (2022)
Carbon-based	Improved H_2O holding and nutrient uptake in soil, increasing plant tolerance to drought and nutrient stress	Wang et al. (2014)
Manganese oxide	Enhanced nutrient availability in soil and reduced stress such as salinity and heavy metal toxicity	Ahmed et al. (2021)
Cerium oxide	Reduced oxidative stress in plants by catalyzing the breakdown of ROS	Ma et al. (2016), Prakash et al. (2021)

fixation (Ivanov et al. 2017). A series of physiological, molecular, and biochemical changes brought on by salinity stress have an immediate impact on plant development. Improved ionic stress and decreased soil osmotic potential are two typical consequences of salt stress on plants. On the other hand, plant turgidity is lost in response to drought stress. The increased protoplasmic dehydration leads to irreversible problems in cellular metabolism. The turgor loss reduces leaf size, ion accumulation, and the yield of photosynthetic products. It also damages membrane integrity, protein structure, and cellular metabolism. As a result of water molecules occupying the air spaces in the soil and affecting the exchange of gases in the soil, roots in wet soil become hypoxic (Singh et al. 2022).

Sufficient evidence suggests that most abiotic stresses have caused multiple biological alterations in response to NP treatment. The application of nanoparticles enhances the photosynthetic rate, chloroplast protection, O_2 evolution, and total chlorophyll content. According to reports, the application of TiO_2 NPs improved the concentration of chlorophyll, net photosynthesis, transpiration rate, and water conductivity in tomato and wheat plants. TiO_2 has photocatalytic qualities that protect chloroplasts, increase the number of light-harvesting complexes, and accelerate the evolution of O_2 (Al-Khayri et al. 2023). As per previous research, NPs alter hormone levels, gene expression, defense mechanisms, antioxidant enzyme activity, and ion homeostasis in plants, thereby affecting their capacity to withstand salinity stress (Soni et al. 2024). These effects could vary depending on the kind, size, and concentration of NPs used, as well as the surrounding environment and plant species. Shorter apoplastic barriers in the roots of canola plants (*Brassica napus*) are caused by CeO_2 nanoparticles, leading to improved Na^+ transfer to shoots and decreased Na^+ accumulation. Moreover, abiotic stressors can upset ionic homeostasis in cells by reducing or switching up the uptake of particular ions (Rehman et al. 2024). For example, when exposed to salinity stress, sodium ions replace potassium or calcium ions. These ions primarily cause cytosolic toxicity, which disrupts a variety of physiological processes such as signaling, nutrient translocation, and photosynthesis. It has been demonstrated that a key factor in determining stress tolerance is the high K^+ to Na^+ ratio, which is changed by salt stress and can be improved by applying NPs (Balasubramaniam et al. 2023).

13.6 Genetic Manipulation

Metal nanoparticles have shown considerable potential as tools in agricultural research, giving unique opportunities to investigate their effects on molecular processes and plant gene expression. To fully realize metal nanoparticles' promise for crop development and sustainable agriculture, it is critical to understand how they influence plant gene expression and molecular pathways.

13.6.1 Effects of Nanoparticles on Plant Gene Expression

According to studies, metal nanoparticles significantly affect plant signaling and have an impact on a variety of biological pathways and processes. The impact of gold nanoparticles on plant receptor–ligand interactions and signaling pathways has been investigated (Pérez-de-Luque 2017). Gold nanoparticles functionalized with auxin hormone analogs could regulate plant growth and development by triggering particular signaling pathways. Gold nanoparticles' impact on plant signaling was highlighted by the alterations in gene expression and physiological reactions they caused (Ferrari et al. 2021). The effect of silver nanoparticles (AgNPs) on plant signaling has also been studied. Silver nanoparticles activated signaling cascades involved in stress tolerance and increased the expression of genes that respond to stress. According to this, plants may be able to respond adaptively to environmental stressors by using silver nanoparticles as signaling inducers (Abasi et al. 2022). Moreover, it has been demonstrated that metal nanoparticles influence signaling molecules and pathways associated with plant defense mechanisms. For example, it has been reported that plants produce more defense-related signaling molecules like salicylic acid and jasmonic acid when exposed to copper nanoparticles (CuNPs). The activation of plant defense is partly attributed to this modulation of signaling pathways (Francis et al. 2024).

Examining the impact of metal nanoparticle exposure on transcription and transcript processing in plants is essential. Their distinct physicochemical characteristics make it possible to precisely manipulate and analyze the regulation of gene expression and RNA metabolism. For the purpose of analyzing their potential uses in agriculture, environmental monitoring, and risk assessment, it is essential to comprehend how metal nanoparticles affect these processes (Abideen et al. 2022). The capacity of gold nanoparticles to affect RNA processing mechanisms and contribute to plant transcriptomic variability was demonstrated by notable changes in the splicing patterns of particular genes. The aluminum oxide nanoparticles affected tobacco plant RNA decay, finding that some RNA molecules were more quickly broken down and that there might have been modifications to the gene expression profiles. This implies that metal nanoparticles may function in plant posttranscriptional regulation (Francis et al. 2024). It is crucial to consider several variables, including the size, concentration, length of exposure, and plant species, as these can show how metal nanoparticles affect plant transcription and transcript processing.

13.6.2 Effects of Nanoparticles on Plant Molecular Processes

The impact of metal nanoparticles on plant translation has been studied. The silver nanoparticles affected *Arabidopsis thaliana* translation efficiency. The results showed modifications in particular protein levels, indicating that silver nanoparticles can influence plant protein synthesis (Geng et al. 2022). Metal nanoparticles also impact plant posttranslational protein modifications. The musk melon protein phosphorylation patterns were affected by iron oxide nanoparticles. The study demonstrated how iron oxide nanoparticles could affect cellular functions and protein signaling by revealing changes in the phosphorylation state of specific proteins. Furthermore, it has been demonstrated that metal nanoparticles affect other posttranslational changes in plants, such as acetylation, methylation, and glycosylation (Wang et al. 2019).

The underlying genetic and molecular processes that control how metal nanoparticles affect plants have not yet been thoroughly investigated. Metal nanoparticles have special qualities that make them suitable for plant molecular processes and gene expression modulators. The seed priming with cerium oxide nanoparticles affected plants' reactions to salinity stress. They discovered that stress-related genes involved in ion transport, antioxidant defense, and stress signaling pathways had different expression levels after seed priming with CeO_2 NPs (Parmar et al. 2017). Certain metabolites that are known to improve plant stress tolerance also accumulate as a result of CeO_2 NPs being present during seed priming. Additionally, under salinity stress conditions, CeO_2 NPs reduced oxidative stress in plant cells by acting as scavengers of reactive oxygen species (ROS). These results underscore CeO_2 NPs' potential as a tool for improving plant stress tolerance and shed light on the molecular mechanisms by which plants become more resilient to salinity stress (Khalid et al. 2022).

NPs cause biological alterations and have an impact on vital components like proteins once they are inside the cells and subcells. AgNPs also damage plant DNA and cause the production of reactive oxygen species (ROS) (Syu et al. 2014). The genetic mechanisms are linked to plants' reaction to metal nanoparticles. The silver nanoparticles affected the transcription of genes and pathways in *Arabidopsis thaliana*, finding differences in their regulation (Geng et al. 2022).

Copper, gold, and silver nanoparticles, among others, have been shown to influence gene expression, activate signaling pathways, and boost signaling molecule synthesis in plants. They influence gene expression profiles, alternative splicing patterns, and RNA stability. Metal nanoparticles also influence protein levels, posttranslational changes, and signaling. These findings show their potential as significant tools for modifying plant molecular processes, increasing environmental responses, and expanding agro and biotechnology applications (Sharma et al. 2019; Kang et al. 2022).

13.7 Phytotoxicity of Nanomaterials in Plants

The toxicity of nanoparticles, however, has not been adequately characterized. The common feature of the toxicity of these nanoparticles is connected with genotoxicity, cell membrane rupture, and the discharge of hazardous chemicals (Ruttkay-Nedecky et al. 2017). Concentrations, on the other hand, have an important influence on both the good and negative aspects of nanoparticles. Higher concentrations can have a major influence on the plant's physiological processes. The persistent use of nanoparticles, particularly in agronomy, may increase their presence in soil and crops (Rajput et al. 2021), resulting in adverse consequences on plants. The bigger the size of nanoparticles, the more hazardous they are in nature. It has also been discovered that crystal structure influences nanotoxicity (Singh et al. 2021).

Nanoparticles can change plant membranes and defense mechanisms directly, as well as indirectly, causing clogging and reactive oxygen species generation. Copper oxide nanoparticles can hinder seed and pollen germination in different *Arabidopsis thaliana* ecotypes; additionally, biomass buildup is parameter-dependent. Following nanoparticle treatment, two genetic materials were differentially expressed: two genes controlling root development and the formation of reactive oxygen species, showing a link between root growth suppression and oxidative stress (Nisha Raj et al. 2021). Zinc nanoparticles' deleterious effects on plant biomass are mostly due to their absorption at the root surface. Silver nanoparticles have been shown to diminish plant biomass in wheat roots, which develop slowly and branch regularly. Antimicrobial drugs containing silver nanoparticles are widely used. Their widespread usage will inevitably result in nanoparticle buildup in the environment, and hence in plants. Nanoparticles at low absorption levels promote growth in Chinese cabbage, whereas high concentrations hinder growth (Ameen et al. 2021). Six advanced plant species such as radish, rape, ryegrass, lettuce, maize, and cucumber are adversely affected by five different forms of nanoparticles that have been displayed to be detrimental to seed growth and root growth: multiwalled carbon nanotubes, aluminum, alumina, zinc, and zinc oxide (Ahmad et al. 2022). The obstruction transpired in the seed breeding stage as opposed to the seed soaking stage, and it was shown that inhibition happened when soybean plants were cultivated in soil that had nanocerium oxide supplements. Rhizobacteria may mitigate the harmful effects of cadmium pollution on barley plants by attaching cadmium ions from the soil and reducing cadmium's permeability in the soil. *Azospirillum brasilense* can mitigate the harmful effects of saline conditions on jojoba roots. The bacteria reduced the influence of salinity on the plant's roots' ability, indicating that *A. brasilense* is more resistant to salt stress. *Bradyrhizobium elkanii* generates rhizobitoxin, which has two effects. Stress-induced ethylene production has a negative influence on nodulation and inhibits ethylene synthesis (Gonzalez et al. 2015). Salicylic acid prevented the growth of root hair cells by lowering several phenotypic impacts and posttranscriptional mechanisms related to nanoparticles. Nanoparticles' negative impacts are intimately linked to hormone levels, and appropriate levels of specific hormones may minimize nanoparticle detrimental effects (Balasubramaniam et al. 2023). Biofortified staple crops have fewer essential nutrients and vitamins per day than industrially improved foods, yet they can meet an individual's daily micronutrient requirements. Unlike commercial or industrially produced fortified foods, they allow rural residents to obtain nutrient-dense foods from their local community (Bouis and Saltzman 2017). Studies have reported adverse impacts of nanofertilizers, including increased oxidative stress and delayed development. Similarly, Ag nanoparticles boosted germination in the early stages but then reduced seedling development (Verma et al. 2022). Toxicology can depend on a variety of factors, including concentration rise above a threshold (e.g., silicon oxide, magnesium, tungsten, and aluminum) and ion release (e.g., zinc and copper oxide nanoparticles) (Naz et al. 2020).

13.8 Exploring the Potential Benefits and Risks of Nanoparticles in Agriculture

Nanoparticles in agriculture have demonstrated encouraging results in a variety of applications, including crop enhancement, control of diseases, and cleanup of the environment. However, in order to ensure its safe and long-term deployment, it is critical to weigh both the possible advantages and hazards. Metal nanoparticles offer potential advantages and dangers in agriculture, affecting crop yield, health of the soil, and environmental sustainability (Banerjee et al. 2021).

13.8.1 Benefits

Metal nanoparticles have antibacterial capabilities and might be used to manage plant diseases. Research has shown that they are efficient against a wide spectrum of plant diseases. For example, silver nanoparticles have been proven to prevent fungal diseases such as *Fusarium* spp., *Botrytis cinerea*, and *Alternaria solani* (Alghuthaymi et al. 2021). The nanoparticles might replace traditional fungicides and reduce the need for chemical treatments. Metal nanoparticles such as copper (Cu), silver (Ag), and zinc oxide (ZnO) are efficient against a wide range of plant diseases, making them useful for disease control (Tudi et al. 2021). Metal nanoparticles have several mechanisms of action in reducing plant diseases, can help prevent pathogen resistance, and contribute to long-term disease management methods (Mittal et al. 2020). Agricultural environmental rehabilitation can benefit from the application of metal nanoparticles, especially in soil and water systems. Pesticides, heavy metals, and organic contaminants have all been removed with their help (Chaud et al. 2021). For example, arsenic-contaminated soil has been remedied with iron nanoparticles, which effectively lower the concentration of arsenic and improve the quality of the soil. These applications help lessen the negative effects that pollutants have on agricultural ecosystems (Al-Juthery et al. 2021).

13.8.2 Risks

Metal nanoparticles may be advantageous, but they can also be harmful to ecosystems and organisms that are not the intended targets. According to studies, beneficial soil bacteria, earthworms, and aquatic species may be toxically affected by several metal nanoparticles, including silver and copper nanoparticles (Rehman et al. 2024). These nanoparticles may upset natural equilibrium and have a detrimental effect on biodiversity in agricultural ecosystems (Prasad et al. 2018). Concerns have also been raised about the possible buildup of metal nanoparticles in the food chain and the potential consequences for human health. Studies have indicated that plants can absorb and relocate metal nanoparticles, which have led to concerns regarding the possibility of these particles building up in plant components that humans can consume. For instance, it has been demonstrated that copper nanoparticles can build up in lettuce leaves and may have an effect on food safety (Francis et al. 2024).

Studies have demonstrated that metal nanoparticles can interact and change with soil constituents, changing how they behave and their ability to move across the environment. Metal nanoparticles in soils have the potential to change nutrient cycling processes and soil structure by interacting with organic matter and soil particles (Ameen et al. 2021). For example, research has shown that at concentrations as low as 0.1 mg/kg, silver nanoparticles (AgNPs) can dramatically change the microbial communities in soil, affecting the fertility and health of the soil (Pallavi et al. 2016).

Similarly, it has been demonstrated that copper nanoparticles (CuNPs) at 50 mg/kg impede plant root elongation and decrease soil enzyme activities, suggesting possible hazards to plant growth and soil health (Xiao et al. 2022). Metal nanoparticles in water bodies have the potential to disturb aquatic ecosystems and impact aquatic organisms. For instance, the EC50 values for suspensions of 60 nm and 300 nm AgNPs were 1.0 μg Ag/L and 1.4 μg Ag/L, respectively, for *Daphnia magna*, a crucial indicator species in marine settings. For *Oryzias latipes* suspensions of 60 nm and 300 nm AgNPs, the 96-h LC50 values were 28 μg Ag/L and 67 μg Ag/L, respectively (Francis et al. 2024).

Similarly, metal nanoparticles in sediments may affect the stability of the sediment and the organisms that live there. Ag nanoparticles, for example, have been reported to lower the survival and reproduction rates of the benthic organism *Lumbriculus variegatus* at concentrations of 1098 mg/kg, suggesting possible ecological hazards (Ramskov et al. 2015).

Determining their long-term effects on ecosystems requires an understanding of their environmental fate. These quantitative investigations demonstrate the necessity of carefully evaluating metal nanoparticle concentrations and persistence in different environmental compartments in order to reduce possible dangers (Wang et al. 2014).

13.9 Conclusions and Future Prospects

The application of metal nanoparticles in agriculture offers a promising alternative, offering fresh answers to the many problems facing the sector. Extensive study has given us valuable insights into the adaptability and underlying principles of metal nanoparticles in agricultural settings. Nanoparticles contain a variety of intriguing features that might revolutionize agricultural practices. They enhance nutrient management by allowing for accurate distribution and optimization of plant nutrition while reducing dependency on traditional fertilizers. Furthermore, nanopesticides incorporating metal nanoparticles offer a more sustainable approach to disease management, decreasing the environmental risks linked with chemical pesticides. The use of nanosensors enables continuous evaluation of crop health and environmental conditions, encouraging precision agriculture and efficient resource utilization. Furthermore, nanofertilizers with controlled release methods increase nutrient effectiveness while minimizing environmental impact and complying with sustainable agriculture principles. Using nanoparticles in nanoremediation techniques has the potential to considerably enhance soil and water quality, hence mitigating the harmful effects of pollutants in agricultural systems.

Following an in-depth analysis of the current research, it is widely accepted that NPs may considerably increase plant growth and development by mitigating a variety of abiotic challenges. Given the bio-stimulant potential of NPs, a fundamental knowledge of how they function and alter plant metabolism is required. The actual capability of NPs in agronomic situations may be realized by significant research into biological, molecular, and biochemical effects on plants. The use of nanoparticles in sustainable agronomy is static in its early stages due to information gaps surrounding the precise target of NPs inside plant cells, biomolecular interactions, and environmental danger assessments.

Despite these potential qualities, some research gaps need to be filled. Long-term effects, ecotoxicity assessments, nanoparticle behavior in agricultural ecosystems, plant absorption and translocation, and standardization of characterization approaches are among the major concerns that demand additional investigation. Furthermore, consideration of socioeconomic ramifications and public image is critical for the effective implementation of nanoparticle-based agricultural solutions. We can build the framework for safe and sustainable agricultural methods if we thoroughly grasp the advantages, hazards, and limits of metal nanoparticles in agriculture. We hope to discover solutions that increase agricultural output, better nutrient management and disease control, and encourage environmental sustainability via ongoing research and collaboration.

References

Aamir Iqbal M (2020) Nano-fertilizers for sustainable crop production under changing climate: a global perspective. In: Hasanuzzaman M, Carvalho Minhoto Teixeira Filho M, Fujita M, Assis Rodrigues Nogueira T (eds) Sustainable crop production. IntechOpen, pp 1–12. https://doi.org/10.5772/intechopen.89089

Abasi F, Raja NI, Mashwani ZUR, Amjad MS, Ehsan M, Mustafa N, Haroon M, Proćków J (2022) Biogenic silver nanoparticles as a stress alleviator in plants: a mechanistic overview. Molecules 27:3378. https://doi.org/10.3390/molecules27113378

Abbasi R, Shineh G, Mobaraki M, Doughty S, Tayebi L (2023) Structural parameters of nanoparticles affecting their toxicity for biomedical applications: a review. J Nanopart Res 25:43. https://doi.org/10.1007/s11051-023-05690-w

Abideen Z, Hanif M, Munir N, Nielsen BL (2022) Impact of nanomaterials on the regulation of gene expression and metabolomics of plants under salt stress. Plants 11:691. https://doi.org/10.3390/plants11050691

Ahmad A, Hashmi SS, Palma JM, Corpas FJ (2022) Influence of metallic, metallic oxide, and organic nanoparticles on plant physiology. Chemosphere 290:133329. https://doi.org/10.1016/j.chemosphere.2021.133329

Ahmed B, Rizvi A, Ali K, Lee J, Zaidi A, Khan MS, Musarrat J (2021) Nanoparticles in the soil–plant system: a review. Environ Chem Lett 19:1545–1609. https://doi.org/10.1007/s10311-020-01138-y

Akhtar N, Ilyas N, Meraj TA, Pour-Aboughadareh A, Sayyed RZ, Mashwani Z-R, Poczai P (2022) Improvement of plant responses by nanobiofertilizer: a step towards sustainable agriculture. Nano 12:965. https://doi.org/10.3390/nano12060965

Alghuthaymi MA, Almoammar H, Rai M, Said-Galiev E, Abd-Elsalam KA (2015) Myconanoparticles: synthesis and their role in phytopathogens management. Biotechnol Biotechnol Equip 29:221–236. https://doi.org/10.1080/13102818.2015.1008194

Alghuthaymi MA, Kalia A, Bhardwaj K, Bhardwaj P, Abd-Elsalam KA, Valis M, Kuca K (2021) Nanohybrid antifungals for control of plant diseases: current status and future perspectives. J Fungi 7:48. https://doi.org/10.3390/jof7010048

Al-Juthery HWA, Lahmod NR, Al-Taee RAHG (2021) Intelligent, nano-fertilizers: a new technology for improvement nutrient use efficiency (article review). IOP Conf Ser Earth Environ Sci 735:012086. https://doi.org/10.1088/1755-1315/735/1/012086

Al-Khayri JM, Rashmi R, Surya Ulhas R, Sudheer WN, Banadka A, Nagella P, Aldaej MI, Rezk AA-S, Shehata WF, Almaghasla MI (2023) The role of nanoparticles in response of plants to abiotic stress at physiological, biochemical, and molecular levels. Plants 12:292. https://doi.org/10.3390/plants12020292

Altammar KA (2023) A review on nanoparticles: characteristics, synthesis, applications, and challenges. Front Microbiol 14:1155622. https://doi.org/10.3389/fmicb.2023.1155622

Ameen F, Alsamhary K, Alabdullatif JA, ALNadhari S (2021) A review on metal-based nanoparticles and their toxicity to beneficial soil bacteria and fungi. Ecotoxicol Environ Saf 213:112027. https://doi.org/10.1016/j.ecoenv.2021.112027

Atrak K, Ramazani A, Taghavi Fardood S (2020) Green synthesis of $Zn_{0.5}Ni_{0.5}AlFeO_4$ magnetic nanoparticles and investigation of their photocatalytic activity for degradation of reactive blue 21 dye. Environ Technol 41:2760–2770. https://doi.org/10.1080/09593330.2019.1581841

Avellan A, Yun J, Morais BP, Clement ET, Rodrigues SM, Lowry GV (2021) Critical review: role of inorganic nanoparticle properties on their foliar uptake and *in Planta* translocation. Environ Sci Technol 55:13417–13431. https://doi.org/10.1021/acs.est.1c00178

Balasubramaniam T, Shen G, Esmaeili N, Zhang H (2023) Plants' response mechanisms to salinity stress. Plants 12:2253. https://doi.org/10.3390/plants12122253

Banerjee SA, Acharya K, Chakraborty N (2021) Nanotechnology: an Emerging Hope in Crop Improvement Letters in Applied NanoBioScience 10(4) 2784-2803. https://doi.org/10.33263/LIANBS104.27842803

Bhalla N, Jolly P, Formisano N, Estrela P (2016) Introduction to biosensors. Essays Biochem 60:1–8. https://doi.org/10.1042/EBC20150001

Bouis HE, Saltzman A (2017) Improving nutrition through biofortification: a review of evidence from harvestplus, 2003 through 2016. Glob Food Secur 12:49–58. https://doi.org/10.1016/j.gfs.2017.01.009

Campaña AL, Saragliadis A, Mikheenko P, Linke D (2023) Insights into the bacterial synthesis of metal nanoparticles. Front Nanotechnol 5:1216921. https://doi.org/10.3389/fnano.2023.1216921

Cao X, Wang C, Luo X, Yue L, White JC, Elmer W, Dhankher OP, Wang Z, Xing B (2021) Elemental sulfur nanoparticles enhance disease resistance in tomatoes. ACS Nano 15:11817–11827. https://doi.org/10.1021/acsnano.1c02917

Chaud M, Souto EB, Zielinska A, Severino P, Batain F, Oliveira-Junior J, Alves T (2021) Nanopesticides in agriculture: benefits and challenges in agricultural productivity, toxicological risks to human health and environment. Toxics 9:131. https://doi.org/10.3390/toxics9060131

Chouhan D, Dutta A, Kumar A, Mandal P, Choudhuri C (2022) Application of nickel chitosan nanoconjugate as an antifungal agent for combating Fusarium rot of wheat Abstract Scientific Reports 12(1). https://doi.org/10.1038/s41598-022-18670-2

Droby S, Gonzalez-Estrada RR, Avila-Quezada G, Durán P, Manzo-Sánchez G, Hernandez-Montiel LG (2022) Microbial antagonists from different environments used in the biocontrol of plant pathogens. In: Kumar A (ed) Microbial biocontrol: food security and post-harvest management. Springer, Cham, pp 227–244. https://doi.org/10.1007/978-3-030-87289-2_9

Elmer WH, White JC (2016) The use of metallic oxide nanoparticles to enhance growth of tomatoes and eggplants in disease-infested soil or soilless medium. Environ Sci Nano 3:1072–1079. https://doi.org/10.1039/C6EN00146G

Elmer W, De La Torre-Roche R, Pagano L, Majumdar S, Zuverza-Mena N, Dimkpa C, Gardea-Torresdey J, White JC (2018) Effect of metalloid and metal oxide nanoparticles on *Fusarium* wilt of watermelon. Plant Dis 102:1394–1401. https://doi.org/10.1094/PDIS-10-17-1621-RE

Ferrari E, Barbero F, Busquets-Fité M, Franz-Wachtel M, Köhler H-R, Puntes V, Kemmerling B (2021) Growth-promoting gold nanoparticles decrease stress responses in *Arabidopsis* seedlings. Nano 11:3161. https://doi.org/10.3390/nano11123161

Francis DV, Abdalla AK, Mahakham W, Sarmah AK, Ahmed ZFR (2024) Interaction of plants and metal nanoparticles: exploring its molecular mechanisms for sustainable agriculture and crop improvement. Environ Int 190:108859. https://doi.org/10.1016/j.envint.2024.108859

Fritea L, Banica F, Costea T, Moldovan L, Dobjanschi L, Muresan M, Cavalu S (2021) Metal nanoparticles and carbon-based nanomaterials for improved performances of electrochemical (bio) sensors with biomedical applications. Materials 14:6319. https://doi.org/10.3390/ma14216319

Fu L, Wang Z, Dhankher OP, Xing B (2020) Nanotechnology as a new sustainable approach for controlling crop diseases and increasing agricultural production. J Exp Bot 71:507–519. https://doi.org/10.1093/jxb/erz314

Gangwar J, Kadanthottu Sebastian J, Puthukulangara Jaison J, Kurian JT (2023) Nanotechnological interventions in crop production—a review. Physiol Mol Biol Plants 29:93–107. https://doi.org/10.1007/s12298-022-01274-5

Geng M, Li L, Ai M, Jin J, Hu D, Song K (2022) Recent advances in metal-based nanoparticle-mediated biological effects in *Arabidopsis thaliana*: a mini-review. Materials 15:4539. https://doi.org/10.3390/ma15134539

Ghosh SK, Bera T (2021) Unraveling the mechanism of nanoparticles for controlling plant pathogens and pests. In: Advances in Nano-fertilizers and Nano-pesticides in agriculture. Elsevier, pp 415–436

Gonzalez AJ, Larraburu EE, Llorente BE (2015) *Azospirillum brasilense* increased salt tolerance of jojoba during in vitro rooting. Ind Crop Prod 76:41–48. https://doi.org/10.1016/j.indcrop.2015.06.017

Hasanuzzaman M, Inafuku M, Nahar K, Fujita M, Oku H (2021) Nitric oxide regulates plant growth, physiology, antioxidant defense, and ion homeostasis to confer salt tolerance in the mangrove species, *Kandelia obovata*. Antioxidants 10:611. https://doi.org/10.3390/antiox10040611

Hazarika A, Yadav M, Yadav DK, Yadav HS (2022) An overview of the role of nanoparticles in sustainable agriculture. Biocatal Agric Biotechnol 43:102399. https://doi.org/10.1016/j.bcab.2022.102399

Hotessa Halake N, Muda Haro J (2022) Role of nanobiotechnology towards agri-food system. J Nanotechnol 2022:1–7. https://doi.org/10.1155/2022/6108610

Ivanov AG, Velitchkova MY, Allakhverdiev SI, Huner NPA (2017) Heat stress-induced effects of photosystem I: an overview of structural and functional responses. Photosynth Res 133:17–30. https://doi.org/10.1007/s11120-017-0383-x

Kang H, Fan T, Wu J, Zhu Y, Shen W-H (2022) Histone modification and chromatin remodeling in plant response to pathogens. Front Plant Sci 13:986940. https://doi.org/10.3389/fpls.2022.986940

Khalid MF, Iqbal Khan R, Jawaid MZ, Shafqat W, Hussain S, Ahmed T, Rizwan M, Ercisli S, Pop OL, Alina Marc R (2022) Nanoparticles: the plant saviour under abiotic stresses. Nano 12:3915. https://doi.org/10.3390/nano12213915

Khan M, Mohammad F (2014) Eutrophication: challenges and solutions. In: Ansari A, Gill S (eds) Eutrophication: causes, consequences and control. Springer, Dordrecht. https://doi.org/10.1007/978-94-007-7814-6_1

Koka JA, Wani AH, Bhat MY (2019) Evaluation of the antifungal activity of magnesium oxide (MgO) and iron oxide (FeO) nanoparticles on rot-causing fungi. J Drug Deliv Ther 9:173–178. https://doi.org/10.22270/jddt.v9i2-s.2479

Konappa N, Krishnamurthy S, Arakere UC, Chowdappa S, Akbarbasha R, Ramachandrappa NS (2021) Nanofertilizers and nanopesticides: recent trends, future prospects in agriculture. In: Advances in nano-fertilizers and nano-pesticides in agriculture. Elsevier, pp 281–330

Kumari S, Raturi S, Kulshrestha S, Chauhan K, Dhingra S, András K, Thu K, Khargotra R, Singh T (2023) A comprehensive review on various techniques used for synthesizing nanoparticles. J Mater Res Technol 27:1739–1763. https://doi.org/10.1016/j.jmrt.2023.09.291

Kutawa AB, Ahmad K, Ali A, Hussein MZ, Abdul Wahab MA, Adamu A, Ismaila AA, Gunasena MT, Rahman MZ, Hossain MI (2021) Trends in nanotechnology and its potentialities to control plant pathogenic fungi: a review. Biology 10:881. https://doi.org/10.3390/biology10090881

Liu C, Zhou H, Zhou J (2021) The applications of nanotechnology in crop production. Molecules 26:7070. https://doi.org/10.3390/molecules26237070

Ma X, Wang Q, Rossi L, Ebbs SD, White JC (2016) Multigenerational exposure to cerium oxide nanoparticles: Physiological and biochemical analysis reveals transmissible changes in rapid cycling Brassica rapa NanoImpact 146–54. https://doi.org/10.1016/j.ecoenv.2021.112403

Mahapatra DM, Satapathy KC, Panda B (2022) Biofertilizers and nanofertilizers for sustainable agriculture: phycoprospects and challenges. Sci Total Environ 803:149990. https://doi.org/10.1016/j.scitotenv.2021.149990

Melotto M, Underwood W, He SY (2008) Role of stomata in plant innate immunity and foliar bacterial diseases. Annu Rev Phytopathol 46:101–122. https://doi.org/10.1146/annurev.phyto.121107.104959

Mgadi K, Ndaba B, Roopnarain A, Rama H, Adeleke R (2024) Nanoparticle applications in agriculture: overview and response of plant-associated microorganisms. Front Microbiol 15:1354440. https://doi.org/10.3389/fmicb.2024.1354440

Mishra P, Singh A, Mishra RR, Ong S, Prasad SM (2018) Application of nanotechnology to enhance the nutrient quality of food crops and agricultural production. In: Nanomaterials in plants, algae, and microorganisms. Elsevier, pp 453–472

Mishra N, Jiang C, Chen L, Paul A, Chatterjee A, Shen G (2023) Achieving abiotic stress tolerance in plants through antioxidative defense mechanisms. Front Plant Sci 14:1110622. https://doi.org/10.3389/fpls.2023.1110622

Mittal D, Kaur G, Singh P, Yadav K, Ali SA (2020) Nanoparticle-based sustainable agriculture and food science: recent advances and future outlook. Front Nanotechnol 2:579954. https://doi.org/10.3389/fnano.2020.579954

Mohapatra B, Chamoli S, Salvi P, Saxena SC (2023) Fostering nanoscience's strategies: a new frontier in sustainable crop improvement for abiotic stress tolerance. Plant Nano Biol 3:100026. https://doi.org/10.1016/j.plana.2023.100026

Naz S, Gul A, Zia M (2020) Toxicity of copper oxide nanoparticles: a review study. IET Nanobiotechnol 14:1–13. https://doi.org/10.1049/iet-nbt.2019.0176

Nisha Raj S, Anooj ES, Rajendran K, Vallinayagam S (2021) A comprehensive review on the regulatory invention of nano pesticides in agricultural nanoformulation and food system. J Mol Struct 1239:130517. https://doi.org/10.1016/j.molstruc.2021.130517

Pallavi MCM, Srivastava R, Arora S, Sharma AK (2016) Impact assessment of silver nanoparticles on plant growth and soil bacterial diversity 3 Biotech 6:254. https://doi.org/10.1007/s13205-016-0567-7

Parmar N, Singh KH, Sharma D, Singh L, Kumar P, Nanjundan J, Khan YJ, Chauhan DK, Thakur AK (2017) Genetic engineering strategies for biotic and abiotic stress tolerance and quality enhancement in horticultural crops: a comprehensive review 3 Biotech 7:239. https://doi.org/10.1007/s13205-017-0870-y

Prakash V, Peralta-Videa J, Tripathi DK, Ma X, Sharma S (2021) Recent insights into the impact fate and transport of cerium oxide nanoparticles in the plant-soil continuum Ecotoxicology and Environmental Safety 221112403. https://doi.org/10.1016/j.ecoenv.2021.112403

Pérez-de-Luque A (2017) Interaction of nanomaterials with plants: what do we need for real applications in agriculture? Front Environ Sci 5. https://doi.org/10.3389/fenvs.2017.00012

Prasad A, Astete CE, Bodoki AE, Windham M, Bodoki E, Sabliov CM (2018) Zein nanoparticles uptake and translocation in hydroponically grown sugar cane plants. J Agric Food Chem 66:6544–6551. https://doi.org/10.1021/acs.jafc.7b02487

Rahman MH, Hasan MN, Nigar S, Ma F, Aly Saad Aly M, Khan MZH (2021) Synthesis and characterization of a mixed nanofertilizer influencing the nutrient use efficiency, productivity, and nutritive value of tomato fruits. ACS Omega 6:27112–27120. https://doi.org/10.1021/acsomega.1c03727

Rajput VD, Singh A, Minkina T, Rawat S, Mandzhieva S, Sushkova S, Shuvaeva V, Nazarenko O, Rajput P, Komariah VKK, Singh AK, Rao M, Upadhyay SK (2021) Nano-enabled products: challenges and opportunities for sustainable agriculture. Plants 10:2727. https://doi.org/10.3390/plants10122727

Ramanand K, Nagarajan A, Suflita JM (1993) Reductive dechlorination of the nitrogen heterocyclic herbicide Picloram. Appl Environ Microbiol 59:2251–2256. https://doi.org/10.1128/aem.59.7.2251-2256.1993

Ramskov T, Forbes VE, Gilliland D, Selck H (2015) Accumulation and effects of sediment-associated silver nanoparticles to sediment-dwelling invertebrates. Aquat Toxicol 166:96–105. https://doi.org/10.1016/j.aquatox.2015.07.002

Rasheed A, Li H, Tahir MM, Mahmood A, Nawaz M, Shah AN, Aslam MT, Negm S, Moustafa M, Hassan MU, Wu Z (2022) The role of nanoparticles in plant biochemical, physiological, and molecular responses under drought stress: a review. Front Plant Sci 13:976179. https://doi.org/10.3389/fpls.2022.976179

Rastogi A, Tripathi DK, Yadav S, Chauhan DK, Živčák M, Ghorbanpour M, El-Sheery NI, Brestic M (2019) Application of silicon nanoparticles in agriculture. 3 Biotech 9:90. https://doi.org/10.1007/s13205-019-1626-7

Rehman A, Khan S, Sun F, Peng Z, Feng K, Wang N, Jia Y, Pan Z, He S, Wang L, Qayyum A, Du X, Li H (2024) Exploring the nano-wonders: unveiling the role of nanoparticles in enhancing salinity and drought tolerance in plants. Front Plant Sci 14:1324176. https://doi.org/10.3389/fpls.2023.1324176

Ruttkay-Nedecky B, Krystofova O, Nejdl L, Adam V (2017) Nanoparticles based on essential metals and their phytotoxicity. J Nanobiotechnology 15:33. https://doi.org/10.1186/s12951-017-0268-3

Sagar NA, Kumar N, Choudhary R, Bajpai VK, Cao H, Shukla S, Pareek S (2022) Prospecting the role of nanotechnology in extending the shelf-life of fresh produce and in developing advanced packaging. Food Packag Shelf Life 34:100955. https://doi.org/10.1016/j.fpsl.2022.100955

Sembada AA, Lenggoro IW (2024) Transport of nanoparticles into plants and their detection methods. Nano 14:131. https://doi.org/10.3390/nano14020131

Shang Y, Hasan MK, Ahammed GJ, Li M, Yin H, Zhou J (2019) Applications of nanotechnology in plant growth and crop protection: a review. Molecules (Basel, Switz) 24:E2558. https://doi.org/10.3390/molecules24142558

Sharma D, Kanchi S, Bisetty K (2019) Biogenic synthesis of nanoparticles: a review. Arab J Chem 12:3576–3600. https://doi.org/10.1016/j.arabjc.2015.11.002

Shukla M, Al-Busaidi KT, Trivedi M, Tiwari RK (2018) Status of research, regulations and challenges for genetically modified crops in India. GM Crops Food 9:173–188. https://doi.org/10.1080/21645698.2018.1529518

Singh RP, Handa R, Manchanda G (2021) Nanoparticles in sustainable agriculture: an emerging opportunity. J Control Release 329:1234–1248. https://doi.org/10.1016/j.jconrel.2020.10.051

Singh P, Choudhary KK, Chaudhary N, Gupta S, Sahu M, Tejaswini B, Sarkar S (2022) Salt stress resilience in plants mediated through osmolyte accumulation and its crosstalk mechanism with phytohormones. Front Plant Sci 13:1006617. https://doi.org/10.3389/fpls.2022.1006617

Singh A, Agrawal S, Rajput VD, Ghazaryan K, Yesayan A, Minkina T, Zhao Y, Petropoulos D, Kriemadis A, Papadakis M, Alexiou A (2024) Nanoparticles in revolutionizing crop production and agriculture to address salinity stress challenges for a sustainable future. Discov Appl Sci 6:317. https://doi.org/10.1007/s42452-024-06009-7

Siwach A, Verma PK (2021) Synthesis and therapeutic potential of imidazole-containing compounds. BMC Chem 15:12. https://doi.org/10.1186/s13065-020-00730-1

Skiba E, Pietrzak M, Glińska S, Wolf WM (2021) The combined effect of ZnO and CeO_2 nanoparticles on *Pisum sativum* L.: a photosynthesis and nutrients uptake study. Cells 10: 3105. https://doi.org/10.3390/cells10113105

Soni S, Jha AB, Dubey RS, Sharma P (2024) Nanowonders in agriculture: unveiling the potential of nanoparticles to boost crop resilience to salinity stress. Sci Total Environ 925:171433. https://doi.org/10.1016/j.scitotenv.2024.171433

Syu Y, Hung J-H, Chen J-C, Chuang H (2014) Impacts of size and shape of silver nanoparticles on *Arabidopsis* plant growth and gene expression. Plant Physiol Biochem 83:57–64. https://doi.org/10.1016/j.plaphy.2014.07.010

Tudi M, Daniel Ruan H, Wang L, Lyu J, Sadler R, Connell D, Chu C, Phung DT (2021) Agriculture development, pesticide application and its impact on the environment. Int J Environ Res Public Health 18:1112. https://doi.org/10.3390/ijerph18031112

Verma KK, Song X-P, Joshi A, Tian D-D, Rajput VD, Singh M, Arora J, Minkina T, Li Y-R (2022) Recent trends in nano-fertilizers for sustainable agriculture under climate change for global food security. Nano 12:173. https://doi.org/10.3390/nano12010173

Waadt R, Seller CA, Hsu P-K, Takahashi Y, Munemasa S, Schroeder JI (2022) Plant hormone regulation of abiotic stress responses. Nat Rev Mol Cell Biol 23:680–694. https://doi.org/10.1038/s41580-022-00479-6

Wang X, Liu X, Chen J, Han H, Yuan Z (2014) Evaluation and mechanism of antifungal effects of carbon nanomaterials in controlling plant fungal pathogen. Carbon 68:798–806. https://doi.org/10.1016/j.carbon.2013.11.072

Wang Z, Xu L, Zhao J, Wang X, White JC, Xing B (2016) CuO nanoparticle interaction with *Arabidopsis thaliana*: toxicity, parent-progeny transfer, and gene expression. Environ Sci Technol 50:6008–6016. https://doi.org/10.1021/acs.est.6b01017

Wang Y, Wang S, Xu M, Xiao L, Dai Z, Li J (2019) The impacts of γ-Fe_2O_3 and Fe_3O_4 nanoparticles on the physiology and fruit quality of muskmelon (*Cucumis melo*) plants. Environ Pollut 249:1011–1018. https://doi.org/10.1016/j.envpol.2019.03.119

Wang X, Xie H, Wang P, Yin H (2023) Nanoparticles in plants: uptake, transport and physiological activity in leaf and root. Materials 16:3097. https://doi.org/10.3390/ma16083097

Wijerathna-Yapa A, Ramtekey V, Ranawaka B, Basnet BR (2022) Applications of *in vitro* tissue culture technologies in breeding and genetic improvement of wheat. Plants 11:2273. https://doi.org/10.3390/plants11172273

Withers PJA, Van Dijk KC, Neset T-SS, Nesme T, Oenema O, Rubæk GH, Schoumans OF, Smit B, Pellerin S (2015) Stewardship to tackle global phosphorus inefficiency: the case of Europe. Ambio 44:193–206. https://doi.org/10.1007/s13280-014-0614-8

Wu H, Li Z (2022) Nano-enabled agriculture: how do nanoparticles cross barriers in plants? Plant Commun 3:100346. https://doi.org/10.1016/j.xplc.2022.100346

Xiao Y, Ma J, Xian J, Peijnenburg WJGM, Du Y, Tian D, Xiao H, He Y, Luo L, Deng O, Tu L (2022) Copper accumulation and physiological markers of soybean (*Glycine max*) grown in agricultural soil amended with copper nanoparticles. Ecotoxicol Environ Saf 229:113088. https://doi.org/10.1016/j.ecoenv.2021.113088

Yee MS-L, Hii L-W, Looi CK, Lim W-M, Wong S-F, Kok Y-Y, Tan B-K, Wong C-Y, Leong C-O (2021) Impact of microplastics and nanoplastics on human health. Nano 11:496. https://doi.org/10.3390/nano11020496

Zulfiqar F, Navarro M, Ashraf M, Akram NA, Munné-Bosch S (2019) Nanofertilizer use for sustainable agriculture: advantages and limitations. Plant Sci 289:110270. https://doi.org/10.1016/j.plantsci.2019.110270

Next-Generation Crop Improvement: Social, Ethical, and Regulatory Perspectives on GMO Commercialization and Future Challenges

14

Neha Boora, Yogesh Ahlawat, Darshna Chaudhary, Pawan K. Jaiwal, and Shilpi Kundu

Abstract

The introduction of genetically engineered (GE) crops into commercial agriculture has triggered significant debate concerning their social, ethical, and regulatory implications. While next-generation crop improvement technologies, including CRISPR-based genome editing and synthetic biology, offer promising solutions for food security, climate resilience, and sustainable farming, their adoption remains constrained by public perception, ethical concerns, and complex regulatory frameworks. This chapter explores the evolving landscape of genetically modified organisms (GMO) commercialization, analyzing how government policies, risk assessments, and intellectual property laws shape market accessibility. Additionally, it examines societal attitudes, ethical dilemmas related to biodiversity and consumer rights, and corporate influence on seed distribution. Emerging challenges, such as cross-border regulatory inconsistencies, misinformation, and the rise of non-GMO labeling trends, are discussed in the context of

N. Boora
West Virginia State University, Institute, WV, USA

Y. Ahlawat (✉)
University Center for Research and Development, Chandigarh University, Mohali, Punjab, India

Allied Health Sciences, Datta Meghe Institute of Higher Education and Research, Wardha, Maharashtra, India
e-mail: ykahlawa@mtu.edu

D. Chaudhary (✉) · P. K. Jaiwal
Center for Biotechnology, Maharshi Dayanand University, Rohtak, Haryana, India
e-mail: darshnarajan.cbt@mdurohtak.ac.in

S. Kundu
Department of Agricultural Extension & Information System, Faculty of Agriculture, Sher-e-Bangla Agricultural University, Dhaka, Bangladesh

Y. Ahlawat et al. (eds.), *Next-Generation Strategies for Crop Improvement*,
https://doi.org/10.1007/978-981-95-0309-4_14

their impact on global food systems. Future perspectives emphasize the need for transparent communication, inclusive policy-making, and adaptive regulatory models that balance innovation with biosafety. Addressing these multidimensional challenges is crucial for leveraging next-generation GM crops to drive sustainable agricultural transformation while ensuring public trust and ethical responsibility.

Keywords

GMO crops · Genome editing · Bioethics · Regulatory frameworks · Commercialization · Agricultural sustainability

14.1 Introduction

Plant breeding has been a key method for crop domestication for approximately 10,000 years, involving the selection of desirable traits from wild relatives across generations (Smýkal et al. 2018). Earlier, farmers domesticated the crops based on visible phenotypes like enhanced harvest and productivity. The modern plant breeding emerged through Darwin's scientific principles of hybridization and natural selection, followed by Mendel's laws of inheritance (Priyadarshan 2019). The evolution of continuous research and work in plant breeding has led to many discoveries as illustrated in Fig. 14.1, such as recombinant DNA technology and gene editing (CRISPR-Cas9). The Green Revolution of the 1960s played a pivotal role in enhancing the productivity of major cereal crops, offering substantial benefits to global agriculture and food systems (Pingali 2012). However, it also led to negative environmental impacts due to the overuse of synthetic fertilizers and

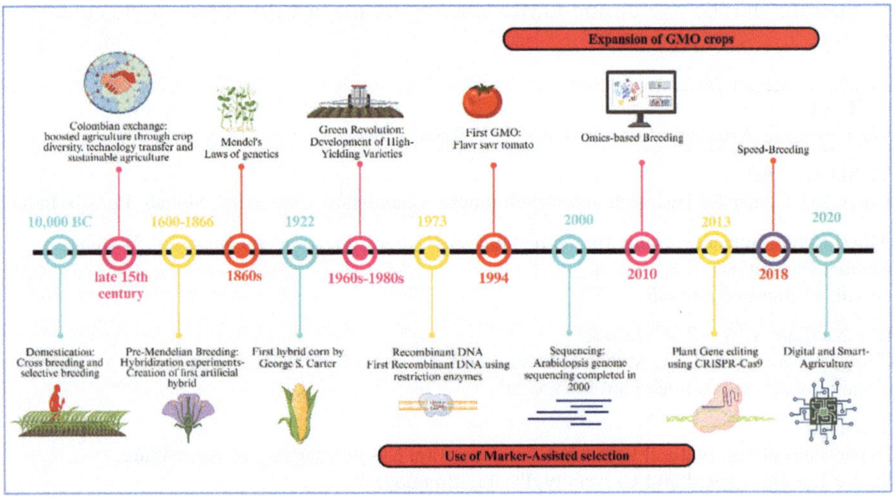

Fig. 14.1 From domestication to genetic engineering: the journey of crop improvement

pesticides (Pingali 2012). Foundational breeding practices have since evolved to include greater genomic diversity by incorporating favorable traits from wild relatives and land races into cultivated lines (Swarup et al. 2021). This is achieved through interspecific crosses aimed at combining desirable agronomic characteristics and selecting superior genotypes (Swarup et al. 2021). Despite this progress, conventional breeding still faces considerable challenges, including genetic drag, erosion of genetic resources, hybridization constraints, and time-intensive selection procedures (Abberton et al. 2016). On average, developing a new crop variety with target traits may take 10–20 years, making the process resource demanding and slow (Mba et al. 2012).

The revolution in crop improvement started with the DNA sequencing methods, such as those developed by Sanger and by Maxam and Gilbert, which laid the groundwork for genetic analysis. The introduction of polymerase chain reaction (PCR), along with the development of reliable DNA-modifying enzymes and the emergence of fluorescent-based automated sequencing systems, has enabled the efficient sequencing of numerous plant genomes (Singh et al. 2022).

Another discovery that helped breeders achieve results in the short term was transgenic breeding; the first transgenic plant was created in 1983 by Fraley et al. (Gasser and Fraley 1992), which was resistant to antibiotics. In 1994, the Flavr Savr tomato developed by Calgene (United States) became the first genetically modified crop to receive approval from the US Food and Drug Administration (FDA) for human consumption. This tomato utilized antisense technology to suppress the polygalacturonase enzyme, which delayed ripening and enhanced resistance to decay (Kramer and Redenbaugh 1994). After a gene responsible for a beneficial trait is discovered, it can be isolated and incorporated into the genome of another plant. Such genetically altered plants that include genes from different species are termed transgenic (Swarup et al. 2021). On average, it takes nearly a decade to develop new genetically engineered crop varieties (Priyadarshan 2019).

Producing new varieties of crops through genetic engineering takes about 10 years on average, which is less than conventional breeding. Significant advancements in molecular marker technologies since the 1980s have transformed plant breeding approaches and improved crop development strategies (Nadeem et al. 2018). Technologies such as high-throughput genotyping and advanced phenotyping platforms have enabled comprehensive studies like genome-wide association studies (GWAS), which are instrumental in identifying genetic components underlying complex plant traits (Xiao et al. 2022).

In parallel, the field of genome editing has evolved rapidly, surpassing traditional molecular marker tools in popularity. Modern editing techniques, including zinc finger nucleases (ZFNs) and transcription activator-like effector nucleases (TALENs), have facilitated precise and efficient gene modifications. These methods are valued for their accessibility, reliability, and cost-effectiveness in plant genetic engineering (Shan et al. 2018). The emergence of CRISPR-associated CAS9 systems has further revolutionized the field by providing a simpler and more versatile genome editing platform (Cong et al. 2013). Crops developed using genetic engineering are often referred to as genetically modified organisms (GMOs) (Cong et al. 2013). However, the term "genetic modification" is frequently misapplied, especially when discussing GMOs. Whether in organic or conventional agricultural systems, plants are typically altered at the genetic level to introduce beneficial traits (Azadi and Ho 2010). Traditional breeding alters genetic composition by selecting plants based on visible characteristics over generations, whereas genetic engineering modifies specific genes directly to achieve desired traits (Gepts 2002).

Recent innovations, such as genomic selection, high-throughput phenotyping (HTP), and lowercase, are increasingly being adopted to accelerate crop improvement (Raman 2017). These tools enhance the precision and efficiency of breeding by enabling better prediction of genetic potential. The integration of genomics, Environics, and phenomics has contributed significantly to modern breeding frameworks and forms the core of what is known as the plant breeding triangle (Crossa et al. 2021).

Next-generation crop improvement refers to the use of new and advanced tools like precise modification of genome, transcriptome, or proteome using tools like high-throughput phenotyping and big data analysis (Fig. 14.2) to increase traits like

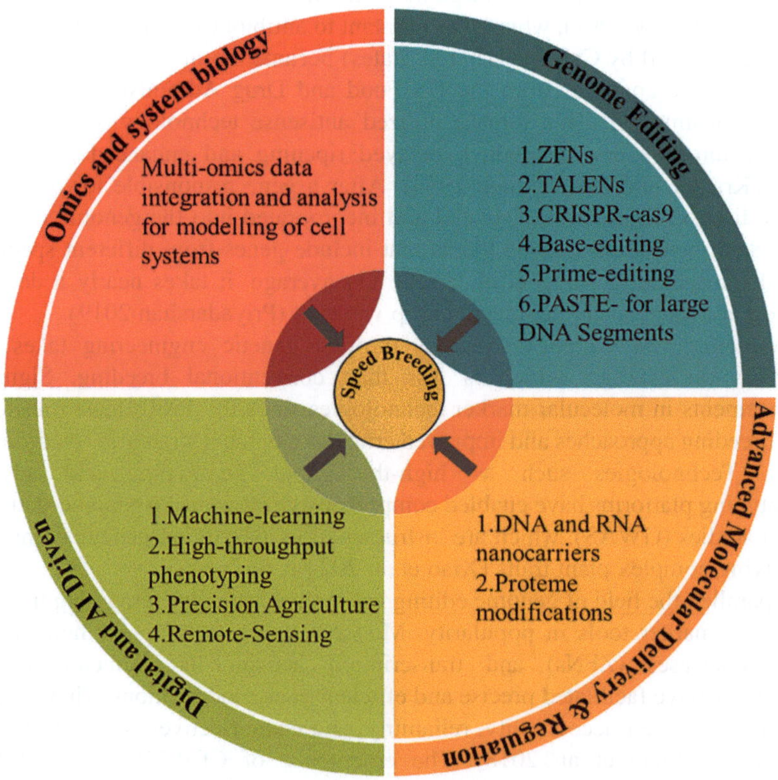

Fig. 14.2 Modern strategies driving next-generation crop improvement

yield and biotic and abiotic stress tolerance. Next-generation crop improvement refers to the precise modifications of genome, epigenome, epi-transcriptome, and proteome using modern gene editing biotechnological advances like CRISPR-Cas, using AI and machine learning which can improve the yield, resistance to biotic and abiotic factors, nutritional value, and other desired traits depending on the crop species in less time period as compared to conventional or first-generation crop improvement (Razzaq et al. 2021).

New genome editing tools now offer precise capabilities for altering specific regions of the genome, making it possible to introduce mutations or edit almost any target gene with high efficiency (Razzaq et al. 2021).

These technological advancements have led to what is referred to as engineered design breeding (Hartung and Schiemann 2014). Looking ahead, the integration of biotechnology with information technology (BT + IT) is expected to shape the next era of intelligent crop design. This approach combines hybrid breeding strategies informed by big data analytics and predictive modeling with biological methods enhanced by artificial intelligence and synthetic evolutionary techniques. As these systems evolve, future breeding programs will incorporate advanced gene editing, large-scale trait discovery, improved variety development, and streamlined biological breeding systems—ultimately forming a holistic framework for intelligent crop improvement (Zhang et al. 2024).

14.1.1 Significance of Genetically Modified Organisms (GMOs) in Agriculture

As reported by the World Health Organization (WHO) in 2023, approximately 2.33 billion individuals across the globe experienced moderate to severe levels of food insecurity. Of these, more than 864 million people endured severe food shortages, occasionally going an entire day or more without eating (Azadi and Ho 2010). With the global population anticipated to grow to 9.1 billion by 2050, an increase of 34% compared to current levels, developing nations are expected to bear most of this growth (Shan et al. 2018). To satisfy the growing food demand, agricultural output must rise by nearly 70%, all while addressing persistent environmental and sustainability concerns (Cong et al. 2013). Genetically modified organisms (GMOs) have revolutionized modern agriculture through enhancing crop yields, reducing production costs, and promoting sustainable farming practices. Since their adoption more than 20 years ago, genetically modified (GM) crops have contributed to improved economic gains for farmers and helped reduce the environmental impact of agriculture (Raman 2017). As of October 2024, GM crops are cultivated, imported, or evaluated in approximately 75 countries, and 13 types of GM crops are currently approved for use worldwide. More than 30 nations have formally authorized the cultivation of GM crops (Turnbull et al. 2021; Haslberger 2006).

14.1.2 Economic Impact of GMO Adoption (1996–2020) (Brookes 2022)

- GM crop adoption increased farm income globally by $261.3 billion, equivalent to an average income rise of $112 per hectare.
- Around 72% of these gains were driven by increased yield and productivity, while 28% resulted from cost savings.
- Over 17 million small farmers benefited economically, supporting more than 65 million people.
- Increased net farm income through reduced labor and input costs.
- Helped lift smallholder farmers from poverty by enhancing yields and reducing losses from pests and environmental stressors.

14.1.2.1 Impact on Yield and Productivity (Brookes and Barfoot 2020)

- Insect-resistant (IR) corn: 16.5% average yield increase
- Insect-resistant (IR) cotton: 13.7% average yield increase
- GM biotechnology has contributed to an additional global production of:
  - 278 million tons of soybeans
  - 498 million tons of corn
  - 32.6 million tons of cotton lint
  - 14 million tons of canola

Case Studies: Impact of GM Crops

I. Bt cotton in India
   - Bt cotton has increased profitability by 50% and yield by 24% compared to conventional cotton.
   - Farmers saw a net profit increase of ₹1877 per acre (US$38) and a yield gain of 126 kg per acre.
   - This led to an 18% increase in annual consumption expenditure, improving the living standards of Bt cotton growers (Kathage and Qaim 2012).

II. GM canola in Australia
   - Canola farming faced significant challenges from blackleg disease and invasive weeds such as charlock and wild radish, which degraded oil quality.
   - The development of Roundup Ready® by Monsanto and InVigor® by Aventis CropScience (now Bayer CropScience) was completed in the early 2000s. Roundup Ready® plants were engineered with a modified EPSP synthase and glyphosate oxidoreductase gene, enabling resistance to glyphosate herbicides.
   - GM canola adoption rose from 4% in 2009 to 14% in 2014, improving crop resilience and reducing losses (Raman 2017).

14.1.2.2 Environmental Benefits of GMOs (Brookes and Barfoot 2020)

Overall, studies indicate that GMOs have a positive environmental impact and can work as a valuable tool in agriculture to mitigate the effects of climate change while promoting sustainability.

- Conserving biodiversity: GM crops have saved 183 million hectares of land by improving productivity between 1996 and 2020.
- Reducing pesticide use: 748.6 million kg of active pesticide ingredients have been prevented from entering the environment. From 1996 to 2017, insecticide application decreased by over 80% among US corn and cotton. It has also been found that conservation tillage is easy for farmers who are growing herbicide-tolerant (HT) crops.
- Farmers who cultivate GM crops have significantly decreased the environmental impact of their crop protection practices by 19%, as assessed using the Environmental Impact Quotient (EIQ), which measures the ecological effects of crop protection.
- Decreased fuel use and tillage reduced 39 billion kg of CO_2 emissions, equivalent to removing 25.9 million cars from the road for a year.

14.2 Social Considerations

- **Public Perception of GMOs**
- Recently, studies have examined public sentiment toward genetically modified organisms (GMOs), revealing significant skepticism and concern.
- A 2023 study by Manreet et al. analyzed public perception of GMOs between 2019 and 2021, finding that only 14% of sentiments were positive, while the majority were either negative or neutral. Notably, *disgust* was the predominant public emotion toward GMO (31%) (Sohi et al. 2023). Similar findings emerged in a nationwide Chinese consumer study by Kai Cui et al., where only 11.9% of respondents had a favorable view, with the rest displaying neutral or negative opinions (Cui and Shoemaker 2018). Correlation analysis in this study indicated a direct relationship between GMO knowledge and acceptance; greater knowledge was associated with lower opposition. Based on the survey conducted by the Pew Research Center between 2019 and 2020, approximately half of the people across 20 countries believed GMOs were unsafe to eat. Gender differences were also observed, with women generally more skeptical of GMO safety than men. Interestingly, education level had little effect on public opinion, as skepticism persisted across different educational backgrounds. However, individuals who had taken at least three science courses showed a slight increase in acceptance of GM foods. Trust in scientific information about GMOs remains relatively low. The same Pew Research Center survey found that only 35% of US adults trust scientists to provide accurate information. In the MENAT region (Middle East, North Africa, and Turkey), a 2024 review by Khalid Ramadan Medani further confirmed that negative attitudes toward GMOs outweighed positive ones

(Medani et al. 2024). These findings highlight a persistent global divide in public perception, where misinformation, limited scientific literacy, and emotional responses continue to shape opinions on GMO safety and benefits.
- **Perceived Risks of GMOs:** Public concerns about GMOs often stem from ethical, health, and environmental considerations.
- **Health Risks and Food Safety Concerns:** Kai Cui et al. highlighted public skepticism regarding the safety of GM foods (Cui and Shoemaker 2018). According to Kunyanga et al., people perceive various health risks associated with GMOs, including the potential transfer of toxins, allergens, and antibiotic resistance. Concerns include immune suppression, allergenic reactions, cancer risks, and teratogenic effects. Some believe that GM foods may result in nutritional loss (Catherine et al. 2024).
- **Environmental and Ecological Risks:** The public also expresses concerns about potential unknown effects on human health, environmental harm, and biodiversity loss (Bawa and Anilakumar 2013).
- **Ethical and Religious Concerns:** According to Khalid et al., many individuals perceive GM foods as morally unacceptable, unsafe, and controversial (Medani et al. 2024). Some participants reported that GMOs contradict their ethical and societal values, while others viewed them as incompatible with their religious principles. Kunyang et al. reported that over 50% of respondents in their study believed that genetic modification equates to "playing God" and is at odds with their spiritual beliefs (Catherine et al. 2024).
- **Perceived Benefits of GMOs:** Despite widespread concerns, some people recognize the potential advantages of GMOs, particularly in food security and agricultural efficiency.
  - **Food Security:** Kunyanga et al. reported that 57% of respondents in Kenya and Africa viewed GM foods as essential for addressing food security (Catherine et al. 2024).
  - **Agricultural Benefits:** The majority of respondents believed GMOs contributed to increased crop resistance to drought and pathogens, reduced pesticide use, higher crop yields, and improved plant protection.
  - **Consumer Benefits:** Many also acknowledged that GM foods enhance shelf life, taste, flavor, appearance, and nutritional quality. Additionally, some people support GM foods for being more affordable.

14.2.1 Misinformation and the Role of Media in Shaping Perceptions

- Public perception of genetically modified organisms (GMOs) is shaped by a variety of factors, including psychological, social, cultural, personal, and economic influences. These may involve aspects such as scientific understanding, lifestyle choices, income level, religious views, and general attitudes. Negative public sentiment, often fueled by misleading media narratives, has contributed to strict regulatory frameworks surrounding GM crops (Woźniak-Gientka et al. 2022). The impact of misinformation was analyzed by Mark Lynas between

2019 and 2021, revealing that 9% of GMO-related reports published during this period contained misleading or false information. These reports predominantly carried negative or neutral sentiments, reflecting the influence of anti-GMO activist networks. Such misinformation reached a global audience of approximately 250 million readers, significantly shaping public discourse (Lynas et al. 2025). Negative media coverage has also contributed to widespread misconceptions about GMO safety. Paulina's study highlights how extreme claims, such as GMOs being responsible for allergic reactions in humans or causing internal organ damage, gained unexpected support, demonstrating the power of media in shaping public beliefs (Kubisz et al. 2021). The study by Mark Lynas underscores the profound effect misinformation has on GMO adoption. Despite no scientifically verified health risks reported since 1996, negative media narratives fuel public skepticism, leading to restrictive regulations or outright bans on GM crops. This issue is particularly concerning in regions like Africa, where misinformation rates are as high as 20%, despite the pressing need for improved food security.

- In a 2023 study, Dmitry analyzed conversations about GMOs on Twitter (now X), focusing on posts shared between January 2020 and December 2022 using the Twitter API. The study revealed that 30.92% of tweets during this timeframe expressed negative sentiments. Despite overwhelming scientific consensus supporting GMO safety and benefits, misinformation on social media platforms has significantly influenced public sentiment. False narratives—such as claims linking GMOs to corporate control, efforts of depopulation, and harmful health effects—have fueled skepticism. Additionally, GMO-related discussions on Twitter often intersect with other controversial topics, including vaccines, COVID-19, Monsanto, and Bill Gates, further amplifying public distrust (Erokhin and Komendantova 2023). The widespread misconceptions surrounding GMOs have tangible consequences. Public skepticism influences policymaking, leading to restrictive GMO regulations and bans. Research by Fernbach et al. (2019) suggests that the most vocal opponents of GMOs often believe they are highly knowledgeable about the topic, yet their actual scientific literacy is relatively low (Fernbach et al. 2019).
- The policy decisions driven by misinformation have real-world implications. Studies indicate that bans on GM crops result in increased pesticide use, lower crop yields, and higher greenhouse gas emissions (Brookes et al. 2017). For instance, resistance to GMOs in Europe has been linked to an additional 33 million tons of CO_2-equivalent emissions (Kovak et al. 2022). Similarly, in sub-Saharan Africa, restrictive policies largely shaped by misinformation have slowed agricultural advancements, limiting opportunities to enhance food security and improve farmer livelihoods (Gbadegesin et al. 2022).

Ethical Concerns Regarding GMOs

The term "*ethics*" originates from the Greek word "*ethos*," meaning character. Ethics, or moral philosophy, involves the exploration of values and principles that help distinguish between right and wrong, as well as what is deemed acceptable or unacceptable behavior. Establishing an ethical framework requires addressing ethical dilemmas, which can be analyzed using ethical paradigms. These paradigms serve as foundational principles that guide moral judgments, defining the values and considerations that shape our understanding of right and wrong (Baggini and Fosl 2024).

The increasing presence of genetically modified organisms (GMOs) in today's food supply has intensified ethical debates surrounding their use. Many modern food products are either genetically modified or contain genetically modified ingredients, raising questions about their moral implications. Ethical concerns related to GMOs stem from cultural and societal values associated with food, influencing the ongoing discourse on biotechnology and genetic modifications (Reiss 2002).

Ethics extends beyond objective facts, aiming to assess how individuals should act toward themselves, each other, and the environment. However, ethical issues are often complex and open to interpretation. Many ethicists support Ronald Sandler's viewpoint that technological advancements should serve to enhance human welfare in fair and sustainable ways, while remaining within ethical boundaries. On the other hand, deep ecologists argue that ethical considerations should not be centered solely on human well-being but should also prioritize the environment and other life forms (Thiroux and Krasemann 2009).

What Are Ethical Paradigms?

Ethical paradigms serve as guiding principles for making moral judgments. They provide a foundational perspective on what is considered right or wrong and the reasoning behind these classifications. These paradigms reflect core values that shape how individuals and societies determine ethical behavior, how people should be treated, and ultimately, how life should be lived. Essentially, ethical paradigms act as a framework through which individuals interpret the world, offering a basis for assessing the morality of actions, policies, and, in this context, specific products and processes (Baggini and Fosl 2024).

Key ethical paradigms include:

1. Consequentialism: Focuses on the outcomes of actions, determining morality based on the consequences they produce
2. Autonomy and Consent Ethics: Emphasizes individual rights, personal freedom, and the importance of informed consent in ethical decision-making
3. Virtue/Tradition Ethics: Centers on moral character, virtues, and cultural traditions, recognizing that ethical values may differ across communities and influence how individuals choose to live their lives

Consequentialism: When evaluating any technology, it is essential to consider its consequences—whether it will lead to beneficial or harmful outcomes (Reiss 2002). In the United States, agricultural policies have historically been shaped by key consequentialist objectives, which include: (Smýkal et al. 2018) ensuring sufficient food production to support a growing, predominantly urban population (Quantity); (Priyadarshan 2019) providing food that meets safety and nutritional standards (Quality); and (Pingali 2012) maintaining food affordability for consumers while ensuring that farmers receive adequate financial returns to sustain their livelihoods (Price). These three goals, collectively called the QQP framework, made the basis for justifying policies and technologies supporting agricultural sustainability. The underlying principle is to increase benefits for the maximum number of people by facilitating access to safe, affordable, and readily available food. To meet the QQP criteria, advancements in agricultural technology must focus on improving productivity and efficiency. Integrating high-tech machinery and innovative farming practices plays a crucial role in achieving these objectives. Genetically modified organisms (GMOs) or genetically modified (GM) crops align with the QQP framework by enhancing food production while ensuring affordability and safety. For example, herbicide-resistant crops, such as Roundup Ready varieties, reduce the reliance on expensive chemical herbicides while maintaining or increasing crop yields. Similarly, Bt crops, engineered to resist pests, lower the cost of plant protection by reducing the need for chemical pesticides (Burkhardt 2001). When evaluating GM crops within the QQP framework, their approval becomes ethically justifiable, as they contribute to meeting consumer needs for accessible, safe, and affordable food while supporting sustainable agricultural practices. But other concerns that people want other than QQP include the following.

Environmental concerns: Biodiversity is a key value that is important for sustainability, and it spans three levels: ecosystems (landscapes), the variety of species present in ecosystems, along with the genetic variation found within and among different plant cultivars. The widespread cultivation of GMOs, particularly in monocultures, raises concerns about the potential negative impact on biodiversity (Reiss 2002). This can occur through mechanisms such as outcrossing GM traits to wild relatives or the unintended effects of modified traits, like the potential harmful effects of Bt crops on insects that are not intended targets. However, the impact on biodiversity is complex and must be evaluated within the context of the specific GM traits, crops, and the environment. One issue that has been raised is the impact of modern crop varieties on the centers of biodiversity. Although the likelihood of introducing traits—such as herbicide tolerance—from commercial fields into local landraces may be relatively low, there is concern that disease resistance genes developed through scientific breeding could disseminate more rapidly in these populations, potentially impacting natural genetic diversity. It is important to emphasize, however, that there is no fundamental distinction between traits produced through genetic modification and those generated via conventional mutation breeding (Gregorowius et al. 2012). Gene flow to wild relatives is a natural process, but with GMOs, there is concern that cross-pollination could transfer genetically modified traits to wild plants, resulting in the creation of "superweeds" or negatively affecting local flora. This process is part of the broader concern over GMOs' effects on genetic diversity.

In the case of plant protection, GMOs, such as Bt crops, have been shown to reduce pesticide use, but the rise of herbicide-resistant weeds presents a significant issue. These resistant weeds could increase herbicide usage, potentially undoing the benefits of reduced pesticide use. This emphasizes the need for a balanced approach in evaluating the environmental concerns of GMOs (Noack et al. 2024). The environmental effects of GMOs are not straightforward and must be assessed on a case-by-case basis. While there are clear benefits, such as reduced pesticide use, there are also risks to biodiversity and plant protection.

Off-Target Effects In plant breeding, introducing desirable traits through crossing can sometimes produce offspring with unfavorable characteristics. These are typically eliminated during the selection phase, a common practice in both traditional breeding and domestication. Notably, conventional random mutagenesis tends to produce more unintended outcomes, both harmful and beneficial, than precise gene-editing methods like CRISPR-Cas. Any unintended changes from gene editing are similarly identified and removed through standard selection procedures (Troadec et al. 2019).

Socioeconomic Concerns: Socioeconomic concerns related to GMOs are often discussed in three categories (Thompson 1997, 2007): the impact on small farms, developing countries' economies, and the scientific community.

1. Impact on Small Farms: The breeder's exemption in plant breeding permits unrestricted use of protected varieties for developing new ones. This provision has traditionally played a key role in maintaining diversity and innovation within the plant breeding sector, but in GM seeds, there have been reported cases of monopolistic practices by large multinational corporations, raising concerns about the survival of small farms. These companies dominate the market, posing a threat to the income of small-scale farmers.
2. Impact on Developing Countries: In developing nations, the commercialization of GMOs could exacerbate inequalities, as large farmers are more likely to benefit from advanced technology and increased seed costs, while smallholder farmers may struggle to access such resources.
3. Impact on the Scientific Community: Increasing commercialization may also shift research priorities, leading to a focus on corporate interests rather than publicly beneficial objectives.

Patenting Issues: Patenting GMOs raises concerns about monopolies held by large corporations, which critics argue undermine the sanctity of life (Uzogara 2000). One of the most contentious aspects of patenting is the shift of seeds from a shared resource to proprietary products. As in the case of terminator technology or Genetic Use Restriction Technologies (GURTs), where genetically engineered plants will produce sterile seeds, farmers need to buy again for the next year, as they are unable to re-plant the harvested seeds. The introduction of Genetic Use Restriction Technologies (GURTs) prevents crops from germinating without an external chemical inducer, ensuring that farmers cannot reuse farm-saved seeds (Corcoran 2008).

Effect on Human Health: Assessing the human health impact of GMOs requires a thorough evaluation of the context, including the features of genetic modification and its potential consequences. For example, Golden Rice is a genetically modified crop developed to address vitamin A deficiency, a major cause of childhood blindness in populations that rely heavily on rice as a staple food. This rice variety has been engineered to accumulate elevated amounts of beta-carotene in the grain (Burkhardt 2001). Likewise, a genetically engineered potato was developed by the J.R. Simplot Company. Similarly, a genetically engineered potato developed by J.R. Simplot Company is designed to reduce the presence of a suspected carcinogen that forms when potatoes are fried and to make the potato more resistant to bruising during transport. These GMOs are not inherently harmful to human health, but must undergo rigorous risk assessments, just as conventional food products do.

This emphasizes the importance of vigilance in risk assessments, with a focus on both food and environmental safety. In agriculture, as in medicine, many consequentialists find GM technology ethically acceptable, as it aligns with addressing human needs and preferences.

14.2.2 Autonomy/Consent Ethics

The concept of autonomy has long been debated in ethical discussions, particularly regarding what makes individuals deserving of personal sovereignty. It centers on the idea of respecting and protecting personal autonomy, which is translated into ethical rules or duties that guide actions. One perspective holds that autonomy is grounded in respecting people's rights, specifically the legitimate claims individuals have regarding how others should act toward them. The autonomy/consent paradigm asserts that self-determination means people have inviolable rights, demanding that they be given the choice in how they act and are treated. A person has the right to take risks on their own terms, if it does not endanger others (Thiroux and Krasemann 2009).

This ethical view focuses less on the broader goals of systems, such as those concerned with agriculture and food production, as consequentialists do. Instead, it emphasizes the right of individuals to make informed decisions. The autonomy/consent paradigm raises the question of whether the food we consume is safe and demands transparency in food production. Consumers are entitled to be informed about the products they buy, including any potential risks to their health. Although agencies like the USDA, EPA, and FDA maintain that there is no risk associated with food, the method of modification, whether through conventional or modern breeding techniques, should not matter, but consumer concerns over the question of the long-term health impacts remain. This situation places consumers at risk, contrary to their informed consent. Another key principle in autonomy ethics is the right of individuals to make an informed decision about consuming genetically modified foods. If someone prefers to avoid GM food, they should have the freedom to do so. This implements the ethical argument for mandatory labeling of GM foods (Burkhardt 2001).

At the farm level, concerns arise about the decreasing availability of non-GM crop seeds. Since farmers often make planting decisions based on market expectations, including consumer demand, the growth of GM crop production, and potential market shifts such as possible boycotts in the EU, may limit their ability to grow non-GM varieties. This creates a monopoly in the GMO seed market, controlled by large multinational corporations, and restricts farmers' freedom to plant non-GMO crops (Corcoran 2008).

14.2.3 Virtue Ethics

Virtue ethics is based on a person who is he/she? Who he/she wants to become or what kind of life he/she should live, it talks about a person's moral character who is carrying out an action instead of focusing on deontology or consequentialism. Virtue ethics in agriculture evaluates biotechnology, especially GMOs, through ethical traditions like Agrarianism and Nature-ism (Burkhardt 2001).

Agrarianism views farming not just as an industry but as a way of life. It places value on qualities such as patience, self-reliance, and community. For Agrarians, the traditional family farm is a place where human values and virtues are practiced and passed down to future generations. Active participation in and a sense of ethical ownership within an agricultural community are considered vital virtues (Thompson 2010b). Agrarians typically oppose GMOs because they perceive these technologies as a threat to traditional farming, which supports small, family-run farms, and because they tend to favor large agribusinesses over rural communities. Agrarians argue that government policies, particularly in the United States, often prioritize large-scale, industrial farming operations, which are believed to be more efficient at producing food for a predominantly urban population. Due to their core beliefs about modern agriculture, many agrarian groups oppose genetic modification based on ethical concerns. This opposition is particularly prominent among small-scale traditional farming communities in Europe and among peasant movements in parts of Africa, Latin America, East Asia, and India.

Nature-ism contends that genetic engineering, especially when it crosses species boundaries, disrupts natural ecosystems. This view is also supported by some religious groups, who argue that scientists engaging in GM research are "playing God." In its broader secular interpretation, Nature-ism argues that transgenic technology should be avoided, as altering species boundaries could have unpredictable ecological consequences (Kirkham 2024). For proponents of Nature-ism, understanding the delicate balance that sustains ecosystems means recognizing that humans have a responsibility not to interfere with these natural processes. At its core, this perspective holds that humans should refrain from manipulating species and ecosystems in ways that could cause disruption (Reiss 2002).

14.2.4 Regulatory Aspects of GMO Crops

14.2.4.1 Global Regulatory Frameworks

Regulations regarding genetically modified organisms (GMOs) vary across countries, depending on different stages of GMO development, such as cultivation, import, production, international trade, and use of genetically modified food and feed items. These regulations are shaped by the specific risks associated with each phase. Typically, GMO evaluation may fall under different regulatory agencies, depending on the characteristics of the end product, although many align their standards with the Cartagena Protocol. This protocol provides the definition of GMOs followed by many nations (Turnbull et al. 2021).

In 1988, the Biotechnology Science Coordinating Committee (BSCC), consisting of representatives from various US agencies, asked the National Academy of Sciences-National Research Council (NRC) to review the scientific evidence related to genetically modified microorganisms and plants and provide criteria for evaluating the risks involved. The NRC's work led to the first set of regulations for the GM crops' environmental release. In 1993, the Organisation for Economic Co-operation and Development (OECD) developed guidelines for the industrial application of GM organisms, laying the groundwork for evaluating environmental impacts and ensuring the safety of GM foods and feeds. In 1995, the United Nations Environment Programme (UNEP) released international safety guidelines for biotechnology. UNEP's collaboration with the Global Environment Facility (GEF) led to a project aimed at helping countries develop National Biosafety Frameworks to align with the Cartagena Protocol. Currently, 77 countries are involved in this initiative. Other sources for GMO regulation include BINAS, ICGEB, and AGBIOS databases (Degrassi et al. 2003), which offer country-specific regulatory information. The Cartagena Protocol on Biosafety, adopted in 2000 and enforced in 2003, aims to safeguard biodiversity and protect human health against potential hazards associated with living modified organisms (LMOs) created using modern biotechnology (Newell and Mackenzie 2000).

Different regulatory approaches are employed to oversee GMOs:

1. **Product-Based Regulation:** This system focuses on the plant varieties themselves, disregarding the techniques used to produce them. The regulation targets the final product, rather than the methods of production, assuming that any associated risks are directly linked to the plant's characteristics. Countries like the United States, Russia, Argentina, Australia, and Chile use this approach, which remains consistent with World Trade Organization (WTO) rules, as it restricts trade based on genetic engineering methods (Sprink et al. 2016).
2. **Process-Based Regulation:** In contrast to product-based systems, this approach focuses on the methods used to create the GM plant. It recognizes that the processes involved may affect potential risks, considering that genes are not typically transferred between species in nature (Sprink et al. 2016).

3. **Regulation of Plants with Novel Traits (PNTs):** The regulatory system of Canada evaluates plants that have novel traits, which may result from either traditional breeding or genetic engineering. These traits may be absent in the species or present in a significantly altered form (Ahmad et al. 2021).

Risk Assessment and Safety Standards The safety of GM foods and feeds is based on years of human and animal consumption of similar foods, which are generally considered safe (Delaney 2015). This principle, known as "familiarity," supports the idea that GM food safety cannot be determined by a single study due to the complex nature of food composition, which can vary widely depending on environmental and agricultural conditions. A comparative approach, known as "substantial equivalence," is commonly used in safety assessments. This involves comparing GM crops with conventional counterparts to identify any significant changes in composition that could pose risks (Séralini et al. 2011). The safety assessment process involves organizations such as the Codex Alimentarius Commission, Food and Agriculture Organization (FAO), World Health Organization (WHO), and the OECD, which collectively propose a comparative evaluation of GMOs using various data sources to weigh the evidence of potential risks (Garcia-Alonso 2013).

An overview of the key parameters for GM crop safety assessment is provided in Fig. 14.3, covering:

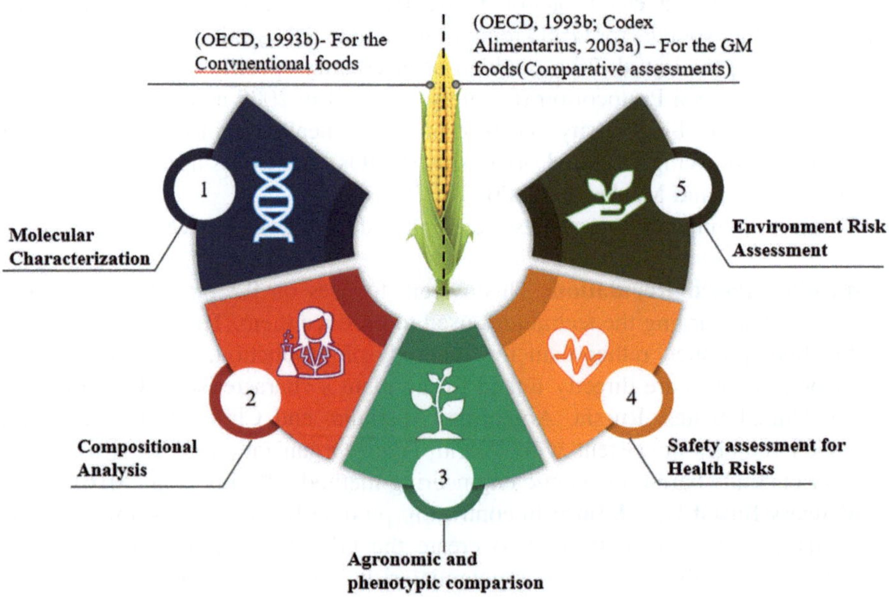

Fig. 14.3 Key evaluation criteria for GM food safety and risk assessment

1. **Molecular Characterization:** This analysis examines the DNA inserted, its location in the genome, and the expressed material to identify potential unintended effects. It ensures that the transgene sequence remains intact and has not caused disruptions in essential genes (Haslberger 2006).
2. **Compositional Assessment:** This involves the evaluation of the GM crop compared to its conventional counterpart to determine if the modifications have resulted in significant differences that could affect safety. If such differences are biologically relevant, they are assessed further for potential risks. But it will not always be a strict approach; for example, if the modification has been done in the metabolic processes, then the metabolic profile will vary significantly. So comparing with the comparator is not very useful, and a complex approach needs to be established to find the effects (Brune et al. 2013).
3. **Phenotypic and Agronomic Comparison:** These studies evaluate whether genetic modifications have resulted in any changes in phenotypic or agronomic traits that could potentially harm the environment. The selection of components and phenotypic or agronomic traits to analyze is partly guided by the specific introduction of the novel trait into the GM plant. However, a broader approach is typically adopted to detect any potential unintended effects of genetic modification (EFSA Panel on Genetically Modified Organisms (GMO) 2015).
4. **Safety Assessments for GMO Crops:**
   - The evaluation of GMO crop safety involves comprehensive assessments that examine food composition, the impact of newly introduced substances such as proteins, and a detailed analysis of genetic modifications. These measures ensure that any approved GM food maintains the same safety standards as conventional food, including potential long-term effects (Huang 2017). The epithelial lining of the small intestine is the main site where nutrients from food are absorbed into the bloodstream. While dietary protein intake is an essential physiological function, a small subset of food proteins can induce allergic reactions or toxicity. These adverse effects may manifest within the digestive tract or extend to other bodily systems, such as in cases of oral allergy syndrome (Kiela and Ghishan 2016).
   - A key aspect of evaluating protein safety involves examining its stability against enzymatic digestion, particularly by pepsin and pancreatic proteases (Codex Alimentarius Commission 2003). Typically, dietary proteins are broken down into peptides and amino acids that enter systemic circulation. This digestion process influences the exposure to intact proteins, which is critical in determining potential allergenic or toxic effects. Regulatory guidelines, including those from Codex Alimentarius, emphasize in vitro digestion studies to evaluate protein safety. Generally, proteins that are fully digested in the gastrointestinal tract have minimal potential for absorption in their intact form, reducing the risk of allergenicity or toxicity. However, proteins that resist digestion may remain in the gut longer, potentially triggering immune responses or toxicity due to prolonged exposure. Protein digestibility alone does not confirm allergenicity or toxicity (Codex Alimentarius Commission 2003). However, proteins that withstand digestive enzymes might persist

longer in the gut, increasing their interaction with the immune system. Most nutrient absorption into the bloodstream occurs through the epithelial cells lining the small intestine. Consequently, digestion resistance studies are considered supplementary evidence in assessing allergenicity and toxicity as part of a weight-of-evidence strategy. In this framework, factors such as protease resistance, structural properties, and potential immunogenicity are evaluated together. The digestion and absorption of proteins are influenced by several physiological and external factors. Interactions between digested food and intestinal epithelial cells, as well as the integrity of the epithelial barrier, determine nutrient transport (EFSA Panel on Genetically Modified Organisms (GMO) et al. 2021). Protein digestibility can also be influenced by factors such as dietary fiber, fats, polyphenols, and the conditions under which the food is processed. Due to the complexity of these interactions, replicating the entire digestive process using static in vitro methods remains a challenge. Thus, in vitro digestibility assays should be used with caution within risk assessment frameworks.

- **Toxicity Assessment**
- The approach to toxicity testing depends on the nature of the introduced proteins or metabolites. If a substance has a well-established history of safe human consumption at expected dietary levels, toxicity assessment can rely on existing studies rather than extensive new testing. This "bridging" approach leverages prior safety data to streamline risk evaluation (Kuiper et al. 2001). For novel proteins and secondary metabolites, more detailed toxicity studies are necessary. The selection of tests is guided by the structure and function of the introduced substance, ensuring a risk-proportionate assessment (Codex Alimentarius Commission 2003).
- Key toxicological testing approaches include:
  - Acute Toxicity Studies: Conducted using purified proteins to assess immediate toxic effects at high doses.
  - Chronic Toxicity and Carcinogenicity Studies: Performed for substances with long-term exposure risks to detect systemic or carcinogenic effects.
  - Reproductive and Developmental Toxicity Studies: Implemented when concerns arise regarding reproductive health or fetal development.
  - Although single-dose oral toxicity tests may not reveal potential long-term effects from continuous exposure, existing studies in animal models have not shown any dietary proteins to be directly associated with teratogenic, mutagenic, or carcinogenic outcomes.
- **Regulatory and Risk Assessment Considerations**: Regulatory bodies, including the European Commission's Scientific Steering Committee, advise that novel proteins lacking a well-established safety history should be subjected to repeated dose toxicity testing in animal models. If early results raise any safety concerns, additional in-depth studies are necessary to thoroughly evaluate potential risks.
- Ultimately, the safety assessment of recombinant proteins in GMOs is based on their similarity to naturally occurring dietary proteins. If an introduced protein shares structural and functional characteristics with known safe

proteins, the assessment process can be expedited. However, novel proteins require thorough toxicological evaluations to ensure consumer safety and regulatory compliance.
- **Allergenicity Assessment**
- Assessing the allergenic potential of newly introduced proteins in genetically modified crops is essential to ensure food safety. Key considerations include evaluating whether allergenic sources that are required to produce the recombinant protein have the potential to induce sensitization or exhibit cross-reactivity with known allergens. Additionally, genetic modifications should not elevate endogenous allergenic protein levels in the crop (Ladics 2019).
- A weight-of-evidence strategy is used to evaluate allergenicity, which involves integrating multiple lines of evidence, including (EFSA Panel on Genetically Modified Organisms (GMO) et al. 2021):
  - Comparing the amino acid sequence of the introduced protein with established allergen databases to identify potential similarities
  - IgE-binding assays to identify immunogenic epitopes
  - Proteolytic stability tests to determine protein degradation in simulated gastric fluid
- Computational tools help evaluate sequence similarities, with an eight-amino acid sequence match commonly used to minimize false-positive allergen predictions. Resistance to digestion is another indicator of allergenic potential, though exceptions exist.
- Current allergenicity assessment methods include targeted serum screening and animal models. While no fully validated animal model for food allergenicity is available, rodent-based immune response studies provide supplementary data. This multi-tiered assessment strategy ensures that novel proteins are thoroughly evaluated before their incorporation into the food supply.

5. Environmental Safety Risk Assessment

Environmental Risk Assessment (ERA) of genetically modified organisms (GMOs) is an essential procedure aimed at identifying potential environmental hazards associated with GM crop cultivation. This process is guided by international standards, including the Cartagena Protocol on Biosafety, biosafety recommendations from the Organisation for Economic Co-operation and Development (OECD), and regulations set by the European Food Safety Authority (EFSA), all of which contribute to a robust and well-defined framework for ERA (Haslberger 2006).

The primary goal of ERA is to evaluate and understand the potential risks that GM plants may pose to the receiving environment. These risks could be direct or indirect, occur immediately or over time, and can include long-term cumulative impacts. The ERA process is highly case-specific, meaning that each GM plant is evaluated individually to determine any potential environmental impact before it is released into the environment. One key aspect of the ERA process is defining the "endpoints," which are critical ecological entities that need protection. These

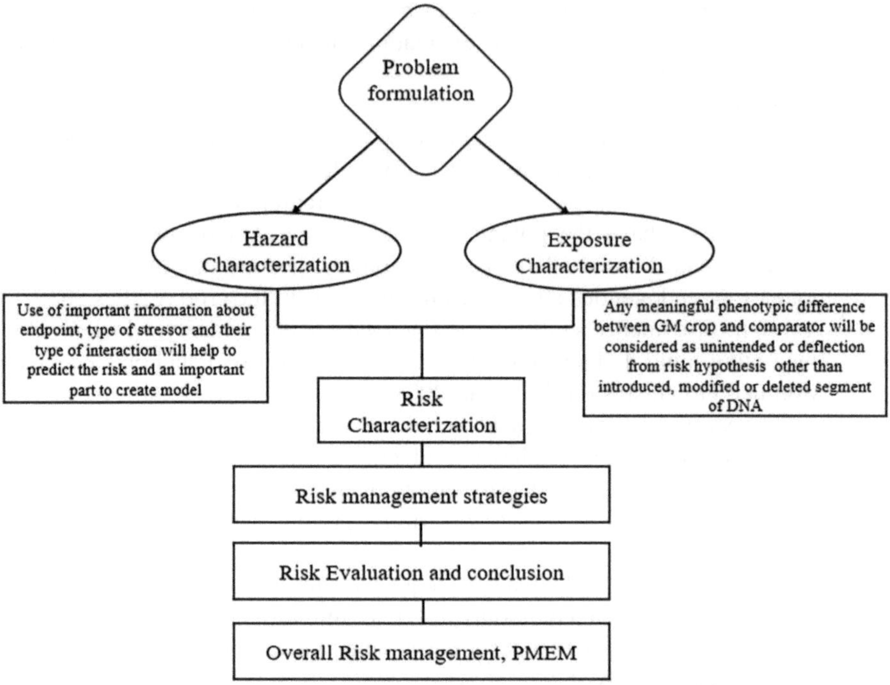

Fig. 14.4 Basic elements for the environment risk assessment model

endpoints can include various components of the environment, such as plant or animal species, soil quality, or water systems. Once the endpoints are identified, the relationship between the environmental stressor and the GM plant is studied. This includes analyzing potential exposure pathways and the likelihood of harm to these ecological entities (Hilbeck et al. 2011).

A conceptual model is then constructed, drawing on available data to help generate hypotheses and predictions about the potential risks, as demonstrated in Fig. 14.4. This model aids in understanding how the GM crop might interact with the environment and informs risk predictions. In ERA, the final step involves prioritizing the necessary tests and data collection methods to accurately assess the risks, ensuring that all relevant factors are considered before any decisions are made regarding the release of GMOs (EFSA Panel on Genetically Modified Organisms (GMO) 2010).

Different countries may adapt international frameworks based on their specific environmental and regulatory needs, but these guidelines generally align with the principles outlined in the Cartagena Protocol and other international agreements. In environmental protection, while there is no equivalent body to the Codex Alimentarius Commission, organizations like EFSA and the OECD have played a pivotal role in developing risk-assessment protocols for GM crops, ensuring that they are evaluated while reducing the risk of harm to biodiversity, ecosystems, and human health. Unintended Effects of Genetic Modifications: Any unintended effects identified during the risk assessment process to determine if they could potentially present a hazard, even if they were not part of the intended genetic modification. The risk evaluation process considers both intended and unintended effects (Garcia-Alonso 2013).

14.2.4.2 Labeling and Consumer Rights

The debate over GMO labeling policies includes transparency, consumer choice, and regulatory standards. Public confidence is a key factor influencing the acceptance and success of new technologies, particularly genetically modified organisms (GMOs). Various factors, including the credibility of information sources, the handling of misinformation, and the transparency of marketing efforts, shape trust (Bain and Dandachi 2014). Closing the disconnect between scientific findings and public opinion is essential to reduce skepticism about the environmental and health effects of GMOs. Key factors influencing consumer acceptance and trust include:

Labeling

Labeling remains a central point of discussion in the acceptance of GMOs. While some scientists argue that labeling is unnecessary if GMO products maintain the same nutritional and safety standards as conventional foods, others emphasize consumers' right to make informed choices.

The US Food and Drug Administration (FDA) has not mandated labeling for genetically modified organisms (GMOs) since 1992 if the product exhibits no significant nutritional or safety differences.

In 2001, the FDA introduced voluntary guidance allowing companies to label their products as GMO-free, provided they met federal standards. However, the FDA maintained its stance from 1992, stating that no safety risks were associated with genetically modified foods.

Voluntary labeling enables companies to seek third-party verification for a "non-GMO" certification, which enhances consumer trust. One of the most recognized organizations in this space is the non-GMO Project (Bain and Dandachi 2014), which certifies products in Canada and the United States. Labeling thresholds vary globally; for example, if a product contains less than 0.9% genetically modified material, then it is considered non-GMO in the European Union.

In 2014, some US states mandated GMO labeling in 2016, which prompted Congress to pass the National Bioengineered Food Disclosure Standard (Hallman 2018). While initially voluntary, mandatory labeling took effect on January 1, 2022, requiring food packaging to include text or a seal indicating "bioengineered" or "derived from bioengineering." However, the choice of terminology, substituting "bioengineered" for the more widely recognized "GMO" sparked public criticism, as did the use of small text, which some argued minimized consumer awareness.

Impact of Labeling on Consumer Behavior

Studies by the American Marketing Association and others have analyzed how different labeling regimes affect consumer preferences:

- Consumers are more likely to prefer products labeled "non-GMO" over those labeled as containing GMOs.
- Label format influences purchasing decisions, particularly among individuals without strong opinions on GMOs.
- Willingness to pay increases for non-GM products when mandatory labeling is enforced.

In Vermont, after mandatory GMO labeling laws were enacted, sales of genetically engineered foods declined by 5.9%. In contrast, the sales of non-GMO products grew by 2.5%, while organic food sales rose by 1.7%, as organic certification standards do not permit the use of GMOs.

Yeh et al. reported that consumers tended to favor products without labels after initially encountering items labeled as genetically modified. However, the reverse was not observed—viewing non-GMO-labeled products first did not significantly affect the perception of unlabeled products (Yeh et al. 2019).

14.2.4.3 Regulatory Challenges in Emerging Biotechnologies

Emerging biotechnologies have revolutionized numerous sectors, from healthcare and agriculture to environmental sciences, offering innovative solutions to address worldwide issues. However, the swift development of these technologies presents unique and evolving regulatory challenges. This chapter explores key hurdles regulators face in managing new technologies, focusing on safety, ethical considerations, intellectual property, international compliance, and how regulations adapt to emerging biotechnologies such as gene editing tools like CRISPR (Son and Park 2022).

1. **Coordinating the Coexistence of GE, Conventional, and Organic Agriculture**

A major regulatory issue in modern agriculture is ensuring the balance of genetically engineered (GE), conventional, and organic farming systems. The accidental occurrence of GE traits in non-GE and organic crops presents economic and trade challenges, particularly for markets with strict regulations regarding GE content. In the United States, regulatory agencies like the USDA primarily focuses on technical assessments. Still, nonregulatory bodies, involving agencies such as the Federal Grain Inspection Service and the Agricultural Marketing Service, also play significant roles in addressing market concerns and promoting coexistence. However, the existing governance frameworks still face difficulties in resolving issues related to identity preservation and organic certification, highlighting the need for clearer guidelines and better coordination (Greene et al. 2016).

2. **Mandatory Labeling of GE Foods**

The mandatory labeling of genetically engineered (GE) foods remains a contentious and multifaceted issue, shaped by varying societal values rather than scientific safety concerns. While some countries require labeling based on consumer demand for transparency, others, including the United States, have adopted mainly voluntary labeling initiatives, such as the non-GMO Project (Bain and Dandachi 2014). These initiatives have gained popularity due to growing consumer preference for clear labeling. Still, they also raise questions about the balance between market forces, consumer rights, and the role of regulatory bodies in shaping public policy. A more consistent global approach to labeling could help improve consumer trust and acceptance (Hallman 2018).

3. **Post-market Regulatory Oversight**

Unlike other regulatory systems, the US Department of Agriculture's Animal and Plant Health Inspection Service (APHIS) lacks authority for post-market monitoring once a GE crop is deregulated. The lack of monitoring after products reach the market has contributed to challenges such as the emergence of weeds resistant to glyphosate. On the other hand, agencies like the Environmental Protection Agency (EPA) impose post-market conditions on certain GM crops, such as Bt crops and herbicide-resistant varieties, to mitigate environmental risks. Expanding post-market regulatory authority would enhance risk management, encourage responsible commercialization, and contribute to long-term environmental sustainability (Yeh et al. 2019).

4. **Public Trust, Transparency, and Participation**

A major regulatory challenge lies in balancing scientific expertise with public accountability. Different countries approach risk assessment and decision-making differently, with some relying on politically accountable bodies for final approvals (Nielsen et al. 2021). In the United States, regulatory agencies often prioritize technical risk assessments, insulating decisions from public influence. Rising public skepticism toward expert-led approaches has prompted increased demands for transparency and more inclusive participation in policy decisions. Initiatives to promote stakeholder engagement, public dialogue, and inclusive decision-making could help enhance public trust in the regulatory process and biotechnology innovations (Wolf 2021).

5. **International Regulatory Harmonization**

The divergence in regulatory standards across countries creates significant trade barriers and complicates the global market for GE products. While some countries impose stringent approval processes that consider socioeconomic factors, others, like the United States, focus primarily on safety assessments. This disparity in regulatory approaches increases the complexity for companies seeking to market their products internationally (Turnbull et al. 2021). Aligning to international regulatory standards while respecting national policy choices is critical for facilitating global trade, promoting innovation, and guaranteeing fair access to the advantages offered by biotechnology.

6. **Cost Implications of Stringent Regulations**

Stringent regulatory requirements can substantially increase the cost of bringing genetically modified (GM) foods to the market. Lengthy approval processes, extensive safety testing, and compliance with evolving regulatory standards can create financial burdens for biotechnology companies (Bernauer et al. 2011). These expenses are frequently reflected in increased prices for consumers, potentially limiting access to GM products and slowing the pace of innovation. Striking a balance between regulatory oversight and efficient approval processes is essential to ensuring that the benefits of biotechnology remain accessible and affordable for all stakeholders.

7. **Adaptability of Regulatory Frameworks and Management of Time Lags**

A major challenge in regulating biotechnology lies in the speed at which new technologies emerge, frequently surpassing the capacity of current regulatory systems to keep up. As new technologies such as gene editing tools like CRISPR emerge, it becomes increasingly important for policymakers to adopt flexible, adaptive regulations that can keep pace with innovation while maintaining rigorous oversight to ensure safety and sustainability. Effective management of time lags between technological developments and regulatory updates is essential to foster innovation while protecting public health and the environment (Son and Park 2022).

8. **Intellectual Property and Commercialization**

Patent regulations and intellectual property rights are central to bringing biotechnology innovations to market, providing legal protection for novel genetic sequences, biopharmaceuticals, and other biotechnological products. However, there is an ongoing challenge in balancing the need to encourage innovation through intellectual property protection with the potential risk of monopolies that could hinder accessibility and equitable distribution of biotechnology benefits (Barragán-Ocaña et al. 2019). Policymakers need to thoughtfully address the intricate balance between protecting intellectual property and ensuring public access, aiming to encourage innovation while avoiding restrictions on competition or access to critical technologies.

14.2.4.4 Commercialization of GMO Crops

According to the 2023 report by the World Health Organization (WHO), approximately 2.33 billion people worldwide experienced moderate to severe food insecurity. Of these, over 864 million faced severe shortages, sometimes going on a full day or longer without food. The global population is projected to reach 9.1 billion by 2050, representing a 34% increase compared to current levels, with most of this growth expected to occur in developing countries. To sustain this rapidly expanding, more urbanized, and wealthier population, food production needs to rise by 70%, excluding food used for biofuels. However, considering environmental and agricultural challenges, this goal must be approached carefully. The commercialization of genetically modified (GM) crops is becoming increasingly important in addressing current global agricultural problems such as food security, climate resilience, and sustainable farming. In comparison to conventional breeding techniques, GM technology allows for more precise and rapid improvements in crop traits. Factors like climate change are increasingly challenging traditional farming methods, limited arable land, and overreliance on synthetic fertilizers and pesticides, leading to volatility and significant environmental stress. Nevertheless, the adoption of GM crops is hindered by the limitations of conventional breeding methods and regulatory hurdles.

Status of GM Crop Commercialization Worldwide

Over the past two decades, more than 18 million farmers (up to 90% of whom are small-scale or resource-poor) in 30 countries have benefited from planting biotech crops. In 2023, according to AgBioInvestor, global GM crop acreage rose by 1.9% compared to 2022, reaching a record of 206.3 million hectares. A total of 27 countries cultivated 11 distinct genetically modified crop varieties, with soybeans being the most widely grown, covering 100.9 million hectares of land, representing the first instance in which the crop exceeded 100 million hectares. Other notable GM crops included maize, cultivated on 69.3 million hectares, and cotton, grown on 24.1 million hectares. Several countries reported notable growth in GM crop adoption, including Pakistan (up by 33.3%), Vietnam (22.2%), Paraguay (8.2%), and Bolivia (7.0%), largely due to expanded planting areas and increased uptake. On the other hand, decreases were observed in countries such as Myanmar (down by 36.7%), China (7.9%), the Philippines (3.4%), Australia (3.1%), and India (2.3%). Recent data from ISAAA reveal that 30 countries have now approved GM crop cultivation, including three African nations—Kenya, Ghana, and Burkina Faso.

The United States remains the leading country regarding GM crop acreage, followed by Brazil. However, the total area dedicated to GM crops in the United States decreased slightly compared to the previous year.

Role of Companies in GM Crop Commercialization

The early breakthroughs that led to GM technology were achieved mainly by public sector scientists, who then collaborated with private sector researchers to develop the technology further. Currently, the development of genetically modified crops involves key contributions from public institutions, private companies, and philanthropic organizations (Dowd-Uribe et al. 2024). Protecting these advancements relies heavily on intellectual property rights, including patents related to GM technologies and crop varieties. These patents allow companies to demand royalty payments for using their proprietary technologies, while also making the underlying scientific knowledge available for public access. This system is similar to drug development, where patents protect discovery- and invention-dependent products, including non-GM crops. Numerous non-GM crop varieties are also subject to licensing restrictions that prevent farmers from saving seeds, requiring them to buy new seeds for each planting cycle.

GM crop research receives funding from national agencies and philanthropic organizations like the Bill & Melinda Gates Foundation, where patents are held for public benefit. The collaboration between the public, private, and charitable sectors has proven beneficial in advancing GM crop development.

Dominance of Multinational Corporations in GM Crop Markets

The global market for genetically modified crops is largely controlled by six leading seed (Table 14.1) and agrochemical corporations. Prior to Bayer's acquisition of Monsanto in 2018, Bayer ranked as the second-largest agrochemical firm, whereas Monsanto held the top position in the seed and biotechnology sector. As of 2022, these six companies collectively held 58% of the global seed market and 78% of the pesticide market. Bayer and Corteva Agrisciences alone accounted for approximately 40% of the global commercial market share (ETC Group 2022).

Table 14.1 GM crops approved for commercialization/planting and importation (Source: https://www.isaaa.org/gmapprovaldatabase/default.asp)

Crop	Improved trait/traits	Developed by	Method of trait introduction
Alfalfa	Herbicide tolerance	Monsanto, Forage Genetics International	Plant transformation facilitated by *Agrobacterium tumefaciens*
Canola (*Brassica napus*)	Herbicide tolerance, male sterility	DuPont, Monsanto, Bayer Crop Science, BASF	Genetic transformation of plants using *Agrobacterium tumefaciens* or by delivering DNA through microparticle (gene gun) bombardment of plant cells or tissues
Cotton	Herbicide and insect tolerance	Dow AgroSciences LLC Monsanto Syngenta	Plant transformation using *Agrobacterium tumefaciens*, along with conventional breeding methods such as cross-hybridization and selection that include transgenic donor lines
Maize (corn)	Herbicide tolerance, drought stress, male sterility, mannose metabolism	DuPont (Pioneer Hi-Bred International Inc.) Monsanto, BASF, Syngenta	Plant transformation using *Agrobacterium tumefaciens*, along with conventional breeding methods such as cross-hybridization and selection that include transgenic donor lines

(continued)

Table 14.1 (continued)

Crop	Improved trait/traits	Developed by	Method of trait introduction
Potato	Herbicide tolerance, insect resistance, lowered free asparagine/reducing sugars, antibiotic resistance, disease resistance	Russian academy of biosciences, Monsanto J.R. Simplot Co., Michigan State University	Plant transformation facilitated by *Agrobacterium tumefaciens*
Papaya	Viral disease resistance, modifies starch, antibiotic resistance, insect tolerance, lowered free asparagine, reduced black spot, lowered reducing sugars, disease resistance	Cornell University, University of Hawaii, South China Agricultural University, Centre Bioengineering Russian Bioacademy of sciences, Monsanto, BASF, J.R. Simplot Co.	Genetic transformation of plants using *Agrobacterium tumefaciens* or by delivering DNA through microparticle (gene gun) bombardment of plant cells or tissues
Rice	Enhanced provitamin A content (Golden Rice)	International Rice Research Institute	Plant transformation facilitated by *Agrobacterium tumefaciens*
Soybean	Herbicide tolerance, modified oil/fatty acid content	BASF, Dow AgroSciences LLC	Trait introduction via gene gun delivery and breeding through crosses with transgenic donor lines
Sugarbeet	Herbicide tolerance	Bayer Crop Science, Monsanto, Novartis Seeds	Plant transformation facilitated by *Agrobacterium tumefaciens*
Sugarcane	Insect tolerance + drought stress tolerance	PT Perkebunan Nusantara XI (Persero), Centro de Tecnologia Canavieira (CTC)	Plant transformation facilitated by *Agrobacterium tumefaciens*
Squash	Viral disease resistance	Seminis Vegetable Seeds, Monsanto	Plant transformation facilitated by *Agrobacterium tumefaciens*
Wheat	Herbicide tolerance and drought stress tolerance	Monsanto, Bioceres S. A.	Genetic transformation of plants using *Agrobacterium tumefaciens* or by delivering DNA through microparticle (gene gun) bombardment of plant cells or tissues
Tomato	Modified flower color, antibiotic resistance, and delayed fruit softening	Norfolk Plant Sciences, Monsanto	Plant transformation facilitated by *Agrobacterium tumefaciens*
Apple	Antibiotic resistance, nonbrowning	Okanagen Speciality fruits incorporated	Plant transformation facilitated by *Agrobacterium tumefaciens*

Why Do Multinational Companies Play a Central Role?

1. **Cost and Time Constraints**: Developing a new plant biotechnology-derived trait from discovery to commercialization costs an estimated $115 million, according to AgBioInvestor. Additionally, the typical timeline to bring a genetic trait to market is about 16.5 years (2017–2022), excluding the time needed to develop stacked trait varieties, common in commercial seed products. The process begins with the discovery phase, where candidate genes with beneficial characteristics such as resistance to herbicides or insects are selected. If these traits demonstrate sufficient biological activity, they move to the genetic construction and testing phase, where they are optimized and incorporated into specific crops for field trials. Based on the success of these trials, the most commercially viable traits are selected for introduction into elite germplasm. In the final stage, regulatory approval is sought by submitting applications to relevant authorities in the countries where the genetically modified crop is intended to be grown and traded. Intellectual property protections, including patents, are also secured at this stage.
2. **Patents and Market Control**: Most patents for GM crop technologies are held by multinational companies such as Bayer and Corteva. These companies control nearly 80% of the patents related to genetically engineered crops, significantly strengthening their market position.
3. **Challenges Faced by Public Sector Research**: Publicly funded institutions often face significant barriers when developing biotech crops due to the significant expenses and lengthy process associated with bringing GM varieties to market (Dowd-Uribe et al. 2024). This has limited the ability of public sector organizations to compete with multinational corporations in the GM crop space.

14.2.4.5 Intellectual Property and Patents

Intellectual property (IP) encompasses a set of legal rights that protect the creations of individuals, such as patents, trademarks, copyrights, and trade secrets. These laws are designed to safeguard inventors' control over their creations, but over time, IP has evolved into a tool for corporations to establish dominance and consolidate control within markets.

- **Plant Patent Act (1930)**: Enacted in 1930, this US federal law grants patent protection for new and distinct varieties of plants that reproduce asexually (Daus 1967).
- **Plant Variety Protection Act (1970)**: This act protects plant breeders, allowing them exclusive rights over newly developed plant varieties. In 1985, the act was expanded to include protection for potatoes and other tuber crops (Butler and Marion 1984).
- **Diamond v. Chakrabarty (1980)**: This landmark Supreme Court case established that patent protection could extend to engineered microorganisms rather than naturally occurring ones, paving the way for biotech patents (Robinson and Medlock 2005).

The growth of intellectual property protections has encouraged private firms to increase their investment in seed biotechnology research and development. However, this has led to a concentration of patent ownership in the hands of a few major corporations (Stein 2005). In a report by Sky Chadde in Investigate Midwest, it was found that Bayer and Corteva Agrisciences now control around 80% of patents related to genetically modified crops, resulting in a monopolistic market environment. This concentration of power raises concerns about the long-term effects on global food security and the economic well-being of farming communities.

The practice of patenting genetic material has sparked significant controversy. Large corporations, such as Bayer and Corteva, have increasingly sought to patent plant genetic traits, whether naturally occurring or modified through genetic engineering. By claiming exclusive rights over these traits, they assert ownership over all plants that express them, whether they are genetically modified or not. For example, Corteva has filed for around 1430 patents related to GMOs, while Bayer/Monsanto has filed 119. These companies argue that new GMOs are as safe as naturally occurring plants, challenging regulatory processes that require safety testing and labeling of GMOs.

Moreover, critics argue that the concentration of seed patents in a few corporations threatens biodiversity and could exacerbate environmental risks. The concentration of agricultural innovation in the hands of a few corporations prompts critical concerns regarding food security, equitable access to farming technologies, and the ethical impact of privatizing control over the future of food systems (Barker et al. 2013).

14.2.4.6 Market Dependency on Genetically Modified (GM) Crops: Associated Risks

The increasing reliance on genetically modified (GM) crops in agriculture introduces a variety of risks, some of which are outlined below:

1. **Monopoly Control and Market Dominance**

A limited number of large biotechnology companies exert significant control over the global GMO seed market, reducing farmers' options and driving up seed prices (Fuglie and MacDonald 2023). This concentration of power by companies like Bayer-Monsanto has serious implications for smallholder farmers.

- In several regions, particularly the United States, the swift uptake of GM crops has been observed, such as GE corn and soybeans, which has often been attributed to high demand. However, much of this adoption is due to major seed suppliers' gradual phase out of non-GMO varieties, leaving farmers with few choices beyond GM seeds.
- Between 2000 and 2010, the market share of GM soybeans grew significantly, with the price of seeds skyrocketing by up to 230%. For example, by 2010, the price of Monsanto's Roundup Ready 2 soybeans had increased by 143% compared to its cost in 2001.

- According to the Economic Research Service of the US Department of Agriculture, seed prices increased significantly between 1995 and 2011, with soybean costs rising by 325 percent and corn by 259 percent. This rise occurred alongside the widespread adoption of genetically engineered crops (Barker et al. 2013).

In other regions like Spain, Germany, and Brazil, the introduction of genetically modified (GM) crops has resulted in fewer seed varieties available to farmers. In Spain, for example, non-GMO maize cultivars have mainly been replaced by GE varieties. Likewise, in Brazil, non-GE soybean acreage has declined sharply since the introduction of GM soybeans (Van Eenennaam and Young 2014).

In South Africa, a similar trend has emerged, where just a few companies predominantly control the soybean market, severely limiting farmers' access to non-GM options.

2. **Escalating Costs and Financial Dependence**

A major concern surrounding GM crops is the cost of purchasing new seeds each year due to intellectual property rights and licensing agreements. Unlike traditional seeds, many GM seeds are patented and sold under terms that prevent farmers from saving and replanting them. As a result, farmers must buy fresh seeds annually, which can place a significant financial burden on them (Shiva 2016). Large multinational corporations like Bayer-Monsanto and Corteva dominate the seed market, imposing licensing fees and restrictive contracts on farmers. While these companies claim that these costs reflect their investment in research and development, they can be prohibitive for smallholder farmers in developing countries.

3. **Loss of Biodiversity**

The large-scale use of genetically modified crops has sparked concerns regarding biodiversity decline. Crops with limited genetic variation may be more susceptible to pests, diseases, and changes in climate, which could pose ecological threats.

- Bt crops provide resistance to specific types of insects, which can lead to the loss of natural competitors and the spread of secondary pests, which affects the balance of the ecosystem, and prolonged cultivation of Bt crops without including crop rotation results in resistance to Bt in insects (Noack et al. 2024).
- An additional threat to biodiversity arises from the possibility of gene transfer between genetically modified crops and their wild counterparts. Horizontal gene transfer (HGT) refers to the movement of genes between different organisms in a nonreproductive manner. This could lead to unintended consequences if GM traits spread to wild plant populations or other crops (Philips et al. 2022). A notable example of this occurred with GM rice, where a variety developed by Bayer (LLRICE601) was found to have contaminated non-GM rice fields, causing trade restrictions and regulatory challenges (Thompson 2010a).

4. **Resistance to Pests and Weeds**

The introduction of herbicide-tolerant crops, such as those modified to withstand glyphosate, resulted in the emergence of resistant weed species. As herbicide-resistant weeds proliferate, farmers may face increased costs and difficulties in managing these weeds. Furthermore, the spread of transgenes from GM crops to wild relatives can disrupt natural ecosystems and reduce genetic diversity. The rise of herbicide-tolerant weeds, such as those found in fields planted with GM crops, highlights the importance of careful management and regulation of GM technology. Over-reliance on herbicide-resistant crops may contribute to the evolution of superweeds, further complicating weed management efforts (Qiu 2013).

5. **Trade Restrictions and Market Limitations**

Some countries have implemented strict regulations or outright bans on GMO imports, which create difficulties for farmers in developing countries to take advantage of GM crops. These trade barriers, particularly in regions like the European Union, have forced farmers in exporting nations to rely on conventional crops, which may be less resilient and productive than their GM counterparts. Developing countries face significant hurdles in adopting GM technology, including the lack of regulatory frameworks to assess and approve GM crops. This lack of clarity often results in delays or outright bans on GM crops, limiting the potential benefits for farmers (Sadikiel Mmbando 2024).

Biotechnological Approaches for the Future of Food Security

Conventional plant breeding techniques alone are insufficient to meet the growing global demand for food, making technological advancements essential to ensuring food security. Conventional breeding methods require a significant amount of time for crop improvement. Despite efforts to enhance yield through various breeding programs, they often fail to meet the increasing global food demand. Traditional breeding alone is not sufficient to enhance crop diversity at the genomic level. Over time, several advancements have been introduced to improve yield and efficiency, such as marker-assisted selection, transgenic approaches, genomic selection using genome-wide markers, high-throughput phenotyping, large-scale sequencing, and genomics-based techniques. The introduction of speed breeding has further accelerated crop improvement, while the introduction of CRISPR-Cas genome editing has marked a new era of transformation in contemporary agriculture (Ahmar et al. 2020).

Biotechnological advances have revolutionized agriculture by providing powerful tools to enhance crop improvement and meet the increasing global food demand. These innovations encompass a wide array of techniques that enable precise genetic modifications in crops, surpassing the limitations imposed by natural reproductive systems. The innovation of GM crop varieties has led to enhanced resistance against pests, tolerance to herbicides, and improved nutritional content, among other benefits. Additionally, breakthroughs in genome editing, particularly CRISPR-

Cas9, have introduced unparalleled precision in modifying specific DNA sequences within plant genomes, opening new avenues for crop enhancement (Liu et al. 2017). Beyond direct genetic modifications, biotechnological approaches have significantly accelerated traditional breeding through marker-assisted selection, which utilizes genetic markers to identify plants with desirable traits. This approach accelerates breeding, allowing enhanced crop varieties to be produced more quickly than with traditional methods.

The integration of plant breeding and biotechnology has significantly enhanced crop improvement by decreasing the time required for genetic advancements and enabling the enhancement of complex traits (Adlak et al. 2023). Although genome editing is not a new concept, earlier techniques such as transcription activator-like effector nucleases (TALENs) and zinc finger nucleases (ZFNs) were used before the emergence of more efficient tools like CRISPR. Conventional breeding now increasingly relies on biotechnology and bioinformatics to enhance precision and efficiency. Additionally, the growing accessibility and affordability of DNA sequencing have revolutionized crop improvement by providing deeper insights into genome structure and organization. This advancement has facilitated the development of novel approaches, including synthetic biology, evolutionary biology, and site-specific genetic editing, opening new possibilities for agricultural innovation.

Important Biotechnological Innovations for Crop Improvement

1. Zinc Finger Nucleases (ZFNs): Zinc finger nucleases (ZFNs) are engineered proteins consisting of a C_2H_2 zinc-finger (ZF) DNA-binding domain fused to a FokI restriction enzyme nuclease domain. The C_2H_2 zinc-finger domain is composed of approximately 30 amino acids, featuring two antiparallel β-sheets and an α-helix, with a zinc ion coordinated by two cysteine residues in the β-sheets and two histidine residues in the α-helix. Due to their unique structure, ZFNs integrate the advantages of both components—the specificity and adaptability of zinc-finger proteins (ZFPs) for DNA binding and the targeted cleavage ability of the FokI nuclease. This modularity allows for independent optimization of them. The presence of distinct DNA-binding and catalytic domains in ZFNs allows for easier retargeting to various genomic sites, enabling their adaptation to different applications and enhancing their efficiency in genetic engineering. (Carroll 2011).
2. TALENs: Transcription activator-like effector nucleases (TALENs) are genome-editing tools derived from TALE proteins, which originate from the bacterial plant pathogen *Xanthomonas* and related genera. These proteins consist of an N-terminal domain, a central DNA-binding domain, and a C-terminal domain that includes an activator domain. In their natural role, TALE proteins are injected into host cells via the type III secretion system, where they bind to specific DNA sequences and activate genes that facilitate pathogen spread. TALENs are engineered by fusing the DNA-binding domain of TALEs to a FokI nuclease, enabling precise genome modifications by introducing double-strand breaks at target sites. Their ease of design and high specificity have made TALENs widely used in genetic engineering, although they have been largely replaced by

CRISPR-Cas9 due to its greater efficiency and simplicity (Joung and Sander 2013).

3. RNAi: RNA interference (RNAi) is a highly conserved gene regulatory mechanism that occurs in many eukaryotic organisms. It involves the silencing of specific genes by short double-stranded RNA (dsRNA) molecules, preventing gene expression at the post-transcriptional level. RNAi can be triggered endogenously, where an organism's own genetic sequences produce regulatory RNA molecules, or exogenously, through the introduction of synthetic dsRNA into a cell. This process is primarily mediated by small interfering RNAs (siRNAs), which are ~21 nucleotides long. These siRNAs are derived from long dsRNA molecules that are cleaved by the RNase III enzyme Dicer. Once formed, siRNAs associate with the RNA-induced silencing complex (RISC), which then targets matching messenger RNA (mRNA) sequences, resulting in their degradation and suppression of gene expression. This mechanism, known as RNA interference (RNAi), is also called RNA silencing, gene silencing, or post-transcriptional gene silencing (PTGS), or quelling, plays a crucial role in natural gene regulation, antiviral defense, and genome stability. Additionally, it has been used as a powerful tool in genetic research, functional genomics, and therapeutic applications, including disease treatment and pest-resistant crops (Pathak and Gogoi 2016).
4. Cisgenesis and Intragenesis: Cisgenesis and intragenesis are two approaches in modern biotechnology that involve genetic modifications within sexually compatible species. These methods offer alternatives to traditional transgenic modification by maintaining natural gene pools. The concept of cisgenesis was first introduced by Jochemsen and Schouten (2000) in their book *Toetsen en begrenzen*, emphasizing that genes or gene elements must originate from the same species. The modern definition of cisgenesis was internationally established in 2006 (Schouten et al. 2006), expanding the concept to include genes from sexually compatible species. In cisgenesis, a gene is transferred in its native form, including its promoter, introns, and terminator, without modifications to its coding sequence. On the other hand, intragenesis, introduced by Rommens (2004), involves the isolation of specific genetic elements from a plant, recombining them outside the organism, and inserting the resulting construct into another plant capable of natural crossbreeding. Unlike cisgenesis, intragenesis allows for the rearrangement of genetic elements, such as combining a promoter from one gene with the coding region of another, leading to customized gene expression (Mujjassim et al. 2019).
5. Marker-Assisted Selection (MAS): Marker-assisted selection (MAS) has revolutionized plant breeding by enabling the precise selection of desirable traits through molecular markers. The concept of using genetic markers in plant breeding gained prominence in the early 1980s, when isozyme markers were first used to accelerate the introgression of monogenic traits from exotic germplasm into cultivated varieties (Tanksley and Rick 1980; Tanksley 1983). A breakthrough came with the introduction of restriction fragment length polymorphism (RFLP) markers, which allowed for more precise identification of genetic

loci associated with important crop traits. Beckmann and Soller (1986) demonstrated the first practical applications of RFLP markers in crop improvement, particularly in marker-assisted backcrossing for transferring qualitative traits. The development of marker-assisted selection (MAS) for quantitative traits was pioneered by Lande and Thompson (1990) (Lande and Thompson 1990), who provided theoretical frameworks for its application. Since then, MAS has played an important role in breeding programs worldwide, enhancing the efficiency, accuracy, and speed of crop improvement by reducing the reliance on phenotypic selection. Today, MAS is widely used to generate improved yield potential, disease-resistant, and climate-resilient crops, making it a powerful tool in modern agriculture. With advancements in genome sequencing and high-throughput genotyping, the efficiency of MAS continues to improve, laying the foundation for more accurate and environmentally sustainable crop improvement methods (Choudhary et al. 2008).

6. CRISPR-Cas9: CRISPR-Cas9: CRISPR systems are renowned for their simplicity, high programmability, and precise targeting capabilities, making them one of the most widely used genome-editing tools. The CRISPR system consists of two essential components: a guide RNA (gRNA), which is designed to complement a specific DNA sequence, and a CRISPR-associated protein (Cas). This complex searches for a protospacer adjacent motif (PAM) and binds to the corresponding target DNA sequence. Traditional CRISPR-Cas genome editing relies on nucleases like Cas9, which not only target specific DNA sequences but also generate double-strand breaks (DSBs) in the DNA. These breaks serve as targets for the plant's inherent DNA repair systems, including homology-directed repair (HDR) and nonhomologous end joining (NHEJ). Recent innovations in CRISPR technology, including base editing, enable accurate base changes without generating double-strand breaks or relying on donor DNA. This innovation bypasses the error-prone repair pathways associated with traditional methods. Although base editing and prime editing are currently more efficient for making small, single-base modifications, CRISPR-based tools are continually advancing, and CRISPR technology continues to evolve, enabling larger-scale genetic modifications, including alterations at the kilobase level and beyond. However, editing large DNA segments remains difficult as a result of certain limitations of cell DNA repair mechanisms and the restricted editing potential of current systems. As technology progresses, CRISPR-based tools are expected to address these challenges and expand their applicability for larger genomic modifications (Bao et al. 2019; Li et al. 2024).
7. Synthetic Biology: Synthetic biology (SynBio) is the field that involves the engineering of biological systems to create complex, biologically based or biologically inspired entities that exhibit functions not found in nature. SynBio allows for the rational and systematic design of biological systems (Serrano 2007). It includes approaches aimed at designing and constructing new biological elements (such as enzymes, genetic circuits, and cells) or redesigning existing biological systems to generate new or enhanced functions. This can occur in two main subfields: (Smýkal et al. 2018) using existing biological components to

create novel combinations that are not found in nature, and (Priyadarshan 2019) developing synthetic components to replicate natural functions or create entirely new ones. As synthetic biology has advanced, it has adopted key engineering concepts, including mechanisms like "switch" and "rewire," along with iterative processes such as the "design, test, simulate, learn" cycle (Sargent et al. 2022). This field significantly expands the capabilities of genetic engineering, allowing for faster transgenic materials development involving intricate genetic alterations, which is particularly advantageous in the creation of elite crop varieties. For instance, the development of C4 rice involved the introduction of five genes from the NADP-ME biochemical subtype (Sargent et al. 2022). This transformation, which could have taken several years using conventional genetic engineering methods (involving individual gene insertions and subsequent stacking), was accomplished within 6 to 12 months using SynBio techniques, such as Golden Gate cloning (Sargent et al. 2022).

8. Systems Biology: A systems biology approach integrates multiple omics data, modeling, and prediction of cellular functions to understand the flow of biological information underlying complex traits. The vast data generated from omics experiments are analyzed using advanced software and analytical methods to unravel the complexity of biological systems. This emerging field is often referred to as integrative biology or systems biology. The core of systems biology is to combine different layers of omics data to build models that explain the behavior of biological systems. These models have predictive power, allowing scientists to forecast the outcomes of mutations or modifications in specific biological pathways (Pazhamala et al. 2021).

   With the advancement of CRISPR technology, scientists have made significant strides in gene editing for vegetable and horticultural plants (Hamdan et al. 2023). Muto and Matsumoto utilized the CRISPR/Cas9 system with a single guide RNA (sgRNA) to modify the GLABRA1 (GL1) orthologs, RsGL1a and RsGL1b, which are known to be involved in leaf trichome formation in radish (Muto and Matsumoto 2022; Hamdan et al. 2023;). Cao et al. applied a multiplex CRISPR/Cas9 approach in soybean to target RS2 and RS3 genes responsible for raffinose synthesis (Hamdan et al. 2023). Their study demonstrated that two-component transcriptional unit (TCTU) and single transcriptional unit (STU) systems, using tRNA as the cleavage site, exhibited higher editing efficiency compared to other systems (Cao et al. 2022; Hamdan et al. 2023). Wang et al. provided a comprehensive overview of CRISPR/Cas technology applications in maize, focusing on gene function, germplasm development, enhanced yield, specialty corns, plant architecture, stress response, haploid induction, and male sterility (Wang et al. 2020). Cheng et al. showed the potential of CRISPR/Cas9 genome editing in accelerating disease resistance in barley. They modified Protein Disulfide Isomerase Like 5-1 (PDIL5-1), which plays a crucial role in the plant's response to viral infections (Cheng et al. 2022; Hamdan et al. 2023). Jasrotia et al. examined interactions between wheat and aphids by studying transcriptome profiles in two genotypes: one susceptible (A-9-30-1) and one tolerant (HD2967) (Hamdan et al. 2023). Their findings indicated that aphid infestation triggered strong expression of defense-related genes in the tolerant variety compared to the susceptible one. This insight into wheat's defense responses may contribute to the development of pest-resistant cultivars (Jasrotia et al. 2021; Hamdan et al. 2023). Modern crop breeding now depends on the rapid generation of genetic variants, and the shift from traditional random mutagenesis to precise, targeted mutagenesis marks a pivotal milestone in this field (Li et al. 2024). Utilizing advanced targeted genetic modification (TGM) methods, combined with innovative plant delivery systems, is now central to crop improvement efforts and is considered vital for addressing future agricultural demands (Li et al. 2024).

14.3 Conclusion

This chapter provides insights into next-generation crop improvement technologies' social, ethical, and regulatory dimensions, particularly GMOs, highlighting their significance in ensuring future food security. GMOs role in enhancing agricultural productivity through herbicide-tolerance, insecticide-tolerance, and stacked-trait crops along with their contributions to environmental sustainability and global regulations are key aspects discussed in this chapter. However, challenges such as consumer acceptance, regulatory concerns, potential environmental impacts, patent-related issues, and market monopolization by multinational corporations remain critical factors that influence the adoption and policy decisions surrounding GMOs. Emerging innovations in next-generation crop improvement, such as synthetic biology, gene editing (CRISPR-Cas9), and RNAi, are advancing rapidly. However, there remains a gap between technological advancements and market acceptance due to consumer skepticism and stringent regulatory frameworks. Bridging this gap requires strict yet balanced regulatory assessments and effective dissemination of scientifically validated information, ensuring that GMOs and related technologies are perceived more accurately in the public domain, potentially leading to less restrictive policies. The future of next-generation crop improvement depends on a balanced approach that embraces scientific advancements while proactively addressing societal, ethical, and regulatory concerns to promote sustainable agricultural development.

References

Abberton M, Batley J, Bentley A, Bryant J, Cai H, Cockram J, Costa de Oliveira A, Cseke LJ, Dempewolf H, De Pace C, Edwards D (2016) Global agricultural intensification during climate change: a role for genomics. Plant Biotechnol J 14(4):1095–1098

Adlak T, Tiwari S, Rathore MS, Tripathi N, Tiwari PN, Tripathi MK (2023) Biotechnological approaches for genetic improvement of crops. In: Cutting edge research in biology, vol 7, pp 63–84

Ahmad A, Munawar N, Khan Z, Qusmani AT, Khan SH, Jamil A, Ashraf S, Ghouri MZ, Aslam S, Mubarik MS, Munir A (2021) An outlook on global regulatory landscape for genome-edited crops. Int J Mol Sci 22(21):11753

Ahmar S, Gill RA, Jung KH, Faheem A, Qasim MU, Mubeen M, Zhou W (2020) Conventional and molecular techniques from simple breeding to speed breeding in crop plants: recent advances and future outlook. Int J Mol Sci 21(7):2590

Azadi H, Ho P (2010) Genetically modified and organic crops in developing countries: a review of options for food security. Biotechnol Adv 28(1):160–168

Baggini J, Fosl PS (2024) The ethics toolkit: a compendium of ethical concepts and methods. Wiley

Bain C, Dandachi T (2014) Governing GMOs: the (counter) movement for mandatory and voluntary non-GMO labels. Sustainability 6(12):9456–9476

Bao A, Burritt DJ, Chen H, Zhou X, Cao D, Tran LSP (2019) The CRISPR/Cas9 system and its applications in crop genome editing. Crit Rev Biotechnol 39(3):321–336

Barker D, Freese B, Kimbrell G (2013) Seed giants vs. US farmers: a report by the Center for Food Safety & save our seeds

Barragán-Ocaña A, Reyes-Ruiz G, Olmos-Peña S, Gómez-Viquez H (2019) Production, commercialization, and intellectual property of transgenic crops in Latin America: a state of the art review. J Agribus Dev Emerg Econ 9(4):333–351

Bawa AS, Anilakumar KR (2013) Genetically modified foods: safety, risks and public concerns—a review. J Food Sci Technol 50(6):1035–1046

Beckmann JS, Soller M (1986) Restriction fragment length polymorphisms and genetic improvement of agricultural species. Euphytica 35(1):111–124

Bernauer T, Tribaldos T, Luginbühl C, Winzeler M (2011) Government regulation and public opposition create high additional costs for field trials with GM crops in Switzerland. Transgenic Res 20:1227–1234

Brookes G (2022) Farm income and production impacts from the use of genetically modified (GM) crop technology 1996–2020. GM Crops Food 13(1):171–195

Brookes G, Barfoot P (2020) GM crop technology use 1996-2018: farm income and production impacts. GM Crops Food 11(4):242–261

Brookes G, Taheripour F, Tyner WE (2017) The contribution of glyphosate to agriculture and potential impact of restrictions on use at the global level. GM Crops Food 8(4):216–228

Brune PD, Culler AH, Ridley WP, Walker K (2013) Safety of GM crops: compositional analysis. J Agric Food Chem 61(35):8243–8247

Burkhardt J (2001) The GMO debates: taking ethics seriously. In Proceedings of the 2001 national public policy education conference. Access at: www.farmfoundation.org/nppecindex.htm

Butler LJ, Marion BW (1984) Has the plant variety protection act achieved its legislative intent? J Agron Educ 13(2):110–114

Cao L, Wang Z, Ma H, Liu T, Ji J, Duan K (2022) Multiplex CRISPR/Cas9-mediated raffinose synthase gene editing reduces raffinose family oligosaccharides in soybean. Front Plant Sci 13: 1048967

Carroll D (2011) Genome engineering with zinc-finger nucleases. Genetics 188(4):773–782

Catherine KN, Mugiira BR, Muchiri NJ (2024) Public perception of genetically modified organisms and the implementation of biosafety measures in Kenya. Adv Agric 2024(1):5544617

Cheng C, Kan J, Li S, Jiang C, He X, Shen H, Xu R, Li B, Feng Z, Yang P (2022) Mutation of barley HvPDIL5-1 improves resistance to yellow mosaic virus disease without growth or yield penalties. Front Plant Sci 13:1018379

Choudhary K, Choudhary OP, Shekhawat NS (2008) Marker assisted selection: a novel approach for crop improvement. Am-Eurasian J Agron 1(2):26–30

Codex Alimentarius Commission (2003) Guideline for the conduct of food safety assessment of foods derived from recombinant-DNA plants. CAC/GL 45(2003):1–13

Cong L, Ran FA, Cox D, Lin S, Barretto R, Habib N, Hsu PD, Wu X, Jiang W, Marraffini LA, Zhang F (2013) Multiplex genome engineering using CRISPR/Cas systems. Science 339(6121): 819–823

Corcoran S (2008) Terminator seeds, the right to food, corporate responsibility, business and human rights. In M. Gibney & S. Skogly (Eds.), Universal Human Rights and Extraterritorial Obligations 203–226. University of Pennsylvania Press

Crossa J, Fritsche-Neto R, Montesinos-Lopez OA, Costa-Neto G, Dreisigacker S, Montesinos-Lopez A, Bentley AR (2021) The modern plant breeding triangle: optimizing the use of genomics, phenomics, and enviromics data. Front Plant Sci 12:651480

Cui K, Shoemaker SP (2018) Public perception of genetically-modified (GM) food: a nationwide Chinese consumer study. npj Sci Food 2(1):10

Daus DG (1967) Plant patents: a potentially extinct variety. Econ Bot 21:388–394

Degrassi G, Alexandrova N, Ripandelli D (2003) Databases on biotechnology and biosafety of GMOs. Environ Biosaf Res 2(3):145–160

Delaney B (2015) Safety assessment of foods from genetically modified crops in countries with developing economies. Food Chem Toxicol 86:132–143

Dowd-Uribe B, Rock JS, Spreadbury T, Chiril P, Uminsky D (2024) Bridging the gap? Public–private partnerships and genetically modified crop development for smallholder farmers in Africa. Plants People Planet 6(2):437–451

Dudek CM (2015) GMO food regulatory frameworks in the US and the EU. In: The new and changing transatlanticism. Routledge, pp 214–231

EFSA Panel on Genetically Modified Organisms (GMO) (2010) Guidance on the environmental risk assessment of genetically modified plants. EFSA J 8(11):1879

EFSA Panel on Genetically Modified Organisms (GMO) (2015) Guidance on the agronomic and phenotypic characterisation of genetically modified plants. EFSA J 13(6):4128

EFSA Panel on Genetically Modified Organisms (GMO), Naegeli H, Bresson JL, Dalmay T, Dewhurst IC, Epstein MM, Firbank LG, Guerche P, Hejatko J, Moreno FJ, Mullins E (2021) Statement on in vitro protein digestibility tests in allergenicity and protein safety assessment of genetically modified plants. EFSA J 19(1):e06350

Erokhin D, Komendantova N (2023) GMO discussion on twitter. GM Crops Food 14(1):1–13

ETC Group (2022) Food barons 2022: crisis profiteering, digitalization and shifting power. ETC Group, Springfield. Available online at: https://etcgroup.org/content/foodbarons-2022. Accessed 7 Dec 2022

Fernbach PM, Light N, Scott SE, Inbar Y, Rozin P (2019) Extreme opponents of genetically modified foods know the least but think they know the most. Nat Hum Behav 3(3):251–256

Fuglie K, MacDonald JM (2023) Expanded intellectual property protections for crop seeds increase innovation and market power for companies. The Economics of Food, Farming, Natural Resources, and Rural America, Amber Waves

Garcia-Alonso M (2013) Safety assessment of food and feed derived from GM crops: using problem formulation to ensure "fit for purpose" risk assessments. Collection Biosaf Rev 8(01):72–101

Gasser CS, Fraley RT (1992) Transgenic crops. Sci Am 266(6):62–69

Gbadegesin LA, Ayeni EA, Tettey CK, Uyanga VA, Aluko OO, Ahiakpa JK, Okoye CO, Mbadianya JI, Adekoya MA, Aminu RO, Oyawole FP (2022) GMOs in Africa: status, adoption and public acceptance. Food Control 141:109193

Gepts P (2002) A comparison between crop domestication, classical plant breeding, and genetic engineering. Crop Sci 42(6):1780–1790

Greene, C., Wechsler, S.J., Adalja, A. and Hanson, J., 2016. Economic issues in the coexistence of organic, genetically engineered (GE), and non-GE crops

Gregorowius D, Lindemann-Matthies P, Huppenbauer M (2012) Ethical discourse on the use of genetically modified crops: a review of academic publications in the fields of ecology and environmental ethics. J Agric Environ Ethics 25:265–293

Hallman WK (2018) Consumer perceptions of genetically modified foods and GMO labeling in the United States. In: Consumer perception of food attributes, pp 44–61

Hamdan MF, Hensel G, Alok A, Tan BC (2023) Genome editing and biotechnological advances for crop improvement and future agriculture. Front Plant Sci 14:1132821

Hammond B, Kough J, Herouet-Guicheney C, Jez JM, ILSI International Food Biotechnology Committee Task Force on the Use of Mammalian Toxicology Studies in the Safety Assessment of GM Foods (2013) Toxicological evaluation of proteins introduced into food crops. Crit Rev Toxicol 43(sup2):25–42

Hartung F, Schiemann J (2014) Precise plant breeding using new genome editing techniques: opportunities, safety and regulation in the EU. Plant J 78(5):742–752

Haslberger AG (2006) Need for an "integrated safety assessment" of GMOs, linking food safety and environmental considerations. J Agric Food Chem 54(9):3173–3180

Hilbeck A, Meier M, Römbke J, Jänsch S, Teichmann H, Tappeser B (2011) Environmental risk assessment of genetically modified plants-concepts and controversies. Environ Sci Eur 23:1–12

https://www.psu.edu/news/agricultural-sciences/story/mandatory-labeling-genetically-engineered-foods-may-reduce-customer

Huang K (2017) Safety assessment of genetically modified foods. Springer, Singapore

Jasrotia P, Bhardwaj AK, Katare S, Yadav J, Kashyap PL, Kumar S, Singh GP (2021) Tillage intensity influences insect-pest and predator dynamics of wheat crop grown under different conservation agriculture practices in rice-wheat cropping system of indo-Gangetic plain. Agronomy 11(6):1087

Jochemsen H. ed., (2000) Toetsen en begrenzen: Een ethische en politieke beoordeling van de moderne biotechnologie. Buijten and Schipperheijn.

Joung JK, Sander JD (2013) TALENs: a widely applicable technology for targeted genome editing. Nat Rev Mol Cell Biol 14(1):49–55

Kathage J, Qaim M (2012) Economic impacts and impact dynamics of Bt (bacillus thuringiensis) cotton in India. Proc Natl Acad Sci 109(29):11652–11656

Kiela PR, Ghishan FK (2016) Physiology of intestinal absorption and secretion. Best Pract Res Clin Gastroenterol 30(2):145–159

Kim Y, Kim S, Arora N (2022) GMO labeling policy and consumer choice. J Mark 86(3):21–39

Kirkham N (2024) A virtue ethics interpretation of the 'argument from nature' for both humans and the environment. Philosophies 9(1):19

Kovak E, Blaustein-Rejto D, Qaim M (2022) Genetically modified crops support climate change mitigation. Trends Plant Sci 27(7):627–629

Kramer MG, Redenbaugh K (1994) Commercialization of a tomato with an antisense polygalacturonase gene: the FLAVR SAVR™ tomato story. Euphytica 79:293–297

Kubisz P, Dalton G, Majewski E, Pogodzińska K (2021) Facts and myths about GM food—the case of Poland. Agriculture 11(8):791

Kuiper HA, Kleter GA, Noteborn HP, Kok EJ (2001) Assessment of the food safety issues related to genetically modified foods. Plant J 27(6):503–528

Ladics GS (2019) Assessment of the potential allergenicity of genetically-engineered food crops. J Immunotoxicol 16(1):43–53

Lande R, Thompson R (1990) Efficiency of marker-assisted selection in the improvement of quantitative traits. Genetics 124(3):743–756

Li B, Sun C, Li J, Gao C (2024) Targeted genome-modification tools and their advanced applications in crop breeding. Nat Rev Genet 25(9):603–622

Liu H, Ding Y, Zhou Y, Jin W, Xie K, Chen LL (2017) CRISPR-P 2.0: an improved CRISPR-Cas9 tool for genome editing in plants. Mol Plant 10(3):530–532

Lynas M, Adams J, Conrow J (2025) Misinformation in the media: global coverage of GMOs 2019-2021. GM Crops Food 16(1):18–27

Mba C, Guimaraes EP, Ghosh K (2012) Re-orienting crop improvement for the changing climatic conditions of the 21st century. Agric Food Secur 1:1–17

Medani KR, Neill A, Garrod G, Ojo M, Hubbard C (2024) Societal perceptions and attitudes towards genetically modified (GM) crops, feed, and food products in the Middle East, North Africa, and Turkey (MENAT) region: a systematic literature review. Food Qual Prefer 117: 105148

Mujjassim NE, Mallik M, Rathod NKK, Nitesh SD (2019) Cisgenesis and intragenesis a new tool for conventional plant breeding: a review. J Pharmacogn Phytochem 8:2485–2489

Muto N, Matsumoto T (2022) CRISPR/Cas9-mediated genome editing of RsGL1a and RsGL1b in radish (*Raphanus sativus* L.). Front Plant Sci 13:951660

Nadeem MA, Nawaz MA, Shahid MQ, Doğan Y, Comertpay G, Yıldız M, Hatipoğlu R, Ahmad F, Alsaleh A, Labhane N, Özkan H (2018) DNA molecular markers in plant breeding: current status and recent advancements in genomic selection and genome editing. Biotechnol Biotechnol Equip 32(2):261–285

Newell P, Mackenzie R (2000) The 2000 Cartagena protocol on biosafety: legal and political dimensions. Glob Environ Chang 10(4):313–317

Nielsen J, Eckstein L, Nicol D, Stewart C (2021) Integrating public participation, transparency and accountability into governance of marketing authorisation for genome editing products. Front Polit Sci 3:747838

Noack F, Engist D, Gantois J, Gaur V, Hyjazie BF, Larsen A, M'Gonigle LK, Missirian A, Qaim M, Sargent RD, Souza-Rodrigues E (2024) Environmental impacts of genetically modified crops. Science 385(6712):eado9340

Pathak K, Gogoi B (2016) RNA interference (RNAi): application in crop improvement: a review. Agric Rev 37(3):245–249

Pazhamala LT, Kudapa H, Weckwerth W, Millar AH, Varshney RK (2021) Systems biology for crop improvement. Plant genome 14(2):e20098

Philips JG, Martin-Avila E, Robold AV (2022) Horizontal gene transfer from genetically modified plants–regulatory considerations. Front Bioeng Biotechnol 10:971402. [online]

Pingali PL (2012) Green revolution: impacts, limits, and the path ahead. Proc Natl Acad Sci 109(31):12302–12308

Priyadarshan, P.M., 2019. PLANT BREEDING: classical to modern

Qiu J (2013) Genetically modified crops pass benefits to weeds. Nature (Aug. 16, 2013), http://perma.cc/5G3T-UHFC

Raman R (2017) The impact of genetically modified (GM) crops in modern agriculture: a review. GM crops & food 8(4):195–208

Razzaq A, Kaur P, Akhter N, Wani SH, Saleem F (2021) Next-generation breeding strategies for climate-ready crops. Front Plant Sci 12:620420

Reiss MJ (2002) Introduction to ethics and bioethics. In: Bioethics for scientists, pp 1–17

Robinson D, Medlock N (2005) Diamond v. Chakrabarty: a retrospective on 25 years of biotech patents. Intell Prop Technol Law J 17(10):12–15

Rommens CM, Humara JM, Ye J, Yan H, Richael C, Zhang L, Perry R, Swords K (2004) Crop improvement through modification of the plant's own genome. Plant Physiology 135(1):421–431

Sadikiel Mmbando G (2024) The adoption of genetically modified crops in Africa: the public's current perception, the regulatory obstacles, and ethical challenges. GM Crops Food 15(1): 185–199

Sargent D, Conaty WC, Tissue DT, Sharwood RE (2022) Synthetic biology and opportunities within agricultural crops. J Sustain Agric Environ 1(2):89–107

Schouten HJ, Krens FA, Jacobsen E (2006) Cisgenic plants are similar to traditionally bred plants: international regulations for genetically modified organisms should be altered to exempt cisgenesis. EMBO reports 7(8):750–753

Séralini GE, Mesnage R, Clair E, Gress S, De Vendômois JS, Cellier D (2011) Genetically modified crops safety assessments: present limits and possible improvements. Environ Sci Eur 23:1–10

Serrano L (2007) Synthetic biology: promises and challenges. Mol Syst Biol 3(1):158

Shan Q, Baltes NJ, Atkins P, Kirkland ER, Zhang Y, Baller JA, Lowder LG, Malzahn AA, Haugner JC III, Seelig B, Voytas DF (2018) ZFN, TALEN and CRISPR-Cas9 mediated homology directed gene insertion in Arabidopsis: a disconnect between somatic and germinal cells. J Genet Genomics 45(12):681

Shiva V (2016) Defending farmers' seed freedom. ANTYAJAA Indian J Women Soc Change 1(2): 205–220

Singh A, Ramakrishna G, Kaila T, Saxena S, Sharma S, Gaikwad AB, Abdin MZ, Gaikwad K (2022) Next-generation sequencing technologies: approaches and applications for crop improvement. In: Genomics of cereal crops. Springer, New York, pp 31–94

Smýkal P, Nelson MN, Berger JD, Von Wettberg EJ (2018) The impact of genetic changes during crop domestication. Agronomy 8(7):119

Sohi M, Pitesky M, Gendreau J (2023) Analyzing public sentiment toward GMOs via social media between 2019–2021. GM Crops Food 14(1):1–9

Son S, Park SR (2022) Challenges facing CRISPR/Cas9-based genome editing in plants. Front Plant Sci 13:902413

Sprink T, Eriksson D, Schiemann J, Hartung F (2016) Regulatory hurdles for genome editing: process-vs. product-based approaches in different regulatory contexts. Plant Cell Rep 35:1493–1506

Stein H (2005) Intellectual property and genetically modified seeds: the United States, trade, and the developing world. Northwest J Technol Intellect Prop 3(2):151

Swarup S, Cargill EJ, Crosby K, Flagel L, Kniskern J, Glenn KC (2021) Genetic diversity is indispensable for plant breeding to improve crops. Crop Sci 61(2):839–852

Tanksley SD, Rick C (1980) Isozymic gene linkage map of the tomato: applications in genetics and breeding. Theoretical and Applied Genetics 58(2):161–170

Tanksley SD (1983) Molecular markers in plant breeding. Plant Molecular Biology Reporter 1 (1):3–8

Thiroux JP, Krasemann KW (2009) Ethics: theory and practice. Prentice Hall, Upper Saddle River, p 480

Thompson PB (ed) (2007) Food biotechnology in ethical perspective, vol 10. Springer

Thompson J (2010a) Spatial economic equilibrium analysis of rice trade: impact of GM contamination on United States exports on prices and trade flows. University of Arkansas

Thompson PB (2010b) The agrarian vision: sustainability and environmental ethics. University Press of Kentucky

Troadec MB, Pagès JC, Scientific Committee of the High Council for biotechnology, France (2019) Where are we with unintended effects in genome editing applications from DNA to phenotype: focus on plant applications. Transgenic Res 28:125–133. Springer

Turnbull C, Lillemo M, Hvoslef-Eide TA (2021) Global regulation of genetically modified crops amid the gene edited crop boom–a review. Front Plant Sci 12:630396

Uzogara SG (2000) The impact of genetic modification of human foods in the 21st century: a review. Biotechnol Adv 18(3):179–206

Van Eenennaam AL, Young AE (2014) Prevalence and impacts of genetically engineered feedstuffs on livestock populations. J Anim Sci 92(10):4255–4278

Wang K, Gong Q, Ye X (2020) Recent developments and applications of genetic transformation and genome editing technologies in wheat. Theor Appl Genet 133(5):1603–1622

Wolf C (2021) Public trust and biotech innovation: a theory of trustworthy regulation of (scary!) technology. Soc Philos Policy 38(2):29–49

Woźniak-Gientka E, Tyczewska A, Twardowski T (2022) Public opinion on biotechnology and genetic engineering in the European Union: polish consumer study. Biotechnologia 103(2):185

Xiao Q, Bai X, Zhang C, He Y (2022) Advanced high-throughput plant phenotyping techniques for genome-wide association studies: a review. J Adv Res 35:215–230

Yeh DA, Gómez MI, Kaiser HM (2019) Signaling impacts of GMO labeling on fruit and vegetable demand. PLoS One 14(10):e0223910

Zhang Y, Huang G, Zhao Y, Lu X, Wang Y, Wang C, Guo X, Zhao C (2024) Revolutionizing crop breeding: next-generation artificial intelligence and big data-driven intelligent design. Engineering 44:245

Epigenetic Modifications and Synthetic Biology Approaches for Crop Improvement 15

Mokhtar Guerfel, J. Essemine, and Yogesh Ahlawat

Abstract

Epigenetics offers crucial and essential biological information that could be directly used for crop enhancement through promoting modern agricultural breeding strategies. Herein, we highlighted the potential epigenetic mechanisms underlying major agronomic traits in crop improvement. Moreover, we reviewed various synthetic biology (SynBio) approaches that may improve several desirable agronomic traits to meet the demand for staple food of an ever-growing worldwide population. We studied recent technologies in plant SynBio, which regulate gene expression at multiple levels, and focused on their future directions. Hence, this chapter provides useful new prospects aimed at harnessing the potential of epigenetics and SynBio in crop improvement for the sake of the worldwide food security.

Keywords

Abiotic stresses · Epigenetics · Synthetic biology · Crop plant · Gene expression

M. Guerfel (✉)
Applied College, Shaqra University, Afif, Saudi Arabia
e-mail: mokhtar_guerfel@su.edu.sa

J. Essemine
Bioinformatics Section, Department of Informatics, University of Quebec at Montreal, Pavillon President-Kennedy, Montreal, QC, Canada

Y. Ahlawat (✉)
University Centre for Research and Development, Chandigarh University, Mohali, Punjab, India

Allied Health Sciences, Datta Meghe Institute of Higher Education and Research, Wardha, Maharashtra, India
e-mail: ykahlawa@mtu.edu

Y. Ahlawat et al. (eds.), *Next-Generation Strategies for Crop Improvement*,
https://doi.org/10.1007/978-981-95-0309-4_15

15.1 Introduction

In recent years, significant progress has been made in understanding how epigenetic mechanisms contribute to plant responses to environmental challenges, particularly various stressors (Mladenov et al. 2021). These environmental stressors can hinder plant development and are increasingly recognized as major threats to both global food security and sustainable agricultural practices (Altieri et al. 2015). Projections suggest that, due to factors such as rapid population growth, climate change, and the loss of arable land, current crop production may not be sufficient to meet global food demands by 2050 (Bailey-Serres et al. 2019). Notably, environmental stresses have been shown to reduce crop yields by up to 96.5% (Thiebaut et al. 2020). Given the rising global demand for food, there is an urgent need to improve crop productivity. This enhancement relies on an interdisciplinary approach, drawing from fields such as plant physiology, biotechnology, epigenetics, and genetics (Gupta and Salgotra 2022). Crop performance is continually challenged by both biotic and abiotic stresses, many of which are governed by complex traits that cannot be easily managed through conventional breeding or traditional genetic engineering techniques. To address these challenges and improve yield potential—especially in the face of climate change—more sophisticated trait development strategies are required. Innovative methods, including synthetic biology (SynBio) and epigenetic modifications, offer promising solutions for developing desirable agronomic traits in current crop varieties, supporting future food and feed demands.

Due to their sessile nature, plants are constantly exposed to a broad spectrum of environmental stressors. To adapt and survive under such challenging and rapidly changing conditions—such as seasonal fluctuations and circadian cycles—plants have evolved an array of complex and integrated mechanisms. These include morphological, physiological, biochemical, molecular, genetic, and epigenetic responses that together contribute to their resilience. A growing body of research highlights the role of epigenetic variation in driving changes in physiological and cellular traits through alterations in gene expression patterns. Such epigenetic changes do not modify the DNA sequence itself but instead involve the activation or repression of specific genes, resulting in functional changes at the cellular level (Moore 2015; Vaiserman and Koliada 2017; Talarowska 2020). These regulatory processes occur swiftly and can operate at both transcriptional and posttranscriptional levels. Ultimately, epigenetic regulation serves as a critical component of the plant's adaptive toolkit, facilitating long-term adjustments to persistent or recurring environmental stresses (Stagé et al. 2016).

In recent decades, efforts to enhance crop yields have primarily depended on the continuous generation and selection of genetic variation, particularly through conventional breeding strategies that focus on elite cultivars with favorable alleles and high-performing traits under diverse environmental conditions. Plant breeders have traditionally employed synthetic selection using both natural and induced variation to integrate promising traits into new varieties.

Today, however, there is a pressing need to explore innovative approaches that complement or surpass the limits of traditional breeding methods (Alvarez-Venegas and De-la-Pe˜na 2016). Emerging technologies—such as synthetic biology (SynBio) and epigenomic engineering—offer promising avenues for increasing crop adaptability and productivity. Unlike genetic modifications that alter the DNA sequence, epigenetic variation influences gene expression through heritable biochemical modifications, including DNA methylation, histone modification, and the regulation by noncoding RNAs (Gawra et al. 2023). These modifications, orchestrated by specific classes of enzymes responsible for writing, erasing, and interpreting epigenetic marks, can influence gene activity and plant phenotype without permanent changes to the genome.

Stoffel et al. (2020) identify key epigenetic mechanisms—including chromatin remodeling, DNA methylation, histone modifications, and the action of noncoding RNAs—as central to the regulation of gene expression patterns in plants. The targeted manipulation of these epigenetic processes presents a powerful tool for plant genetic improvement (Alvarez-Venegas et al. 2014). Furthermore, advancements in epigenetic editing techniques raise the potential to regulate phenotypic plasticity, enabling the development of climate-resilient, high-performing crops suited for sustainable agriculture (Kumar 2019).

Synthetic biology further broadens the scope of crop improvement by enabling the design and construction of novel biological components—such as genetic circuits, enzymes, or entire genomes—or by redesigning existing systems to achieve enhanced functions. SynBio facilitates the artificial introduction of genetic variation through the insertion of multiple well-characterized genes in a single transformation event, thereby accelerating the development of cultivars with complex, elite traits. This technology also supports the assembly and transfer of synthetic genomes into target cells or organisms, opening up new possibilities for trait engineering and innovation in crop science.

In this chapter, we provide an overview of recent advancements in plant epigenome editing tools and examine their impact on crop improvement. We explore key epigenetic regulatory mechanisms—DNA methylation, histone posttranslational modifications (HPTMs), and noncoding RNAs—and assess the extent to which these modifications are stable and heritable. Finally, we highlight studies that demonstrate the practical applications of SynBio in enhancing plant plasticity and developing agronomically valuable traits.

15.2 Applications of Epigenetic Modifications for Crop Enhancement

Currently, one of the primary objectives in agricultural research is to develop crop cultivars with enhanced agronomic traits, including increased yield, improved tolerance to biotic stresses, and resistance to abiotic stresses. These traits are typically governed by complex genetic networks, which often present significant challenges to crop improvement efforts (Mosammaparast and Shi 2010). Despite advances in genetic modification, there remains limited evidence demonstrating substantial yield

Table 15.1 Examples of recent studies of harnessing epigenetic modifications for crop enhancement

No.	Plant species	Epigenetic modification	Improvement	References
1	*Arabidopsis*	DNA methylation	Drought stress tolerance	Ramirez-Prado et al. (2019))
2	Rice	DNA methylation	Cadmium stress tolerance	Feng et al. (2020)
3		DNA methylation	Increased nutritional value	Liu et al. (2018)
4		Histone modification	Enhanced nitrogen use efficiency	Wu et al. (2020)
5	Cotton	DNA methylation	Methylation at H3K9me2 controls fiber differentiation	Wang et al. (2016a, b)
6	Pineapple	DNA methylation	Somatic embryogenesis enhancement	Luan et al. (2020)
7	Pigeon pea	DNA methylation	High heterosis	Sinha et al. (2020)
8	Sugar beet	DNA methylation	Cold acclimation, hormonal pathway genes, and flowering	Hebrard et al. (2015)
9	Watermelon	DNA methylation	Resistance to viral contamination	Sun et al. (2019)
10		Histone modification	Carotenoid biosynthesis	Anwar et al. (2021)
11	Wheat	Histone modification	Fungal pathogen resistance	Kong et al. (2020)
12		miRNA	Development of anther	Wang et al. (2012)
13	Banana	Histone modification	Binding of MaHDA6 to the MaERF11/15 promoters results in ripening	Fu et al. (2018)
14		Histone modification	Impeding ripening	Han et al. 2016
15	Cotton	miRNA	Fiber development	Guan et al. (2014)
16		Histone modification	Fiber initiation	Kumar et al. (2018)
17	Barley	miRNA	Grain density and cleistogamous flowering	Houston et al. (2013)
18	Soybean	miRNA	Increased yield and stability	Raju et al. (2018)

improvements in major crops through traditional genetic engineering approaches (Paul et al. 2020).

Epigenetics offers a promising alternative, capable of influencing both qualitative and quantitative traits, thereby accelerating crop improvement significantly (see Table 15.1). Studying epigenetic mechanisms not only elucidates how modifications to DNA structure can regulate gene expression but also opens new avenues for precise gene-editing techniques aimed at producing and selecting desirable phenotypes efficiently and specifically. These epigenetic regulations operate throughout plant development and enable adaptation to environmental conditions. Understanding the molecular foundations of these adaptive responses holds considerable agronomic value, as identifying epigenetic factors affecting traits such as yield, biomass, flowering time, fruit ripening, and abiotic stress resistance can lead to targeted and efficient crop enhancement.

Epigenetic tools have emerged as valuable resources in modern breeding programs. One key advantage is the ability to manipulate gene expression by targeting specific epigenetic regulators—through mutation or deletion—resulting in novel epigenetic variants with potential agronomic benefits. Importantly, epigenetic modifications represent reversible and heritable changes in gene activity that occur without altering the underlying DNA sequence. These modifications encompass DNA methylation, histone modifications, and RNA-based mechanisms.

15.2.1 DNA Modification

DNA methylation is a crucial epigenetic mechanism primarily involved in gene silencing and is predominantly found in heterochromatic regions such as telomeres and pericentromeric areas (Zhang et al. 2018). Additionally, methylation occurring near gene regulatory regions—typically promoters and terminators—is strongly associated with gene repression. Liu and He (2020) described DNA methylation in plants as the addition of a methyl group to the fifth carbon of cytosine's pyrimidine ring or the sixth position of adenine's purine ring. Among epigenetic modifications, DNA methylation is the most extensively studied and has been implicated in the regulation of chromatin structure, a phenomenon often referred to as "Cell Memory" (Alvarez-Venegas et al. 2014). This methylation, by modifying the DNA molecule, influences biological functions such as the silencing and reactivation of transposable elements and the regulation of gene expression across both plants and animals. Importantly, DNA methylation is reversible and can be actively removed by specific demethylase enzymes. In plants, active demethylation is mainly governed by DNA glycosylases operating through a base-excision repair pathway (Penterman et al. 2007).

Recent research by Jiang et al. (2020) confirmed that DNA methylation serves as a heritable epigenetic mark crucial for stabilizing gene silencing during plant development. It is essential for processes such as transposable element (TE) silencing, genomic imprinting, X-chromosome inactivation, and maintaining genome stability (Jiang et al. 2020). Gang et al. (2018) further emphasize that inheritance of DNA methylation patterns is a key mechanism by which plants preserve genomic integrity. Additionally, DNA methylation contributes to gene expression differences between parental genomes in the endosperm, a process known as genomic imprinting (Parashar et al. 2020). Epigenetic modifications broadly influence gene expression regulation and serve as a vital source of phenotypic variation, positively impacting plant growth and agronomic traits (Banerjee and Roychoudhury 2017). Some epigenetic changes persist beyond the original inducing stimuli and can be transmitted to subsequent generations as stable epialleles, offering a novel source of variation that can be harnessed in breeding programs, especially in crops with limited genetic diversity. Because these epigenetic marks are heritable, DNA methylation can be effectively utilized in breeding both seed-propagated and vegetatively propagated crops.

DNA methylation also plays a pivotal role in the heritable adaptation of stress tolerance. For example, Zhang et al. (2011) demonstrated in rice (*Oryza sativa*) that the genes P5CS and d-OAT, involved in osmotic stress response, exhibited methylation in mother plants but not in the immediate offspring exposed to stress, indicating that demethylation regulates their expression. While many stress-induced epigenetic modifications arise in response to the stress itself, some are maintained as stress memory, enabling stable gene expression patterns across generations. Both abiotic and biotic stresses are major drivers of heritable changes in DNA methylation (Kong et al. 2020). Tissue-specific differential methylation of HKT genes in wheat under salt stress further confirms the role of DNA methylation in regulating gene expression under adverse conditions (Kumar et al. 2017).

In tomato, Lang et al. (2017) found that global DNA methylation decreases during fruit ripening, a process linked to the upregulation of a DNA demethylase gene, SlDML2. Suppressing or knocking out SlDML2 delays ripening, illustrating the enzyme's regulatory role. Similarly, in cultivated cotton, DNA demethylation activates the COL2 gene, which triggers photoperiod sensitivity, potentially expanding cotton cultivation to diverse geographical areas (Wang et al. 2017). These examples underscore the practical impact of epigenetic modifications on agriculturally relevant traits.

Further studies have highlighted the role of DNA methylation in diverse biological processes such as fiber differentiation and lipid biosynthesis in cotton, mediated through an H3K9me2-dependent pathway influencing reactive oxygen species (ROS) dynamics (Wang et al. 2016a, 2016b). Naturally occurring epialleles also significantly influence traits like vitamin E accumulation in tomato fruits (Quadrana et al. 2014). In *Brassica rapa*, DNA demethylation has been associated with increased seed protein content, while in oilseed rape, it correlates with decreased oil content and enhanced protein levels (Amoah et al. 2012; Long et al. 2011). Dynamic epigenetic regulation partially controls fruit ripening, with citrus plants exhibiting increased global DNA methylation during this process (Huang et al. 2019). Understanding these mechanisms has facilitated the development of late-maturing varieties and improved fruit quality in several crop species (Liu et al. 2015, 2016; Osorio et al. 2020).

Collectively, these findings highlight the significant potential of epigenetic modifications, particularly DNA methylation, in enhancing agronomic traits and provide a strong foundation for their application in crop improvement programs.

15.2.2 Histone Modification Associated with Crop Traits

Histone modifications represent a key layer of epigenetic regulation and typically occur as posttranslational alterations on the N-terminal tails of histone proteins. These modifications vary depending on the functional groups added to specific amino acids and include acetylation, methylation, phosphorylation, ubiquitination, sumoylation, and glycosylation (Gibney and Nolan 2010). Such chemical changes influence chromatin architecture, shifting it between more condensed (inactive) and relaxed (active) states, thereby controlling gene transcription (Pfluger and Wagner 2007). Among these, histone methylation is especially prominent and has been well characterized in plants, where it occurs on lysine and arginine residues. These histone marks are dynamically regulated: acetylation is modulated by histone acetyltransferases and histone deacetylases, while methylation is controlled by histone methyltransferases and demethylases (Yu et al. 2020). These enzymes directly influence chromatin status and transcriptional activity (Jenuwein and Allis 2001; Yu et al. 2020).

The manipulation of such modifications offers a novel strategy for enhancing crop yield through epigenetic editing. Plants possess various histone variants and modifying enzymes that play vital roles in regulating gene expression under diverse environmental conditions. The impact of a histone modification is determined not only by the type of modification but also by its distribution across the gene region—referred to as the "histone modification landscape."

Research has highlighted the relevance of histone acetylation and methylation in crop development and stress responses. In rice, for example, grain size and flowering efficiency—key yield-related traits—are regulated through histone modifications (Shi et al. 2015). Epigenetic alterations across the genome also enhance the expression of drought-responsive genes (Li et al. 2013), while histone acetylation in *Arabidopsis* has been linked to drought stress tolerance (Zheng et al. 2016). In maize, epigenetic and morphological changes under salt stress have been attributed to histone acetyltransferase-mediated gene activation, which promotes cell enlargement to mitigate salinity effects (Li et al. 2014a, 2014b).

More recently, histone modifications have been implicated in the reprogramming of genes involved in both pollination-dependent and -independent fruit set, expanding our understanding of floral-to-fruit developmental transitions (Hu et al. 2021). Abiotic stress factors such as extreme temperatures, salinity, drought, and heavy metals severely affect plant growth and yield (Chang et al. 2020), and histone modifications play an essential role in modulating stress responses, fruit quality, and reproductive development. In perennial crops, low-temperature-related epigenetic changes—specifically an increase in the repressive mark H3K27me3 and a decrease in the activating mark H3K4me3—have been shown to regulate dormancy and its release (Leida et al. 2012).

Unlike DNA methylation, histone modifications are not typically inherited across generations. However, they can be highly effective in clonally propagated crops, which constitute over 60% of globally important crops, such as potatoes, yams, sorghum, and taro (Meyer et al. 2012; Latutrie et al. 2019). Clonal propagation facilitates the fixation of favorable epigenetic traits, making it a valuable tool in breeding programs targeting such crops.

Collectively, the findings from these studies underscore the significance of histone modification in regulating growth-related traits. These regulatory pathways provide a valuable foundation for developing epigenetic-based strategies to improve crop performance and meet the increasing demand for food under variable environmental conditions.

15.2.3 Noncoding RNAs (ncRNAs)

Noncoding RNAs (ncRNAs), which include both small (sRNAs, typically 20 to 24 nucleotides) and long forms (lncRNAs, generally over 200 nucleotides), are classified based on their size, function, or chromosomal origin. These molecules are dispersed throughout the genome and have emerged as key elements in the regulation of gene expression, maintenance of genome integrity, and defense mechanisms against foreign genetic materials. Both small and long ncRNAs are involved in the epigenetic regulation of gene activity by influencing chromatin structure—either activating or silencing genes.

Among the small ncRNAs (sncRNAs) in plants, two primary types are recognized: short interfering RNAs (siRNAs) and microRNAs (miRNAs), both of which contribute to gene silencing at the posttranscriptional level. These RNA classes differ in their origin and processing: siRNAs usually derive from exogenous double-stranded RNAs, whereas miRNAs are produced from endogenous stem-loop precursors (Guleria et al. 2011; Rajewsky et al. 2017). Increasing evidence highlights the essential role of ncRNAs in plant development and responses to environmental stresses (Campalans et al. 2004). They are now being explored as promising tools for molecular breeding strategies aimed at enhancing traits such as stress tolerance, herbicide resistance, and yield improvement.

Notably, ncRNAs play an important role in plant defense by mediating RNA silencing mechanisms that target pathogens. Their involvement in plant immunity, especially under stress conditions, makes them potential candidates for genetic improvement. Jha et al. (2020) reported distinct expression patterns of ncRNAs under various stress conditions, while Zhang et al. (2016) identified changes in the expression of specific miRNAs during drought stress at the seedling stage—17 downregulated and 16 upregulated miRNAs. These regulatory molecules also contribute to disease resistance by modulating genes involved in producing defense proteins, secondary metabolites, and activating phytohormone signaling in response to pathogens.

The flexibility of gene expression in plants under changing conditions is largely mediated by both transcriptional and posttranscriptional regulation involving miRNAs and other ncRNAs (Wang et al. 2017; Song et al. 2019). Although miRNAs are small (typically 21 to 24 nucleotides), they play critical roles by targeting messenger RNAs for degradation or translational repression (Wu 2013). Their regulatory influence has been studied extensively in model organisms like

Arabidopsis as well as in major cereal crops including rice, wheat, maize, and barley (Bhogireddy et al. 2021). Despite the broad range of miRNA expression profiles under stress, only a limited number of miRNA–target pairs have been identified to directly modulate specific stress responses.

siRNAs, another important class of small RNAs, mediate gene silencing through two main mechanisms: transcriptional silencing via DNA and histone methylation, and posttranscriptional silencing through mRNA degradation (Borges and Martienssen 2015). These 20 to 24 nucleotide double-stranded RNAs include subtypes such as heterochromatic siRNAs (hc-siRNAs), trans-acting siRNAs (tasiRNAs), long siRNAs (lsiRNAs), and repeat-associated siRNAs (ra-siRNAs) (Zhang et al. 2019). Their biogenesis often depends on RNA-dependent RNA polymerases (RDRs), which synthesize the dsRNA precursors (Song et al. 2019).

Recent findings suggest that some miRNAs may also regulate their own targets during translation (Bordersen et al. 2015; Prathiba et al. 2017), and in certain cases, may play minor roles in DNA methylation (Cuerda-Gil and Slotkin 2016). Kawakatsu and Ecker (2019) proposed a feedback loop between siRNA activity and DNA methylation, suggesting that this mechanism helps distribute stable epialleles across the genome. Furthermore, miRNAs are involved in key biological processes such as organ morphogenesis, transition from vegetative to reproductive phases, response to environmental stimuli, and phytohormone signaling (Zhao et al. 2017). Interestingly, studies have revealed that hybrid plants can exhibit distinct small RNA expression profiles compared to their parental lines (He et al. 2010), indicating a role for ncRNAs in hybrid vigor or heterosis.

Given their diverse regulatory roles, ncRNAs have become a promising focus for research into trait inheritance and genetic enhancement. Over the past two decades, significant advancements have been made in understanding their structure, functions, and potential applications in agriculture. However, to fully exploit their utility in crop improvement, more comprehensive studies are needed to clarify the mechanisms through which ncRNAs influence heritable traits and stress adaptation.

15.3 SynBio as an Epibreeding Strategy

Synthetic biology (SynBio) has emerged as a transformative approach in plant science, offering powerful strategies to enhance growth traits, increase stress resilience, and boost crop productivity (Table 15.2). By engineering or redesigning metabolic pathways, SynBio enables the development of plants better equipped to handle environmental challenges such as drought, salinity, and temperature extremes (Yang et al. 2020). Moreover, SynBio provides a platform for generating genetically modified crops with improved physiological responses, enhanced tolerance to abiotic stress, and optimized yield performance (Jagadeesh Chandra Bose et al. 2024).

One of SynBio's most significant contributions lies in its ability to overcome key limitations in traditional breeding methods, including interspecific incompatibility, genetic drag (the co-introduction of unfavorable traits), and difficulties in stacking multiple alleles required for complex phenotypes (Kitashiba and Nasrallah 2014;

Table 15.2 Examples of recent studies of synthetic biology approaches for crop improvement

Technology	Description	References
Gene drives	Facilitating the inheritance of detrimental alleles, including those that induce sterility or lethality, within insect pest populations	Bier (2022)
Golden Gate	Simultaneous, directional in vitro assembly of multiple DNA fragments enables gene stacking for complex traits and facilitates modular cloning, allowing precise exchange of promoter and terminator elements	Engler et al. (2009)
CRISPR-Cas9	Targeted in vivo gene editing is an efficient method for silencing, modifying, or enhancing specific genes, as well as integrating transgenes at precise genomic locations	Mao et al. (2013)
Regulated promoters	Regulated promoters enable temporal control of gene expression, activating or repressing downstream genes in response to specific conditions, such as environmental stress or developmental stages	Khan et al. (2017) and Schreiber and Tissier (2017)

Langridge and Fleury 2011; Varshney et al. 2014; Lyzenga et al. 2021). SynBio offers novel tools to bypass these barriers, allowing precise control over genetic elements and pathway functions to achieve desired agronomic outcomes.

The urgency of developing innovative crop improvement strategies has never been more pressing, particularly in the face of global food security concerns and accelerating climate change. SynBio enables the de novo design of physiological traits, entire metabolic networks, and complex regulatory circuits to enhance plant adaptation and yield stability. These capabilities are increasingly relevant to the development of climate-resilient crops, especially as traditional breeding techniques struggle to keep pace with emerging agricultural challenges.

Practical applications of SynBio extend across multiple fields, notably in agriculture and biotechnology. For instance, the creation of plants with enhanced drought tolerance or improved photosynthetic efficiency—through the engineering of Rubisco or ATP synthase enzymes—represents a significant step forward in sustainable crop production. SynBio also offers the tools to develop and fine-tune complex traits, such as stress sensing and signaling, by precisely editing genes and pathways involved in environmental responses.

Furthermore, SynBio is redefining plant breeding by enabling not only the insertion of foreign DNA, as in conventional transgenic approaches, but also the editing and redesign of native genomes and biosynthetic pathways (Rao and Wang 2021; Kaul et al. 2021). This dual capability allows researchers to generate crops with tailored traits, including improved growth, resistance to stresses, and higher nutritional quality.

Advances in imaging and molecular diagnostics, driven by SynBio, now allow real-time visualization of fluorescent proteins within plant canopies, facilitating the monitoring of physiological processes and providing valuable insights for optimizing productivity (Rigoulot et al. 2021). Additionally, SynBio has been successfully applied to engineer microbial systems for the production of valuable bioactive compounds, such as resveratrol, which can benefit both plant performance and human health (Song et al. 2021). These developments illustrate the expanding potential of SynBio not only in crop enhancement but also in therapeutic molecule production.

The integration of artificial promoters that respond to specific stimuli—chemical, biotic, or abiotic—further enhances the control over gene expression in engineered plants, offering new avenues for precise regulation of desired traits (Schreiber and Tissier 2017). Looking ahead, SynBio is poised to play a central role in improving photosynthetic capacity, optimizing nutrient use efficiency, and increasing the nutritional value of crops.

By leveraging computational tools and SynBio principles, researchers are now equipped to design and implement environmentally sustainable agricultural technologies. These innovations mark a shift toward a new paradigm in plant biotechnology, one that combines precision, efficiency, and ecological sensitivity in meeting the demands of a growing global population.

15.4 Conclusion and Future Thrust

As demonstrated throughout this chapter, both epigenetic mechanisms and synthetic biology (SynBio) represent transformative approaches in the enhancement of agronomic traits. Epigenetic editing, in particular, offers novel avenues for introducing phenotypic diversity into crop improvement programs, expanding the scope beyond what is achievable through classical breeding or targeted mutagenesis. This includes the modulation of tissue-specific expression, organ development, and environmentally responsive traits—areas where traditional methods often fall short. By leveraging epigenetic knowledge, researchers can manipulate key molecular processes to develop crops with improved adaptability, resilience, and productivity, especially in the context of climate change.

Among the primary strategies for exploiting epigenetic regulation is the targeted mutation or removal of specific epigenetic regulators, resulting in new epigenetic variants with potential agronomic value. DNA methylation, histone modifications, and noncoding RNAs (ncRNAs) have emerged as critical regulators of gene expression, contributing to phenotypic plasticity and stress memory in plants. These memory traits enable plants to respond more effectively to recurring stresses, a capacity highly valuable under increasingly volatile environmental conditions.

In parallel, SynBio offers a complementary platform for crop improvement. It enables the precise engineering of genetic and metabolic pathways to modulate plant development and stress responses. Using advanced genome editing tools, such as CRISPR-based systems, it is now possible to induce or reverse specific epigenetic marks—like DNA methylation or histone modifications—to activate beneficial genes or suppress the expression of detrimental ones. This level of control offers a promising strategy for improving key traits such as yield, stress tolerance, and nutritional quality.

Furthermore, SynBio facilitates the integration of novel genetic circuits, the introduction of genes with desirable functions, and the application of RNA-based technologies, including the use of small RNAs as gene regulators or biopesticides. These innovations accelerate the breeding cycle and expand the toolkit available for developing elite crop varieties. Importantly, both naturally occurring and artificially induced epigenetic diversity can be harnessed as a reservoir of selectable traits, significantly contributing to phenotypic variability within breeding populations.

Despite the exciting potential of these technologies, several obstacles remain. Challenges related to the stability and heritability of epigenetic modifications, regulatory hurdles, technical limitations in delivery systems, and the complexity of multigene trait control continue to restrict their widespread adoption in mainstream breeding programs.

Nonetheless, the convergence of epigenetics and synthetic biology marks a critical step forward in modern crop science. A deeper understanding of the molecular underpinnings and intergenerational transmission of epigenetic marks will be essential for realizing the full potential of these technologies. In the near future, their integration into crop breeding pipelines could revolutionize our ability to develop resilient, high-yielding, and sustainable crops tailored for the demands of a changing global environment.

References

Altieri MA, Nicholls CI, Henao A, Lana MA (2015) Agroecology and the design of climate change-resilient farming systems. Agron Sustain Dev 35(3):869–890. https://doi.org/10.1007/s13593-015-0285-2

Alvarez-Venegas R, De-la-Pe˜na C (2016) Editorial: recent advances of epigenetics in crop biotechnology. Front Plant Sci 7(March):413. https://doi.org/10.3389/fpls.2016.00413

Alvarez-Venegas R, De-la-Peña C, Casas-Mollano JA (2014) Epigenetics in plants of agronomic importance: fundamentals and applications: transcriptional regulation and chromatin remodelling in plants. Springer. https://doi.org/10.1007/978-3-319-07971-4

Amoah S, Kurup S, Rodriguez Lopez CM, Welham SJ, Powers SJ, Hopkins CJ et al (2012) A Hypomethylated population of *Brassica rapa* for forward and reverse epi-genetics. BMC Plant Biol 12(1):193. https://doi.org/10.1186/1471-2229-12-193

Anwar S, Brenya E, Alagoz Y, Cazzonelli CI (2021) Epigenetic control of carotenogenesis during plant development. Crit Rev Plant Sci 40(1):23–48. https://doi.org/10.1080/07352689.2020.1866829

Bailey-Serres J, Parker JE, Ainsworth EA, Oldroyd GED, Schroeder JI (2019) Genetic strategies for improving crop yields. Nature 575:109–118

Banerjee A, Roychoudhury A (2017) The gymnastics of epigenomics in rice. Plant Cell Rep 37:25–49

Bhogireddy S, Mangrauthia SK, Kumar R, Pandey AK, Singh S, Jain A, Budak H, Varshney RK, Kudapa H (2021) Regulatory non-coding RNAs: a new frontier in regulation of plant biology. Funct Integr Genom 21:313–330

Bier E (2022) Gene drives gaining speed. Nat Rev Genet 23(1):5–22. https://doi.org/10.1038/s41576-021-00386-0

Bordersen P, Sakvarelidze-Achard L, Baruun-Rasmussen M, Dunoyer P, Borges F, Martienssen RA (2015) The expanding world of small RNAs in plants. Nat Rev Mol Cell Biol 16:727–741. https://doi.org/10.1038/nrm4085

Borges F, Martienssen RA (2015) The expanding world of small RNAs in plants. Nat Rev Mol Cell Biol 16(12):727–741. https://doi.org/10.1038/nrm4085

Campalans A, Kondorosi A, Crespi M Enod40 (2004) A short open reading frame–containing mRNA, induces cytoplasmic localization of a nuclear RNAbinding protein in *Medicago truncatula*. Plant Cell 16 (4): 1047 1059 https://doi.org/10.1105/tpc.019406

Chang YN, Zhu C, Jiang J, Zhang H, Zhu JK, Duan CG (2020) Epigenetic regulation in plant abiotic stress responses. J Integr Plant Biol 62(5):563. https://doi.org/10.1111/jipb.12901

Cuerda-Gil D, Slotkin RK (2016) Non-canonical RNA-directed DNA methylation Nat Plants 2: 16163. https://doi.org/10.1038/nplants.2016.163

Feng SJ, Liu XS, Ma LY, Khan IU, Rono JK, Yang ZM (2020) Identification of epigenetic mechanisms in paddy crop associated with lowering environmentally related cadmium risks to food safety. Environ Pollut 256:113464. https://doi.org/10.1016/j.envpol.2019.113464

Fu CC, Han YC, Guo YF, Kuang JF, Chen JY, Shan W et al (2018) Differential expression of histone deacetylases during banana ripening and identification of MaHDA6 in regulating ripening-associated genes. Postharvest Biol Technol 141:24–32. https://doi.org/10.1016/j.postharvbio.2018.03.010

Gang WG, Liu K, Shen T, Shi J, Liu B, Han M, Peng M, Fu H, Song Y, Zhu J, Dong A, Ni T (2018) Position-specific intron retention is mediated by the histone methyltransferase SDG725. BMC Biol 16. https://doi.org/10.1186/s12915-018-0513-8

Gawra J, Valdivieso A, Roux F, Laporte M, de Lorgeril J, Gueguen Y, Saccas M, Escoubas J.-M, Montagnani C, Destoumieux-Garzón D et al (2023) Epigenetic variations are more substantial than genetic variations in rapid adaptation of oyster to Pacific oyster mortality syndrome. Sci Adv 9: eadh8990

Gibney ER, Nolan CM (2010) Epigenetics and gene expression. Heredity 105:4–13

Guan X, Pang M, Nah G, Shi X, Ye W, Stelly DM et al (2014) miR828 and miR858 regulate homoeologous MYB2 gene functions in *Arabidopsis* trichome and cotton fibre development. Nat Commun 5:3050. https://doi.org/10.1038/ncomms4050

Guleria P, Mahajan M, Bhardwaj J, Yadav SK (2011) Plant small RNAs: biogenesis, mode of action and their roles in abiotic stresses. GPB 9(6):183–199. https://doi.org/10.1016/S1672-0229(11)60022-3

Gupta C, Salgotra RK (2022) Epigenetics and its role in effecting agronomical traits. Front Plant Sci 13:925688

Han YC, Kuang JF, Chen JY, Liu XC, Xiao YY, Fu CC et al (2016) Banana transcription factor MaERF11 recruits histone deacetylase MaHDA1 and represses the expression of MaACO1 and expansions during fruit ripening. Plant Physiol 171:1070–1084. https://doi.org/10.1104/pp.16.00301

He G, Zhu X, Elling AA, Chen L, Wang X, Guo L, Liang M, He H, Zhang H, Chen F, Qi Y, Chen R, Deng XW (2010) Global epigenetic and transcriptional trends among two rice subspecies and their reciprocal hybrids. Plant Cell 22:17–33. https://doi.org/10.1105/tpc.109.072041

Hebrard C, Peterson D, Willems G, Delaunay A, Jesson B, Lefebvre M et al (2015) Epigenomics and bolting tolerance in sugar beet genotypes. J Exp Bot 67:207–225. https://doi.org/10.1093/jxb/erv449

Houston K, McKim SM, Comadran J, Bonar N, Druka I, Uzrek N et al (2013) Variation in the interaction between alleles of HvAPETALA2 and microRNA172 determines the density of grains on the barley inflorescence. Proc Natl Acad Sci USA 110:16675–16680. https://doi.org/10.1073/pnas.1311681110

Hu G, Huang B, Wang K, Frasse P, Maza E, Djari A, Benhamed M, Gallusci P, Li Z, Zouine M, Bouzayen M (2021) Histone posttranslational modifications rather than DNA methylation underlie gene reprogramming in pollination-dependent and pollination-independent fruit set in tomato. New Phytol 229:902–919

Huang H, Liu R, Niu Q, Tang K, Zhang B, Zhang H et al (2019) Global increase in DNA methylation during orange fruit development and ripening. Proc Natl Acad Sci USA 116(4): 1430–1436. https://doi.org/10.1073/pnas.1815441116

Jagadeesh Chandra Bose K, Sharma S, Sarwan J, Uddin N (2024) Methods from the field of synthetic biology that aim to improve plant growth and resistance to stress through the use of genetic engineering. In: Metabolomics, proteomics and gene editing approaches in biofertilizer industry. https://doi.org/10.1007/978-981-97-2910-4_6

Jenuwein T, Allis CD (2001) Translating the histone code. Science 293:1074–1080

Jha UC, Nayyar H, Jha R, Khurshid M, Zhou M, Mantri N et al (2020) Long non-coding RNAs: emerging players regulating plant abiotic stress response and adaptation. BMC Plant Biol 20(1): 1–20. https://doi.org/10.1186/s12870-020-02595-x

Jiang J, Ding AB, Liu F, Zhong X (2020) Linking signaling pathways to histone acetylation dynamics in plants. J Exp Bot 71:5179–5190. https://doi.org/10.1093/jxb/eraa202

Kaul S, Choudhary M, Gupta S, Dhar MK (2021) Engineering host microbiome for crop improvement and sustainable agriculture. Front Microbiol 12:635917. https://doi.org/10.3389/fmicb.2021.635917

Kawakatsu T, Ecker JR (2019) Diversity and dynamics of DNA methylation: Epigenomic resources and tools for crop breeding. Breed Sci 69:191–204

Kitashiba H, Nasrallah JB (2014) Self-incompatibility in *Brassicaceae* crops: lessons for interspecific incompatibility. Breed Sci 64:23–37

Kong L, Liu Y, Kong L, Liu Y, Wang X, Chang C (2020) Insight into the role of epigenetic processes in abiotic and biotic stress response in wheat and barley. Int J Mol Sci 21. https://doi.org/10.3390/ijms21041480

Kumar S (2019) Epigenetics and epigenomics for crop improvement. Curr Opin Adv Biotech Micro 14:555879. https://doi.org/10.19080/AIBM.2019.14.555879

Kumar S, Singh AK, Mohapatra T (2017) Epigenetics: history, present status and future perspective. Indian J Genet Plant Breeding 77:445–463. https://doi.org/10.5958/0975-6906.2017.00061.X

Kumar V, Singh B, Singh SK, Rai KM, Singh SP, Sable A, et al. (2018) Role of GhHDA5 in H3K9 deacetylation and fiber initiation in Gossypium hirsutum. Plant J 95(6):1069–1083. https://doi.org/10.1111/tpj.14011

Lang Z, Wang Y, Tang K, Tang D, Datsenka T, Cheng J, Zhang Y, Handa AK, Zhu JK (2017) Critical roles of DNA demethylation in the activation of ripening-induced genes and inhibition of ripening-repressed genes in tomato fruit. Proc Natl Acad Sci 114:E4511–E4519

Langridge P, Fleury D (2011) Making the most of 'omics' for crop breeding. Trends Biotechnol 29: 33–40

Latutrie M, Gourcilleau D, Pujol B (2019) Epigenetic variation for agronomic improvement: an opportunity for vegetatively propagated crops. Am J Bot 106:1281–1284. https://doi.org/10.1002/ajb2.1357

Leida C, Conesa A, Ll'acer G, Badenes ML, Ríos G (2012) Histone modifications and expression of *DAM6* gene in peach are modulated during bud dormancy release in a cultivar-dependent manner. New Phytol 193:67–80

Li T, Chen X, Zhong X, Zhao Y, Liu X, Zhou S et al (2013) Jumonji C domain protein JMJ705-mediated removal of histone H3 lysine 27 trimethylation is involved in defense-related gene activation in rice. Plant Cell 25(11):4725–4736. https://doi.org/10.1105/tpc.113.118802

Li H, Yan S, Zhao L, Tan J, Zhang Q, Gao F et al (2014a) Histone acetylation associated up-regulation of the cell wall related genes is involved in salt stress induced maize root swelling. BMC Plant Biol 14(1):1–14. https://doi.org/10.1186/1471-2229-14-105

Li RQ, Huang JZ, Zhao HJ, Fu HW, Li YF, Liu GZ, Shu QY (2014b) A down-regulated epi-allele of the genomes uncoupled 4 gene generates a xantha marker trait in rice. Theor Appl Genet 127(11):2491–2501. https://doi.org/10.1007/s00122-014-2393-9

Liu L, Zhang Y, Tan H, Wang Y, Tong Y (2018) Coordinate regulation of photosynthesis, chlorophyll biosynthesis, and seed oil body formation by a single transcription factor in Arabidopsis. Plant Cell 30(10):2463–2481. https://doi.org/10.1105/tpc.18.00425
Liu J, He Z (2020) Small DNA methylation, big player in plant abiotic stress responses and memory. Front Plant Sci 11:595603. https://doi.org/10.3389/fpls.2020.595603
Liu R, How-Kit A, Stammitti L, Teyssier E, Rolin D, Mortain-Bertrand A, Halle S, Liu M, Kong J, Wu C, Degraeve-Guibault C, Chapman NH, Maucourt M, Hodgman TC, Tost J, Bouzayen M, Hong Y, Seymour GB, Giovannoni JJ, Gallusci P (2015) A demeter-like DNA demethylase governs tomato fruit ripening. Proc Natl Acad Sci 112:10804–10809
Liu D-D, Zhou L-J, Fang M-J et al (2016) Polycomb-group protein SlMSI1 represses the expression of fruit-ripening genes to prolong shelf life in tomato. Sci Rep 6:31806. https://doi.org/10.1038/srep31806
Long Y, Xia W, Li R, Wang J, Shao M, Feng J et al (2011) Epigenetic QTL mapping in Brassica napus. Genetics 189(3):1093–1102. https://doi.org/10.1534/genetics.111.131615
Luan A, Chen C, Xie T, He J, He Y (2020) Methylation analysis of CpG islands in pineapple SERK1 promoter. Genes 11(4):425. https://doi.org/10.3390/genes11040425
Lyzenga WJ, Pozniak CJ, Kagale S (2021) Advanced domestication: harnessing the precision of gene editing in crop breeding. Plant Biotech J 19:660–670
Meyer RS, DuVal AE, Jensen HR (2012) Patterns and processes in crop domestication: an historical review and quantitative analysis of 203 global food crops. New Phytol 196:29–48. https://doi.org/10.1111/j.1469-8137.2012.04253.x
Mladenov V, Fotopoulos V, Kaiserli E, Karalija E, Maury S, Baranek M (2021) Deciphering the epigenetic alphabet involved in transgenerational stress memory in crops. Ijms 22(13):7118. https://doi.org/10.3390/ijms22137118
Moore D (2015) The developing genome: an introduction to behavioral epigenetics. Pitzer Faculty Books. Retrieved from. https://scholarship.claremont.edu/pitzer_facbooks/16
Mosammaparast N, Shi Y (2010) Reversal of histone methylation: biochemical and molecular mechanisms of histone demethylases. Annu Rev Biochem 79(1):155–179. https://doi.org/10.1146/annurev.biochem.78.070907.103946
Osorio S, Carneiro RT, Lytovchenko A et al (2020) Genetic and metabolic effects of ripening mutations and vine detachment on tomato fruit quality. Plant Biotechnol J 18:106–118. https://doi.org/10.1111/pbi.13176
Parashar NC, Parashar G, Nayyar H, Sandhir R (2020) Differential DNA methylation in regulation of deacetylvindoline-4-Oacetyl transferase (DAT) gene in Catharanthusroseus. J Plant Biochem Biotechnol 30:326. https://doi.org/10.1007/s13562-020-00592-7
Paul MJ, Watson A, Griffiths CA (2020) Linking fundamental science to crop improvement through understanding source and sink traits and their integration for yield enhancement. J Exp Bot 71(7):2270–2280. https://doi.org/10.1093/jxb/erz480
Penterman J, Zilberman D, Jin HH, Ballinger T, Henikoff S, Fischer RL (2007) DNA demethylation in the Arabidopsis genome. Proc Natl Acad Sci USA 104(16):6752–6757. https://doi.org/10.1073/pnas.0701861104
Pfluger J, Wagner D (2007) Histone modifications and dynamic regulation of genome accessibility in plants. Curr Opin Plant Biol 10:645–652
Prathiba KY, Usha S, Suchithra B, Jyothi MN, Devaraj VR, Nageshbabu R (2017) Computational identification of miRNAs and their targets from Niger (Guizotia abyssinica). J Appl Biol Biotechnol 5(2):53–58. https://doi.org/10.7324/JABB.2017.50208
Quadrana L, Almeida J, Asis R, Duffy T, Dominguez PG, Bermudez L et al (2014) Natural occurring epialleles determine vitamin E accumulation in tomato fruits. Nat Commun 5:4027. https://doi.org/10.1038/ncomms5027
Rajewsky N, Jurga S, Barciszewski J (2017) Plant epigenetics, 536
Raju SKK, Shao MR, Sanchez R, Xu YZ, Sandhu A, Graef G, Mackenzie S (2018) An epigenetic breeding system in soybean for increased yield and stability. Plant Biotechnol J 16:1836–1847. https://doi.org/10.1111/pbi.12919

Ramirez-Prado JS, Latrasse D, Rodriguez-Granados NY, Huang Y, Manza- Mianza D, Brik-Chaouche R et al (2019) The Polycomb protein LHP1 regulates *Arabidopsis thaliana* stress responses through the repression of the MYC2-dependent branch of immunity. Plant J 100(6): 1118–1131. https://doi.org/10.1111/tpj.14502

Rao MJ, Wang L (2021) CRISPR/Cas9 technology for improving agronomic traits and future prospective in agriculture. Planta 254(4, 68). https://doi.org/10.1007/s00425-021-03716-y

Rigoulot SB, Schimel TM, Lee JH, Sears RG, Brabazon H, Layton JS, Li L, Meier KA, Poindexter MR, Schmid MJ (2021) Imaging of multiple fluorescent proteins in canopies enables synthetic biology in plants. Plant Biotechnol J 19(4):830–843

Schreiber T, Tissier A (2017) Generation of dTALEs and libraries of synthetic TALE-activated promoters for engineering of gene regulatory networks in plants. Methods Mol Biol 1629:185–204

Shi J, Dong A, Shen WH (2015) Epigenetic regulation of Rice flowering and reproduction. Front Plant Sci 5:1–13

Sinha P, Singh VK, Saxena RK, Kale SM, Li Y, Garg V, Meifang T, Khan AW, Kim KD, Chitikineni A, Saxena KB, Sameer Kumar CV, Liu X, Xu X, Jackson S, Powell W, Nevo E, Searle IR, Lodha M, Varshney RK (2020) Genome-wide analysis of epigenetic and transcriptional changes associated with heterosis in pigeonpea. Plant Biotechnol J 18:1697–1710

Song X, Li Y, Cao X, Qi Y (2019) MicroRNAs and their regulatory roles in plantenvironment interactions. Annu Rev Plant Biol 70:489–525. https://doi.org/10.1146/annurev-arplant-050718-100334

Song J-T, Woo D-U, Lee Y, Choi S-H, Kang Y-J (2021) The semi-supervised strategy of machine learning on the gene family diversity to unravel resveratrol synthesis. Plan Theory 10(10):2058

Stagé C, Moreau S, Sallet E, Lefort G, Auriac MC, Remblière C et al (2016) Reprogramming of DNA methylation is critical for nodule development in *Medicago truncatula*. Nat Plants 2(11): 1–10. https://doi.org/10.1038/nplants.2016.166

Stoffel M, Gardini E, Ehrenthal JC, Abbruzzese E, Ditzen B (2020) Evaluation of stress management and stress prevention using epigenetic markers. Verhaltenstherapie 30:18. https://doi.org/10.1159/000506323

Sun Z, Su C, Yun J, Jiang Q, Wang L, Wang Y, Cao D, Zhao F, Zhao Q, Zhang M, Zhou B, Zhang L, Kong F, Liu B, Tong Y, Li X (2019) Genetic improvement of the shoot architecture and yield in soya bean plants via the manipulation of GmmiR156b. Plant Biotechnol J 17:50–62

Talarowska M (2020) Epigenetic mechanisms in the neurodevelopmental theory of depression. Depress Res Treat 1–9. https://doi.org/10.1155/2020/6357873

Thiebaut F, Hemerly AD, PCG F (2020) A role for epigenetic regulation in the adaptation and stress responses of non-model plants. Front Plant Sci 10. https://doi.org/10.3389/fpls.2019.00246

Vaiserman AM, Koliada AK (2017) Early-life adversity and long-term neurobehavioral outcomes: epigenome as a bridge? Hum Genomics 11:34. https://doi.org/10.1186/s40246-017-0129-z

Varshney RK, Terauchi R, Mccouch SR (2014) Harvesting the promising fruits of genomics: applying genome sequencing technologies to crop breeding. PLoS Biol 12:e1001883

Wang Y, Tang H, Debarry JD, Tan X, Li J, Wang X (2012) MCScanX: a toolkit for detection and evolutionary analysis of gene synteny and collinearity. Nucleic Acids Res 40(7):e49. https://doi.org/10.1093/nar/gkr1293

Wang M, Wang P, Tu L, Zhu S, Zhang L, Li Z et al (2016a) Multi-omics maps of cotton fibre reveal epigenetic basis for staged single-cell differentiation. Nucleic Acids Res 44(9):4067–4079. https://doi.org/10.1093/nar/gkw238

Wang X, Xin C, Cai J, Zhou Q, Dai T, Cao W, Jiang D (2016b) Heat priming induces trans-generational tolerance to high temperature stress in wheat. Front Plant Sci 7(April):501. https://doi.org/10.3389/fpls.2016.00501

Wang W, Zhang X, Deng F, Yuan R, Shen F (2017) Genome-wide characterization and expression analyses of superoxide dismutase (SOD) genes in *Gossypium hirsutum*. BMC Genomics 18:376

Wu G (2013) Plant microRNAs and development. J Genet Genom 40(5):217–230

Wu K, Wang S, Song W, Zhang J, Wang Y, Liu Q (2020) Enhanced sustainable green revolution yield via nitrogen-responsive chromatin modulation in rice. Science 367:eaaz2046. https://doi.org/10.1126/science.aaz2046

Yang K, Gong XL, Li GC, Huang LQ, Ning C, Wang CZ (2020) A gustatory receptor tuned to the steroid plant hormone brassinolide in Plutella xylostella (*Lepidoptera: Plutellidae*). elife 9: e64114

Yu J, Su D, Yang D, Dong T, Tang Z, Li H, Han Y, Li Z, Zhang B (2020) Chilling and heat stress-induced physiological changes and microRNA-related mechanism in sweetpotato (*Ipomoea batatas* L.). Front Plant Sci 11:687

Zhang M, Xu C, Von Wettstein D, Liu B (2011) Tissue-specific differences in cytosine methylation and their association with differential gene expression in *Sorghum bicolar*. Plant Physiol 156: 1955–1966. https://doi.org/10.1104/pp.111.176842

Zhang F, Luo X, Zhou Y, Xie J (2016) Genome-wide identification of conserved microRNA and their response to drought stress in Dongxiang wild rice (*Oryza rufipogon Griff.*). Biotechnol Lett 38(4):711–721. https://doi.org/10.1007/s10529-015-2012-0

Zhang Q, Ma C, Zhang Y, Gu Z, Li W, Duan X, Wang S, Hao L, Wang Y, Wang S, Li T (2018) A single-nucleotide polymorphism in the promoter of a hairpin RNA contributes to Alternaria alternata leaf spot resistance in apple (*Malus × domestica*). Plant Cell 30:1924–1942

Zhang M, Zhang X, Guo L, Qi T, Liu G, Feng J (2019) Single-base resolution methylomes of cotton CMS system reveal epigenomic changes in response to high-temperature stress during another development. J Exp Bot 71(3):951–969. https://doi.org/10.1093/jxb/erz470

Zhao D, Chong K, Palanivelu R (2017) Editorial: molecular and cellular plant reproduction. Front Plant Sci 8:199. https://doi.org/10.3389/fpls.2017.00199

Zheng Y, Ding Y, Sun X, Xie S, Wang D, Liu X (2016) Histone deacetylase HDA9 negatively regulates salt and drought stress responsiveness in *Arabidopsis*. J Exp Bot 67(6):1703–1713. https://doi.org/10.1093/jxb/erv562

MIX
Papier aus verantwortungsvollen Quellen
Paper from responsible sources
FSC® C105338
FSC
www.fsc.org

If you have any concerns about our products,
you can contact us on
ProductSafety@springernature.com

In case Publisher is established outside the EU,
the EU authorized representative is:
Springer Nature Customer Service Center GmbH
Europaplatz 3, 69115 Heidelberg, Germany

Printed by Libri Plureos GmbH
in Hamburg, Germany